IET ENERGY ENGINEERING SERIES 152

Modern Power Electronic Devices

Other volumes in this series

Volume 1 **Power Circuit Breaker Theory and Design** C.H. Flurscheim (Editor)
Volume 4 **Industrial Microwave Heating** A.C. Metaxas and R.J. Meredith
Volume 7 **Insulators for High Voltages** J.S.T. Looms
Volume 8 **Variable Frequency AC Motor Drive Systems** D. Finney
Volume 10 **SF_6 Switchgear** H.M. Ryan and G.R. Jones
Volume 11 **Conduction and Induction Heating** E.J. Davies
Volume 13 **Statistical Techniques for High Voltage Engineering** W. Hauschild and W. Mosch
Volume 14 **Uninterruptible Power Supplies** J. Platts and J.D. St Aubyn (Editors)
Volume 15 **Digital Protection for Power Systems** A.T. Johns and S.K. Salman
Volume 16 **Electricity Economics and Planning** T.W. Berrie
Volume 18 **Vacuum Switchgear** A. Greenwood
Volume 19 **Electrical Safety: A guide to causes and prevention of hazards** J. Maxwell Adams
Volume 21 **Electricity Distribution Network Design, 2nd Edition** E. Lakervi and E.J. Holmes
Volume 22 **Artificial Intelligence Techniques in Power Systems** K. Warwick, A.O. Ekwue and R. Aggarwal (Editors)
Volume 24 **Power System Commissioning and Maintenance Practice** K. Harker
Volume 25 **Engineers' Handbook of Industrial Microwave Heating** R.J. Meredith
Volume 26 **Small Electric Motors** H. Moczala *et al.*
Volume 27 **AC–DC Power System Analysis** J. Arrillaga and B.C. Smith
Volume 29 **High Voltage Direct Current Transmission, 2nd Edition** J. Arrillaga
Volume 30 **Flexible AC Transmission Systems (FACTS)** Y.-H. Song (Editor)
Volume 31 **Embedded Generation** N. Jenkins *et al.*
Volume 32 **High Voltage Engineering and Testing, 2nd Edition** H.M. Ryan (Editor)
Volume 33 **Overvoltage Protection of Low-Voltage Systems, Revised Edition** P. Hasse
Volume 36 **Voltage Quality in Electrical Power Systems** J. Schlabbach *et al.*
Volume 37 **Electrical Steels for Rotating Machines** P. Beckley
Volume 38 **The Electric Car: Development and future of battery, hybrid and fuel-cell cars** M. Westbrook
Volume 39 **Power Systems Electromagnetic Transients Simulation** J. Arrillaga and N. Watson
Volume 40 **Advances in High Voltage Engineering** M. Haddad and D. Warne
Volume 41 **Electrical Operation of Electrostatic Precipitators** K. Parker
Volume 43 **Thermal Power Plant Simulation and Control** D. Flynn
Volume 44 **Economic Evaluation of Projects in the Electricity Supply Industry** H. Khatib
Volume 45 **Propulsion Systems for Hybrid Vehicles** J. Miller
Volume 46 **Distribution Switchgear** S. Stewart
Volume 47 **Protection of Electricity Distribution Networks, 2nd Edition** J. Gers and E. Holmes
Volume 48 **Wood Pole Overhead Lines** B. Wareing
Volume 49 **Electric Fuses, 3rd Edition** A. Wright and G. Newbery
Volume 50 **Wind Power Integration: Connection and system operational aspects** B. Fox *et al.*
Volume 51 **Short Circuit Currents** J. Schlabbach
Volume 52 **Nuclear Power** J. Wood
Volume 53 **Condition Assessment of High Voltage Insulation in Power System Equipment** R.E. James and Q. Su
Volume 55 **Local Energy: Distributed generation of heat and power** J. Wood
Volume 56 **Condition Monitoring of Rotating Electrical Machines** P. Tavner, L. Ran, J. Penman and H. Sedding
Volume 57 **The Control Techniques Drives and Controls Handbook, 2nd Edition** B. Drury
Volume 58 **Lightning Protection** V. Cooray (Editor)
Volume 59 **Ultracapacitor Applications** J.M. Miller

Volume 62 **Lightning Electromagnetics** V. Cooray
Volume 63 **Energy Storage for Power Systems, 2nd Edition** A. Ter-Gazarian
Volume 65 **Protection of Electricity Distribution Networks, 3rd Edition** J. Gers
Volume 66 **High Voltage Engineering Testing, 3rd Edition** H. Ryan (Editor)
Volume 67 **Multicore Simulation of Power System Transients** F.M. Uriate
Volume 68 **Distribution System Analysis and Automation** J. Gers
Volume 69 **The Lightening Flash, 2nd Edition** V. Cooray (Editor)
Volume 70 **Economic Evaluation of Projects in the Electricity Supply Industry, 3rd Edition** H. Khatib
Volume 72 **Control Circuits in Power Electronics: Practical issues in design and implementation** M. Castilla (Editor)
Volume 73 **Wide Area Monitoring, Protection and Control Systems: The enabler for smarter grids** A. Vaccaro and A. Zobaa (Editors)
Volume 74 **Power Electronic Converters and Systems: Frontiers and applications** A.M. Trzynadlowski (Editor)
Volume 75 **Power Distribution Automation** B. Das (Editor)
Volume 76 **Power System Stability: Modelling, analysis and control** B. Om P. Malik
Volume 78 **Numerical Analysis of Power System Transients and Dynamics** A. Ametani (Editor)
Volume 79 **Vehicle-to-Grid: Linking electric vehicles to the smart grid** J. Lu and J. Hossain (Editors)
Volume 81 **Cyber-Physical-Social Systems and Constructs in Electric Power Engineering** S. Suryanarayanan, R. Roche and T.M. Hansen (Editors)
Volume 82 **Periodic Control of Power Electronic Converters** F. Blaabjerg, K. Zhou, D. Wang and Y. Yang
Volume 86 **Advances in Power System Modelling, Control and Stability Analysis** F. Milano (Editor)
Volume 87 **Cogeneration: Technologies, optimisation and implementation** C.A. Frangopoulos (Editor)
Volume 88 **Smarter Energy: from Smart Metering to the Smart Grid** H. Sun, N. Hatziargyriou, H.V. Poor, L. Carpanini and M.A. Sánchez Fornié (Editors)
Volume 89 **Hydrogen Production, Separation and Purification for Energy** A. Basile, F. Dalena, J. Tong and T.N. Veziroğlu (Editors)
Volume 90 **Clean Energy Microgrids** S. Obara and J. Morel (Editors)
Volume 91 **Fuzzy Logic Control in Energy Systems with Design Applications in MATLAB®/Simulink®** İ.H. Altaş
Volume 92 **Power Quality in Future Electrical Power Systems** A.F. Zobaa and S.H.E.A. Aleem (Editors)
Volume 93 **Cogeneration and District Energy Systems: Modelling, analysis and optimization** M.A. Rosen and S. Koohi-Fayegh
Volume 94 **Introduction to the Smart Grid: Concepts, technologies and evolution** S.K. Salman
Volume 95 **Communication, Control and Security Challenges for the Smart Grid** S.M. Muyeen and S. Rahman (Editors)
Volume 96 **Industrial Power Systems with Distributed and Embedded Generation** R Belu
Volume 97 **Synchronized Phasor Measurements for Smart Grids** M.J.B. Reddy and D.K. Mohanta (Editors)
Volume 98 **Large Scale Grid Integration of Renewable Energy Sources** A. Moreno-Munoz (Editor)
Volume 100 **Modeling and Dynamic Behaviour of Hydropower Plants** N. Kishor and J. Fraile-Ardanuy (Editors)
Volume 101 **Methane and Hydrogen for Energy Storage** R. Carriveau and D.S.-K. Ting
Volume 104 **Power Transformer Condition Monitoring and Diagnosis** A. Abu-Siada (Editor)
Volume 106 **Surface Passivation of Industrial Crystalline Silicon Solar Cells** J. John (Editor)
Volume 107 **Bifacial Photovoltaics: Technology, applications and economics** J. Libal and R. Kopecek (Editors)

Volume 108	**Fault Diagnosis of Induction Motors** J. Faiz, V. Ghorbanian and G. Joksimović
Volume 110	**High Voltage Power Network Construction** K. Harker
Volume 111	**Energy Storage at Different Voltage Levels: Technology, integration, and market aspects** A.F. Zobaa, P.F. Ribeiro, S.H.A. Aleem and S.N. Afifi (Editors)
Volume 112	**Wireless Power Transfer: Theory, technology and application** N. Shinohara
Volume 115	**DC Distribution Systems and Microgrids** T. Dragičević, F. Blaabjerg and P. Wheeler
Volume 117	**Structural Control and Fault Detection of Wind Turbine Systems** H.R. Karimi
Volume 119	**Thermal Power Plant Control and Instrumentation: The control of boilers and HRSGs, 2nd Edition** D. Lindsley, J. Grist and D. Parker
Volume 120	**Fault Diagnosis for Robust Inverter Power Drives** A. Ginart (Editor)
Volume 121	**Monitoring and Control Using Synchrophasors in Power Systems with Renewables** I. Kamwa and C. Lu (Editors)
Volume 123	**Power Systems Electromagnetic Transients Simulation, 2nd Edition** N. Watson and J. Arrillaga
Volume 124	**Power Market Transformation** B. Murray
Volume 125	**Wind Energy Modeling and Simulation, Volume 1: Atmosphere and plant** P. Veers (Editor)
Volume 126	**Diagnosis and Fault Tolerance of Electrical Machines, Power Electronics and Drives** A.J.M. Cardoso
Volume 128	**Characterization of Wide Bandgap Power Semiconductor Devices** F. Wang, Z. Zhang and E.A. Jones
Volume 129	**Renewable Energy from the Oceans: From wave, tidal and gradient systems to offshore wind and solar** D. Coiro and T. Sant (Editors)
Volume 130	**Wind and Solar Based Energy Systems for Communities** R. Carriveau and D.S.-K. Ting (Editors)
Volume 131	**Metaheuristic Optimization in Power Engineering** J. Radosavljević
Volume 132	**Power Line Communication Systems for Smart Grids** I.R.S Casella and A. Anpalagan
Volume 139	**Variability, Scalability and Stability of Microgrids** S.M. Muyeen, S.M. Islam and F. Blaabjerg (Editors)
Volume 145	**Condition Monitoring of Rotating Electrical Machines** P. Tavner, L. Ran, C. Crabtree
Volume 146	**Energy Storage for Power Systems, 3rd Edition** A.G. Ter-Gazarian
Volume 147	**Distribution Systems Analysis and Automation, 2nd Edition** J. Gers
Volume 155	**Energy Generation and Efficiency Technologies for Green Residential Buildings** D. Ting and R. Carriveau (Editors)
Volume 157	**Electrical Steels, 2 Volumes** A. Moses, K. Jenkins, Philip Anderson and H. Stanbury
Volume 158	**Advanced Dielectric Materials for Electrostatic Capacitors** Q. Li (Editor)
Volume 159	**Transforming the Grid towards Fully Renewable Energy** O. Probst, S. Castellanos and R. Palacios (Editors)
Volume 160	**Microgrids for Rural Areas: Research and case studies** R.K. Chauhan, K. Chauhan and S.N. Singh (Editors)
Volume 166	**Advanced Characterization of Thin Film Solar Cells** N Haegel and M. Al-Jassim (Editors)
Volume 172	**Lighting Interaction with Power Systems, 2 volumes** A. Piantini (Editor)
Volume 905	**Power System Protection, 4 volumes**

Modern Power Electronic Devices

Physics, applications, and reliability

Edited by
Francesco Iannuzzo

The Institution of Engineering and Technology

Published by The Institution of Engineering and Technology, London, United Kingdom

The Institution of Engineering and Technology is registered as a Charity in England & Wales (no. 211014) and Scotland (no. SC038698).

First published 2020

The Institution of Engineering and Technology
Michael Faraday House
Six Hills Way, Stevenage
Herts, SG1 2AY
United Kingdom

www.theiet.org

British Library Cataloguing in Publication Data
A catalogue record for this product is available from the British Library

ISBN 978-1-78561-917-5 (hardback)
ISBN 978-1-78561-918-2 (PDF)

Typeset in India by MPS Limited

Contents

About the editor

Francesco Iannuzzo received the M.Sc. degree in Electronic Engineering and the Ph.D. degree in Electronic and Information Engineering from the University of Naples, Italy, in 1997 and 2002, respectively. He is primarily specialized in power device modeling.

He is currently a professor of Reliable Power Electronics at the Aalborg University, Denmark, where he is also part of CORPE, the Center of Reliable Power Electronics. His research interests are in the field of reliability of power devices, including mission-profile based life estimation, condition monitoring, failure modeling, and testing up to MW-scale modules under extreme conditions. He is the author or co-author of more than 220 publications on journals and international conferences, three book chapters, and four patents. Besides publication activity, over the past years, he has been contributing 17 technical seminars about reliability at first conferences as ISPSD, EPE, ECCE, PCIM, and APEC.

Prof. Iannuzzo is a senior member of the IEEE (Industry Application Society, Reliability Society, Power Electronics Society, and Industrial Electronics Society). He currently serves as Associate Editor for the *IEEE Journal of Emerging and Selected Topics in Power Electronics*, *Transactions on Industry Applications*, the *EPE Journal*, and *Elsevier Microelectronics Reliability*. He is the vice-chair of the IEEE IAS Power Electronic Devices and Components Committee. In 2018 he was the general chair of the 29th ESREF, the first European conference on the reliability of electronics, and has recently been appointed general chair for the EPE 2023 conference in Aalborg.

Preface

Over the last three decades, Power Electronics has been supporting the development and innovation of an extraordinary number of applications and disciplines, such as robotics, automotive, aeronautics, satellites, industrial automation, the internet of things, as well as personal computers, mobile phones, telecommunications, and almost the entire electricity production and distribution chains. Power Electronic devices are the building blocks in all of these applications but are far away from behaving ideally, that is, as switches that can be considered perfectly on or perfectly off. Many limitations exist, depicting a complex scenario of performance, on the one hand, and demands, on the other, which require designers to master power devices' behaviors in order to avoid trivial and costly design mistakes.

Bearing the above in mind, this book is intended for Power Electronics designers who want to deepen their knowledge and confidence with respect to Power Electronic devices by providing concrete answers and support to fundamental questions inevitably arising during the application-design process. Our effort was directed toward describing devices' behaviors without necessarily introducing theoretical discussions about device fundamentals. Many more authoritative books are available in the literature that can help the reader in that respect. The focus here has been rather to describe the operation principles and then move as quickly as possible to the applications, including practical examples and dependence of device behavior on key parameters, such as temperature. Failure mechanisms are also discussed, wherever relevant. Finally, special attention has been paid to reliability, which is one of the most important, yet difficult, aspects to evaluate in Power Electronics. We tried to cover all the most relevant types of devices used nowadays, giving a very special emphasis to the emerging ones made of wide-bandgap materials, which are silicon carbide (SiC) and gallium nitride (GaN). In addition, we included a chapter on module design, another one specifically on the switching-cell design, and a third one about optimal gate driving strategies. This choice comes from the awareness of technological trends, which aim at minimizing the external device package, if not even removing it, in order to meet the present pushes for reduced parasitic elements. We believe that Power Electronics application designers must become confident with this challenge, especially in the fields where packaged-device-based design is not possible anymore for the lack of space or for power density constraints, and bare dies must be used directly on printed-circuit boards with no intermediate enclosures. In such a case, mastering power circuits' layout is paramount, together with a proper gate circuit design. In the end, the book is nicely integrated with an entire chapter on the prospects in power electronics technology and market, which provides the reader with the trend and a forecast of how the power device sector is about to grow.

The structure we gave to the book is topic-specific, where chapters can be regarded as stand-alone, in order to speed up the reading process, with immediate focus on the design parameters, especially from the user's point of view, for the device chapters, and on concrete solutions/challenges for the others. A brief description of the 14 chapters is given below for the sake of the reader's orientation.

In Chapter 1, I myself introduce the present challenges in Power Electronics, emphasize the relevance of this discipline in today's scenario, and point out the key parameters to pay attention to from the application-design perspective.

In Chapter 2, Lin Liang presents junction diodes, starting from the PN-junction and key concepts such as breakdown voltage and junction capacitance. Then, the most important classification is presented among PiN diodes, fast-recovery diodes, and drift-step recovery diodes, and some typical applications are shown together with instability issues, especially temperature-related. SiC-based junction diodes are also briefly introduced.

In Chapter 3, Jan Vobecky introduces thyristors, the very first Power Electronic devices in time (1952), which recently brought the application segment of high-voltage direct current (HVDC) transmission to the record level of 12 GW, and can also be found in motor control and power quality systems as well as others. Their relative ease of driving, a very limited cost per converted watt, together with a huge current and voltage handling capacity and forward and reverse blocking capability up to 10 kV, make the thyristor market ever-growing. The chapter explains their key characteristics and application aspects and draws the principal technology improvements which led to modern thyristors, together with the limitations and future challenges of this technology.

In Chapter 4, Gerald Deboy presents the metal-oxide-semiconductor field-effect transistor (MOSFET), which is by far the most used component in Power Electronics. Starting from basics, the author moves into details of the super junction principle and the shielded-gate concept, evidencing how tremendous progress through technology nodes has allowed the development of such an incredibly robust and advantageous technology.

In Chapter 5, Munaf Rahimo and Paula Díaz Reigosa present the insulated-gate bipolar transistor (IGBT). No need for introducing what IGBT device is, and how its invention has radically changed the Power Electronics arena. The chapter starts with basic operation principles, both static and dynamic, then presents the technology trends in the field. Afterward, an important discussion around instabilities and failure modes make out of the chapter a precious reference for the reader. In the end, an interesting roadmap with a nice overview of forthcoming improvements concludes the chapter.

In Chapter 6, Eric Carroll presents the most recent high-power turn-off device, i.e. the integrated gate-commutated thyristor (IGCT), which combines the main characteristics of IGBTs, that is, snubberless turn off, and thyristors, that is, high blocking voltage and low on-state voltage drop. The chapter starts from the device ancestor, that is, the gate-turn-off thyristor (GTO) and describes both the static and dynamic characteristics of IGCTs. After that, the gate-driving principle circuit is introduced together with the clamp circuit. Applications, including practical details such as mechanical mounting and circuit simulations, follow the characteristic part, then the author concludes with some prospects.

In Chapter 7, Jens Peter Konrath introduces silicon-carbide (SiC) diodes. After an important classification, needed to clarify the important differences among the several types of diodes in SiC, the author presents the topic of forward-bias ruggedness by means of surge current stability. The impact of the edge termination on the reverse bias reliability is discussed thereafter, followed by a concrete example of operation. Prospects about future development end the chapter.

In Chapter 8, Luca Maresca, Alessandro Borghese, Gianpaolo Romano, Asad Fayyaz, Michele Riccio, Giovanni Breglio, Alberto Castellazzi, and Andrea Irace present the MOSFET device in silicon-carbide technology. After a description of the operation principle, including differences with silicon MOSFETs, an important discussion about instabilities taking place both in short-circuit conditions and avalanche conditions is presented, providing practical and important details about testing procedures, too.

In Chapter 9, Shu Yang and Shaowen Han present gallium-nitride (GaN) metal-insulator-semiconductor field-effect transistors (MIS-FETs). The recent progress and primary challenges in GaN power devices are reviewed and discussed in this chapter. Several kinds of instabilities that may take place at different operating conditions are discussed extensively, providing many cases and examples. An important mention of the problems related to gate driving is also given, with interesting insights into the phenomena taking place during switching operations. The chapter is concluded with some prospects about the near-future improvements of this very promising technology.

In Chapter 10, Giorgia Longobardi provides an overview of GaN transistor applications and describes the frontiers of vertical technology, clearly pointing out the main differences with respect to the high-electron mobility transistor (HEMT) structure. State-of-the-art vertical technologies are thoroughly described, with six possible solutions, each of them presented and discussed together with their advantages and disadvantages.

In Chapter 11, Daohui Li, Xiaoping Dai, and Guoyou Liu present the state-of-the-art fundamentals of module design, including parasitic study and thermal optimization. The second though very important part of the chapter is about reliability estimation and enhancement. The authors discuss wire bonding, solder materials, sintering process, ultrasonic welding, and other important module fabrication technologies from the reliability perspective, pointing out the strengths and the weaknesses of the considered solutions.

In Chapter 12, Eckart Hoene and Kirill Klein present the fundamental principles of switching-cell design, including thermal, electrical, and mechanical interfaces. Afterward, the chapter deals with DC-link design concepts, which are key to avoid trivial mistakes at the layout level. Alternative contacting technologies are also introduced at the end of the chapter.

In Chapter 13, Haoze Luo, Wuhua Li, and myself introduce some important principles to be understood for a correct design of gate driving circuits for IGBTs aimed at improving robustness and reliability. After classifying gate driving circuits according to the basic driving principles, that is, voltage-source and current-source types, fault detection, active gating methods, and active thermal control are introduced as complementary techniques for IGBT gate driving.

Finally, in Chapter 14, Elena Barbarini draws the current trend in the semiconductor market, especially in relation to the electric vehicles and wide-bandgap devices. Development and evolution in the power device packaging sector are also reported, which depict an amazingly evolving scenario. Quite interestingly, silicon IGBTs and MOSFETs are yet expected to grow, basically due to the increasing number of power electronics applications together with the incomparable reliability of those technologies.

Aalborg, Denmark — *Francesco Iannuzzo*
April 2020

Chapter 1
Introduction: Power Electronics challenges

Francesco Iannuzzo[1]

1.1 Power Electronics

The demand for electricity in the new decade is close to 30,000 TWh/year and most of it flows through Power Electronics installations [1,2]. The Power Electronics discipline was officially introduced by William E. Newell during a keynote speech at the PESC conference in 1973. He proposed "[. . .] Power Electronics [as] the interstitial technology in which significant amounts of electrical power flow through and are controlled by electronic devices" [3]. After almost half a century, this concept has originated thousands of diverse applications. Nowadays, in the majority of cases in which electricity has to be adapted in voltage and current or converted to other forms of energy, the use of Power Electronics has become inevitable. This has conferred to Power Electronics a key and enabling role in human subsistence and development. According to a study by Yole Développement in 2019, the Power Electronic market for discretes and modules alone was worth \$17.5 billion in 2018 and is forecasted to be \$21.8 billion by 2024 [4].

The principle schematic of a Power Electronics converter is shown in Figure 1.1. The input power with parameters V_{in}, I_{in} is typically fed into an input filter in order to reduce disturbances (electromagnetic interference, EMI) to the circuits' upstream of the converter. After the input filter, a power processor based on semiconductor devices operates in two phases: first, it draws energy from the input side and stores it temporarily into an energy buffer; afterward, it connects the energy buffer to the output and releases the stored energy to it. By varying the process timing, one achieves different electrical parameters V_{out}, I_{out} at the output side. These two phases are repeated continuously, at a frequency called *switching frequency*. The energy buffer plays a key role in the overall conversion process as input and output power parameters are typically different, e.g. in voltage or frequency, which creates a delay and, in general, nonsynchronized energy fluxes. In order to keep the overall efficiency as high as possible, the energy buffer is made of a conservative element, i.e. a capacitor, as in the figure, or an inductor. The output power is fed into an output filter for similar reasons as the input side. The input and

[1]Department of Energy Technology, Aalborg University, Aalborg, Denmark

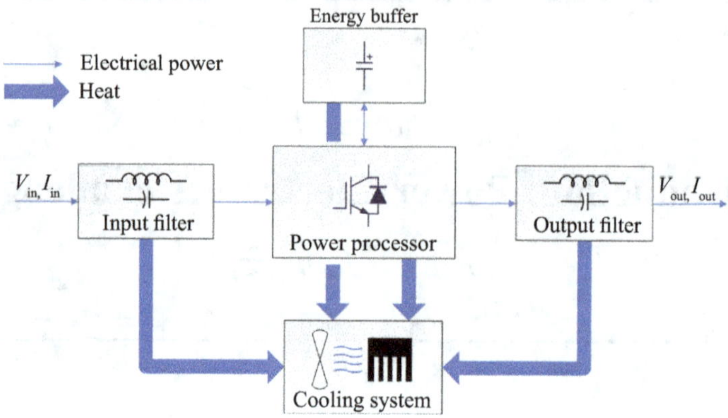

Figure 1.1 Principle schematic of a Power Electronics converter

output filters are typically made up of capacitors and inductors. Very often, it is possible to combine one or both the filter stages with the energy buffer.

Power Electronic converters exhibit typically very high efficiencies, which seldom fall below 80%–90%. Nevertheless, not being ideal machines, they generate power losses that have to be dissipated by a cooling system. Heat is generated in all the parts of the circuit, including interconnections such as printed circuit boards (PCBs) and busbars.

1.2 Power devices: the core of Power Electronics

Unlike filters and the energy buffer, the power processor of Figure 1.1 is made up of the so-called active devices or Power Electronic devices, i.e. semiconductor components that allow the current flow under certain conditions, e.g. when a command signal is applied.

A Power Electronic device operates as a switch, i.e. in two possible states: off-state and on-state. In an ideal Power Electronic device, off-state means that the current is zero and the voltage drop across it can increase arbitrarily. On the other hand, on-state means that the voltage across it is zero and an arbitrary amount of current can flow through it. In addition, commutations between the two states are instantaneous and lossless. Of course, the above conditions are not achievable in reality, so a number of imperfections have to be taken into account. Table 1.1 summarizes a set of these shortcomings and their implications on real applications.

Power Electronic devices can be classified based on their controllability, i.e. the ability on the user's side to switch among the two states. They can be (a) uncontrolled, (b) semi-controllable or (c) (fully-)controllable (see Figure 1.2).

Schottky-barrier diodes (SBDs) and PiN diodes are basically the only devices belonging to the uncontrolled category, which means that the user has no way to force a status change; in other words, their status only depends on the current force

Table 1.1 Principal imperfections of Power Electronic devices

Imperfection	Consequences in real applications
Nonzero current at off-state ("leakage current")	Power losses at off-state
Limited voltage at off-state ("Breakdown voltage")	The device choice is affected by the maximum expected voltage, including overvoltages
Limited current at on-state	The device choice is affected by the maximum expected current, including overcurrents
Nonzero voltage drop at on-state	Power losses at the on-state ("conduction losses")
Nonzero switching times	Switching losses

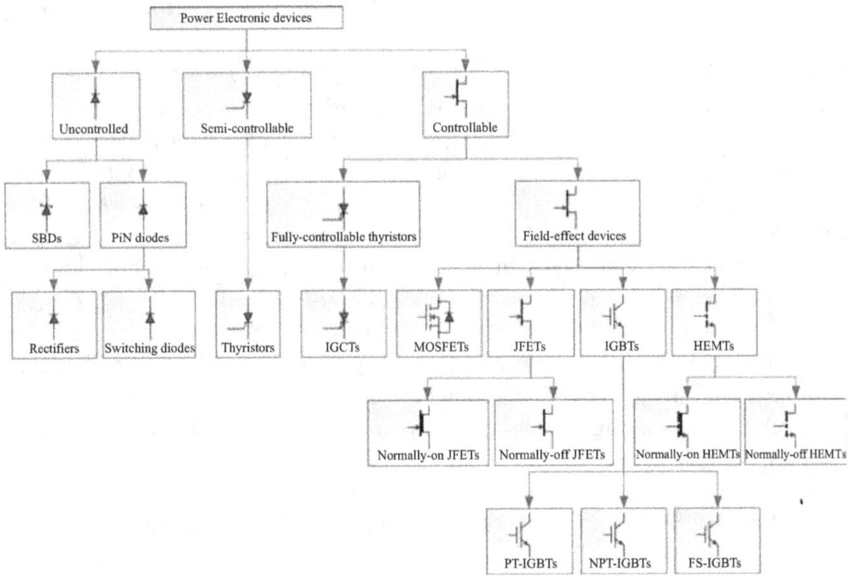

Figure 1.2 Classification of Power Electronic devices

from the external circuit. These two types of devices differ considerably from each other, if one looks into their internal structure. The internal structure of a PiN diode is basically a modified PN junction with an intermediate "intrinsic" layer, wherefrom the name P-i-N. PiN diodes are further classified into rectifiers and switching diodes, based on the applications. In particular, rectifiers operate at 50 or 60 Hz and are optimized for AC/DC conversion from the grid. At such low frequencies, only conduction losses are relevant and switching losses are negligible. Switching diodes are optimized based on the trade-off between on-state and switching losses and are suitable for working at high switching frequencies. SBDs are diodes with a "Schottky junction," from physicist W. Schottky, which is a metal-semiconductor junction. Even though SBDs exhibit quite a large off-state current, they perform very well from both conduction and switching points of view. However, in the case of silicon, their maximum operating voltage is seldom beyond 100 V. On the other hand, thanks to the quick spread of wide-bandgap (WBG) materials, particularly silicon carbide (SiC), SBDs can now reach 1,700 V and above.

In the semi-controllable category, there are basically silicon-controlled rectifiers (SCRs) or thyristors, i.e. rectifiers that are normally off and can be turned on by means of a current pulse fed to a gate terminal. On the other hand, once fired, the gate loses the control capacity and the only way to turn thyristors off is to make the current return to zero. Thyristors were invented in the early 1950s; since then, they revolutionized the world of electronic control, making it possible the practical adoption of switching modulation. Nowadays, however, thyristors have been completely outperformed by fully-controllable Power Electronic devices at low and medium voltage and are mainly used in high-voltage, line-commutated applications.

The first fully-controllable (or just "controllable") Power Electronic devices were power bipolar junction transistors (BJTs), Darlingtons and, later, Triplingtons. However, a special mention goes to the gate turn-off thyristors (GTOs), which originated from thyristors, but with the disruptive feature of turn-off controllability. In the 1990s though, IGBTs (insulated-gate bipolar transistors) and MOSFETs (metal-oxide-semiconductor field-effect transistors) started taking over, thanks to their simplified controllability with zero gate current, which has probably been the most important breakthrough in the evolution of Power Electronic devices. Nowadays, two well-established categories of controllable devices exist, i.e. fully-controllable thyristors and field-effect devices. The former category basically comprises the integrated gate-commutated thyristor (IGCTs) only, which dominates the high-voltage, high-current application sector. In the latter category, MOSFETs, JFETs (junction field-effect transistors), IGBTs and HEMTs (high-electron-mobility transistors) populate the landscape. It is worth pointing out, though, that MOSFETs and IGBTs are the most used devices made of silicon, whereas SiC MOSFETs and JFETs and gallium nitride (GaN) HEMTs are the most used ones made of wide-bandgap materials. JFETs can be further classified based on their conduction state when their gate is biased at 0 V ("normal state"). If they are not conducting, then they are referred to as "normally-off," otherwise they are called "normally-on." A very similar classification holds for HEMTs. It is worth pointing out that normally-on devices are generally deprecated as such, as they require additional circuitry, e.g. during power-up in order to avoid a shoot-through of the converter DC-link. For the sake of completeness though, it is worth mentioning that the so-called "cascode-connection" solves, to a large extent, the above problem by using a high-efficiency silicon MOSFET in series with a high-voltage normally-on device. A different classification holds for IGBTs, based on technological generations. The first one was the punch-through (PT) IGBT. The devices developed in PT technology had fairly low conduction losses but were unstable in temperature, which is a limiting feature in Power Electronics. Successively, non-punch-through (NPT) IGBT could be fabricated also, thanks to a number of technological improvements, which allowed one more degree of freedom in device design and obtaining temperature stability as well. The last generation is the field-stop (FS) IGBT, which is a combination of the former two.

1.3 Wide-bandgap semiconductors

A Power Electronic device is made of a piece of semiconductor material; hence, its electrical performance is inherently related to the physical properties of that specific material. Consider, for instance, the critical electric field, which is the maximum electric field a device can withstand without undergoing avalanche breakdown.* For a given device thickness, the higher the critical field, the higher the blocking voltage. Since the thickness is related to the on-state resistance, it goes without saying that if a semiconductor material exhibits a higher critical electric field than silicon, one can fabricate devices with a higher blocking voltage having the same on-state resistance or with a lower on-state resistance having the same blocking voltage.

Figure 1.3 shows the contours of Baliga figure-of-merit (BFOM), defined as [5–7]:

$$\mathrm{BFOM} = \varepsilon_S \mu_n E_C^3, \tag{1.1}$$

where ε_S and μ_n are the permittivity and the electron mobility of the considered semiconductor material, respectively, and E_C is its critical electric field. For a

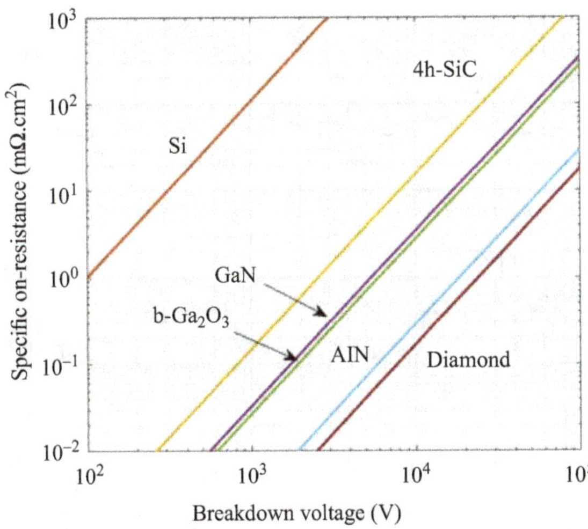

Figure 1.3. Contour plot of Baliga figure-of-merit (BFOM) for silicon and various wide-bandgap semiconductors [7]

*Avalanche breakdown is the phenomenon of uncontrolled current flow occurring when a semiconductor junction is biased over its maximum admissible voltage.

constant value of BFOM, it is possible to show from (1.1) that the square root of breakdown voltage and the specific on-resistance† have a constant product, i.e. form a negative-gradient straight line on a log-log plot. Figure 1.3 clearly displays the enormous potential of WBG materials. If one compares 4H–SiC, which is the most used crystalline form of SiC, with traditional silicon at the same specific on-resistance, one obtains a factor of ≅ 25 in breakdown voltage. Similarly, if one compares them at the same breakdown voltage, a factor of ≅ 500 can be theoretically achieved. Although these observations are based on a theoretical figure of merit, they give the first impression of how large the benefits of adopting WBG semiconductors are.

The advantages of WBG materials with respect to silicon also come from their thermal properties. More specifically, the thermal resistance of a given device is inversely proportional to the thermal conductivity of the semiconductor material. The higher the thermal conductivity, the lower the thermal resistance, the higher the admissible losses not to exceed a given temperature limit. The latter point allows processing higher power at the same temperature, or the same power at a lower temperature, thus improving, e.g. the conversion efficiency.

Table 1.2 summarizes the key differences among silicon and the principal WBG semiconductors, in particular, three types of SiC and GaN [8]. The first and most important entry is the *bandgap*. In the energy diagram of a piece of semiconductor material, the bandgap is the forbidden range of energy values for electrons. This introduces a gap in the energy levels and forces the lower ("valence") band, which is responsible for the chemistry of the material, to stay apart from the

Table 1.2 Comparison of principal wide-bandgap material characteristics with the ones of silicon [8]

Property	**Si**	**4H–SiC**	**6H–SiC**	**3C–SiC**	**Diamond**	**ZnO**	**AlN**	**GaN**
Bandgap (eV)	1.1	3.2	3.0	2.3	5.45	3.37	6.2	3.42
Dielectric constant	11.8	10	9.7	9.6	5.5	8.66	8.5	8.9
Critical electric field (MV/cm)	0.3	2.0	2.4	1.2	5.6	4.0	1.8	3.3
Thermal conductivity (W/cm-K)	1.5	4.5	4.5	4.5	20	0.6–1.16	2	2.4
Saturated electron velocity (10^7 cm/s)	1.0	2.0	2.0	2.0	2.7	3.0	1.5	2.7
Electron mobility (cm^2/V-s)	1,350	720	370	900	1,900	210	300	1,300 >2,000 (2DEG)[a]

[a]2D-electron gas conduction mode.
Note: Dielectric constant is a unitless, dimensionless quantity.

†As resistance is inversely proportional to the area, specific on-resistance is defined as [resistance]·[area unit] (Ω·cm^2).

upper ("conduction") band, which holds free electrons and is responsible for the current conduction. The larger this bandgap, the higher the energy jump needed for a valence electron to travel across the forbidden band and to reach the conduction band. The energy jump is directly related to the critical electric field mentioned in the beginning of this section. This is where the WBG materials take their name from. Some other parameters are also very relevant, e.g. the critical electric field and the thermal conductivity, both of them are mentioned earlier in this section. In the case of WBG materials, both quantities are overwhelming with respect to silicon. Electron mobility deserves an important mention, though. As a matter of fact, only some of the WBG semiconductors exhibit higher electron mobility than silicon. Lower electron mobility causes a larger resistivity of pure material. This has some negative implications in the design and operation of bipolar devices, i.e. the ones where two species of carriers contribute to the conduction process, such as thyristors and IGBTs. In fact, the reduced electron mobility prompts device designers to compensate for it with an enhanced carrier concentration, which in turn makes transients slower for bipolar devices made of WBG semiconductors.

Along with the many advantages, there are some significant disadvantages of WBG semiconductors. The main drawback is essentially the manufacturing cost, as the fabrication of ingots of pure material cannot be done with the same techniques as for silicon, such as Czochralski and float zone [9,10]. Nevertheless, the cost is unanimously predicted to decrease significantly in the next 5 years and the related market is forecast to boom. According to the IEEE–PELS International Technology Roadmap for Wide Bandgap Power Semiconductors (ITRW, 2019 Edition), "[the] total market for SiC power devices by 2023 will be over $1.5 billion with a compound annual growth rate (CAGR) of 31% for next five years, and [...] GaN device business to reach around $423 million by 2023 with a CAGR of 55%. Projections by PowerAmerica are that the total market for WBG devices in Power Electronics will be over $11 billion by 2027" [11].

1.4 Operational range

The limitations introduced in Section 1.2 have pushed Power Electronic device designers over the past years to develop large product categories, in particular regarding blocking voltage, which best suit specific application requirements. Figure 1.4 shows a qualitative distribution of sample Power Electronics applications with respect to the operating voltage, the switching frequency and the converter power [12]. It is worth noting that the switching frequency is roughly inversely proportional to the square of the operating voltage, which comes directly from a trade-off intrinsic to semiconductor physics. Although the diagram of Figure 1.4 is merely qualitative, Power Electronics applications can be in a first approximation distributed along a straight line. Silicon Power Electronic devices (solid ovals) covering such applications range from MOSFETs and SBDs, through IGBTs and FRDs, to SCRs, IGCTs and rectifiers when voltage moves from few volts to tens of kilovolts. WBD devices (dashed

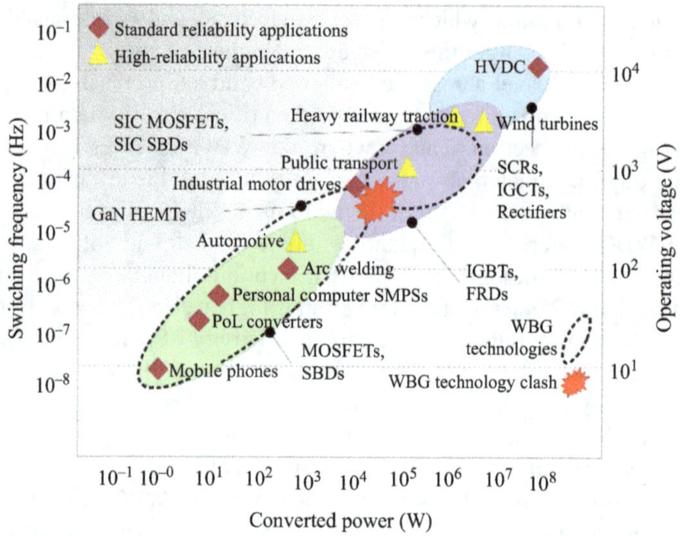

Figure 1.4 Application range of modern Power Electronic devices [12]

ovals), i.e. GaN HEMT, SiC MOSFETs and SBDs are emerging very quickly; they cover a more and more significant part of the whole application range. In particular, the center region around 600 V and beyond is a clash area for the two new technologies.

From Figure 1.4, a rule of thumb can be directly inferred for both silicon and WBG materials. For silicon, below 200–300 V (and powers up to few kilowatts), MOSFETs and SBDs should be used, whereas IGBTs and FRDs should be used above this limit. For WBG materials, GaN HEMTs should be used up to 500–600 V, whereas SiC MOSFETs and SiC SBDs should be used above this limit. Beyond several kilovolts, and especially at high currents, hence powers, silicon still dominates the arena with no reasonable alternative (rectifiers, IGBTs, SCRs and IGCTs). This conclusion is just as preliminary as it can be, as many design parameters and management decisions in the product development process can divert attention toward a neighboring device or material, although this reflection can still be considered as a good basis for starting a design process.

1.5 Temperature, reliability and other challenges

The Power Electronics application design process can be very long and, in general, can require many iterations to ensure that all the functional specifications and all design constraints are met. Undoubtedly, one of the greatest challenges to face is to confidently ensure that the maximum junction temperature in each device does not exceed the admissible level, nor gets too far below it, with a consequent loss of money. In fact, this is a complex problem, for three main reasons:

1. The instantaneous temperature depends on the instantaneous power loss, which can be very challenging to be worked out, as an accurate current and voltage prediction of a given device relies on its accurate electrical modeling, which pushes forward the issue to find simulation models accurate enough in all the given operating conditions. This is especially true in nonhard switching applications, such as soft-switching and resonant topologies. On top of that, losses depend, in turn, upon temperature; therefore, an accurate temperature estimation requires two-domain modeling.
2. The thermal stack is made up of different materials with different geometries that may not be known at design time. This requires an extra design effort and, as mentioned in the beginning of this section, several design iterations.
3. Wear of components occurring during real life is also relevant as it eventually affects their electrical and thermal characteristics, ending up in altering the junction temperature profile and, in particular, its maximum.

The maximum junction temperature is not the only problem to cope with during the Power Electronic design process. Reliability definitely deserves special attention. In a nutshell, a converter working at time zero may not be functional anymore after a short period of time. The main reason is that components wear out during the operational life and a parametric drift can occur until specifications are no longer met. Predicting the expected operational life, and ensuring that the converter will reach it, is a very complicated task. It goes without saying that the wear rate of components depends on temperature too. This gives the first impression of how complex the design process can be.

In general, many other operating conditions that affect the expected life of converters should be taken into account. Among others, we can list up:

- Humidity
- Salinity
- Vibration
- Human factor/imperfect maintenance actions
- Radiation
- EMI

In Figure 1.4, some examples of high-reliability applications are highlighted.

In general, a robust Power Electronic design is an iterative and multidisciplinary process, where no perfect recipes can be proposed. Definitely, a deep knowledge of the principle of operation of Power Electronic devices can avoid trivial mistakes and a huge waste of time and money.

1.6 Summary

Power Electronics has nowadays become an enabling discipline and has originated thousands of diverse applications. The core of a Power Electronic converter is the Power Electronic device. Power Electronic devices can be classified as uncontrolled, semi-controllable and fully-controllable. According to this classification, and because of the nonideal behavior, e.g. in limited blocking voltage, nonzero on-state voltage drop and noninfinite switching speed, many different kinds of power devices have populated the Power Electronics arena in the past five decades.

Recently, WBG semiconductors emerged as a very promising technology to fabricate efficient Power Electronic devices, but many fabrication problems are still to be solved in order to lower the cost at a competitive level with traditional silicon. However, all the market forecasts predict a boom of WBG technologies in Power Electronics in the next 5–10 years.

The design of Power Electronics converters is a complex and tricky process, involving several disciplines, such as device operation theory and thermal management, which require design engineers to gain vast and cross-disciplinary expertise. Among the many different challenges, correct and accurate junction temperature estimation and reliability prediction are paramount.

Note to the reader

This chapter is part of an edited book. To cite or use this chapter as a reference, please use the following format:

Iannuzzo F. "Introduction: Power Electronics challenges." In: Iannuzzo F (ed.), *Modern Power Electronic Devices: Physics, applications, and reliability*. Stevenage, UK: IET; 2020. pp. 1–11.

References

[1] IEA, "World Energy Outlook 2019–Electricity." [Online]. Available: https://www.iea.org/reports/world-energy-outlook-2019/electricity.

[2] IRENA, "Global Energy Transformation: A Roadmap to 2050." [Online]. Available: http://www.irena.org/publications/2018/Apr/Global-Energy-Transition-A-Roadmap-to-2050.

[3] W. E. Newell, "Power Electronics-Emerging from Limbo [Conference Keynote Address]," *PESC Rec. - IEEE Annu. Power Electron. Spec. Conf.*, vol. 1973-Jan, pp. 6–12, 1973.

[4] A. Villamor, "Status of the Power Electronics Industry 2019." [Online]. Available: https://www.i-micronews.com/products/status-of-the-power-electronics-industry-2019/.

[5] B. J. Baliga, "Semiconductors for high-voltage, vertical channel field-effect transistors," *J. Appl. Phys.*, vol. 53, no. 3, pp. 1759–1764, 1982.

[6] B. J. Baliga, *Fundamentals of Power Semiconductor Devices*. Cham: Springer International Publishing, 2019.

[7] J. Y. Tsao, S. Chowdhury, M. A. Hollis, *et al.*, "Ultrawide-bandgap semiconductors: Research opportunities and challenges," *Adv. Electron. Mater.*, vol. 4, no. 1, 2018.

[8] M. A. Fraga, H. Furlan, R. S. Pessoa, and M. Massi, "Wide bandgap semiconductor thin films for piezoelectric and piezoresistive MEMS sensors

applied at high temperatures: An overview," *Microsyst. Technol.*, vol. 20, no. 1, pp. 9–21, 2014.

[9] J. Friedrich, "Methods for bulk growth of inorganic crystals: Crystal growth," *Ref. Modul. Mater. Sci. Mater. Eng.*, 2016.

[10] R. Abbaschian, "Crystal growth," in *Encyclopedia of Materials: Science and Technology*, Elsevier, 2001, pp. 1860–1863.

[11] IEEE PELS, "ITRW International Technology Roadmap for Wide Bandgap Power Semiconductors 2019 Edition," 2019.

[12] F. Iannuzzo, C. Abbate, and G. Busatto, "Instabilities in silicon power devices: A review of failure mechanisms in modern power devices," *IEEE Ind. Electron. Mag.*, vol. 8, no. 3, pp. 28–39, 2014.

Chapter 2
Junction diodes

Lin Liang[1]

2.1 Introduction

In a broad sense, all of the devices that have two electrodes outside could be called diodes, whether the core semiconductor structure has one, two, even three PN junctions, or metal-semiconductor contact barrier inside. But in this chapter, we only discuss the very basic case, i.e., the two-electrode devices that contain one PN junction, while they may also contain several high-low junctions of P^+P or N^+N certainly.

Generally speaking, the one-directional conductivity of the PN junction is applied to the junction diodes, which is in common. But there are great differences in their structures and characteristics in detail. Except for the basic differences of the voltage and current ratings, some have the one-dimensional PiN structure (here, i means intrinsic, though there is usually P-type or N-type dopant of low concentration actually), while some have a complex layout at anode or cathode, and then the two-dimensional effects have to be considered. Some need the extra process for the lifetime control of minority or excess carriers. There is also a big difference for the reverse recovery characteristics, which reflect soft recovery, hard recovery or even super-hard recovery. Though diodes are semiconductor devices that have a very basic structure, the research on them has never been ended.

According to the functions, junction diodes could be used as rectifiers, freewheeling diodes (FWD), clamping diodes, break-over diodes, opening switches, etc. Thereinto, the rectification is the most conventional function of diodes. The freewheeling is the function through which the diodes provide a continual loop for current from current sources constituted by load inductances of the main switches. To limit destructive voltage overshoots caused by large inductances in high-power converters, clamping diodes are frequently used. To protect electronics circuits from voltage spikes and to absorb some of the excess transient energy, break-over or avalanche diodes are used. The opening is a little special, which is the function that diodes cut off the current quickly and output high voltage in the pulsed power systems of the inductive energy storage type.

[1]School of Electrical and Electronic Engineering, Huazhong University of Science and Technology, Wuhan, China

The basic theory of PN junction is introduced briefly first in this chapter. Then, there are three sections of PiN diodes, FRDs (fast recovery diodes) and DSRDs (drift step recovery diodes), just divided by the functions described in the preceding paragraph. In each section, the structure and operation principle, characteristics and parameters, typical applications and instabilities are illustrated, respectively. In addition, some research work of silicon carbide (SiC) devices is also introduced.

2.2 PN junction [1]

2.2.1 Definition and types

In an N-type (or P-type) semiconductor wafer, the P-type (or N-type) doping is added by ion beam implantation with subsequent dopant activation and diffusion (drive-in) or by dopant predeposition and subsequent diffusion. This makes the two connected areas on the wafer to have the electric conduction types of N and P, respectively. The transition region of their interface is called the PN junction.

The PN junctions could be divided into the abrupt junction and graded junction, as shown in Figure 2.1(a) and (b). The characteristics of the abrupt junction are that the doping concentration distributes uniformly inside the P region and N region and the doping type changes abruptly at the interface. The PN junction formed, for example, by the ion implantation followed by rapid thermal annealing (dopant activation) is a typical abrupt junction. The characteristics of the graded junction are that the doping concentration does not distribute uniformly inside the P region and the N region and the doping type changes slowly from one to another by some laws of function near the PN junction with the lowest effective doping concentration at the interface. The PN junction formed by diffusion method frequently used for high-voltage power diodes is a typically graded junction. The real PN junction could be approximated by an abrupt junction or linear graded junction by the doping concentration distribution.

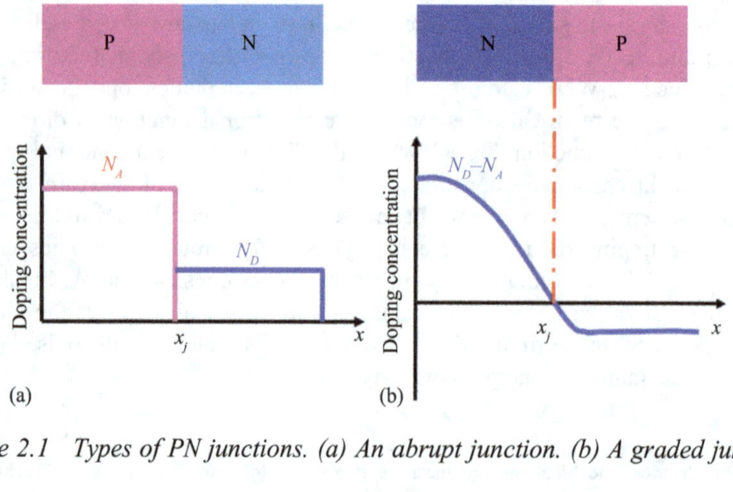

Figure 2.1 Types of PN junctions. (a) An abrupt junction. (b) A graded junction

2.2.2 Equilibrium PN junction

At the interface between the P-type and N-type regions, due to the difference in the carrier concentration of free electrons and holes, the majority carriers (electrons in the N region and holes in the P region) move forward into the opposite area by diffusion. As the mobile majority carriers diffuse away, they leave behind the immobile impurity ions, as shown in Figure 2.2. The created space charge region (SCR) of immobile ionized impurity atoms is composed of a bilayer of opposite charge polarities, the charge of which is not compensated (electrons and holes are gone). As a result, the bilayer has a high electric field in the direction of decreasing electrical potential φ. The potential energy of an electron is then $E_e = -q^*\varphi$, where q is the electron charge. It implies the energy band diagram as well the so-called built-in voltage V_D that points from the N region with positively ionized donors to the P region with negatively ionized acceptors as shown in Figure 2.3.

Together with the diffusion of the majority carriers outside the SCR, the electric filed extracts the minority electrons from the N region and the minority holes from the P region. In other words, there is a drift of electrons and holes in the opposite direction to that of the diffusion. Once the drift current densities of electrons and holes become equal to the diffusion current densities in the opposite direction, the PN junction enters the thermodynamic equilibrium. Without applying an external voltage, there is no current flowing through the PN junction. The corresponding energy band diagram is shown in Figure 2.3. The width of the SCR, which reads $X_m = x_N + x_P$, is set to the value, which provides the equilibrium state for given donor and acceptor concentrations N_D and N_A. As the area density of positive and negative charges at both sides of the bilayer must be equal, we receive the equation: $N_D{}^*x_N = N_A{}^*x_P$. It implies that the SCR is wider at the side of a lower doping concentration. This relation will still remain after application of a negative external voltage between the P and N regions.

The energy band diagram in Figure 2.3 shows the existence of energetic barrier q^*V_D, which maintains the equilibrium between the majority electrons in the N region and minority electrons in the P region and analogically between the majority holes in the P region and minority holes in the N region.

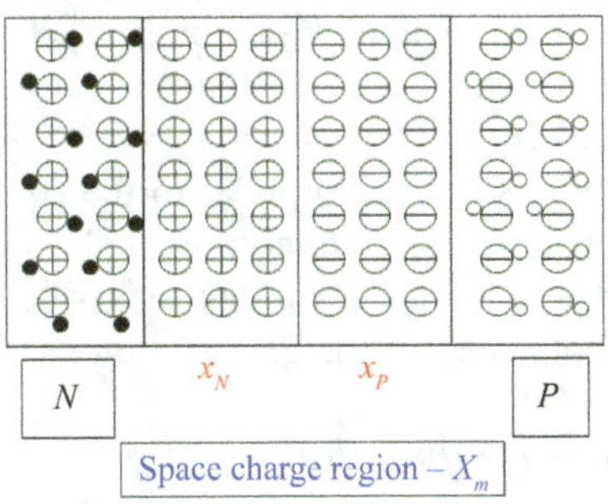

Figure 2.2 Space charge region with $x_N = x_P$ for the case $N_D = N_A$

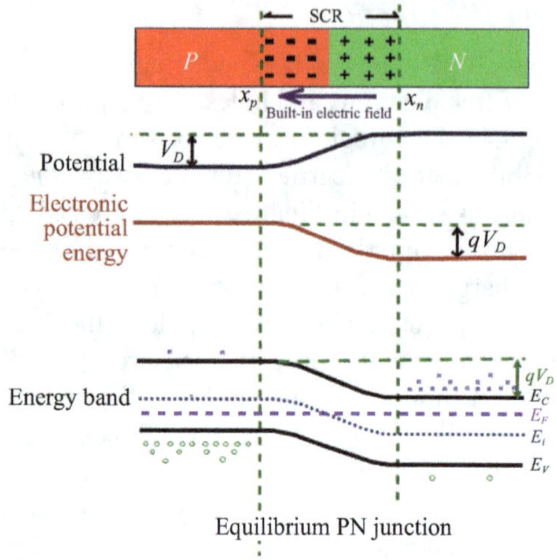

Figure 2.3 Energy band diagram for the equilibrium PN junction

The formulas of the built-in electric potential difference V_D for the abrupt junction and the linear graded junction are shown by (2.1) and (2.2), separately. Thereinto, n_i is the intrinsic carrier concentration, a is the doping concentration gradient, X_m is the width of the SCR and $kT/q = 0.026$ V ($T = 300$ K) is the thermal voltage.

$$V_D = \frac{kT}{q} \ln \frac{N_A N_D}{n_i^2} \tag{2.1}$$

$$V_D = \frac{2kT}{q} \ln \frac{a X_m}{2n_i} \tag{2.2}$$

Equation (2.1) can be rewritten as $V_D = kT\,{*}\ln(n_{N0}/n_{P0})/q$, where n_{N0} and n_{P0} are the concentrations of electrons in the N region and the P region, respectively, at zero applied voltage (analogically for the holes). The built-in voltage is then given by the concentration ratio of the majority and minority carriers at the corresponding edge of the SCR. Every growth of the ratio (n_{N0}/n_{P0}) or (N_D/N_A) by one order of magnitude then results in the growth of V_D by 60 mV. We can also write equation $n_{P0} = n_{N0}{*}\exp(-q{*}V_D/kT)$, which tells us that to transfer the charge of single electron over the potential barrier difference V_D, we need the energy $q{*}V_D$ and that the probability of doing so is proportional to the so-called Boltzmann factor $\exp(-q{*}V_D/kT)$. This probability is relatively low and, therefore, $n_{P0} \ll n_{N0}$. It is the existence of the energetic barrier $q{*}V_D$ that makes so much different electron (hole) concentrations at both sides of the PN junction possible.

2.2.3 *Nonequilibrium PN junction*

The PN junction with applied external voltage is called nonequilibrium PN junction. When the P region is connected to the positive pole and the N region to the negative pole of a power supply, the PN junction is forward biased. With the opposite polarity, the PN junction is reverse biased. As the impedance of the silicon outside the SCR is negligible, the whole applied voltage appears basically on the PN junction. As the polarity of the forward voltage is opposite to that of the V_D, it lowers the original barrier height as well as the width of the SCR. As the applied electric field has the opposite direction to the built-in electric field, the diffusion movement is stronger than the drift. The reverse voltage widens the barrier area and increases the original barrier height. As the applied electric field has the same direction as the built-in electric field, the drift is stronger than the diffusion. With the assumption of the ideal PN junction, its I–V characteristics could be described by the Shockley equation (2.3), where I_0 is the reverse leakage current and V_A is the applied voltage.

$$I = I_0\left(e^{\frac{qV_A}{kT}} - 1\right) \tag{2.3}$$

2.2.4 PN junction breakdown

When the reverse voltage on the PN junction reaches a certain value, an abrupt rise of the reverse current takes place and the PN junction appears at breakdown. The corresponding voltage is called the breakdown voltage. There are two kinds of breakdown mechanisms for the PN junction, which are avalanche breakdown and tunnel breakdown.

The process of the avalanche breakdown is as follows. When the reverse voltage at the PN junction increases, the carriers drifting through the SCR obtain sufficient energy from the electric field and collide with the lattice atoms. After reaching a critical electrical field (about 200 kV/cm in silicon), the received energy is high enough to cause the so-called impact ionization. The valence band electron is raised to the conduction band over the energy bandgap and leave a hole behind. This produces a pair of electron and hole while the original electron which caused the scattering effect is drifting further over the SCR. The newly formed electron and hole are accelerated by the electric field and collide to produce new electrons and holes. This leads to the carrier multiplication and abrupt increase of reverse current and the PN junction breakdown in the end.

The electrical breakdown caused by tunneling effect appears in the devices with very high doping concentration N_D and N_A, (typically, above 10^{18} cm^{-3}). The resulting PN junction is so narrow that it can be easily overcome by tunneling through the energetic barrier (quantum effect). When the reverse voltage at the PN junction increases, the valence band maximum of the P region is higher than the conduction band minimum of the N region. With the small width of the barrier area, the electrons in the valence band of P region pass through the barrier area to reach the conduction band of the N region by a certain probability, which forms the electron-hole pairs. Since this also increases the reverse current of the PN junction, electrical breakdown takes place. In silicon, the diodes subjected to tunneling have a breakdown voltage of less than 8 V and are called Zener diodes. The breakdown voltages above 8 V are achieved via the impact ionization leading to the avalanche multiplication.

Both the avalanche and tunnel breakdowns are nondestructive and reversible effects. If the electrical breakdown is not limited, the current continues to grow, then the thermal runaway can take place and lead to a destruction (unrecoverable). Well-designed state-of-the-art diodes are avalanche safe.

2.2.5 PN junction capacitance

PN junction features barrier and diffusion capacitances.

The SCR of PN junction contains the immobile charge of ionized dopant impurities and is empty of free carriers. As a result, it behaves as an insulator with the permittivity of silicon. When the applied voltage varies periodically, carriers flow in or out of the barrier area periodically, which is equivalent to the charging and discharging of a capacitor. This is called the PN junction barrier capacitance. The barrier capacitance is a nonlinear capacitance. It exists in the forward bias, zero bias and reverse bias of the PN junction. It is related to the applied voltage. The higher the forward voltage is, the larger the capacitance is, because the SCR shrinks. The higher the reverse voltage is, the smaller the capacitance is, because the SCR widens. It is related to the doping concentration profile, which determines how much the SCR extends with applied reverse voltage. One can simplify it to the parallel plate capacitor with capacity $C = \varepsilon_0\varepsilon_r{*}A/X_m$, where A is the area of the PN junction.

With the forward bias, the capacitance effect, which is about the electrical charges outside the SCR with varying applied voltage, is called diffusion capacitance. When the forward voltage V_A is applied to the PN junction, there is charge accumulation of the unbalanced minority carriers outside the SCR. When the V_A increases, the concentration of the injected minority carriers increases, which is equivalent to the capacitance charging. When the V_A decreases, the concentration of the injected minorities decreases, which corresponds to the capacitance discharging.

2.3 PiN diodes [2]

2.3.1 Structures and operation principle

PiN diode has been introduced by R. N. Hall [3] to provide a low forward voltage drop at high currents and a high breakdown voltage at the same time. The nearly intrinsic i layer of high resistivity (in practice called N or P base, depending on the used dopant type) is in the middle, and the high doping P^+ layer and N^+ layer of low resistivity acting as emitters injecting high carrier concentrations into the base are on both sides of the base, as shown in Figure 2.4(a). Thereinto, the middle i layer is always replaced by P layer or N layer of the resistivity, according to the requested breakdown voltage. It is usually called π layer for the high-resistivity p layer and ν layer for the high-resistivity n layer. It has unidirectional conductivity

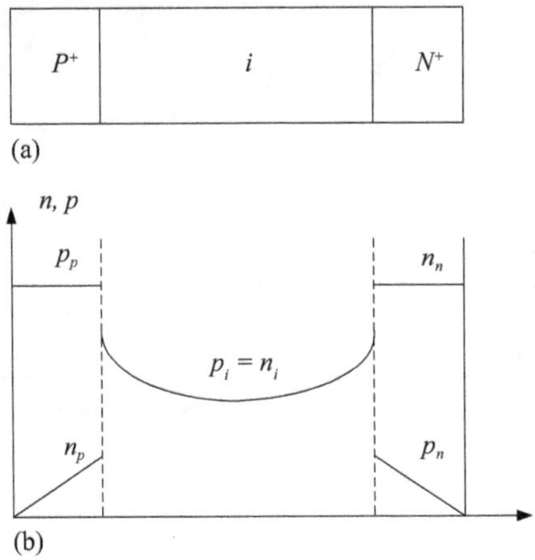

Figure 2.4 PiN diode. (a) Structure. (b) Carrier concentration distribution with forward bias

at low frequency. On the bias of forward voltage, the conduction impedance is low, such as a short circuit. On the bias of reverse voltage, the impedance is high, such as an open circuit.

When the forward voltage is applied to the two terminals of the device, the barrier regions become narrower for P^+i junction and iN^+ junction. A large number of holes and electrons are injected into i region from P^+ and N^+ regions. With a certain concentration gradient, they diffuse toward the center of i region oppositely, with recombination at the same time. When the injected and recombined amount of the electrons and holes equals to each other in a unit time, the electron and hole distributions get stable. Due to the requirement of electric neutrality in the i region, the electrons and holes have the same distribution, as shown in Figure 2.4(b). The conductivity modulation effect is formed by the injected electrons and holes. It increases the conductivity of i region and makes it to be low impedance, which is the ON state of PiN diode. When the reverse voltage is applied on the device, the width of the depletion layer becomes larger for P^+ region and N^+ region and the SCR for the i region grows with increasing magnitude of voltage. When the voltage is increased to a certain value so that the whole i region becomes depleted, the diode has reached its punch-through voltage. The reverse voltage can be further increased until the critical electric field for the impact ionization is reached and the diode undergoes the avalanche breakdown.

2.3.2 Characteristics and parameters

The static characteristics of PiN diode are mainly referred to its voltage–current characteristics, as shown in Figure 2.5. When the forward current I_F increases obviously, the corresponding voltage is the threshold voltage U_{TO}. The voltage on the diode corresponding to I_F is its forward voltage drop U_F. When it sustains the reverse voltage, there is only tiny reverse leakage current.

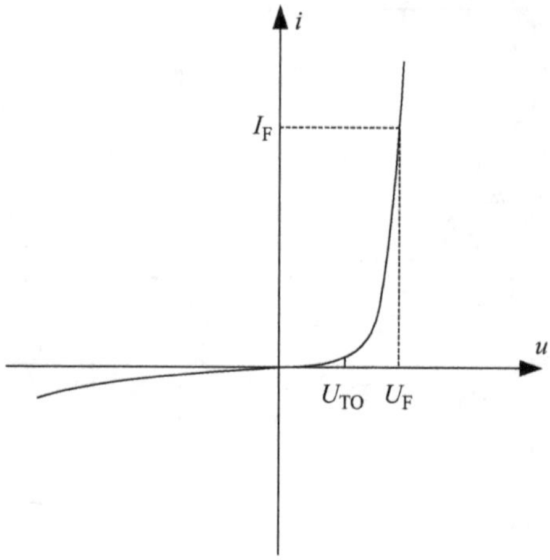

Figure 2.5 Voltage–current characteristics of a PiN diode

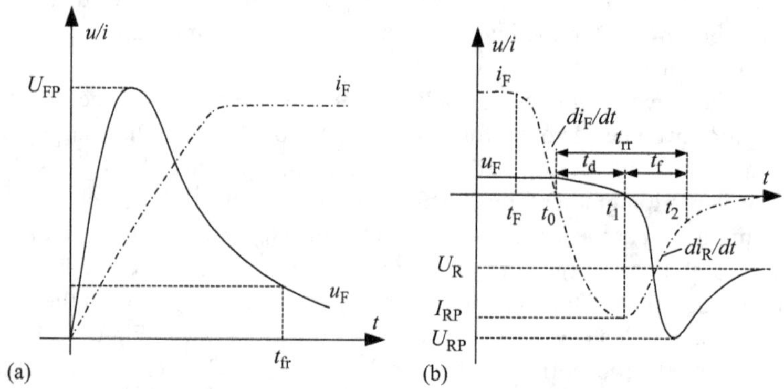

Figure 2.6 Current and voltage waveforms of the switching process of a diode. (a) The turn-on process. (b) The turn-off process

The dynamic characteristics are the variations of the voltage and current characteristics versus time of the diode. Generally, it is related to the junction capacitance and the minority carrier storage effect. Figure 2.6 shows the current and voltage waveforms of the switching process. Figure 2.6(a) shows the turn-on process. An overshoot U_{FP} appears firstly on the forward voltage drop. After some period, it just tends to be a steady-state value. So, there is a forward recovery time t_{fr}. The higher the current rise rate is, the higher the U_{FP} is. In practice, the overshoot U_{FP} of a high-power diode with a wide N base can amount to several hundred volts. This is because the originally low concentration of free carriers in the N base is increased by several orders of magnitude by the injection of electrons and holes from the emitters. As the uncompensated injected carriers cause a high electric field in the base region, there is a force which tends to establish the charge neutrality. The high electric field manifests itself by the voltage overshoot taking place until the charge neutrality is established. The wider base region and the larger rate of current pulse *di*/*dt*, the higher the voltage overshoot. At the end of the forward recovery, the base is flooded by electron-hole plasma with concentration about three orders of magnitude higher than the base doping. This stored charge, which brings us a very low ON-state voltage drop, must be removed during turn-off. As the diode must be recovered from the stored charge, the turn-off process is called the reverse recovery process.

Figure 2.6(b) shows the turn-off process. The instant t_0 corresponds to the current zero-crossing point. At this time, the gradient of carrier concentration close to the emitters changes its polarity and the flow of current is reversed. The instant t_1 is corresponding to the reverse current peak point. At this time instant, the SCR at the anode PN junction is reestablished and the reverse voltage starts to rise toward the DC link voltage of the reverse voltage source. The rise of the reverse current is stopped due to the very high impedance of the SCR. Since then, the amplitude of the current decreases. At this time period, the power (voltage times current) reaches its peak value and the thermal loading of the time is peaking. The integral of dissipated power over the whole recovery process gives us the energy loss called recovery losses E_{rec}. At this instant, t_2 corresponds to the reverse current drops of up to 25% of its peak value. Usually, we define the delay time t_d, current fall time t_f and reverse recovery time t_{rr} as (2.4–2.6). After t_{rr} period, the diode is recovered from the stored charge, the reverse voltage becomes constant and the static OFF state is reached.

$$t_d = t_1 - t_0 \tag{2.4}$$

$$t_f = t_2 - t_1 \tag{2.5}$$

$$t_{rr} = t_d + t_f \tag{2.6}$$

The main parameters are described as follows.

Forward average current $I_{F(AV)}$/rated current: At a specified junction temperature and dissipation condition, it is the average value of maximum allowed half-sine wave current. The rated current should be selected by the principle of root mean square (RMS) equivalence in practice, with enough margins.

Forward voltage drop U_F: At a specified temperature, it is the forward voltage drop corresponding to a specified steady-state forward current.

Reverse repetitive peak voltage U_{RRM}: It is the maximum reverse peak voltage that could be applied to diodes repetitively. Usually, there should be a double margin depending on the application.

Reverse recovery time t_{rr}: As defined in (2.6), it is the sum of delay time and fall time during the turn-off process.

Maximum operation junction temperature T_{Jmax}: It is the maximum temperature at which the diode is allowed to operate. It is usually between 125 and 180 °C for silicon devices.

Surge current I_{FSM}: It is the maximum allowed diode current for one or several pulses typically 1 and 10 ms long the diodes must sustain.

Take PνN diode as an example. The bulk voltage drop U_m on the ν region is expressed as (2.7), where W_n is the width of ν region and L_p is the diffusion length of minority carriers in the ν region.

$$U_{m} = \frac{2kT}{q}\left(\frac{W_{n}}{L_{p}}\right)^{2} \tag{2.7}$$

The junction voltage drop U_J is the sum of voltages at $P^{+}\nu$ and νN^{+} junctions, which has a similar expression as (2.1). If the voltage drop at ohmic contacts is ignored, the forward voltage drop of PνN diode is the sum of the bulk voltage and junction voltage drop, as shown in (2.8).

$$U_{F} = U_{m} + U_{J} \tag{2.8}$$

The breakdown voltage is determined by the avalanche breakdown voltage U_B according to (2.9) and the punch-through breakdown voltage U_{PT}, is given by (2.10), where N_b is the doping concentration in ν region.

$$U_{B} = 5.6 \times 10^{13} N_{b}^{-\frac{3}{4}} \tag{2.9}$$

$$U_{PT} = \frac{qN_{b}W_{n}^{2}}{2\varepsilon\varepsilon_{0}} \tag{2.10}$$

Producers of semiconductor devices assure enough margins between the actual breakdown voltage and the reverse repetitive peak voltage down to the minimal operation temperature (–40 °C or 0 °C) at which is the breakdown voltage lowest due to the negative temperature coefficient of this parameter.

2.3.3 *Typical application [4]*

The typical application of PiN diode is rectification. Rectifier is a kind of AC–DC converter. According to the input power supply types, rectifiers could be divided into single-phase and three-phase. Figure 2.7(a) shows the single-phase bridge uncontrolled rectifier circuit formed of four diodes. During the positive half-cycle of input voltage, the power supply is applied to load by diodes D_1 and D_4. During the negative half-cycle, the power supply is reversely applied to load by diodes D_2 and D_3, i.e., $v_D = |v_S|$, as shown in Figure 2.7(b) and (c). The average output voltage is

$$V_{D} = \frac{1}{\pi}\int_{0}^{\pi}\sqrt{2}V_{S}\sin(\omega t)\,d(\omega t) = 2\frac{\sqrt{2}}{\pi}V_{S} = 0.9\,V_{S} \tag{2.11}$$

This bridge rectifier can be used for the rectification of the AC net voltage, where the input is the 50 or 60 Hz sine wave from a voltage transformer and the

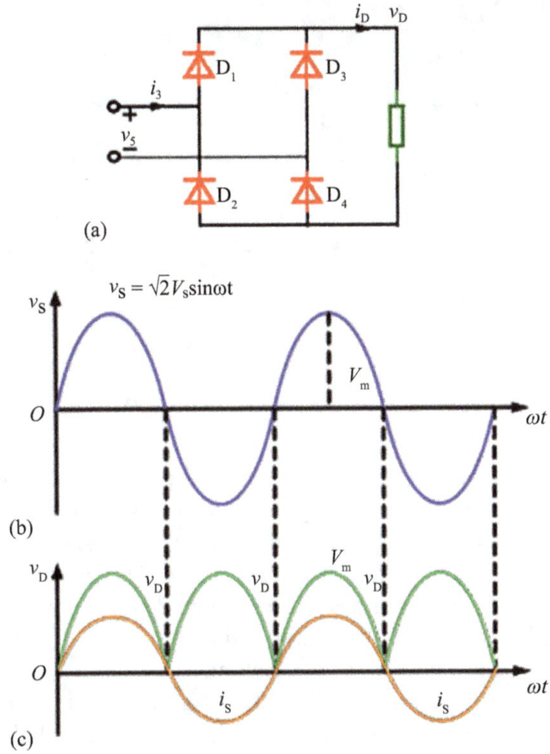

Figure 2.7 Single-phase bridge uncontrolled rectifier circuit. (a) Circuit. (b) Input voltage. (c) Output voltage and input current

load is either RC or LC filter or the input of power factor correction circuit instead of the simplified load impedance shown in Figure 2.7.

2.3.4 *Instabilities*

Figure 2.8 is a sketch map for the safe operation area (SOA) of a low-power PiN diode. The limiting factors include the maximum voltage U_M, the maximum current I_M, the maximal allowed power dissipation P_M ($I = P_M/U$) and the secondary breakdown line, which can limit the operation space at certain types of bipolar devices.

For the voltage-blocking capability, a proper bulk structure design (resistivity and thickness of starting silicon wafer) is necessary, first of all. On top of it, special care must be devoted to the junction termination (JT) structure. The JT appears at the edge of the diode, where the electrical field can leave the silicon material and enter the ambient (air) with much lower electrical strength compared to silicon. At the same time, the diode surface must be clean and protected against humidity and foreign particles, which would otherwise increase the leakage current, cause the charging of the surface and reduce the blocking capability. The surface protection is

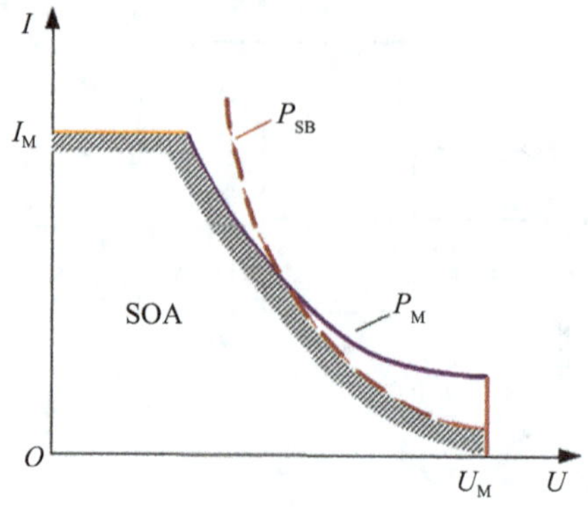

Figure 2.8 Safe operation area (SOA)

called passivation and typical materials used for this purpose are polyimide, silicone rubber, glass, amorphous silicon, silicon-polysilicon, diamond-like carbon, etc. The typical types of JT structures are positive and negative bevels, mesa structures, field limiting rings (FLR), junction termination extension (JTE), and a variation of lateral doping (VLD); some of them are described in more detail in Chapter 3.

To achieve a low forward voltage drop and high surge current I_{FSM}, the doping profile of the anode and cathode must be optimized.

It is worth noting that the forward voltage drop is a temperature-dependent parameter and that this dependence has a big impact on device reliability. As a kind of bipolar device, the diode is historically considered to have a negative temperature coefficient for forward voltage drop, as shown in Figure 2.9 [5]. This is probably given by the fact that the first diodes had strongly injecting highly doped emitters motivated by a low contact resistance and low voltage drop. However, the diodes that have a crossover point between the cold and hot IV curves are preferred. We take the 5SDD 54N4000 rectifier diode from ABB (Asea Brown Boveri) Ltd., for example [6]. As shown in Figure 2.10, the forward I–V characteristics curves with different temperature have a crossover point at about 9 kA, below which the forward voltage drop has a negative temperature coefficient and above which has a positive coefficient. A rough analysis is as follows. The junction voltage drop dominates at low current. The temperature rise makes n_i increase and then the voltage drop decreases. The bulk voltage drop dominates at high current. The temperature rise makes mobility decrease and then the voltage drop increases. The third factor is the temperature dependence of the lifetime of minority carriers, which can make the carrier lifetime increase as well as decrease. More details are given in Chapter 3. The competition between the three temperature effects makes the crossover point possible. As the temperature characteristics of forward voltage drop will influence the reliability of parallel-connected diodes, the existence of the crossing point below the maximal rating current is a must. In general, the lower crossing point current, the better. The only exception is the temperature sensor where parallel cold and hot IV curves are preferred.

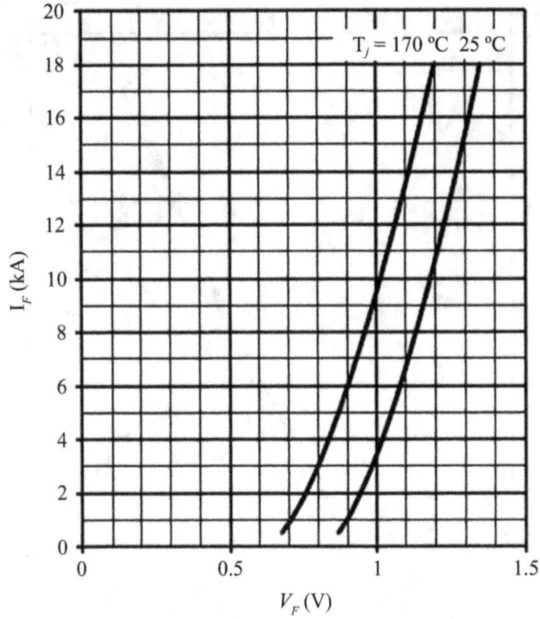

Figure 2.9 Forward I–V characteristics with temperature dependence

2.3.5 SiC PiN diodes [7–34]

Compared with Si material, SiC has 10 times higher breakdown field strength, 3 times higher thermal conductivity, and 2 times higher carrier saturated drift velocity and higher intrinsic temperature (when the intrinsic concentration becomes equal either to N_D or N_A). It is, therefore, to be considered as a better material to fabricate power devices. PiN diode fabricated from the SiC has the same operation principle as Si PiN diode and similar structure. According to (2.8), the forward voltage drop includes the bulk voltage drop and the junction voltage drop. As the wide-bandgap material, due to its much lower intrinsic carrier concentration, the junction voltage drop of SiC devices is about 2 V higher than that of Si devices. Therefore, only when the applied voltage is high enough and the bulk voltage drop is dominant, could the advantages appear for SiC bipolar devices. Generally speaking, the overall performance for SiC PiN diode is better than Si with the voltage rate above 3 kV. A complication of the SiC, which is not present in the silicon, is that the available dopants such as boron, aluminum and nitrogen are located in the band gap at much higher energetic distance from the conduction and valence bands. As a result, they are not all ionized at room temperature which makes the injection of minority carriers more difficult. Most of the commercial diodes are, therefore, based on the Schottky diode concept, where the energetic barrier is formed at the metal-semiconductor (MS) boundary instead of at PN junction. PN junction is usually combined with the Schottky contact to shield the MS contact from a high electric field under reverse bias.

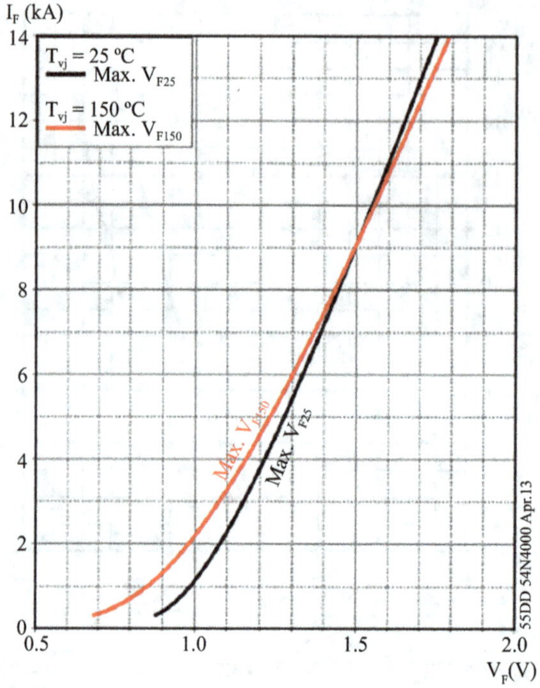

Figure 2.10 Forward I–V characteristics at different temperatures

For SiC PiN diode, the low epitaxial layer defects and the stable junction terminations are two challenges.

In the epitaxial layer defects, the basal plane dislocation (BPD) is the fatal two-dimensional defect that causes the "forward voltage drift" for bipolar SiC devices and, therefore, has been concerned for a long time. The BPD is a kind of dislocation. Basically, dislocations are considered to be caused by shear stress during the material grown process. Electron-hole recombination at the BPD induces stacking faults what causes an irreversible increase of the ON-state voltage drop. The BPDs can also propagate from the substrate to the epitaxial layers. The stable devices are acquired by eliminating the stacking fault nucleation sites.

The minority carrier lifetime also has a close relationship with the epitaxial layer point defects. For N-type 4H-SiC, now $Z_{1/2}$ center is considered to be the carrier lifetime killer, which is the acceptor level of a carbon monovacancy. It could be eliminated by either carbon ion implantation followed by annealing at 1,600–1,700 °C or thermal oxidation at 1,300–1,400 °C. The Kyoto University has reported a 26.9 kV SiC PiN diode with the differential on-resistance of 9.7 mΩ·cm^2. The width of the drift region is 268 μm. From the photoconductance decay measurement, the carrier lifetime is enhanced from 2.4 to 21.6 μs, by thermal oxidation at 1,400 °C for 72 h.

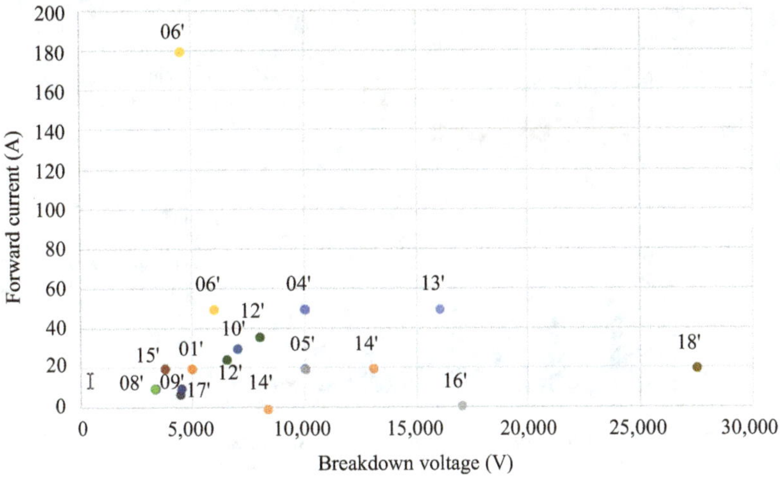

Figure 2.11 Research and development progress for the SiC PiN diodes

The principle for the junction termination technology is the same as at the Si PiN diode – to extend the surface width of the SCR. But from the realization point of view in practice, due to the high hardness of the SiC material the feasibility is bad for the mesa termination by grinding angles. In addition, since the breakdown field strength is 10 times higher than at Si, it is certainly more difficult for reduce the termination electric field. The common technologies are still guard ring, floating field ring and different kinds of the JTE structures.

Figure 2.11 shows the research and development progress for the SiC PiN diodes. No SiC PiN diode products are found from companies. The highest voltage rating reported until now is a 27.5 kV 4H-SiC PiN diode developed by the National Institute of Advanced Industrial Science and Technology from Japan. The structure is shown in Figure 2.12. The termination structure is formed by two-zone JTE and space-modulated JTE. Through carbon ion implantation and thermal oxidation, the carrier lifetime is enhanced. Through controlling the annealing temperature and time, the electron injection from cathode is controlled. Through the design of the anode contact electrode pattern, the hole injection from anode is controlled.

2.4 FRDs (fast recovery diodes)

2.4.1 Structures and operation principle

The FRD is the diode with very short t_{rr}, usually below 5 μs. Actually, the t_{rr} could reach the order of several tens of nanosecond already. The requirement of modern power electronics technology on power diodes includes high blocking voltage, low energy dissipation, fast recovery and recovery voltage waveforms without voltage overshoots. The t_{rr} could be shortened by reducing the stored charge. This can be realized by introducing recombination centers in the forbidden energy band, using

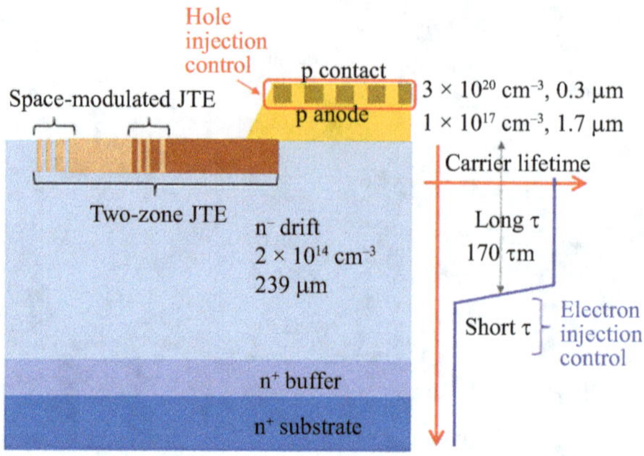

Figure 2.12 Ultra-high-voltage SiC PiN diode structure reported in 2018 [31]

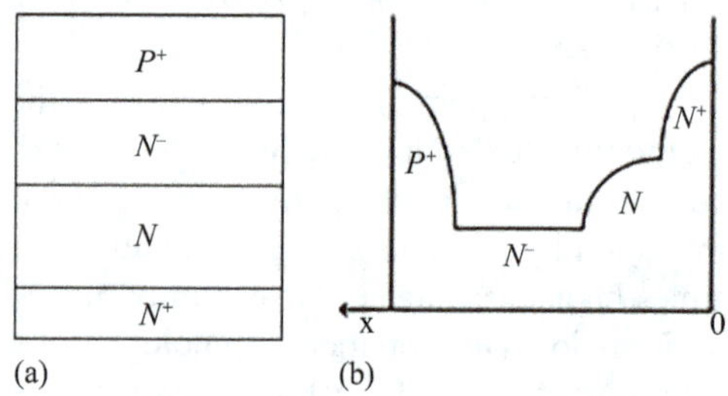

Figure 2.13 PiN diode with a buffer. (a) Structure. (b) Doping profile

methods such as electron or ion irradiation, diffusion of gold or platinum atoms, etc. As a result, the forward voltage drop and leakage current are increased, but the dynamic parameters such as t_{rr}, peak recovery current I_{RM}, recovery losses E_{rec}, softness and the SOA are improved.

There are several design concepts for the FRD, for example: (1) drift layer with soft punch-through buffer, (2) anode layer with special design such as reduced emitter efficiency (a lower anode surface doping), (3) cathode layer with special structures injecting and additional charge at the tail phase of the reverse recovery and/or eliminating the existence of the second SCR at the NN^+ junction, and (4) controlled axial lifetime of excess carriers.

For the first type, the buffer layer of moderate doping concentration is introduced in the drift region, as shown in Figure 2.13 [35]. The buffer could be formed by either epitaxy or diffusion process. For the buffer concentration, it is a uniform distribution for the former one and Gaussian distribution for the latter one. The original intention of setting up the buffer structure in the power devices is to coordinate the off-state and on-state performance. It makes the device with the same voltage rating and a lower ON-state voltage drop by introducing the buffer.

The research on diodes shows that the well-designed buffer layer could also improve the soft recovery characteristics, reduce the voltage oscillations during turn-off and eliminate the snap-off. Buffer has this kind of ability because its relatively high doping concentration prevents the creation of the second SCR at NN^+ junction, which would otherwise lead to abrupt removal of the rest of the stored plasma and generation of the voltage peak $V = -L^* \, di/dt$ at the load and/or stray inductance L.

For the second type, the typical device is self-adjusting P emitter efficiency diode, with the short form of SPEED [36]. It controls the injection efficiency by improving the anode structure. Figure 2.14 shows its structure, in which the anode region is formed by alternating high-doping P^+ region and low-doping P region. The cell size is designed reasonably such that the effective injection is formed only at high current for the high-doping P^+ region, which guarantees the overcurrent ability of the device. Only the low-doping P emitter region works at normal operation. It has low injection efficiency, which could reduce the reverse current peak significantly and soften the recovery characteristics. The robustness of this concept can be questionable, because of enhanced injection from the edges of the highly doped P^+ regions during the fast reverse recovery.

For the third type, structures of controlled injection of backside holes (CIBH) [37] and field charge extraction (FCE) could be put into this category. Actually, it is also a kind of self-adjusting mechanism, which adjusts the injection efficiency of the cathode region during the tail phase of the reverse recovery. Figure 2.15 shows the structures of CIBH and FCE diodes, respectively. For the CIBH structure, the floating P layer is buried in the N^- region and then two PN junctions are added into the cathode region. The PN^+ junction formed by the N^+ emitter and P buried layer is equivalent to an avalanche diode, which has low breakdown voltage and, therefore, could suppress the dynamic avalanche during the reverse recovery process effectively. During the reverse recovery, P buried layer injects holes into N^- region, which makes it have soft recovery characteristics, too. The advantage of CIBH is to get a good trade-off among the forward voltage drop, soft recovery and dynamic ruggedness. For the FCE structure, its cathode structure is formed by the combination of N^+ and P^+ regions. The existence of P^+ region can inject extra carriers from the cathode into the N-base thereby preventing the creation of the cathode SCR. The extra carriers decrease the rate of the current fall (lower di/dt), so the soft recovery performance is achieved.

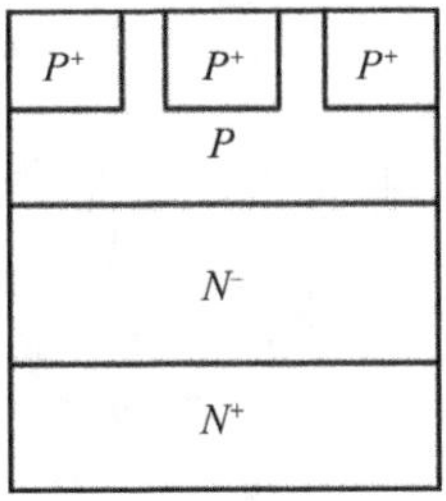

Figure 2.14 Structure of self-adjusting P emitter efficiency diode (SPEED)

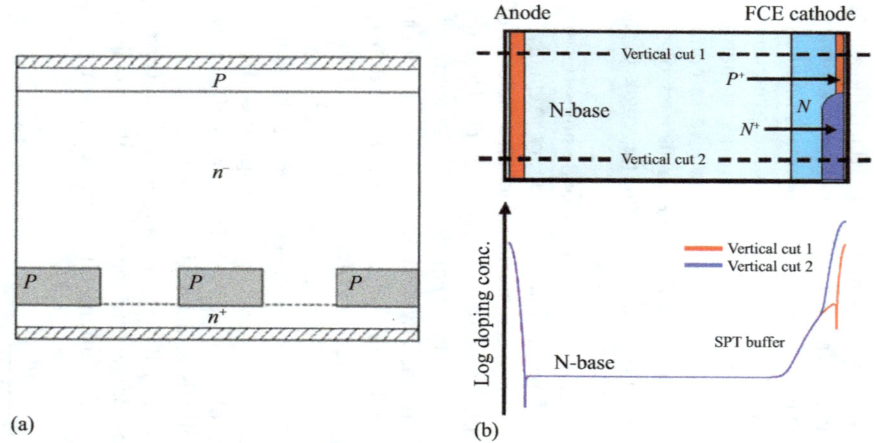

Figure 2.15 Structures of (a) controlled injection of backside holes (CIBH) and (b) field charge extraction (FCE) diodes

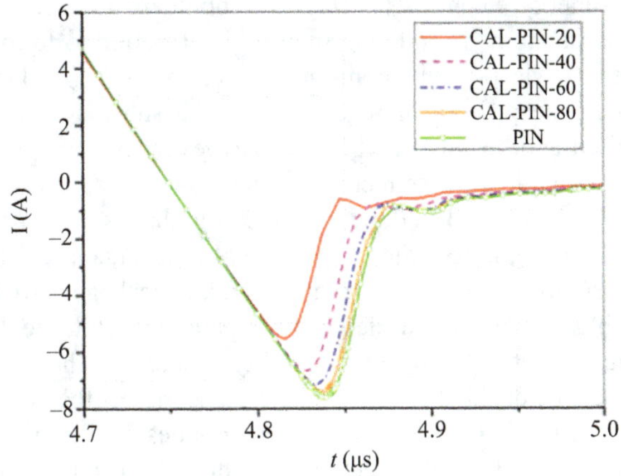

Figure 2.16 Reverse recovery characteristics of CAL diodes with different carrier lifetimes (20, 40, 60 and 80 ns) at the PN junction versus an ordinary PiN diode

For the fourth type, the controllable axial lifetime (CAL) diode is usually fabricated by hydrogen or He^{2+} ion implantation [38]. If the carrier lifetime is adjusted by electron irradiation, the axial distribution of the recombination center could hardly be localized. On the contrary, the ion irradiation (H or He) can be localized by chosen irradiation energy and can be realized after the device processing was completely finished. For the CAL, the key point is to reduce the carrier lifetime at the PN junction, while keeping the carriers in the neutral region beyond PN junction with a longer lifetime. In this way, both the high switching speed and low forward voltage drop could be acquired. Therefore, the recombination center should be as close to the PN junction as possible. Figure 2.16 shows the reverse recovery characteristics of diodes with different carrier lifetime at the PN junction. It could be seen that lower carrier lifetime could reduce the reverse recovery current peak [39].

The introduction of deep energetic levels in the silicon bandgap by the electron and ion irradiation techniques is always related to the increase of leakage current, especially if the radiation defects are placed into the SCR of the PN junction. As a result, the CAL diode has shown a high leakage current, which prevents the device to operate up to the junction temperatures beyond 125 °C. To eliminate this drawback, the Field-Shielded Anode (FSA) concept has been developed [40,41], where the anode SCR is shielded from the region with the radiation defects by P-type buffer. This allowed to qualify high-power chip FRDs with $T_{Jmax} \geq 150$ °C.

2.4.2 Characteristics and parameters

In order to develop the FRD, the technology of minority carrier lifetime control is a frequently used option. The technology of the CAL by He^{2+} ion implantation or the FSA by proton or helium implantation is a relatively new method compared to the electron irradiation, gold (Au) and platinum (Pt) diffusion. For the electron irradiation, high-energy electrons (1–10 MeV) bombard the semiconductor chip, detaching the silicon atoms from the normal lattice position to form vacancies and interstitials. Then, the deep-level recombination centers are formed in the forbidden band of silicon in the process of reaction with the substrate impurities such as oxygen and carbon and between each other. However, as it introduces a uniform lifetime spatial profile the reverse recovery is snappy. The soft recovery characteristics are better for the Au and Pt doping. The energy levels of $E_c - 0.54$ eV for Au and $E_C - 0.21$ eV for Pt play a major role in the N-type silicon. Because the energy level of Au is closer to the center of forbidden band, its leakage current is higher and represents a problem especially at elevated temperatures. But, the Au doping also has the advantage of better temperature stability. Therefore, the selection of deep-level dopant should be decided by the application scopes of the device.

For the parameters of the FRD, the parameters describing the performance of PiN diode in Section 2.3.2 are all effective. In addition, the softness factor *S* should be emphasized here. Seeing Figure 2.6(b), its early definition is

$$S = \frac{t_\mathrm{f}}{t_\mathrm{d}} \tag{2.12}$$

i.e., in the reverse current, the higher the ratio of fall time is, the softer the characteristics are. It is true for most cases, but it may also appear that even if the fall time is long, a high *di*/*dt* turns up at the end of switching off. It is still snap-off. Therefore, the description of softness factor in this way could not define the soft recovery characteristics of diode completely. Another definition is as follows

$$S = \left| \frac{-\frac{di}{dt}\big|_{i=0}}{\left(\frac{di_r}{dt}\right)_{\max}} \right| \tag{2.13}$$

i.e., described by the ratio of the zero-crossing slope of reverse current and the maximum slope of recovery current. As the slope of the current is not the relevant parameter to judge the softness, the definition $S = V_{pkr}/V_{DC\ link}$ is used in the practice. Here, the V_{pkr} is the magnitude of the voltage overshoot of a snappy diode, while the $V_{DC\ link}$ is the DC link voltage of the application circuit – the constant voltage we usually see after the recovery profess passed. It is a common rule that the V_{pkr} lower or equal to the rating breakdown voltage of a given diode is accepted. However, the lower the V_{pkr}, the softer the diode. With increasing $V_{DC\ link}$, the softness of diode is decreasing. This is because the N base regions of the diodes are designed as narrow as possible to minimize the ON-state voltage drop. The diode design is, therefore, subjected to a compromise between the minimal ON-state voltage drop and maximal allowed V_{pkr} on top of other characteristics such as the dynamic SOA.

2.4.3 *Typical application [42]*

The typical application of the FRD is the freewheeling diode (FWD). Taken the chopper circuit shown in Figure 2.17 as an example, the FRD is applied together with the main switch IGBT to provide the freewheeling loop. The basic operation

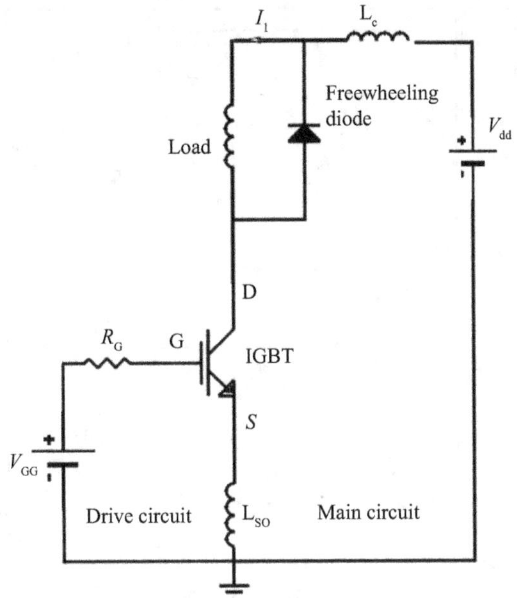

Figure 2.17 FSRD applied as a freewheeling diode (FWD)

process is as follows. When the voltage V_{GG} is applied to the gate of the IGBT, the IGBT is turned on and the current I_1 flows through. When the gate voltage is removed, the IGBT is turned off and the overvoltage caused by the inductive load could damage the IGBT. In order to protect the IGBT, the FWD provides the freewheeling loop. The energy stored in the inductance dissipates in the loop of load and the FWD. But because the resistivity of the loop is low, the current decays slowly. When the gate voltage is applied again, i.e., the next pulse comes, the current in the FWD decreases while the current in the IGBT increases with the same rate. It could be seen why the soft recovery characteristics of the FWD are important. At the turn-on process, turn-on loss of the IGBT has a close relationship with the FWD, because the recovery current of the FWD flows into the IGBT as shown in Figure 2.18.

2.4.4 Instabilities

The instability issues for the PiN diode are all involved in the FRD, including the different failure modes of overvoltage failure, overcurrent failure and overheating failure. For the FRD in the freewheeling application, the failure caused by dynamic avalanche should be paid special attention. The dynamic avalanche occurs during the reverse recovery period when the velocity of holes traversing the N base region saturates due to the high electric field of the anode PN junction. As a result, the hole concentration grows and the electric field as well. When the critical field for impact ionization is reached, the diode enters the dynamic avalanche that happens at much lower voltages than in the case of the static reverse bias. The generated

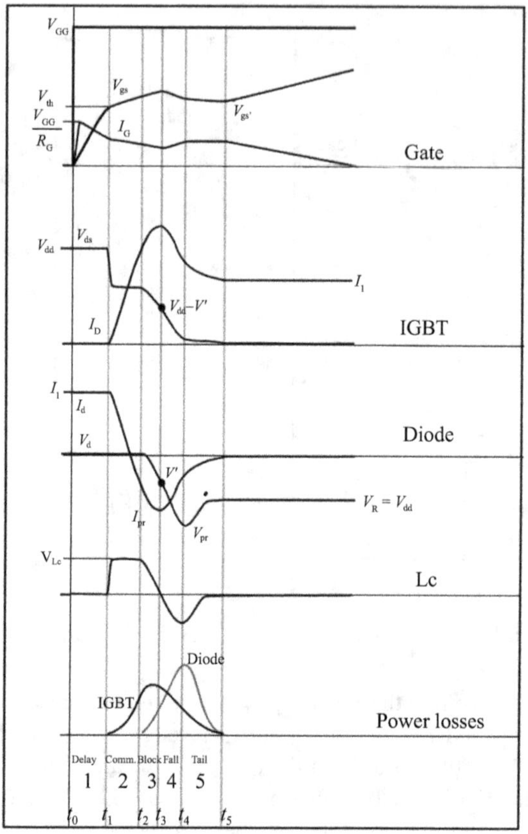

Figure 2.18 IGBT and diode interaction during IGBT turn-on

carriers form current filaments, which are the electrical instabilities leading to the secondary breakdown. The fails after the current density in a narrow filament melt a hole through the silicon wafer. The failure mechanisms could be divided into single PN junction dynamic avalanche failure and double PN/NN$^+$ junction dynamic avalanche current filament failure. The local high-temperature failure would always be caused by the latter one. The CIBH structure mentioned in Section 2.4.1 is to enhance the dynamic avalanche capability of the diode by avoiding the NN$^+$ junction avalanche injection.

For the temperature dependence of the FRD, similar to the PiN diode, the crossing points appear in the forward I–V characteristics curve at low and high temperatures. Here, the forward I–V characteristics for two types of FRD chips from Infineon Technologies are provided as examples in Figure 2.19.

For the temperature dependence of the reverse recovery process, a simulation result is provided, as shown in Figure 2.20. Due to the rise of junction temperature, the minority carrier lifetime increases and the stored charge in the drift region

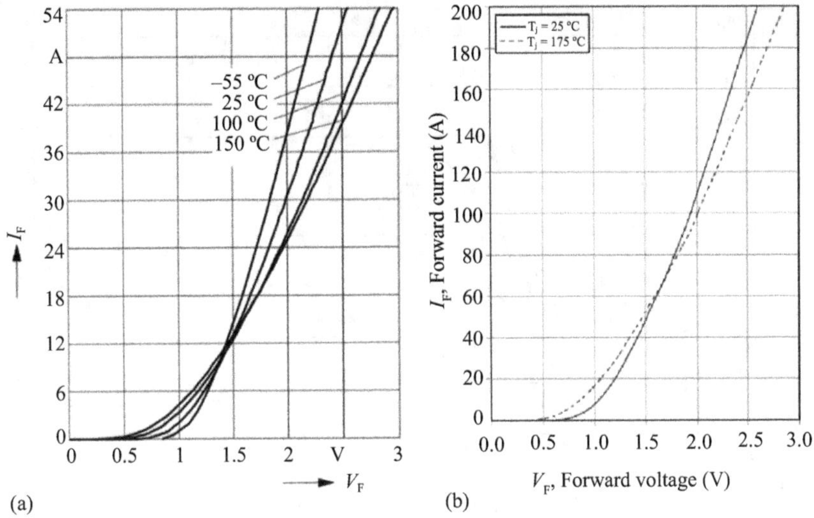

Figure 2.19 Forward I–V characteristics at different temperatures. (a) IDP18E120, 1,200 V/18 A and (b) IDW50E60, 600 V/50 A FSRDs from Infineon Technologies [43,44]

increases as well with the consequence of enhanced forward current. This results in increasing reverse recovery charge, reverse current peak and reverse recovery time. As shown in Figure 2.20(a) and (b), the reverse current chops off at low temperature, which may cause a spike in the reverse voltage. The higher the temperature is, the higher the minority carriers' energy is, making them less likely to be captured by the recombination center and then having a longer lifetime. As a minority carrier device, the minority carrier lifetime directly influences the reverse recovery dynamic process of the FRD.

2.5 DSRDs (drift step recovery diodes)

2.5.1 Structures and operation principle

By contrast, the DSRD (drift step recovery diode) is a kind of diodes with super-hard reverse recovery characteristics. Ioffe Physical Technical Institute in Russia proposed this device. It is capable of producing the maximum voltage rise rate up to 2×10^{12} V/s and the maximum voltage over 1 kV [45]. The DSRD is a semiconductor opening switch used in the field of pulse power technology to generate high-voltage pulses of several kilovolts to tens of kilovolts with a few nanoseconds rising front. The DSRD is a diode-like device with two electrodes and either a three-layer (P^+PN^+) or a four-layer (P^+PNN^+) structure. In general, the DSRD is manufactured by deep Al diffusion to the n-type substrate in order to enable the DSRD to withstand high voltage and cut off large current [46]. In this case, the PN junction is located deep in the wafer. Due to the process characteristics, the PN

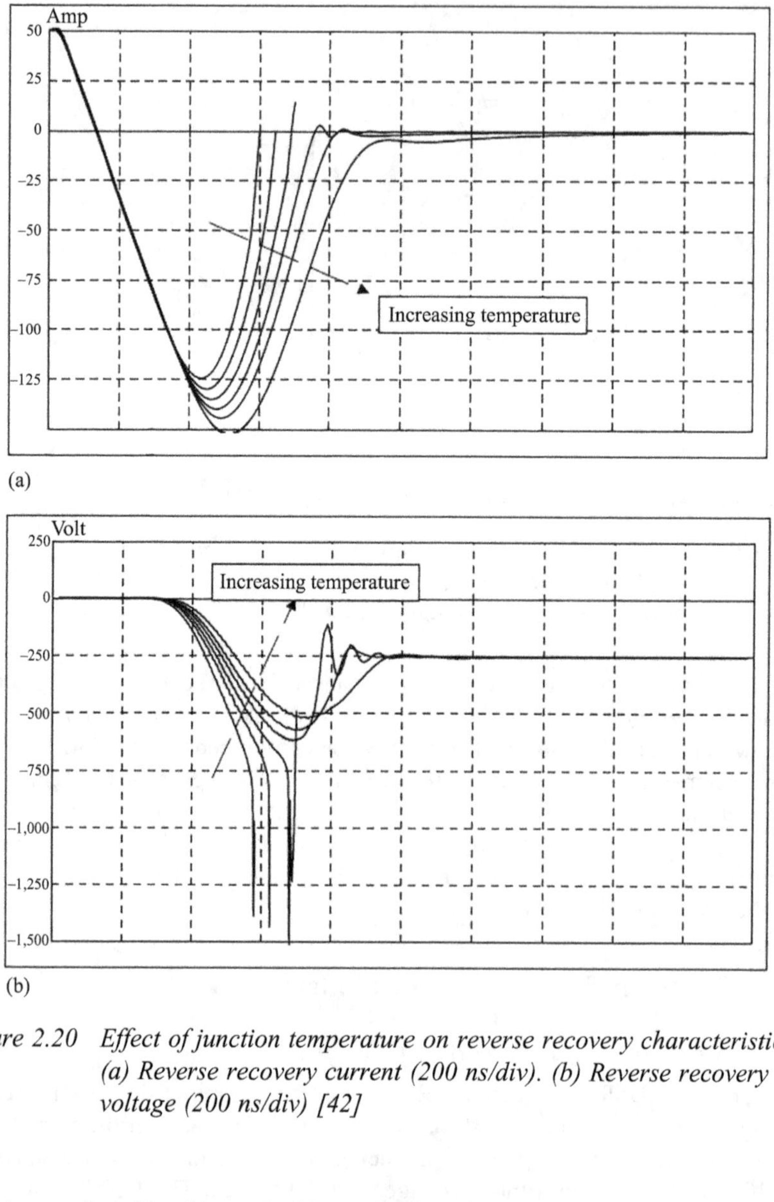

Figure 2.20 Effect of junction temperature on reverse recovery characteristics. (a) Reverse recovery current (200 ns/div). (b) Reverse recovery voltage (200 ns/div) [42]

junction produced by this method is graded junction. The structure parameters and working process are shown in Figure 2.21 [47].

The schematic diagram of the DSRD's working circuit is shown in Figure 2.22 [48]. As per Figures 2.21 and 2.22, its working process can be divided into three stages. Stage I: switch S_1 turn-on. A positive pulse is applied across the DSRD. During this period, the DSRD exhibits a characteristic similar to the forward conduction of the diode. The current I_1 charges the DSRD and the device is filled with

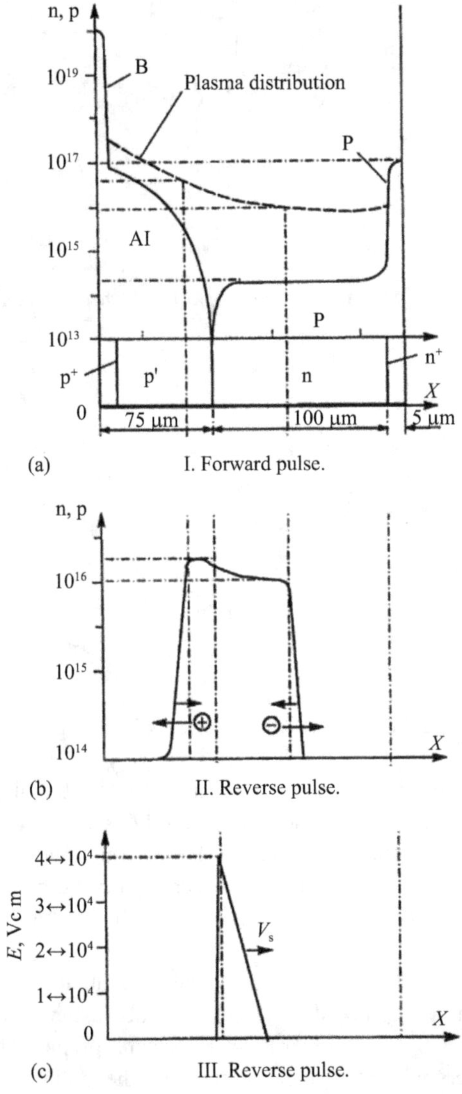

Figure 2.21 The DSRD working process. (a) Plasma distribution. (b) Plasma motion. (c) Space charge region formation

the electron-hole plasma of distribution indicated by the dotted line, as in Figure 2.22(a). The voltage drop across the DSRD is low and there is no current in load *R*. Stage II: switch S_2 is turned on. Current I_2 injects into the DSRD reversely and rapidly extracts the plasma inside the device. In this process, since there are still carriers capable of conducting electricity inside the device, the reverse conducting current continues to increase and still no current flows through the load *R*.

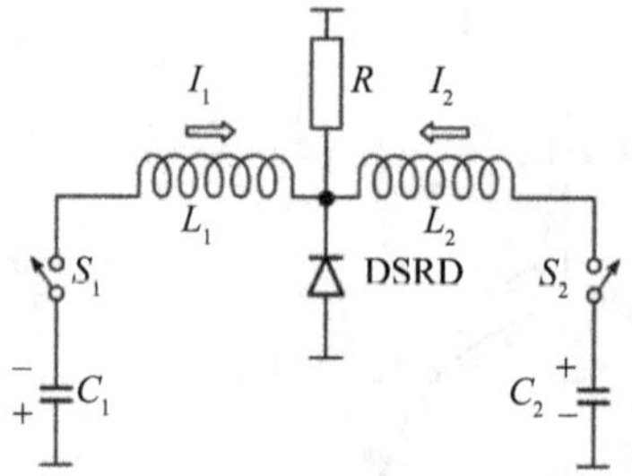

Figure 2.22 A simplified working circuit of the DSRD

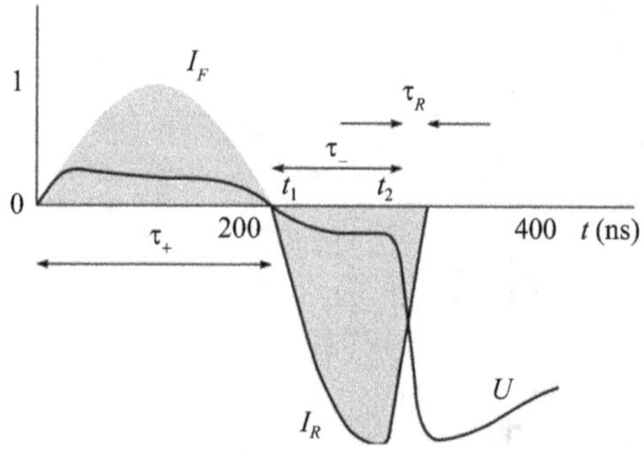

Figure 2.23 Voltage–current curve of the DSRD [48]

Stage III: with the depletion of carriers during the reverse conduction process, the space charge region is rapidly formed inside the DSRD and its impedance rapidly increases, exhibiting the characteristics of the opening switch. At the same time, due to the rapid increase of the DSRD reverse impedance, the reverse current flowing through the DSRD is quickly converted to the load *R*, thus forming a high-voltage pulse.

In addition, with the improvement of the semiconductor process, it has become possible to prepare thick epitaxial layers by the chemical vapor deposition (CVD) [49]. For the first time, Soreq Nuclear Research Center prepared the DSRD based on silicon epitaxial layers with as-grown junctions by the CVD [50]. According to the characteristics of the process, this method can well control the doping concentration of each layer of the device, thereby optimizing the characteristics of the DSRD.

A single silicon-based DSRD device can only operate at a few hundred volts; but due to the special structural characteristics and operating principle of DSRD, it is easy to increase the operating voltage by connecting several or even dozens of devices in series.

2.5.2 Characteristics and parameters

The typical voltage and current curves of the DSRD are shown in Figure 2.23, where I_F is the forward current, I_R is the reverse current and *U* is the voltage across the DSRD. τ_+, τ_- and τ_R are, respectively, the duration of the charge pumping, charge decay and DSRD current interruption and voltage rise stage. In general, τ_R, the duration of DSRD current interruption and voltage rise stage, is 1–2 ns. Usually, a single DSRD operates at only a few hundred volts, so the rate of voltage change during switching is only about 10^{11} V/s. By connecting dozens of DSRD devices in series, it can output high-voltage pulses of thousands to tens of thousands of volts, so that the voltage change rate in the switching process can be increased to 10^{12} V/s, which can be used as the trigger pulse of the superfast ionization fronts. Typical reverse current densities of DSRD devices range from 200 A/cm^2 to about 2 kA/cm^2 [51].

2.5.3 *Typical application*

The DSRD is a semiconductor open-circuit switch with hard reverse recovery characteristics, which can cut off high current quickly in the nanosecond time scale. With this feature, it is often applied as the main switch of the pulse generator in the field of nanosecond pulse power technology to output high voltage with steep front.

As shown in Figure 2.24 [52], it is a DSRD-based ultrashort pulse-generation circuit. Initially, the DSRD is positively biased, nonequilibrium carriers are injected into the base region and the DSRD is in the conductive state. When the forward current decreases to zero, the current flowing through the DSRD changes direction and begins to extract nonequilibrium carriers. When the extraction process ends, the DSRD reverse recovery process begins: majority carriers start to be extracted from the base region at the saturation speed. After a short time, the DSRD resistance rises remarkably. At the same time, a voltage pulse with a very steep leading edge is formed on the load resistor R_{load} and its voltage peak is much higher than the power supply voltage V_1.

Figure 2.25 is also a pulse-generation circuit based on the DSRD [53]. The circuit is divided into three stages. The first stage generates a high voltage through a power MOSFET turning off the current in L_1. The DSRD is used in the second stage to cut off the reverse current to create a 1-ns rise time pulse. The third stage utilizes SAS (silicon avalanche shaper) and its capacitor C_{SAS} to form a pulse with an even faster rise rate.

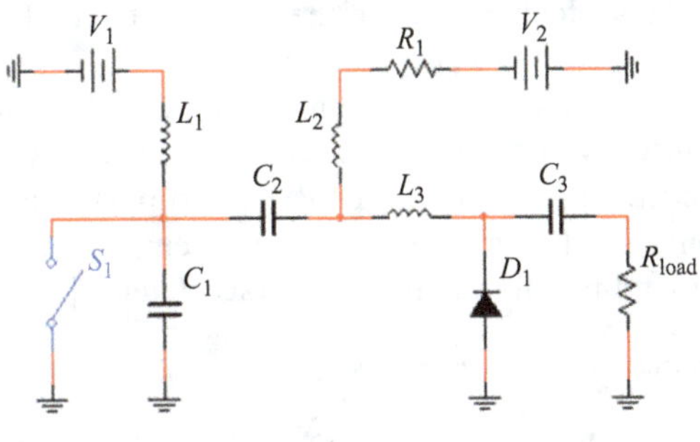

Figure 2.24 Ultrashort pulse-generation circuit based on the DSRD (D1)

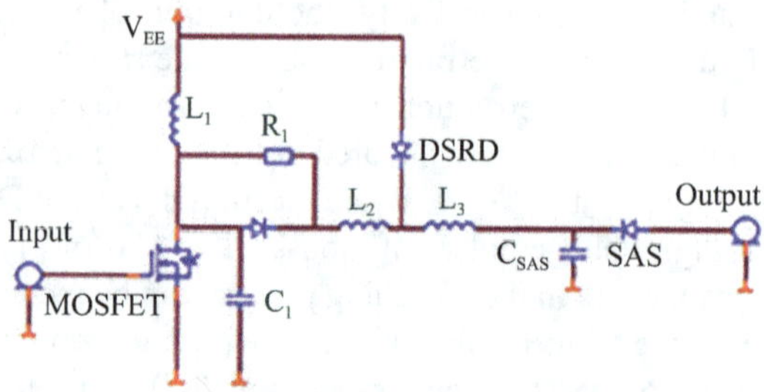

Figure 2.25 Pulse-generation circuit with the DSRD as a trigger of the last stage switch

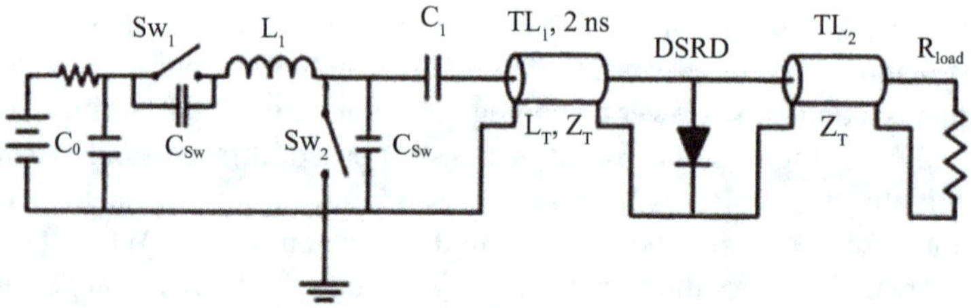

Figure 2.26 The 5-kV modulator circuit diagram based on the DSRD

Initially, the MOSFET is in off-state, no current flows through the circuit and C_1 and C_{SAS} are charged to V_{EE}. Then, the MOSFET is turned on and current flows through L_1 and L_2. L_1 continues to be charged until the MOSFET is turned off. The current flowing through L_2 and R_1 acts as the forward pumping current of the DSRD. Next, the MOSFET is turned off and L_1 discharges to the capacitor C_1 while generating a current that flows through the DSRD through the L_1-diode-L_2-DSRD loop, removing the charges that have been accumulated previously in the DSRD. When the accumulated charges are fully removed, the DSRD quickly cuts off the current, producing a high-voltage pulse that triggers the SAS into conduction. In the last stage, as the SAS turns on, the voltage pulse front generated on the DSRD is further steepened. The switching speed of the DSRD directly determines the leading edge steepness of the output voltage on the load.

Figure 2.26 shows a 5-kV modulator circuit based on the DSRD [54], which is used to generate pulses with a reverse current peak of 200 A. It can be used in Ultrafast Beam Kicker. In addition, the DSRD-based pulse generator can also be used in the ignition system of internal combustion engines [55], electromagnetic pulse radar [56], ground-penetrating radar [57], accelerator [58] and so on.

2.5.4 SiC DSRDs

The SiC DSRD was first reported in 2002 [52]. Due to the advantage of the wide-bandgap material, it is theoretically predicted that the switching speed of SiC

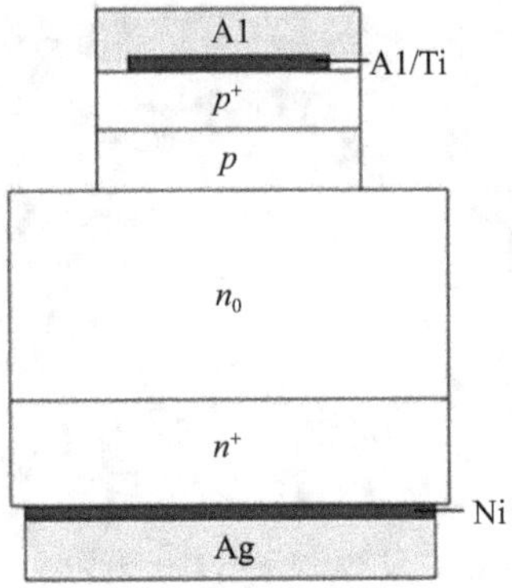

Figure 2.27 A mesa epitaxial p^+-p-n_0-n^+ 4H-SiC DSRD

DSRD can reach 20 times that of Si DSRD. The SiC DSRD has an even better prospect than the Si DSRD in high-voltage pulse generators, since the former could get higher blocking voltage for single chip and better heat dissipation characteristics in the repetitive frequency pulse power area hopefully, etc.

The structure of the SiC DSRD is basically the same as that of the Si DSRD, as shown in Figure 2.27 [59]. The 4H-SiC DSRD is prepared by epitaxial growth on an n-type substrate wafer into a p^+-p-n_0-n^+ 4-layer structure. The parameters of each layer are as follows. The p^+ emitter layer thickness is 2 μm, with a doping concentration of 10^{19} cm^{-3}. The p base region thickness is 5 μm, with a doping concentration of 5×10^{16} cm^{-3}. The n_0 base region thickness is 40 μm, with a doping concentration of 1.5×10^{15} cm^{-3}. The n^+ emitter layer thickness is 1 μm, with a doping concentration of 10^{18} cm^{-3}. By the magnetron sputtering, the p^+ region is deposited with Al/Ti as the anode ohmic contact metal and the n^+ substrate is deposited with Ni as the cathode ohmic contact metal, which is annealed at 950 °C in vacuum to form the ohmic contact. An etch process is used to form a step structure as the termination. A layer of metal Al is predeposited as an etch mask and SF_6 is the etching gas [60].

The top view of a 4H-SiC DSRD device is shown in Figure 2.28 [52]. In order to prevent surface discharge, the periphery of the fabricated diode is covered with high-temperature polyimide. In order to increase the operation voltage and switching speed, SiC DSRDs are also connected in series sometimes. As shown in Figure 2.29, it is a stack of four 4H-SiC DSRDs in series [61]. In 2018, Russia reported that a stack of 2-kV SiC DSRDs was used to output a voltage pulse with a peak value of 30.5 kV and leading edge of 1.6 ns [62].

The SiC DSRD can achieve higher power density than the Si DSRD. The increase of power density may increase the heat dissipation, which causes the junction temperature rise of the SiC DSRD. It may change the effective lifetime of nonequilibrium carriers in the drift region. A necessary condition for the DSRD to quickly switch off is that the forward current–pumping time should be shorter than the carrier lifetime in the base region. Therefore, the influence of temperature on

Figure 2.28 Top view of fabricated 4H-SiC DSRD

Figure 2.29 Four 4H-SiC DSRDs in series

the lifetime of nonequilibrium carriers is important for designing and analyzing the dynamic process of the device.

Studies have shown that [63], as shown in Figure 2.30, for a 1-kV 4H-SiC DSRD with p^+-p^--n^+ structure, as the device temperature rises from 300 to 673 K, the lifetime of the nonequilibrium carriers in the p base region of the DSRD increases from 250 ns to 1.4 μs, a sixfold increase. Lifetime measurements are carried out on the falling edge of the post-injection electromotive force by the open circuit voltage decay (OCVD) method.

In addition, the effect of temperature on dynamic performance is also studied, especially the effect of temperature on the injected charge loss [64]. The effect of temperature on voltage is shown in Figure 2.31. The higher the temperature is, the better the storage of the injected charge is, which leads to an increase in the time taken for the carrier to be withdrawn from the base. Within this temperature range, the magnitude of the generated voltage pulse is reduced from 1,400 V at room temperature to 650 V at 573 K, and the average switching speed is reduced from 2 to 0.7 V/ps.

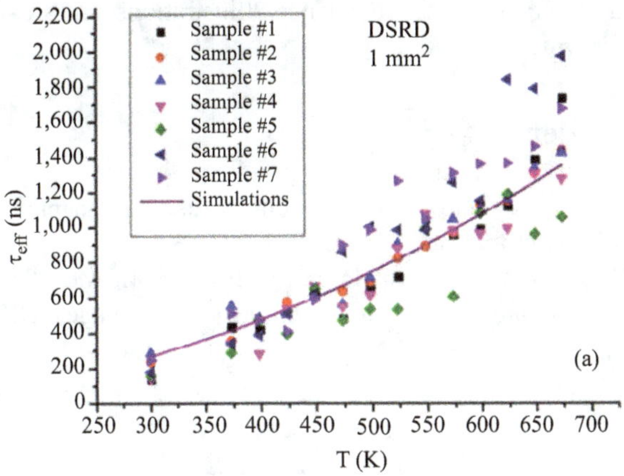

Figure 2.30 Experimentally measured temperature dependence of the effective minority carrier lifetimes in the p base of the 4H-SiC DSRD

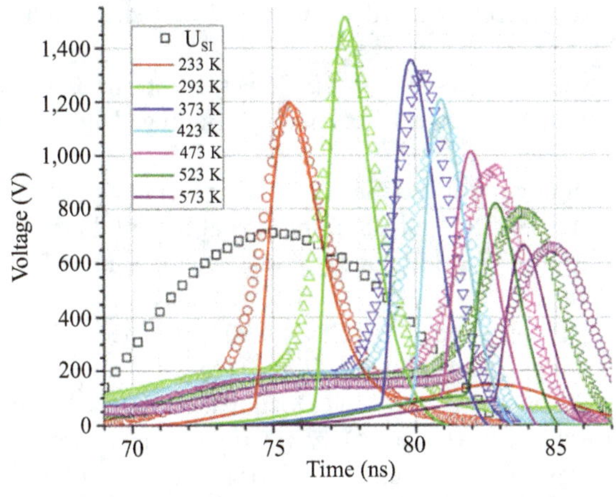

Figure 2.31 Voltage pulses (markers) at different temperatures. Simulations – solid lines; experiments – markers

2.6 Summary

In this chapter, the junction diodes are specified in focus. It is started from the base of PN junction. The concepts of equilibrium and nonequilibrium PN junctions are introduced, together with the ideas of junction breakdown and junction capacitance. Then, according to the different reverse recovery characteristics and applications, junction diodes are divided into three categories, including PiN diodes, FSRDs and DSRDs. Their structures, operation principles and characteristics are described. Some typical applications are illustrated. Instability issues are covered, especially the ones related to temperature dependence. Illustrations for SiC-based junction diodes are also provided.

Note to the reader

This chapter is part of an edited book. To cite or use this chapter as a reference, please use the following format:

Liang L. “Junction diodes.” In: Iannuzzo F (ed.), *Modern Power Electronic Devices: Physics, Applications, and Reliability*. Stevenage, UK: IET; 2020. pp. 13–48.

References

[1] Gang L, Xiaoming H, Tao C. Microelectronic Devices and IC Design. Beijing: Science Press. 2005.

[2] Gang L, Yuehui Y, Jiqun S, *et al*. Semiconductor Devices Power, Sensitive, Photonic, Microwave Devices. Beijing: Electronic Industry Press. 2000.

[3] Hall R N. Power Rectifiers and Transistors. Proceedings of the IRE, 1952, 40 (11):1512–1518.

[4] Jian C, Yong K, Xinbo R, *et al*. Power Electronics, Power Electronics Transformation and Control Technology. Beijing: Higher Education Press. 2011.

[5] ABB Ltd. High current rectifier diodes for welding applications [EB/OL]. (2013-03) [2019-12]. https://search-ext.abb.com/library/Download.aspx?DocumentID=5SYA2013&LanguageCode=en&DocumentPartId=&Action=Launch.

[6] ABB Ltd. Rectifier Diode 5SDD 54N4000 Data Sheets [EB/OL]. (2002-01) [2019-12]. http://search.abb.com/library/Download.aspx?DocumentID=5SYA1171-02&LanguageCode=en&DocumentPartId=&Action=Launch.

[7] Singh R, Hefner A R, Berning D, *et al*. High Temperature Characteristics of 5 kV, 20 A 4H-SiC PiN Rectifiers. 2001:45–48.

[8] Singh R, Irvine K G, Richmond J, *et al*. High-Temperature Performance of 10 Kilovolts, 200 Amperes (Pulsed) 4H-SiC PiN Rectifiers. Materials Science Forum, 2002, 389–393:1265–1268.

[9] Singh R, Cooper J A, Melloch M R, *et al*. SiC Power Schottky and PiN Diodes. IEEE Transactions on Electron Devices, 2002, 49(4):665–672.

[10] Das M K, Sumakeris J J, Hull B A, *et al*. High Power, Drift-Free 4H-SiC PiN Diodes. High Performance Devices, 2004. Proceedings IEEE Lester Eastman Conference on. IEEE, 2004.

[11] Hull B A, Das M K, Sumakeris J J, *et al*. Drift-Free 10-kV, 20-A 4H-SiC PiN Diodes. Journal of Electronic Materials, 2005, 34(4):341–344.

[12] Das M K, Sumakeris J J, Hull B A, *et al*. Drift-Free, 50 A, 10 kV 4H-SiC PiN Diodes with Improved Device Yields. Materials Science Forum, 2005, 483–485:965–968.

[13] Das M K, Hull B A, Richmond J T, *et al.* Ultra High Power 10 kV, 50 A SiC PiN Diodes. International Symposium on Power Semiconductor Devices & ICs. IEEE, 2005.

[14] Hull B A, Das M K, Richmond J T, *et al.* A 180 Amp/4.5 kV 4H-SiC PiN Diode for High Current Power Modules. IEEE International Symposium on Power Semiconductor Devices & ICs. IEEE, 2006.

[15] Hull B A, Das M K, Richmond J, *et al.* High Current 6 kV 4H-SiC PiN Diodes for Power Module Switching Applications. Materials Science Forum. 2006.

[16] Hull B A, Sumakeris J J, Das M K, *et al.* Progress on the Development of 10 kV 4H-SiC Pin Diodes for High Current/High Voltage Power Handling Applications. Materials Science Forum, 2007, 556–557:895–900.

[17] Brosselard P, Camara N, Hassan J, *et al.* 3.3 kV-10A 4H-SiC PiN Diodes. Materials Science Forum. 2008: 991–994.

[18] Brosselard P, Pérez-Tomás A, Hassan J, *et al.* Low Loss, Large Area 4.5 kV 4H-SiC PIN Diodes with Reduced Forward Voltage Drift. Semiconductor Science and Technology, 2009, 24(9):095004.

[19] Tanaka Y, Ohashi H, Sung K M, *et al.* Development of 6kV-Class SiC-PiN Diodes for High-Voltage Power Inverter. International Symposium on Power Semiconductor Devices & ICs. 2010.

[20] Niwa H, Suda J, Kimoto T. 21.7 kV 4H-SiC PiN Diode with a Space-Modulated Junction Termination Extension. Applied Physics Express. 2012, 5(6):1598–1600.

[21] Bolotnikov A V, Losee P A, Matocha K, *et al.* Design, Yield and Process Capability Study of 8 kV 4H-SiC PIN Diodes. Materials Science Forum, 2012, 717–720:953–956.

[22] Elasser A, Agamy M S, Nasadoski J, *et al.* Static and Dynamic Characterization of 6.5-kV 100-A SiC Bipolar PiN Diode Modules. IEEE Transactions on Industry Applications, 2014, 50(1):609–619.

[23] Cheng L, Agarwal A K, O'Loughlin M, *et al.* 16 kV, 1 cm^2, 4H-SiC PiN Diodes for Advanced High-Power and High-Temperature Applications. Materials Science Forum, 2013, 740–742:895–898.

[24] Kaji N, Niwa H, Suda J, *et al.* Ultrahigh-Voltage SiC PiN Diodes with an Improved Junction Termination Extension Structure and Enhanced Carrier Lifetime. Japanese Journal of Applied Physics, 2013, 52(7R):070204.

[25] Okamoto D, Tanaka Y, Mizushima T, *et al.* 13-kV, 20-A 4H-SiC PiN Diodes for Power System Applications. Materials Science Forum, 2014, 778–780:855–858.

[26] Salemi A, Buono B, Hallén A, *et al.* Fabrication and Design of 10 kV PiN Diodes Using On-Axis 4H-SiC. Materials Science Forum, 2014, 778:836–840.

[27] Deng X, Xiao H, Wu J, *et al.* Design, Fabrication and Characterization of Mesa Combined with JTE Termination for High-Voltage 4H–SiC PiN Diodes. Superlattices and Microstructures, 2015, 88:167–173.

[28] Runhua H, Yonghong T, Ling W, *et al.* Development of 17 kV 4H-SiC PiN Diode. Journal of Semiconductors. 2016, Vol.37(No.8):084001.

[29] Sung W, Baliga B J. A Comparative Study 4500-V Edge Termination Techniques for SiC Devices. IEEE Transactions on Electron Devices, 2017, 64(4):1647–1652.

[30] Li J, Xiao C, Xu X, *et al.* High-Voltage 4H-SiC PiN Diodes with the Etched Implant Junction Termination Extension. Journal of Semiconductors, 2017, 38(2):024003.

[31] Nakayama K, Mizushima T, Takenaka K, *et al.* 27.5 kV 4H-SiC PiN Diode with Space-Modulated JTE and Carrier Injection Control. 2018 IEEE 30th International Symposium on Power Semiconductor Devices and ICs (ISPSD). IEEE, 2018.

[32] Skowronski M, Ha S. Degradation of Hexagonal Silicon-Carbide-Based Bipolar Devices. Journal of Applied Physics, 2006, 99(1):011101.

[33] Okada A, Nishio J, Iijima R, *et al.* Dependences of Contraction/Expansion of Stacking Faults on Temperature and Current Density in 4H-SiC p–i–n Diodes. Japanese Journal of Applied Physics, 2018, 57(6).

[34] Kaji N, Niwa H, Suda J, *et al.* Ultrahigh-Voltage SiC p-i-n Diodes with Improved Forward Characteristics. IEEE Transactions on Electron Devices, 2015, 62(2):374–381.

[35] Heinze B, Felsl H P, Mauder A, *et al.* Influence of Buffer Structures on Static and Dynamic Ruggedness of High Voltage FWDs. International Symposium on Power Semiconductor Devices & ICs. IEEE, 2005.

[36] Schlangenotto H, Serafin J, Sawitzki F, *et al.* Improved Recovery of Fast Power Diodes with Self-Adjusting p Emitter Efficiency. IEEE Electron Device Letters, 1989, 10(7):322–324.

[37] Chen M, Lutz J, Domeij M, *et al.* A Novel Diode Structure with Controlled Injection of Backside Holes (CIBH). Power Semiconductor Devices and ICs, 2006. ISPSD 2006. IEEE International Symposium on. IEEE, 2006.

[38] Kopta A, Rahimo M. The Field Charge Extraction (FCE) Diode: A Novel Technology for Soft Recovery High Voltage diodes. International Symposium on Power Semiconductor Devices & ICs. IEEE Xplore, 2005.

[39] Lutz J, Schlangenotto H, Scheuermann U, *et al.* Semiconductor Power Devices Physics, Characteristics, Reliability [M]. Mechanical Industry Press, 2013.

[40] Vobecky J, Hazdra P, Homola J. Optimization of Power Diode Characteristics by Means of Ion Irradiation. IEEE Transactions on Electron Devices, 1996, 43(12):2283–2289.

[41] Matthias S, Vobecky J, Corvasce C, *et al.* Field Shielded Anode (FSA) Concept Enabling Higher Temperature Operation of Fast Recovery Diodes// Power Semiconductor Devices and ICs (ISPSD), 2011 IEEE 23rd International Symposium on. IEEE, 2011.

[42] Rahimo M T. Switching Characteristics of Fast Power Diodes in IGBT Circuits. 1996.

[43] Infineon Technologies. IDP18E120 Data Sheets [EB/OL]. (2009-09) [2019-12]. https://www.infineon.com/dgdl/Infineon-IDP18E120-DS-v02_03-en.pdf?fileId=db3a30432313ff5e01237a90bd2b7c24.

[44] Infineon Technologies. IDW50E60 Data Sheets [EB/OL]. (2014-04) [2019-12]. https://www.infineon.com/dgdl/Infineon-IDW50E60-DS-v02_02-en.pdf?fileId=db3a30433cd75ebf013cf1ddd0fb255a.

[45] Grekhov I V, Efanov V M, Kardo-Sysoev A F, *et al.* Power Drift Step Recovery Diodes (DSRD). Solid-State Electronics, 1985, 28(6): 597–599.

[46] Shinohara S, Kobayashi H, Hasegawa Y, *et al.* High Voltage Charge-Storage Diodes and Their New Applications. IEEE International Symposium on Power Semiconductor Devices & ICs. 1997.

[47] Grekhov I V, Mesyats G A. Physical Basis for High-Power Semiconductor Nanosecond Opening Switches. Plasma Science IEEE Transactions on, 2000, 28(5):1540–1544.

[48] Kozlov V A, Kardo-Sysoev A F, Brylevskii V I. Impact Ionization Wave Breakdown of Drift Step Recovery Diodes. Semiconductors, 2001, 35(5): 608–611.

[49] Gusev A V, Udalov O F, Krasil'nikov V. S., *et al.* Silicon Epilayers Grown by Chemical Vapor Deposition from High-Purity Silane. Inorganic Materials, 2006, 42(8):815–818.

[50] Merensky L M, Shafir I, Sharabani Y, *et al.* Fast Switching of Drift Step Recovery Diodes based on All epi-Si Growth. IEEE International Conference on Microwaves. IEEE, 2010.

[51] Sharabani Y, Rosenwaks Y, Eger D. Mechanism of Fast Current Interruption in p-π-n Diodes for Nanosecond Opening Switches in High-Voltage-Pulse Applications. Physical Review Applied. 2015, 4(1):014015.

[52] Ivanov B V, Smirnov A A, Shevchenko S A, *et al.* High Voltage Subnanosecond Silicon Carbide Opening Switch. 2016 57th International Scientific Conference on Power and Electrical Engineering of Riga Technical University (RTUCON). IEEE, 2016.

[53] Merensky L M, Kardo-Sysoev A F, Flerov A N, *et al.* A Low-Jitter 1.8-kV 100-ps Rise-Time 50-kHz Repetition-Rate Pulsed-Power Generator. IEEE Transactions on Plasma Science, 2009, 37(9):1855–1862.

[54] Benwell A, Burkhart C, Krasnykh A, *et al.* A 5KV, 3MHz Solid-State Modulator based on the DSRD Switch for an Ultra-fast Beam Kicker. Power Modulator and High Voltage Conference (IPMHVC), 2012 IEEE International. IEEE, 2012.

[55] Tropina A A, Kuzmenko A P, Marasov S V, *et al.* Ignition System Based on the Nanosecond Pulsed Discharge. IEEE Transactions on Plasma Science, 2014, 42(12):3881–3885.

[56] Prokhorenko V, Ivashchuk V, Korsun S. Drift Step Recovery Devices Utilization for Electromagnetic Pulse Radiation. Tenth International Conference on Ground Penetrating Radar. IEEE, 2005.

[57] Prokhorenko V, Boryssenko A. Drift Step Recovery Diode Transmitter for High-Power GPR Design. Proceedings of SPIE – The International Society for Optical Engineering, 2000, 4084.

[58] Arntz F O, Kardosysoev A, Krasnykh A, *et al.* SLIM, Short-Pulse Technology for High Gradient Induction Accelerators. IEEE International Pulsed Power Conference, 2009: 1–4.

[59] Ivanov P A, Grekhov I V. Subnanosecond 4H-SiC Diode Current Breakers. Semiconductors, 2012, 46(4):528–531.

[60] Ivanov P A, Grekhov I V. Subnanosecond Semiconductor Opening Switch Based on 4H-SiC Junction Diode. Materials Science Forum. 2013.

[61] Afanasyev A V, Ivanov B V, Ilyin V A, *et al.* Temperature Dependence of Silicon Carbide Drift Step Recovery Diodes Injection Electroluminescence. Journal of Physics: Conference Series, 2016, 741(1):012175.

[62] Ilyin V A, Afanasyev A V, Demin Y S, *et al.* 30 kV Pulse Diode Stack Based on 4H-SiC. Materials Science Forum, 2018, 924:841–844.

[63] Afanasyev A V, Ivanov B V, Ilyin V A, *et al.* Temperature Dependence of Minority Carrier Lifetime in Epitaxially Grown p^+-p^--n^+ 4H-SiC Drift Step Recovery Diodes. Materials Science Forum, 2015, 821–823:632–635.

[64] Smirnov A A, Shevchenko S A, Ivanov B V, *et al.* Investigation of the Temperature Effect on the Dynamic Parameters of Ultrafast Silicon Carbide Current Switches.. Journal of Physics Conference Series, 2018, 1038(1):012036.

Chapter 3

Thyristors

Jan Vobecky[1,2]

3.1 Introduction

Thyristor is a four-layered p-n-p-n structure in which the current flow between anode and cathode is triggered by a current pulse into the gate electrode. The gate trigger current flows from a current or voltage source connected between the gate and cathode, altogether forming the gate (control) circuit. If the impedance of the main circuit connected between the anode and cathode allows the flow of anode current higher than the so-called latching current, the thyristor stays in the forward conduction state (ON-state) even after removing the gate current. In other words, thyristor can be triggered to the ON-state by a current pulse containing a sufficient integral charge and remain ON as long as the anode current is higher than the so-called holding current.

The ON-state of the thyristor is characterised by **the lowest voltage drop V_T** for a given current density from all available semiconductor switches (Figure 3.1). As this implies the lowest energy losses in electronic applications, this represents the most significant advantage of thyristors. The low ON-state losses are usually related to the advantage of a very high **surge** (short circuit) **current** capability. Nowadays, high-power thyristors, with device area up to 140 cm^2, provide the maximal surge current I_{TSM} up to 150 kA for current pulses $t_p = 10$ ms, depending on their voltage class.

Another significant advantage represents the fact that thyristor can be turned ON by a single current pulse (typically, $t_{ON} > 10$ μs) and continue to conduct until the voltage between anode and cathode is reversed. It makes the gate driving circuits simple compared to other switches and lowers the energy consumption necessary to maintain the ON-state. Both these advantages are currently available at commercial thyristors produced from silicon together with reverse and forward blocking capability between 400 V and 10 kV, up to the maximal operation temperature of 125 °C. The magnitude of the maximal DC rating current up to 6,250 A and that of the surge current about 20 times more illustrates the fact that thyristors

[1]ABB Power Grids Switzerland Ltd. Semiconductors, Lenzburg, Switzerland
[2]Microelectronics Department, Faculty of Electrical Engineering, Czech Technical University in Prague, Prague, Czech Republic

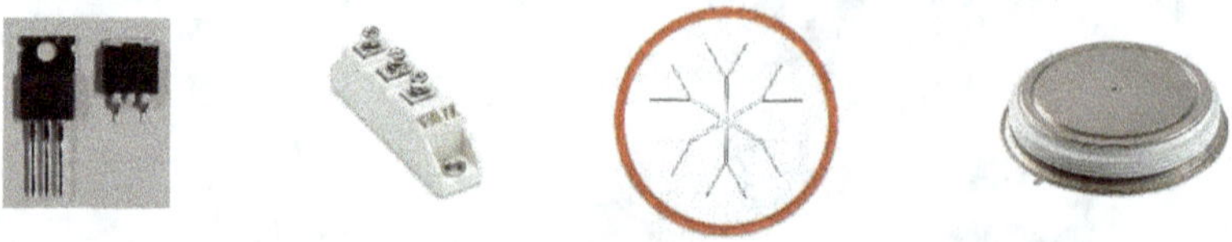

(a) TO220 and DD2PAK (b) SEMIPPAK (c) Thyristor at 150 mm wafer (d) Ceramic housing with thyristor wafer from (c)

Figure 3.1 Thyristors can be found in various packages depending on current and voltage ratings. Small area thyristors are produced as rectangular chips in moulded packages (a–b). Large area thyristors consume the whole wafer (c) or significant part and are packed in hermetic ceramic housings (d) with pressure contact. Not to scale

can **control the largest amount of power** from all existing power devices. One can find thyristors everywhere starting from low-power applications up to the top of a power pyramid.

The classical thyristor structure provides

- blocking of forward and reverse current,
- gate-controlled turn-on into the ON-state with very low V_T and
- turn-off of by circuit commutation (reverse recovery).

There is a broad group of thyristors possessing these features, called phase control thyristor (PCT), which is dedicated to the applications with line frequencies 50 and 60 Hz. Their commutation turn-off time can amount up to 1 ms, depending on voltage and current class. For higher frequencies, there is a group of fast thyristors with commutation turn-off time below 100 μs and significantly higher V_T.

For applications which require a turn-off capability using the gate electrode, there is available a group of turn-off thyristors. Historically first and already abandoned was the gate turn-off (GTO) thyristor followed by the faster and more reliable concept of gate commutated thyristor (GCT) or integrated gate commutated thyristor (IGCT). These devices are described in Chapter 6 of this book.

The importance of understanding thyristor physics is mostly driven by a need to design and apply the devices with the capability of a reliable turn-on and forward and reverse blocking. However, the four-layer p-n-p-n thyristor structure can also be found as a parasitic one in the complementary metal oxide semiconductor (CMOS)-based integrated circuits and other switches such as IGBTs, where it is prone to turn ON (latch-up) and destroy the whole chip. In this case, an exactly opposite sort of knowledge is needed. We need to identify the potential parasitic thyristor structure and introduce design provisions to prevent it from turning ON within the entire technical specification.

Before making a decision to use thyristor in a given application, it is essential to consider the character of output characteristics and its impact on system design and performance in comparison with transistors (MOSFET, IGBT). Figure 3.2 compares the measured output current versus voltage of thyristor and IGBT. The output characteristics of the thyristor are steep (low V_T) up to the maximal rating current and cannot be controlled by the device itself. Transistor switches are also used in the steep part of the output characteristics (resistive region). However, their characteristics can saturate at high collector-emitter voltages V_{CE} at a magnitude depending on the gate control voltage. This is advantageous especially under short circuit conditions when the full supply voltage accidentally appears on the transistor. Well-designed transistors allow a safe operation up to 10 μs under short circuit conditions, thanks to the saturation of the output *I–V* curves. A safe operation is assured by transistor turn-off within this interval to avoid thermal destruction.

Thyristors do not show any saturation of the output current. The output curves are 'triode-like' in analogy to vacuum tubes, the predecessors of semiconductors. However, every system design must assure a safe operation up to the maximal practically possible current, which is the short circuit current. In the thyristor circuits, it dictates an application of external protection components, such as fuses, chokes or reactors (inductances), to limit the maximal current to a level below the maximal thyristor rating, which is the surge current I_{TSM}. The inserted inductances not only limit the maximal current but also reduce the rate of rising the current (*di*/*dt*) during turn-on, which thyristor cannot control contrary to the transistor. In every case, it means the usage of additional components, which increases the system footprint and cost. On the other hand, it allows us **to limit the short circuit current** to a well-defined and reasonably high value that thyristors can survive. It can easily happen in the high-power systems with transistors that

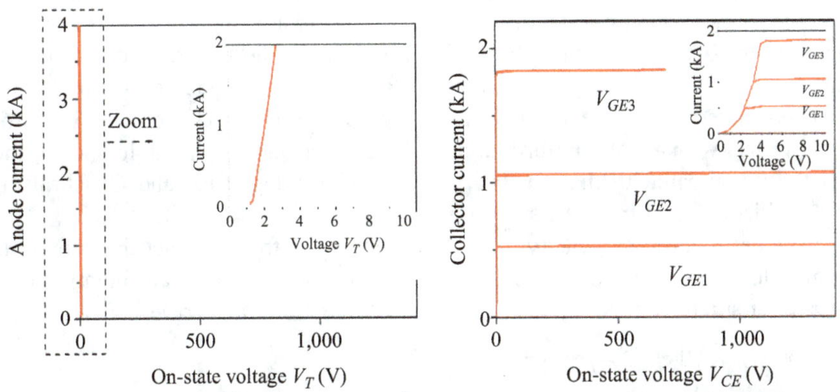

Figure 3.2 Output characteristics of thyristor type switch (left) show no saturation (8.5-kV thyristor example). The output current of transistor type switch (right) can saturate at high voltages depending on driving voltage V_{GE} (1.2-kV IGBT module example)

missing inductances allow a flow of so high short circuit currents that no component can survive and additional protection components must be added anyway, for example, the so-called bypass devices.

3.2 History and current state

Shortly after the invention of p-n junction and transistor in 1947, the complexity and unique features of p-n-p-n structure attracted the scientists at Bell Labs. This resulted in the experimental demonstration of a two-terminal silicon switch produced by the combination of diffusion and alloying [1]. The practical importance of the third electrode (gate) for a simple and well-defined thyristor switching was first understood and demonstrated by the engineers from General Electric in 1957 [2]. The three-terminal p-n-p-n electrically gated switch received the name silicon-controlled rectifier, which was later shortened to the thyristor.

According to the historical records from eyewitnesses, the thyristor played a unique role in the development of silicon technology (diffusion, oxidation, masking, metallisation) [2] (Figure 3.3). Their development enabled the experimental realisation of new design features giving thyristor reproducible and thermally stable electrical parameters (cathode shorts), application variability (Triac, GTO, RCT, LTT) and easiness of usage (amplifying gate) [3], where the individual abbreviations can be found in Figure 3.3. By the end of the 1960s, all relevant thyristor operation, design and processing concepts were known, and nothing kept thyristors from applications up to voltages as high as $\approx$2 kV. Although the quality of silicon wafers in terms of homogeneity of resistivity and contamination content was far away from present standards, the way to a low- and medium-power segment was paved.

Further increase of thyristor voltage ratings was enabled by the implementation of the neutron transmutation doping (NTD) to the modification of silicon ingots prior to wafer cutting. The NTD principle, for the first time described by Lark–Horovitz in 1951 [4], was experimentally proved in 1962 at silicon [5] and commercialised in the late 1970s [6]. The float zoned (FZ) NTD process has provided wafers with a more uniform doping distribution and a tighter tolerance of the resistivity compared to the existing methods of gas-doped FZ and Czochralski-grown silicon doped from the silicon melt.

The excellent homogeneity of NTD wafers stems from the fact that silicon is permeable for neutrons, thus allowing a spatially homogeneous transmutation of a fraction of silicon atoms to phosphorus ones through the (n, γ) reaction

$$^{30}Si(n, \gamma)^{31}Si \rightarrow {}^{31}P + \beta^-$$

For this purpose, the silicon rods originally produced by the float zone method with the resistivity of several kΩ.cm are rotating in the neutron flux. After annealing at $\approx$800 °C to remove the radiation damage, the resulting N-type wafers are doped by the stable ^{31}P. The wafers are free from macroscopic and microscopic resistivity variations with typical radial resistivity variation below 10 per cent and

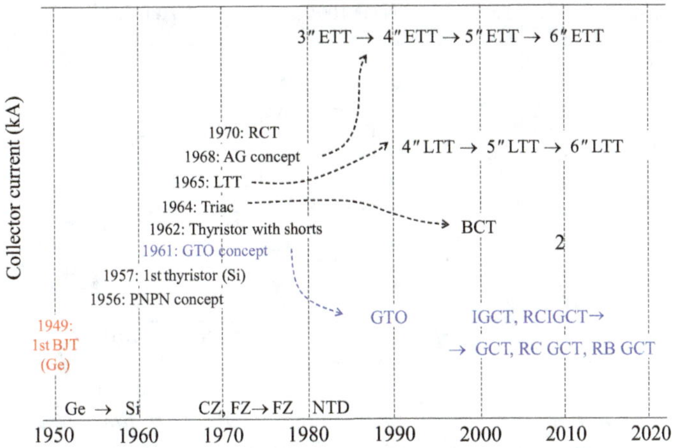

BJT = Bipolar Junction Transistor, PCT = Phase Control Thyristor (50/60 Hz), BCT = Bidirectional Control Thyristor, LTT = Light Triggered Thyristor, ETT = Electrically Triggered Thyristor, RCT = Reverse Conducting Thyristor, RB = Reverse Blocking, GTO = Gate Turn-Off Thyristor, IGCT = Integrated Gate Commutated Thyristor, FZ = Float Zone, NTD = Neutron Transmutation Doping, AG = Amplifying Gate.

Figure 3.3 Evolution of thyristor technology and relevant materials: classical thyristors (black) and gate turn-off thyristor (blue). 3″, 4″, 5″ and 6″ denote the time evolution of thyristor size in inches meaning 75, 100, 125 and 150 mm silicon, respectively, starting wafer diameter

absolute reproducibility of resistivity within a given ingot around 5 per cent. Nowadays, this technology is available on silicon wafers with a diameter of up to 200 mm. For high-voltage thyristors, the resistivity is precisely adjusted in the NTD process between 100 and 500 Ω.cm according to the required voltage class, which is ranging from 2.5 to 10 kV.

It is worth noting that the FZ process itself provides silicon wafers with much fewer oxygen atoms ($<2.10^{16}$ cm^{-3}) in comparison to the ingots pulled using the Czochralski method. During the processing of wafers, the originally electrically neutral (inactive) oxygen atoms can be converted to the electrically active oxygen thermal donors with maximal effect around 450 °C. The risk of an uncontrolled increase of the N-base doping concentration and reduction of the required blocking voltage of final devices is, therefore, reduced with the FZ wafers. Recently, the quality of gas-doped FZ silicon was significantly improved and their usage has increased for cost reasons. Anyway, the NTD process has remained irreplaceable for the high-end thyristors (large area, voltage classes above 6.5 kV).

All the improvements gradually enabled the production of thyristors with diameters up to 150 mm, breakdown voltages up to 10 kV and rating currents up to 6 kA, as summarised in Figure 3.4.

Thyristors with the highest voltage ratings (7.2–9.5 kV) and current ratings (up to 6.25 kA) from Figure 3.4 are used in the load commutated converters (LCC) of high-voltage direct-current (HVDC) transmission systems, where the maximal transmitted power reached 12 GW in 2018. Similarly, powerful thyristors are used

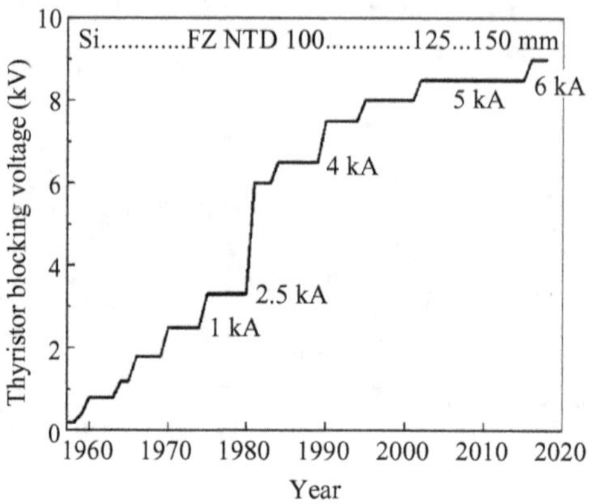

Figure 3.4 Evolution of the thyristor blocking capability in time. The labels show the maximal current ratings achieved in the HVDC systems and the diameter of available FZ NTD silicon wafers

in hydro-pumping power systems. All these devices are utilising full 150-mm silicon wafers and discrete ceramic packages with pole pieces ≈138 mm in diameter. Thyristors of a similar or smaller size and power can be found in the flexible AC transmission systems (FACTS), which are used to control the reactive power in the energy transmission networks.

3.3 The thyristor structure and its two-transistor analogue

The basic thyristor structure is composed of four layers (p-n-p-n), three p-n junctions (J1, J2, J3) and three electrodes (anode, cathode, gate). A practical thyristor structure contains more than four layers for the reason of manufacturability, electrical performance and a low cost of processing (see Figure 3.5). These layers are diffused from both surfaces into the starting phosphorus-doped FZ or FZ NTD silicon wafer. The remaining nondiffused part of the wafer forms the **N-base**, which has the lowest doping level and the biggest thickness. Together with two neighbouring low-doped p-type layers (resulting from the diffusion of aluminium in Figure 3.6, optionally of gallium), two p-n junctions with high breakdown voltages are formed. The **anode junction J1** provides a high reverse blocking voltage V_{RRM} (under negative anode to cathode voltage V_{AC}) and the **blocking junction J2** a high forward blocking voltage V_{DRM} (positive V_{AC}). Actually, the high blocking voltages are provided only by the low-doped p-type regions with the concentration below 1.10^{16} cm^{-3} together with the N-base doped to the order of 10^{13} cm^{-3}. The highly doped p-type and n-type layers P^{++} and N^{++}, respectively, are forming the anode and cathode emitters. They provide the highest possible injection efficiency of holes and electrons in the ON-state, respectively. Thyristors with the lowest ON-state voltage drop V_T are the ones with the surface boron (P^{++}) and phosphorus

(N^{++}) concentrations well above 1.10^{19} cm^{-3}. As the doping concentrations at both sides of the **cathode junction J3** are high, the breakdown voltage of this junction is typically below 5 V (below 25 V at GTO or GCT). Since the only task of the J3 is the turn-on (forward bias of J3), the low breakdown voltage does not constitute a problem provided that the J3 is protected against a negative bias.

Figure 3.7 shows the thyristor representation called a two-transistor analogue, which is useful to explain several features specific just to the thyristor. It helps especially to those familiar with basic transistor circuits. The four layers are split into two transistors, which are connected into positive feedback as is best illustrated by the electrical scheme. Figure 3.8 highlights the fact that the thicknesses of individual p- and n-type layers are very different in practical thyristors.

The two-transistor analogue facilitates understanding the turn-on process and keeping thyristor in the ON state even after the extinction of the gate current. Figure 3.8 shows what happens when the thyristor is connected to a power supply $+V_{supply}$ with a load and the gate current I_G is applied as the base current $I_{B\ NPN}$ (step 1). The NPN transistor turns ON and the gate current $I_G = I_{B\ NPN}$ is amplified to the collector current $I_{C\ NPN} = \beta_{NPN}.I_G$ (step 2), where β_{NPN} is the current amplification factor of the NPN transistor in the common-emitter configuration. The collector-emitter voltage $V_{CE\ NPN}$ decreases and the NPN transistor enters the

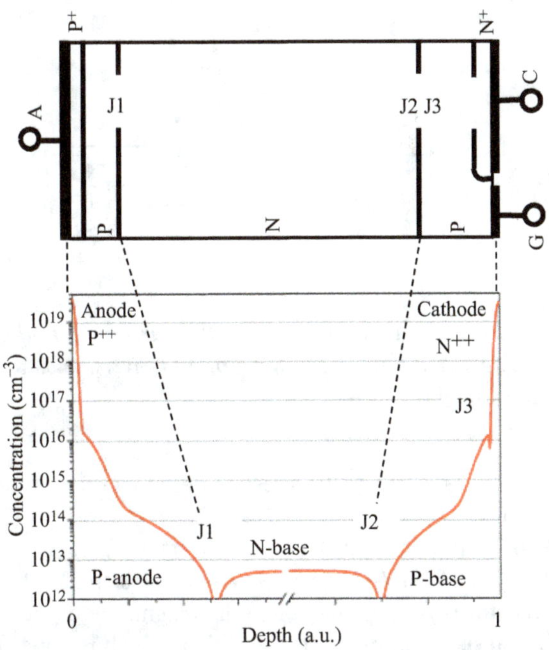

Figure 3.5 A simplified sketch of high-voltage thyristor structure (top) and 1-D doping profile (bottom). Note the interruption of x-axis. J1 = anode junction, J2 = blocking junction and J3 = cathode junction

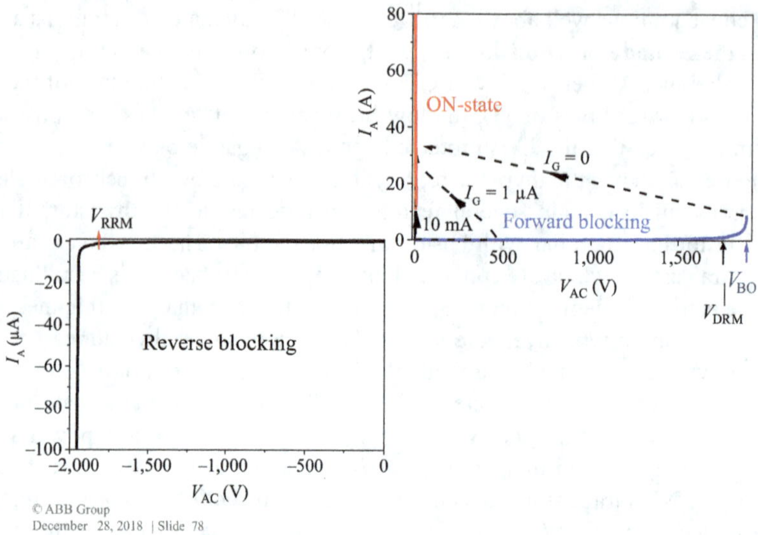

Figure 3.6 Three working regimes of thyristor: Forward blocking (V_{BO} = break-over voltage), reverse blocking and ON-state. Blocking and breakdown voltage not related to the doping profile in Figure 3.5

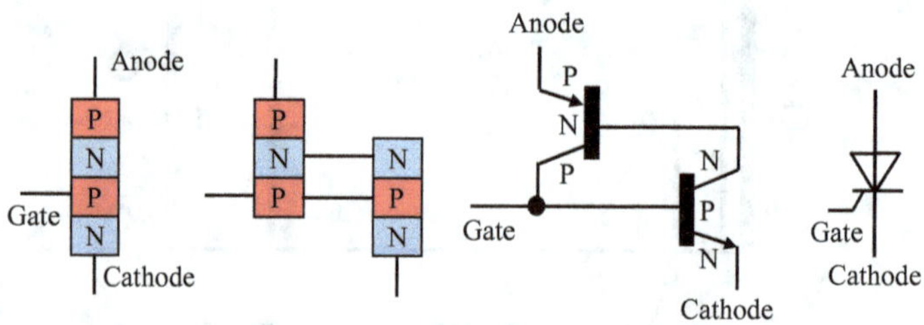

Figure 3.7 A simplified thyristor structure (left) rearranged to the two-transistor analogue scheme (middle). Symbol of a thyristor as used in electrical schemes (right)

saturation: $V_{CE\ NPN} = V_{CE\ sat} \rightarrow 0$. This results in a flow of the base current of the upper PNP transistor $I_{B\ PNP} = I_{C\ NPN} = \beta_{NPN}.I_G$.

The upper PNP transistor turns ON what reflects in $V_{EC\ PNP} = V_{EC\ sat} \rightarrow 0$ and $I_{C\ PNP} = \beta_{PNP}.I_{B\ PNP} = \beta_{PNP}.I_{C\ NPN} = \beta_{PNP}.\beta_{NPN}.I_G$ (step 3). The current amplification of the thyristor is $\beta_T = \beta_{PNP}.\beta_{NPN}$. The magnitude of the current amplification factor of a transistor drops in the saturation, which is not a problem for the thyristor to stay in the ON-state, if $\beta_T \geq 1$, i.e., when the current from transistors' amplification is higher than the original gate current. The ON-state voltage drop of the thyristor V_T is very low, as suggests the appearance of both transistors in the saturation – the state with the lowest possible V_{CE} even smaller than the V_{BE} due to the forward-biased base-collector junction (= blocking junction J2 of the thyristor).

By providing the base drive current for each other the two transistors keep themselves in the saturation and the thyristor in the ON-state. When the gate pulse I_G ceases, it is substituted by the current $I_{C\ PNP}$, the existence of which is maintained by that of the $I_{C\ NPN} = \beta_{NPN} \cdot I_{C\ PNP}$. Positive feedback is closed ($\beta_T > 1$ or $\alpha_{PNP} + \alpha_{PNP} = 1$ if common base amplification factors α are considered) and the thyristor stays safely in the ON state. A thyristor is the only device with this advantageous feature.

The positive feedback can be broken – it is regenerative. However, the flow of thyristor current I_T and, hereby, the positive feedback can be interrupted only by changes in the external circuit. It happens for example by reversing the polarity of V_{supply} or by increasing the load impedance in a way that the flow of $I_{C\ PNP}$ is reduced below a limit originally keeping the positive feedback active.

As the box scheme in Figure 3.8 (left) suggests, the individual transistors do not represent a typical bipolar junction transistors (BJTs) with high current amplification factors ($\beta > 50$) and high transient cut-off frequencies ($f_T >$ 50 MHz). This is because the main prerequisite for these two parameters, namely a narrow base region (narrower than the diffusion length of minority carriers passing this region), is not fulfilled at all. In this respect, the NPN transistor is much better than the PNP one, which reflects in a bigger impact of its transistor effect on thyristors' electrical ratings.

On the other hand, the very competitive ratings of the constituent transistors are the open base breakdown voltages $V_{BR\ CE0}$ (when $I_B = 0$, i.e., when leakage current I_{CE0} flows). This is given by the low-doped and very thick n-type region (N-base of thyristor = N-collector of the NPN and N-base of the PNP transistor) and neighbouring low-doped p-type region (P-base of thyristor = P-base of the NPN and P-collector of the PNP transistor). The high breakdown voltage of the PNP and

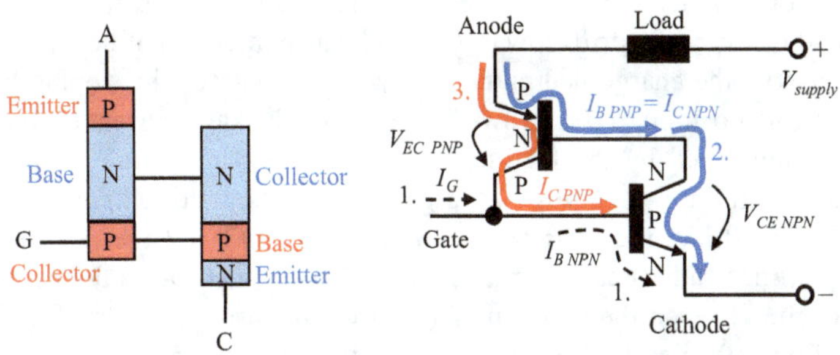

Figure 3.8 Two transistor analogue highlighting the different thicknesses of individual layers (left). Turn-on of thyristor as explained using the two-transistor analogue (right). The growing thickness of the lines represents the amplification of current

NPN transistors allows us to achieve, respectively, the high reverse and forward blocking voltages of the thyristor, which is discussed in Section 3.4.

3.4 Forward and reverse blocking

Based on the same principle as in the P-i-N diode, thyristor provides high blocking voltage and low ON-state voltage drop V_T at the same time. As the low-doped wide N-base is surrounded by relatively low-doped p-type regions at both sides (Figures 3.5 and 3.9), the thyristor can block the flow of current for both polarities of V_{AC}, if the gate current is below its trigger value as shown on Figure 3.6.

Even if the doping profiles forming the junctions J1 and J2 are close to each other due to the diffusion of p-type dopants simultaneously from anode and cathode side, the physical behaviour under reverse and forward blocking is different. Under **reverse blocking** (anode negative, cathode positive, J1 and J3 reverse biased, J2 forward-biased), only the forward-biased junction **J2** (P-base–N-base) **can amplify the leakage current** generated in the space charge regions (SCR) of the reverse-biased J1 and J3. As the SCR of the J3 has a negligible thickness in comparison with J1, the most of leakage current is generated in the wide SCR of the junction J1 (see Figure 3.9). This current is then multiplied at forward-biased J2 and extracted by the J1. The holes flow towards the negatively polarised anode (to the left) and the electrons towards the positively polarised cathode (to the right).

A significant magnitude of electric field $\boldsymbol{E}$ can be found in the SCR of the reverse-biased anode junction J1, while that of the J3 is too narrow to be visible on this scale. At low voltages represented by $V_{AC} = 2$ kV, the region outside the SCR contains only electrons with the concentration $\boldsymbol{n}$ equal to the donor concentration N_D of the N-base. At higher voltages represented by $V_{AC} = 5$ kV in the middle of Figure 3.9, the peak electric field increases up to ≈100 kV/cm. At the position of the arrow outside the SCR in Figure 3.9 (middle), we can see a slight increase of both the electron and hole concentration. The electrons on top of the N_D (dotted line) are the ones generated in the SCR of the J1. They are accelerated by the high electric field of the SCR (drift) towards J2 and cause injection of holes from the P-base to conserve the charge neutrality ($n = p + N_D^+$) – see the sloping hole concentration at the position of the arrow. This way, the leakage current of the reverse-biased J1 is amplified by the forward-biased J2 as evidenced by the increase of reverse current in the *I–V* curves on top graphs (see the bold red point).

At $V_{AC} = 10$ kV, the SCR reaches the junction J2 and the **punch-through effect** (sometimes called reach-through) takes place. Practically all electrons from the SCR of the J1 reach the junction J2 (see the blue arrow), where they cause an amplified flow of holes in the opposite direction and the anode current sharply increases. Besides the punch-through effect, the electric field increases up to 150 kV/cm and causes an increased generation rate of electrons and holes by the **impact ionisation effect**. These carriers also participate in the sharp increase of the anode current.

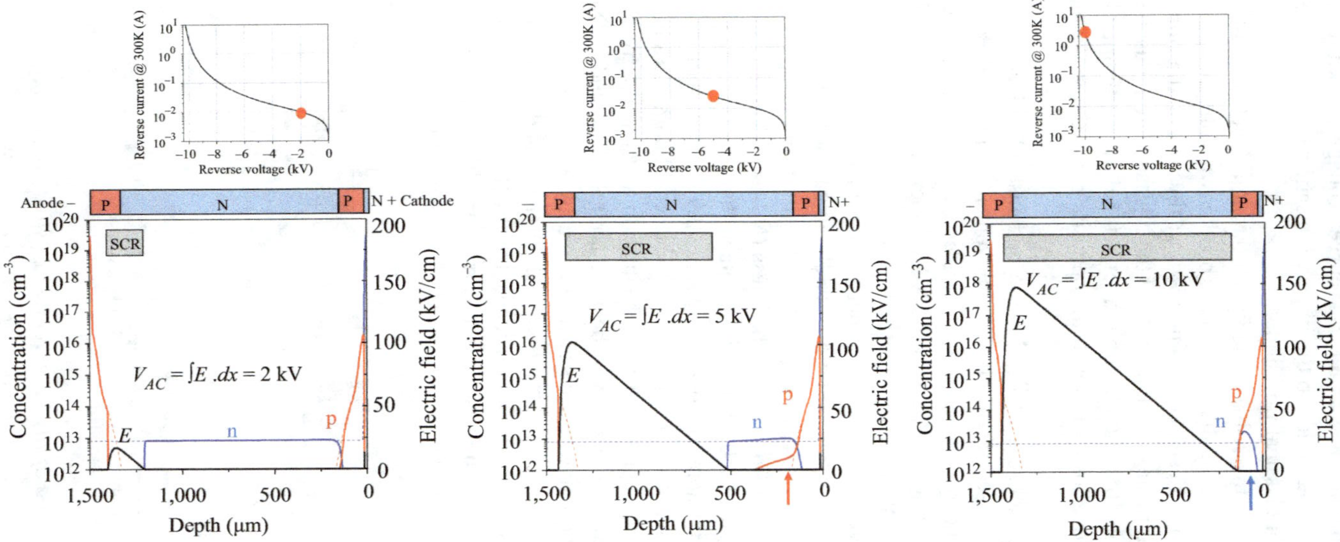

Figure 3.9 Electric field (E), electron (n) and hole concentration (p) of thyristor under the reverse bias of 2, 5 and 10 kV. The space charge region (SCR) of the anode junction J1 expands with the growing reverse bias

If we increased the thickness of the N-base, the punch-through effect would appear either at a higher V_{AC} or the knee of the I–V curve (voltage breakdown) would appear before the punch-through due to the impact ionisation leading to the **avalanche breakdown**, like in a classical p-n diode.

If we reduced the thickness of the N-base, the punch-through effect would be fully responsible for the sharp knee of the I–V curve (breakdown voltage). This implies an important fact from the designer's point of view: thyristor is the **non-punch-through** (NPT) device. It means that the width of the N-base region must be designed thick enough to avoid the punch-through effect below the nominal blocking voltage. The SCR is not allowed to reach the opposite p-n junction under normal operation.

If we did not change the thickness of the N-base and instead reduced the doping concentration N_D (= increased resistivity), the movement of the SCR with growing V_{AC} would increase and the punch-through effect would take over the impact ionisation. If we for the same thickness of the N-base increased the doping concentration N_D (= reduced resistivity), an opposite effect would take place: the impact ionisation would take over. In the thyristor design, we, therefore, strive to find a good balance between the two competing effects, the impact ionisation and punch-through, to maximise the utilisation of silicon as shown in Figure 3.10(a). This is performed by selecting an optimal wafer thickness (N-base thickness) and resistivity, as summarised in Figure 3.10(b), for high-voltage thyristors.

Figure 3.11 shows the distributions analogous to Figure 3.9 for the conditions of **forward blocking**. In this case, the SCR of the reverse-biased J2 dominates the N-base. Because of the symmetric doping profiles, the situation looks qualitatively similar to that of the reverse blocking. However, because both the PNP and NPN transistors operate in the normal active regime characterised by a relatively high current amplification, the leakage current is amplified by both the **J1** and **J3**, where $\alpha_{NPN} > \alpha_{PNP}$. As the increasing voltage V_{AC} widens the SCR of the J2 and decreases the base widths of both the transistors, the current amplification further increases (Early effect). As a result, about two times higher leakage current flows at

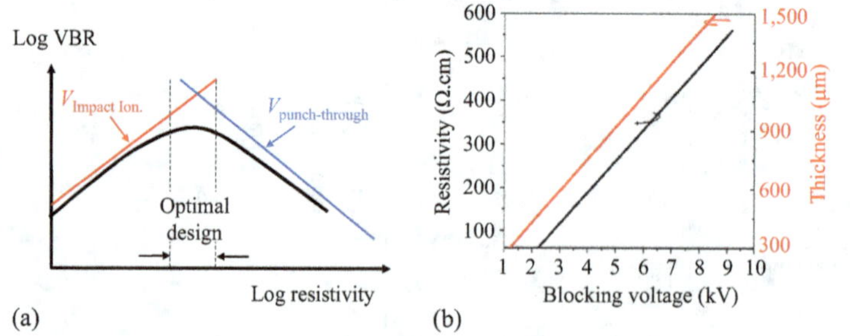

Figure 3.10 (a) Breakdown voltage of thyristor versus resistivity of starting silicon wafer in the light of responsible physical effects. (b) Resistivity and thickness of starting silicon wafer versus breakdown voltage of thyristor

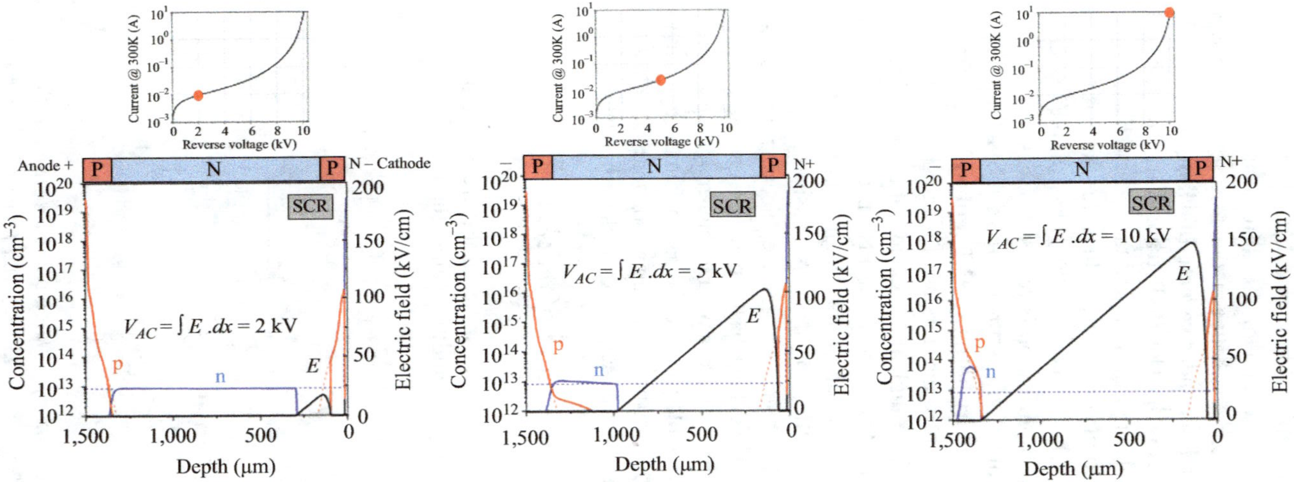

Figure 3.11 Electric field (E), electron (n) and hole concentration (p) of thyristor under the forward bias of 2, 5 and 10 kV. The space charge region (SCR) of the blocking junction J2 expands with growing forward bias

the same voltage compared to the reverse blocking (note the logarithmic scale). The competition between the punch-through and impact ionisation effect is the same likewise its impact on thyristor design depicted in Figure 3.10.

Figure 3.12 gives a graphical visualisation of the differences between the current amplification of the leakage current between the reverse and forward blocking. The polarisation of p-n junctions under **reverse blocking** results in the amplification of the leakage current **only by the PNP transistor**. The forward-biased junction J2 provides the current amplification and the reverse biased junctions J1 and J3 the extraction (collection) of corresponding free carriers. As the effective base width of this transistor, which is equal to the N-base of the thyristor, is wide up to very high voltages, this amplification is relatively weak.

The polarisation of p-n junctions of a thyristor under **forward blocking** results in the amplification of the leakage current **by both the PNP and NPN transistors**. Both the B-E junctions of the constituent transistors are forward biased (*F*) and fulfil one of the necessary conditions for a high amplification of current. The current amplification of the NPN is higher than that of the PNP because of the narrower base region. The second necessary condition for a high amplification factor, a narrow base, is hereby much better fulfilled for the NPN transistor.

One forward-biased junction under thyristor's reverse blocking versus two forward-biased junctions under forward blocking has a practical consequence. At simple thyristor structures, we typically measure a visibly higher leakage current under forward blocking than under reverse blocking. In the past, when the cleanliness of silicon was much lower than today, this has been a strong reason for the application of cathode shorting structures in the thyristor design, which is described later.

3.4.1 Advanced methods for optimisation of blocking capability

The choice of silicon resistivity and thickness illustrated in Figure 3.10 represents the first level of optimisation of blocking capability against the ON-state losses.

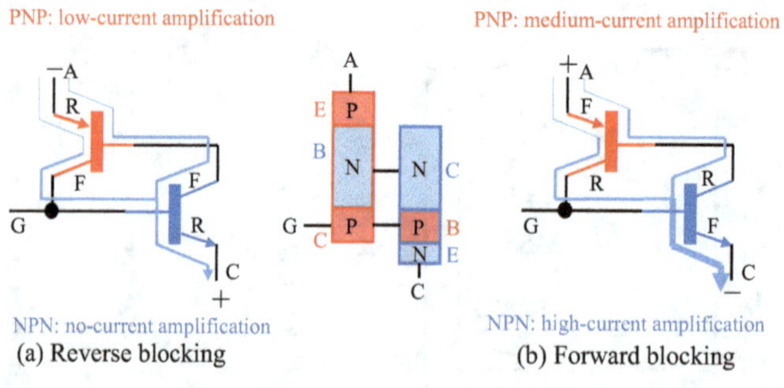

Figure 3.12 Overview of the amplification of leakage current using the two-transistor analogue under the reverse (a) and forward blocking (b). Line thickness is proportional to the magnitude of current amplification. Amplification of current takes place at forward-biased p-n junctions (F)

The result of this optimisation can be evaluated by measuring the breakdown voltage at the lowest operation temperature and the leakage current at the maximal operation temperature. These two parameters can then be related to the achieved voltage drop V_T.

Blocking ratings can be further improved by

1. minimising the punch-through effect and
2. by reducing the excess carrier lifetime (= reducing the amplification of leakage current by the NPN and PNP transistors by reducing their base transport factors).

Figure 3.13(a) shows a possible method to reduce the leakage current and to increase the breakdown voltage by increasing the width of the N-base $w_{N\text{-}base}$ in the bulk of the thyristor compared to the N-base width w_{JT} at junction termination (JT). This method is advantageous for the devices with the JT formed by negative bevel [7].

The improvement is based on the fact that the punch-through effect is suppressed in the internal region of the thyristor with a wider N-base region $w_{N\text{-}base}$. The enlargement of the N-base happens at the expense of the lower thickness of the p-type layer forming the P-base and P-anode. It is not a problem as long as the lower p-type layer thickness supports the required breakdown voltage of the one-dimensional p-n junction in the bulk.

An important feature of this concept is that the internal region represents about 90 per cent of the total device area. At the peripheral region, occupying about 10 per cent of the thyristor area, the N-base region width w_{JT} is much smaller and the punch-through effect can dominate there at voltages close to the breakdown voltage. The requirement of a deeper p-type layer is dictated by design principles of the negative bevel used to minimise the surface electric field at the edge of the silicon wafer (JT).

Figure 3.13(b) shows the gain in the leakage current for a given voltage and, hereby, an increase of the blocking voltage measured at a four-inch thyristor with

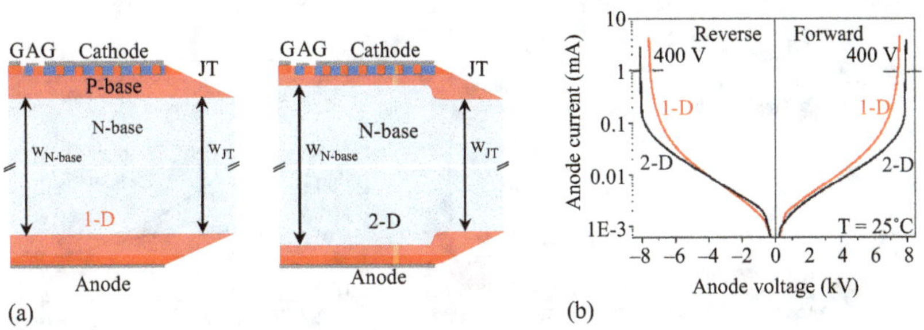

Figure 3.13 (a) Thyristor with the 1-D diffusion profile (1-D) and 2-D diffusion profile (2-D). (b) Forward and reverse blocking I–V curves of thyristors with 1-D and 2-D doping profiles, respectively

an active area of about 60 cm^2. The gain in the blocking voltage is usually invested in the reduction of the wafer thickness providing us a lower V_T. This makes sense mainly in the thyristors for the HVDC valves, where every 100 mV of reduced V_T reduces the energy losses of transmitting power in the GW order by several percents. For example, the latest 8.5-kV thyristor from Figure 3.13(b) provides $V_T < 1.6$ V at $I_T = 1.5$ kA and $T = 90$ °C.

Figure 3.14(a) shows a method to further reduce the leakage current by local reduction of the excess carrier lifetime by proton irradiation (right) [8]. The state-of-the-art production of thyristors utilises mainly the electron irradiation (left) as a mean for a uniform reduction of the lifetime to adjust relevant parameters such as the recovery charge Q_{rr} and the turn-off time described later. The uniformity is given by the electron irradiation energy chosen so high that the electrons penetrate the whole wafer. As a side effect, the electron irradiation improves parameters such as breakdown voltage and leakage current as illustrated in Figure 3.14(b). This is because the irradiation introduces point defects acting as the generation-recombination (g-r) centres, which reduce the carrier lifetime and consequently the amplification of leakage current by the internal NPN and PNP transistors. The g-r centres themselves increase the leakage current of a reverse-biased p-n junction. However, this effect is smaller than the suppression of leakage current by reduced amplification of the internal transistors.

Using the proton irradiation, we can further benefit from introducing the recombination centres locally in the regions where they suppress the leakage current more efficiently. Such a location is the N-base close to the junction J1 and J2, where the amplification of leakage current takes place as shown in Figures 3.9 and 3.11. Since we need to suppress the leakage current under both the forward and reverse bias, the protons must be applied symmetrically through the anode and cathode surfaces.

As shows the upwards pointing arrow in Figure 3.14(b), the leakage current of proton irradiated devices is increased at low voltages in comparison with the electron irradiation. This is the result of introducing the g-r centres by the protons into the SCR of a reverse-biased p-n junction, i.e., into the J2 under forward bias and into the J1 under reverse bias, in the case of the reverse blocking from Figure 3.14(b).

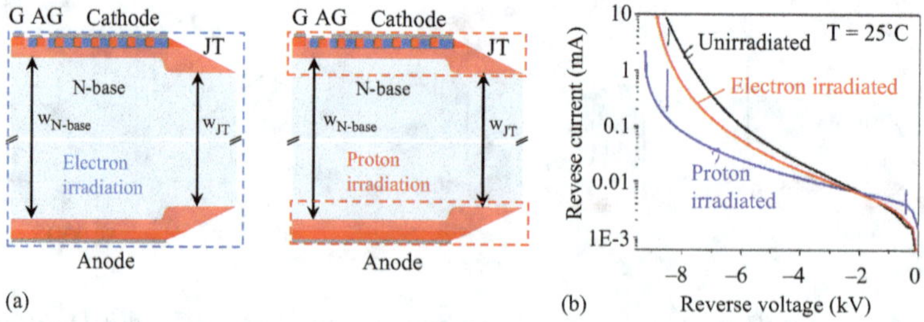

Figure 3.14 (a) Electron irradiation affects the whole volume. The proton irradiation acts only locally. (b) The reduction of leakage current by electron and proton irradiation

Since the g-r centres are placed only in a spatially limited region close to the p-n junction, there is no further increase of the leakage current with increasing voltage. On the contrary, at higher voltages, above ≈2 kV in Figure 3.14(b), the effect of the g-r centres on the opposite side of the thyristor starts to reduce the leakage current. This happens thanks to the reduction of the diffusion length of free carriers (outside the SCR). Fewer carriers then reach the amplifying p-n junction and the leakage current is reduced. This effect adds to the 2-D effect from Figure 3.13(a).

3.4.2 Junction termination

When the electric field at the p-n junction in the bulk (active region of a device) reaches the critical value of the electric field for silicon $E_{crit} \approx 200$ kV/cm, the excessive impact ionisation causes the electrical breakdown. When the p-n junction reaches the surface and the electric field exceeds the dielectric strength E_{max} of the ambient dielectric (air or nitrogen), the dielectric breakdown occurs. The high electric field ionises the molecules and changes them from an insulator to a conductor. Because of the recombination of electrons and ions, sparks occur and the required blocking capability is lost. As the dry air or nitrogen shows only $E_{max} \leq$ 30 kV/cm, the edge of the thyristor wafer or chip must be specially treated against the dielectric breakdown. This part of the thyristor is called junction termination (JT) and represents the most sensitive region regarding the blocking capability and its long-term stability.

Figure 3.15(a) shows the distribution of electric potential and electric field calculated from the Poisson equation for a reverse bias diode without any JT. Although the maximal surface electric field is about five times of the E_{max}, there is no breakdown, because an ideal dielectric is used for the simulation. In a real

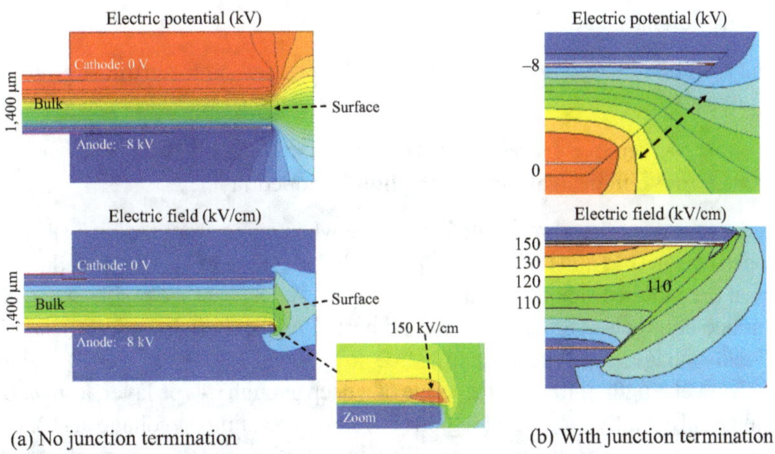

Figure 3.15 The distribution of electric potential and electric field at the edge of a device without (a) and with junction termination (b). The device is upside down in the case (b)

situation, the device would break at the location of the peak electric field, which is located where the anode p-n junction J1 reaches the surface. Such weak points at the edge must be eliminated.

Figure 3.15(b) shows that bevelling can prolong the surface length and bend the potential lines so that the maximal electric field moves from the surface into bulk and the surface electric is lowered. In practice, the silicon surface is passivated by a thin layer with appropriate E_{max} and electrical conductivity (electroactive passivation) in order to eliminate the existence of poor quality native oxides, the effect of surface dangling bonds, charging effects, conducting surface channels analogous to the doping enhancement in MOS devices and to shield the silicon bulk from external charges. These layers can be formed by silicone, doped glass, polyimide, silicon nitride, amorphous silicon, semi-insulating polycrystalline silicon (SIPOS), diamond-like carbon (DLC) and, eventually, by their smart combinations.

The bending of equipotential lines (Figure 3.15) is caused by an unequal change in the amount of electrical charge at the individual sides of the p-n junction. Regardless of how much volume is removed, the charge balance at the p-n junction described by the equation

$$\int N_A^- \times dy \times dz = \int N_D^+ \times dy \times dz \tag{3.1}$$

must be maintained. If we, for example, reduce more volume at the p-doped side than at the n-doped one, the SCR at the p-side expands to compensate for this removal and contracts at the n-side. The expansion of the SCR reduces the surface electric field.

There are two basic options to modify the amount of charge:

(a) selective charge removal and
(b) selective charge addition.

The selective charge removal methods (a) are based on mechanical grinding of the edge of a silicon wafer under an appropriate angle so that

- n. either we remove more materials in the highly doped layer or
- p. we remove more materials in the lightly doped layer.

In the case of *n*, the bevel angle is negative when going from the high (P^+) to the low-doped region (N^-) and this JT concept is called **negative bevel.** As we reduce more volume at the p-doped side than at the n-doped one, the SCR at the p-side expands and at the n-side contracts. To maximise the expansion of the SCR for a sufficiently high breakdown voltage, a p-type layer must be low doped (10^{13}–10^{15} cm^{-3}) and fill a sufficient bevel length. It implies the usage of a deep enough p-type layer, a small bevel angle and a sufficiently long bevel region (1–4 mm depending on voltage class). Under such conditions, the SCR expands along the whole surface and there is no electric field at the steep wall between the top and bottom surface as shown in Figure 3.16.

Usage of the negative bevel in the devices with a shallow p-type anode or P-base layer does not make sense, because it would lead either to a low breakdown

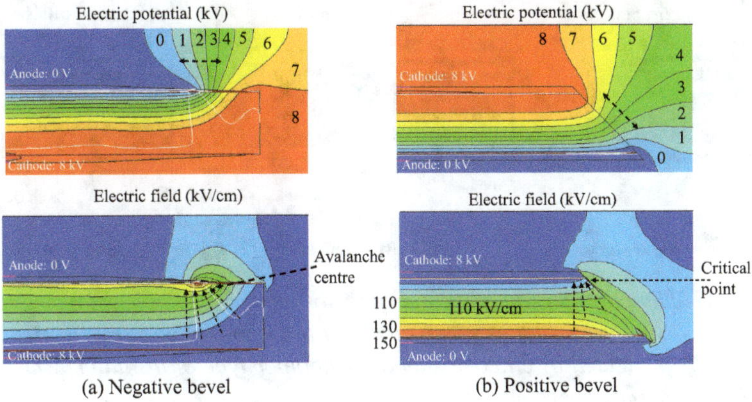

Figure 3.16 The distribution of electric potential and electric field at the edge of a device terminated by negative bevel (a) and positive bevel (b). The distance between the equipotential lines at the surface is increased

voltage or to a very wide bevel consuming too much active area. A typical range of negative angles giving a high breakdown voltage is 1 to 5 angle degrees. The minimal angle giving the highest breakdown voltages is limited by manufacturability to 1.5–2.5 degrees.

Figure 3.16(a) shows that **the avalanche centre** appears **in bulk** where the SCR of the blocking p-n junction stops at the highly doped shoulder (P^+) closer to the surface. It is typically where the concentration of acceptors reaches the level > 2.10^{15} cm^{-3}. To satisfy all design needs of a thyristor, it is advantageous to use the double diffused p-type layer when the negative bevel is used as the JT. The low-doped deep p-type layer provides the maximal breakdown voltage and the high-doped shallower one is used to adjust the injection efficiency in the ON-state.

In the case *p.*, the bevel angle is positive when going from the high (P^+) to the low-doped region (N^-) and this JT concept is called **positive bevel.** As we reduce more volume at the n-doped side than at the p-doped one, the SCR at the n-side expands and at the p-side contracts. As the n-doped side is always low doped (resistivity of the starting wafer), the expansion of the SCR is maximal without any design effort. At the same time, the p-doped layer can be relatively thin just to satisfy the breakdown voltage at one-dimensional p-n junction in the bulk. A typical range of positive angles is 30 to 80 degrees. The preferred range in the practice is 45 to 60 degrees as a compromise between a high breakdown voltage (low angle) and a smaller area consumed by the JT (high angle).

Figure 3.16(b) shows that the equipotential lines are widened at the surface, the electric field is lowered and there is no avalanche centre in the bulk. However, when the SCR reaches the opposite side of the wafer (punch-through), a region with a high electric field can be created by focusing the field lines and reduce the blocking capability. A low-doped region (a buffer) and an electrically strong surface passivation can help to minimise this field.

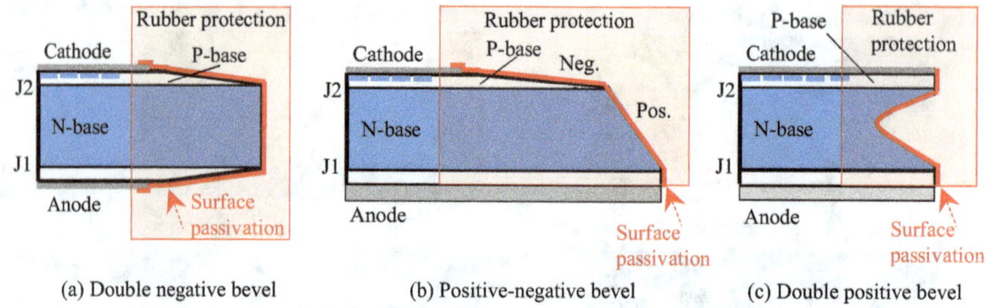

Figure 3.17 Junction termination of a thyristor using various bevel concepts. Silicon surface is passivated and the whole edge is mechanically protected by silicone rubber or another type of insulator, which also avoids sparking

The positive and negative bevel concepts, based on mechanical grinding, are used in discrete thyristors. It is advantageous especially when circular wafers are used. Figure 3.17 summarises the typical combinations used in practice [9]. As we need to terminate two p-n junctions, namely the J1 and J2, a minimum of two bevels must be used at thyristors.

One option is to use the negative bevel (a) on both sides. The advantage is relatively simple processing. The disadvantage is that both two edges are hanging in the air so that it does not make sense to bond the silicon wafer to a molybdenum disk to improve the cooling of the wafer, because the wafer cannot be bonded up to the edge. The combined positive-negative bevel (b) allows us to use the low-temperature bonding (LTB) process to attach the silicon wafer to the molybdenum disk at the anode contact to cool the silicon wafer up to its edge. As the two bevels are running in the same direction, the consumption of the wafer area is higher than in the case (a). The double-side positive bevel (c) offers the lowest area consumption at the price of the most complicated processing of surface and its passivation at a required quality. All the three above-described concepts can be found in high-power thyristors formed by a circular wafer.

The selective charge addition methods (b) are based on creating areas with locally increased p-type doping concentration. As they result in a planar JT, they are mainly used for thyristors processed in the form of rectangular chips. This planar JT can be in the form of concentric rings with constant surface concentration, so-called field rings. Their amount is given by the required blocking voltage. It can also be a single wider ring with a lower and constant doping concentration along the surface (junction termination extension, JTE). The width and doping concentration are given by the required blocking voltage. Eventually, it can be a single ring with outwards decreasing surface concentration and depth (variable lateral doping, VLD), where the length and depth are given by the required blocking voltage.

Figure 3.18(a) shows an example of using the guard ring (GR) concept to stretch the electric field along the silicon surface. The electrical potential is decreasing along the surface at the individual GRs from its maximal value at the p-

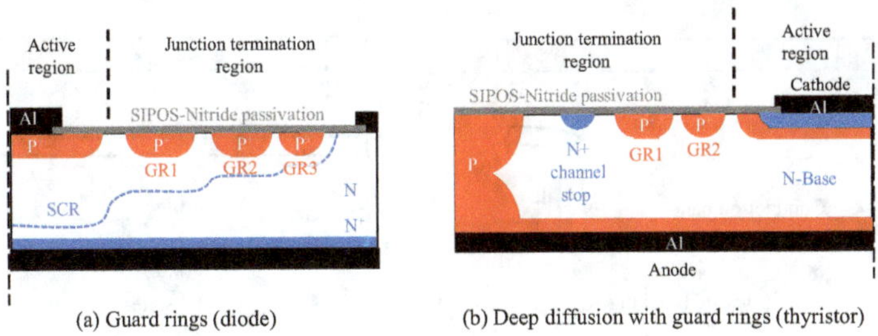

Figure 3.18 Planar junction termination of a diode (a) and thyristor (b) using guard ring concept. Only a part of the active region is shown, as the dashed line shows

n junction down to zero at the edge (right). The dimensions, doping and distance between the GRs must be designed so as to the high electric field never reaches the surface between the GRs and its maximal value is below the critical field for impact ionisation E_{crit}. The surface passivation region must be semi-insulating and connected to the anode metal because it must drain the charge collected by the GRs under reverse bias. We can imagine that as a distributed resistance forming a resistance divider connecting the GRs to the anode metal. Eventually, there can be an N^+ channel stop GR behind the last p-type GR to stop the spreading of the SCR at a predefined position.

Figure 3.18(b) shows a planar JT used for low-voltage thyristors (<2 kV) requiring silicon wafer thickness not exceeding 400 μm. The blocking junction J2 is terminated using the GRs as in the case of the diode including the channel stop. The anode junction J1 is terminated by two low-doped P-sink regions at the left, which are formed by a special aluminium diffusion from both sides of the wafer, which is typically 300–400 μm thick. At the end of production, the thyristor chip is sawn through this region and separated from other chips. The high thickness of the P-type layers results in a low-doping concentration along with its entire depth, which spreads the SCR and reduces the electric field likewise at the JTE. The advantage of this concept is the possibility to use the standard semiconductor processes from the integrated circuit industry.

3.5 Turn-on into the ON-state

A thyristor can be turned on in several ways if the polarity of the anode is positive, compared to the cathode ($V_{AC} > 0$). In this case, the blocking junction J2 is reverse biased and its highly resistive SCR supports practically the whole voltage V_{AC} between the anode and cathode and blocks the flow of anode current. Small leakage current is given by the transport of minority electrons extracted by the electric field from the P-base side of the J2 towards the anode and minority holes extracted from the N-base side of the J2 towards the cathode, respectively, multiplied by the J1 and J3. The carrier transport from J1 and J3 is small as suggests the high-energy barriers in Figure 3.19 (left).

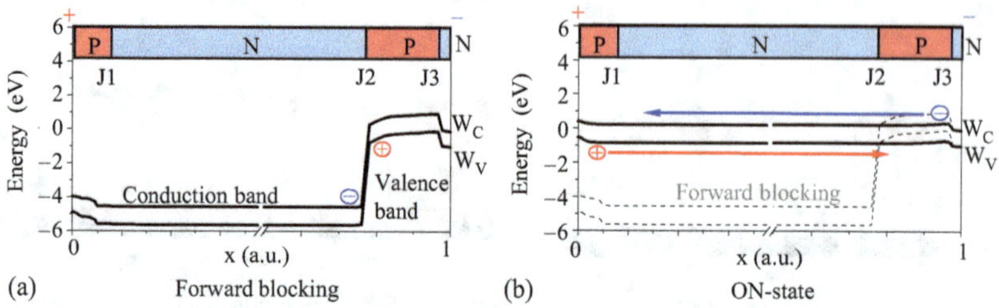

Figure 3.19 Energy band diagram of a thyristor in the forward blocking for $V_{AC} \approx 5$ V (a) and in the ON-state (b). The energy of the electron grows upwards. The energy of holes flows downwards due to the opposite polarity of its electrical charge

To bring the thyristor into the ON-state with a very low V_T, we have to remove the energy barrier of J2 and flood the N- and P-base regions by massive concentrations of free carriers similar to the P-i-N diode. For this purpose, we have to establish a strong injection of holes from the anode and electrons from the cathode. The J1 and J3 must appear under high injection characterised by correspondingly lowered energy barriers as shown in Figure 3.19 (right).

There are about four basic ways to turn thyristor into the ON-state:

1. by current pulse into the gate electrode,
2. by the light pulse into the gate-cathode junction,
3. by application of $V_{AC} \geq V_{BO}$ (break-over voltage) and
4. by application of $dV_{AC}/dt \geq (dV_{AC}/dt)_{crit}$ (critical rate of rise of voltage).

The modes 1 and 2 are the desirable ways of switching, while the 3 and 4 are undesirable as they can cause a failure.

3.5.1 *Turn on by the gate current I_G*

Application of a current pulse into the gate of a thyristor in the forward blocking regime results in the injection of electrons from the cathode emitter into the blocking junction J2. These electrons are extracted by the electric field of the reverse-biased J2 and transported towards the positive anode, where they cause a corresponding injection of holes to conserve the charge neutrality law. The holes injected from the J1 are extracted by the J2 and continue towards the J3. For the same reason as in the case of electrons at the anode, the holes stimulate the injection of electrons from the cathode emitter. These electrons then undergo the same scenario towards the anode and so on. The positive feedback explained already in Section 3.3 is closed and, from now on, the gate current is not needed to maintain the thyristor in the ON-state. The thyristor structure has got into the latch-up. The whole N- and P-bases (including the J2) are swamped by a high concentration of carriers (Figure 3.20) with a spatial distribution such as in the P-i-N diode. Thyristor shows a very low V_T – typically below 2 V for the maximal rating current.

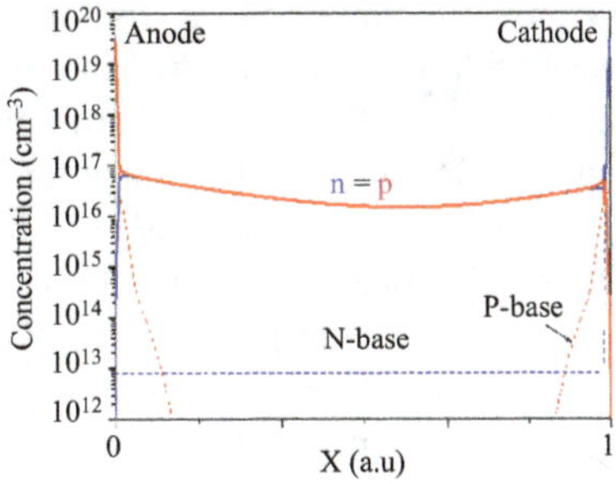

Figure 3.20 Spatial distribution of electron and hole plasma in the ON-state. The electron concentration in the N-base has increased by four orders of magnitude compared to the forward blocking regime

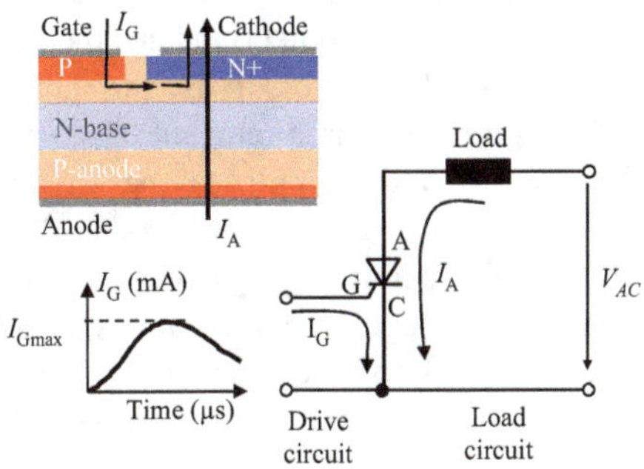

Figure 3.21 Turn-on by current pulse into the gate. 2-D view of the planar gate-cathode structure (top left)

Thyristor stays in the latch-up if the external circuit in Figure 3.21 allows the flow of current higher than the so-called latching current I_L. It is the minimum value of anode current I_A that keeps the device conducting when the gate pulse is removed after the immediate transition from the forward blocking into the ON-state. After a certain period in the ON-state, the thyristor holds in the ON-state, if the anode current from the eternal circuit does not drop below the holding current I_H. The I_H is the minimal anode current I_A necessary to keep the thyristor in the ON-state at $I_G = 0$.

State-of-the-art thyristors with the ON-state currents in the kA range require a large area ranging somewhere between 50 and 150 cm^2. The area of gate amounts typically $\leq$1 cm^2 to facilitate a reliable electrical contact and is placed either in the corner of a chip or in the middle of a circular wafer. The electron-hole plasma distributed between the anode and cathode according to Figure 3.20 is first created around the gate region and spreads laterally outwards the gate with the speed of $\approx$100 μm/μs. The thyristor is fully turned ON after the plasma is spread laterally over the most of the device area, which could require 200–800 μs. To assess the turn ON speed, the gate-controlled turn-on delay time t_d is measured as the time instant between the increase of I_G to 10 per cent of the maximal gate current I_{GM} and the decrease to 90 per cent of the thyristor forward voltage V_{DM}. To speed up, the lateral spread of the turned ON area and to reduce the t_d, the **amplifying gate** (AG) structure inserted between the gate and cathode can be used (see Figure 3.22) [10]. As shown in the detailed electrical scheme in Figure 3.22(c), the AG and the NPN transistor form the Darlington configuration of two NPN transistors. The basic feature of this configuration is the amplification of gate current I_G to that of the cathode $I_C = \beta_{pilot} \cdot \beta_{main} \cdot I_G$, where β is the current amplification factor of a transistor in the common-emitter configuration. A faster growth of cathode current, thanks to the current amplification, results in a faster and more uniform spread of plasma in the lateral direction, reduced t_d (0.2–2.5 μs depending on applied V_D at thyristors up to 150 mm in diameter) and a higher *di*/*dt* capability (*di*/*dt* of anode current not leading to a destruction). The AG structure can be star-shaped structured from the gate region over the whole cathode surface – see the snowflake AG in Figure 3.1(c). The longer the gate-cathode boundary, the faster the turn ON. The AG concept allows the producers of thyristors to use the same or a very similar AG structure independent of the thyristors' area (current class) and thickness (voltage class). This kind of platforming then simplifies the situation for designers of gate units.

After the application of the current I_G, the current flows first through the resistors R_p–R_{AG}–Rm. Once the voltage drop at R_p and R_m reaches 0.7 V, the base-emitter junctions are injecting the amplified gate current. The design rule $R_p > R_m$ assures that the AG and, hereby, the pilot thyristor turns ON before the main

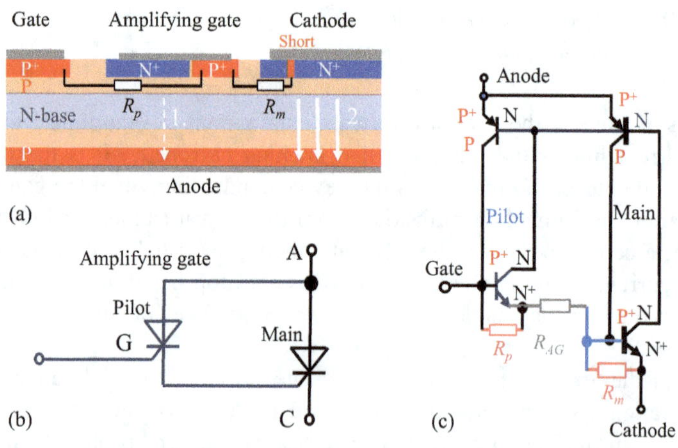

Figure 3.22 *Amplifying gate (AG) structure (a), its simplified electrical scheme (b) and scheme taking into account resistances R_p and R_m relevant for triggering (c). First to turn ON is the AG part (arrow 1) followed by the main thyristor (arrows 2). R_{AG} is the metal resistance connecting the N^+ and P^+ layers of the AG*

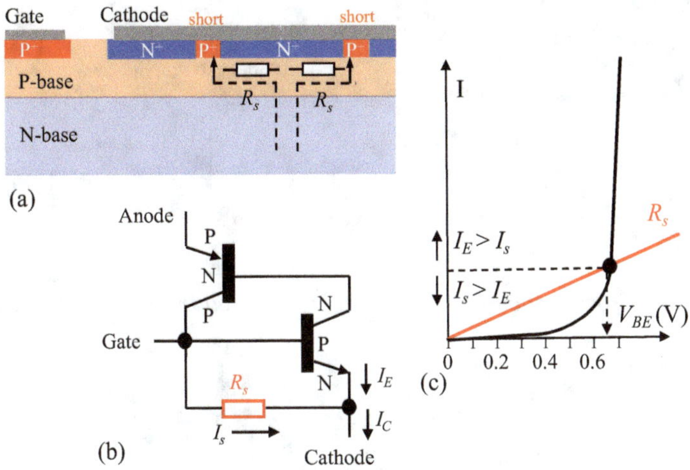

Figure 3.23 The principle of cathode shorting structure (a). Cathode short in the two-transistor analogue (b). The effect of cathode short resistance R_s on the onset of current amplification by the NPN transistor (c)

one [10]. In the opposite case, the pilot thyristor would be useless. A robust thyristor is designed in a way that the AG triggers first under all four modes 1–4 summarised earlier.

The resistor R_m is connected at the cathode side through the cathode shorting structure (Figure 3.22(a)). The cathode short is a circular p-type region 50–500 μm in diameter interrupting the n-type cathode layer. The shorts are usually homogeneously distributed over the cathode area using triangular or hexagonal patterns. The role of the shorts is

- to allow the current from J2 to bypass the base-emitter junction of the NPN transistor, hereby reducing its current gain at leakage and displacement (low) current levels,
- to slow down and make uniform (homogenise) the lateral spread of plasma to avoid local overloading by an excessive current density at turn ON with a high di/dt,
- to increase the critical dV/dt during fast transients of V_{AC} under forward blocking by diverting the displacement current into the cathode electrode without traversing the junction J3 (no amplification of current) and
- to increase the break-over voltage V_{BO} and V_{DRM} (repetitive peak off-state voltage) close to that of the reverse breakdown voltage and V_{RRM} (repetitive peak reverse voltage).

Under forward blocking regime, the leakage or displacement current from the J2 flows preferably into the p-type short (Figure 3.23(a)) as long as the lateral voltage drop on the equivalent short resistance R_s does not reach the threshold voltage $V_{BE} \approx 0.7$ V of the J3 (Figure 3.23(b)). The I–V curve of the parallel

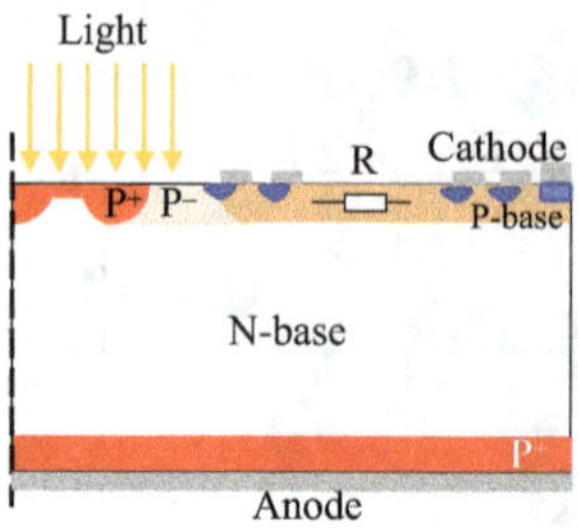

Figure 3.24 Cross-section of a typical light triggered thyristor (not to scale) [9]

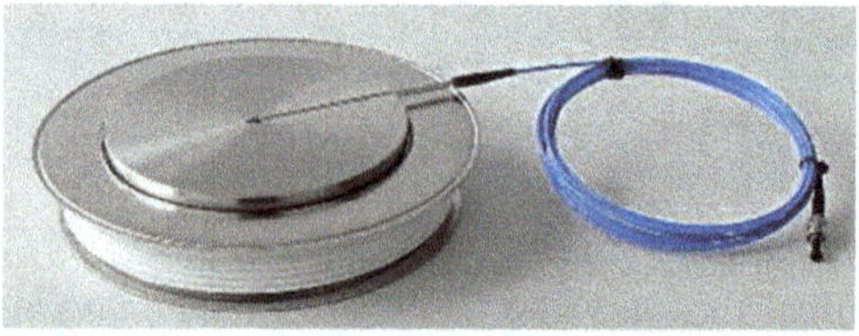

Figure 3.25 Housing with optical cable of the direct light triggered thyristor produced at 150-mm silicon wafer by Infineon [18]

combination of R_s and J3 suggests that the short diverts the current which would be otherwise amplified at J3 as long as $V_{BE} < 0.65$ V. Above this voltage, the current across the thyristor is already high and the current amplification as well, what is favourable for the latch-up and low V_T in the ON-state. The slope of R_s in Figure 3.23(c) can be adjusted by the distance between the shorts and, so, the transition current level from shorting to amplification. A high density of cathode shorts improves dynamic parameters such as the turn-off time and $(dV/dt)_{crit}$. Too many shorts close the AG delay turn ON and reduce the di/dt capability. As the shorts consume the useful cathode area and increase the V_T, it is wise to minimise the diameter of shorts down to a reasonable level.

3.5.2 Turn on by the light pulse

The main difference between the electrically triggered thyristor (ETT) described above and the LTT is the gate structure shown in Figure 3.24 [9]. The triggering of the LTT is initiated by a light pulse of several 10 mW applied to the photosensitive area located within the 1st AG in Figure 3.24. Free carriers are generated there which act as photocurrent that flows through the LTT via the multiple AG structure when a forward blocking voltage V_D is applied. Due to the small physical dimension of about 1 mm, the light-sensitive area has to be protected against a very high di/dt by a current limiting resistor (R), which is implemented into the P-base between 2nd and 3rd AGs. By proper gate and resistance design, very high repetitive and nonrepetitive di/dt ramps, up to 5 kA/µs and 10 kA/µs, respectively, can be realised.

While early concepts of the LTT were based on a light-activated auxiliary thyristor [11,12], nowadays used direct LTTs came up from the end 1970s. With the availability of reliable, high-performance laser diodes in 1997, 8-kV LTTs with integrated protection functions became available. A special ceramic housing with an internal light guide embedded in the encapsulated cover of the housing is shown in Figure 3.25. Direct light triggering is possible through a window in the centre of the encapsulated cover of the hermetically sealed housing. A sophisticated multi-layer structure with sapphire glass is needed for reliability reasons.

3.5.3 *Turn on by overcoming the break-over voltage V_{BO}*

This mode comes into consideration when $I_G = 0$ and we 'slowly' apply $V_{AC} > 0$ (forward blocking) with such a high magnitude that the reverse-biased junction J2 is subject to intensive impact ionisation leading to the avalanche breakdown. The generated current brings both the J1 and J3 into high injection, trigger the positive feedback and causes thyristor to latch-up as described above ($\beta_T > 1$ or $\alpha_{PNP} + \alpha_{PNP} = 1$). In other words, the current from impact ionisation substitutes the current I_G.

Figure 3.26(a) shows the current and voltage waveforms for the case $V_{AC} < V_{BO}$. The sine wave from a voltage source induces the displacement current $I_D \approx C{\cdot}dV/dt$ at the capacitance of the reverse-biased J2. As the dV/dt is small even in the steepest part of the V_{AC} waveform, the displacement current is smaller than the leakage current I_L resulting from the rising V_{AC} at a given temperature. Because of $V_{AC} < V_{BO}$, a full sine wave of voltage is measured. The thyristor stays in a forward blocking regime.

Figure 3.26(b) shows similar waveforms for the case when V_{AC} gets over the V_{BO}. Before reaching the maximum of the voltage sine wave, the thyristor turns ON

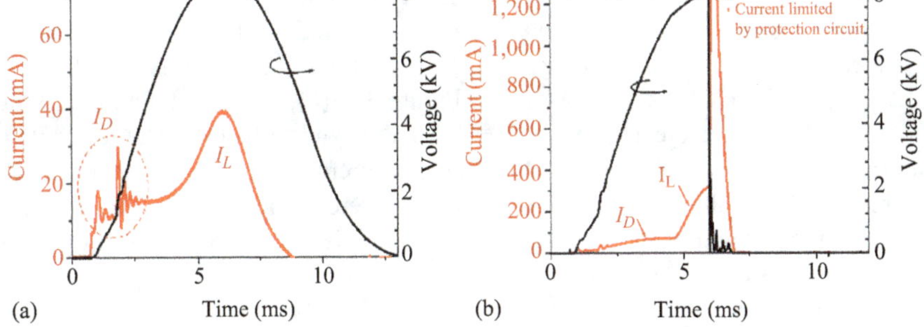

Figure 3.26 (a) Forward blocking test of 8.5-kV thyristor by applying a sine voltage with $f = 6.25$ Hz. $V_D < V_{BO}$. The blocking test passed. (b) Forward blocking test of 8.5-kV thyristor by applying a sine voltage with $f = 6.25$ Hz. $V_D > V_{BO}$. The blocking test failed

due to an excessive current generation at J2 by the impact ionisation. The turned-ON region is localised in the area of the highest impact ionisation, typically somewhere at the wafer edge (junction termination) instead of a uniform spread over the whole area. Consequently, this **turn-ON mode** can be destructive and is, therefore, **undesired**. The reduction of the V_{BO} corresponding to the failure waveform in Figure 3.26(b) can also be caused by a defect resulting from device processing. Such devices are normally scrapped after production testing.

3.5.4 Turn on by a fast rise of the anode voltage (by overcoming the V_{BO} at high dV/dt)

This mode takes place when $I_G = 0$ and we apply $V_{AC} > 0$ (forward blocking) with a high voltage rise time dV/dt. The SCR of the reverse-biased J2 contains immobile ionised acceptors and donors, which constitute the capacitance C decreasing with increasing forward voltage. During dV/dt transient, the charge from the J2 is swept out and the current $I_D \approx C \cdot dV/dt$ is generated. Since the dV/dt typically amounts to 1 to 10 kV/μs, the generated current can trigger the thyristor into the ON-state. In other words, the displacement current substitutes the current I_G.

In this mode, the thyristor can in principle locally trigger at any location depending on the design. A properly designed thyristor triggers around the gate or AG region so that the lateral spread of plasma proceeds uniformly such as during switching by the gate current. In such a case, the thyristor subjected to an extremal dV/dt undergoes so-called protective firing. It turns into the ON-state and avoids destruction. In the next cycle, it is again ready to block. A thyristor with design weak points can fail destructively leaving a locally burnt region at surface metallisation.

Optimal design and placement of cathode shorts are crucial to the achievement of a high dV/dt capability. The highest dV/dt is achieved when the cathode shorts are placed as close to the AG as possible with minimal spacing and the AG contains shorts as well. It is important, especially in the device concepts with low-dose electron irradiation, to reduce the stored charge and even more in the designs without using any lifetime control technique (cost reduction).

Figure 3.27(a) shows the voltage waveforms of 8.5-kV thyristor tested to the full-rated V_D. After the steep voltage rise, the thyristor still blocks forward blocking voltage $V_D = 8.5$ kV. The test passed. In Figure 3.27(b), thyristor fails to hold the $V_D = 8.5$ kV when subjected to $dV/dt = 8.5$ kV/μs and turns ON. The test failed. This thyristor remains functional, but the measurement confirms that its dV/dt capability is lower than 8.5 kV/μs. Turn ON by overcoming the V_{BO} is **undesired**. Such a method would be very impractical in comparison with the triggering by the gate and unreliable.

3.6 Turn off

The classical thyristor connected in the circuit on Figure 3.21 has no own means to turn OFF the anode current I_A. It can turn OFF only after the anode current in the load circuit decreases below the holding current I_H – this process is called **natural**

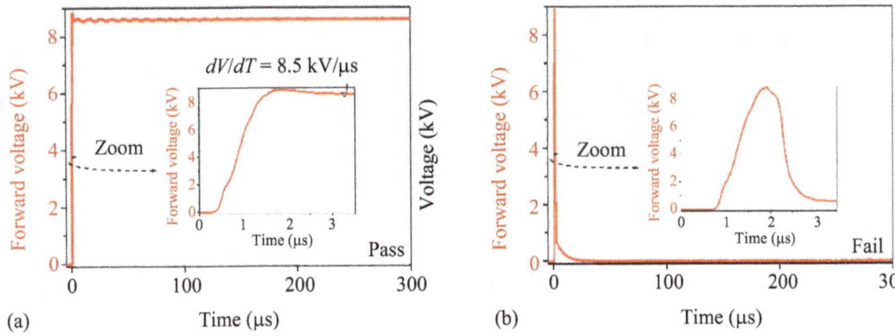

Figure 3.27 (a) dV/dt up to full rated V_D test passed. dV/dt capability is 8.5 kV/µs. Thyristor stays in forwarding blocking. (b) dV/dt up to full rated V_D test failed. The thyristor is not capable to stay in the forward blocking and turns ON

turn OFF. In this case, the thyristor turns OFF after the electron-hole plasma (charge) in the N- and P-base, responsible for the very low V_T shown in Figure 3.20, drops close to the background doping thanks to the recombination of free carriers. Unfortunately, there is not much practical use for this approach.

In the circuits with alternating current (AC), typically with the sinusoidal voltage with the frequency of 50 or 60 Hz, the thyristor can turn OFF after the anode current decreases below the holding current I_H thanks to the reversal of polarity of the V_{AC} at the end of half period. As this happens thanks to an external circuit – it is called **forced** turn OFF. As it is fast enough and well controllable, it is the most used turn OFF principle. The applied reverse voltage stops the injection into the N- and P-base and the stored charge is extracted by the electric field caused by the reversed voltage. The required time before the thyristor can support the forward blocking after the anode current crosses zero is called the **circuit commutated recovery time** t_q.

Figure 3.28 and 3.29 show typical waveforms resulting from the measuring of the t_q. The reverse voltage $V_R = 200$ V is applied to the thyristor in the ON-state with $I_T = 3$ kA. Because of the load inductance L, the current decreases with

$$\frac{di}{dt} = -\frac{V_R}{L} = -\frac{200\ \text{V}}{133\ \mu\text{H}} = -1.5\ \frac{\text{A}}{\mu\text{s}}.$$

The plasma dynamics during turn OFF is shown in Figure 3.30. The anode current drops to zero as long as the plasma concentration in the middle of the N-base becomes higher than close to the emitters. Then, the gradient of plasma concentration changes its polarity (note the arrows) and so does also the anode current.

After the current crosses zero, the thyristor enters the **reverse recovery** process. In this process, the reverse voltage recovers the N- and P-base from the excess charge of the electron-hole plasma originally injected in the ON-state (Figure 3.30 (b)). Once the highly resistive SCR of the J1 is established by the reversed V_{AC}, the thyristor starts to limit the flow of current and the negative I_T reaches its maximum

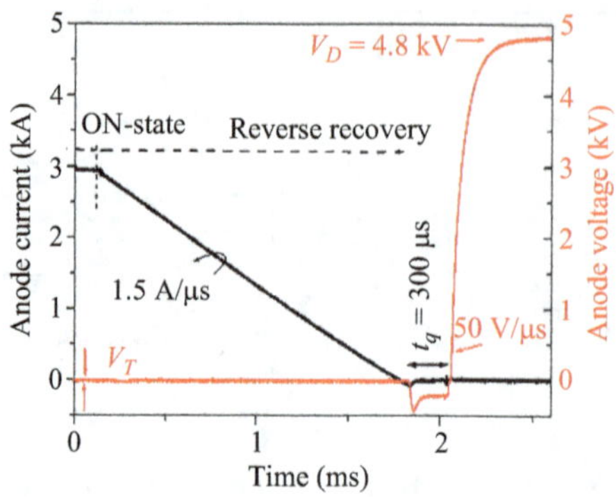

Figure 3.28 Turn OFF with subsequent application of forward voltage V_D. ON-state → reverse recovery → forward blocking. $T = 90$ °C

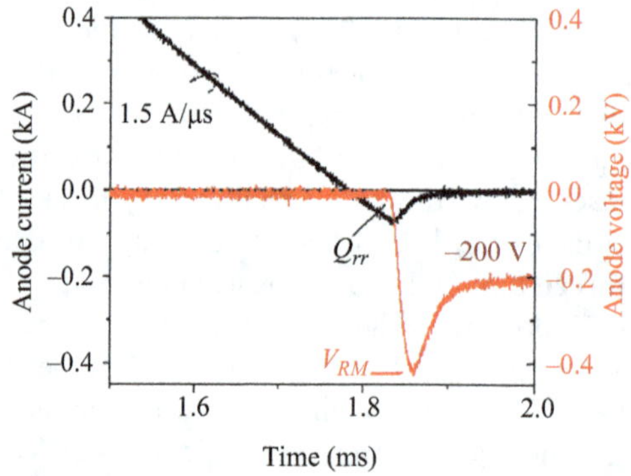

Figure 3.29 Zoom into the reverse recovery process with $V_R = -200$ V

called **the maximal reverse recovery current** I_{RM}. The integrated portion of the curve $I_T = f(t)$ below zero in Figure 3.29 represents **the recovery charge** Q_{rr}. The highly resistive SCR overtakes the applied reverse voltage, which shoots over the DC supply voltage level (200 V, in our case) due to the inductance of the load circuit (see the maximal recovery voltage V_{RM}). To limit this voltage overshoot at higher AC voltages in order to prevent thyristor breakdown, an RC snubber is often connected parallel to the thyristor.

After the voltage overshoot, the plasma removal goes on. According to Figure 3.28, the thyristor needs 300 μs to extract so much plasma that after the application of $V_D = 4.8$ kV with $dV/dt = 50$ V/μs, it stays in the forward blocking regime and does not turn ON.

Figure 3.31(a) shows the distribution of the electric field $\boldsymbol{E}$, electron concentration $\boldsymbol{n}$ and hole concentration $\boldsymbol{p}$ during the reverse recovery (reverse: $t = 2$ ms

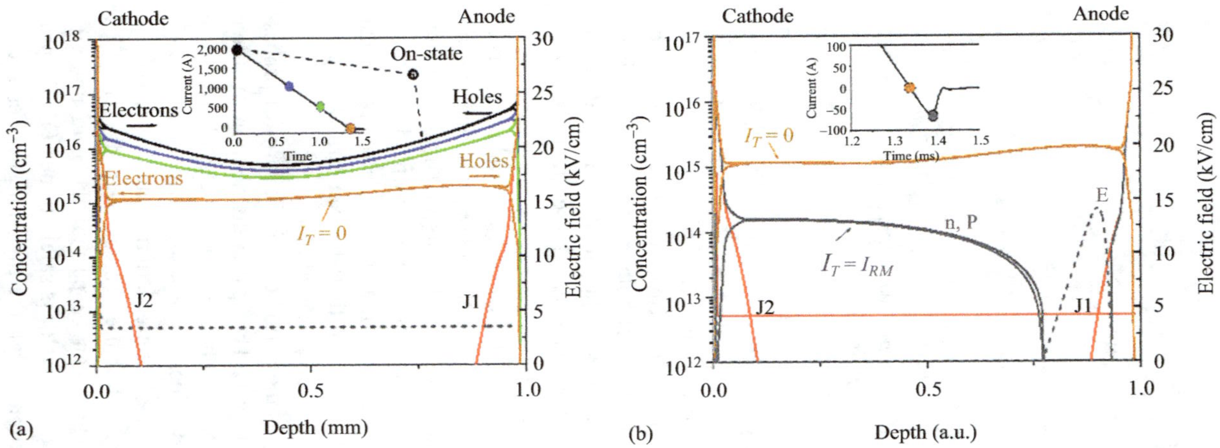

Figure 3.30 *(a) Electron-hole plasma during turn OFF: ON-state → $I_T = 0$. Current I_T is shown in the insert. The gradient of electron and hole concentration changes at $I_T = 0$. (b) Electron-hole plasma during turn OFF: $I_T = 0$ → reverse recovery. The region with high electric field E is the SCR of the reverse-biased J1*

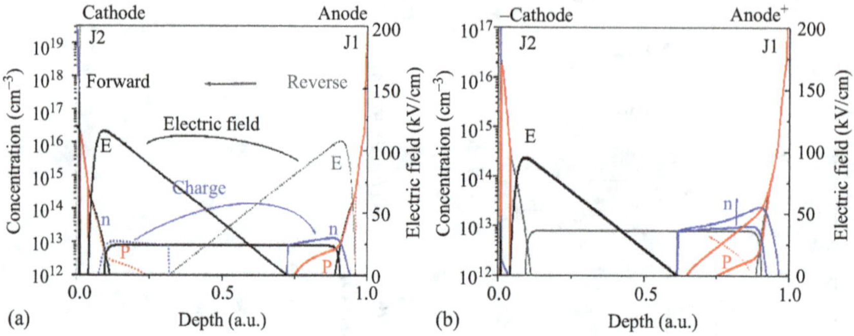

Figure 3.31 (a) Electric field, electron and hole concentration during reverse recovery (reverse) and forward blocking (forward). (b) Electric field, electron and hole concentration during two consecutive time instants of forward blocking leading to latch-up

in Figure 3.29) and after the application of forward voltage V_D (forward: $t = 2.5$ ms in Figure 3.28). Under reverse bias, the electric field at J1 shows the SCR without carriers and the rest of the N-base recovered from the excess electrons and holes. After reversing the V_{AC} back to the forward direction, the blocking junction J2 is reverse biased and depleted from free carriers. The remaining electrons are swept towards the anode, where they cause the injection of holes to conserve the charge neutrality. Their concentration is multiplied there by the junction J1 (PNP transistor). The holes from the N-base are extracted by the reverse-biased J2 and cause the injection of electrons at the J3 (NPN transistor).

Figure 3.31(b) shows a situation of growing concentrations n and p in time by current amplification at J1 and J3, which can precede the subsequent latch-up (not shown here). The forward voltage V_D has already reached its maximal value and the electric field does not change further. The E-lines are shown for two consecutive time instants overlap. However, the electron and hole concentration can grow further as a result of current amplification and thyristor can turn ON (latch-up). This can happen when the forward voltage V_D is applied at time $t < t_q$.

The figures above have shown that t_q is the period during the turn-off process when the state of the device is not known well. The spatial distribution of stored charge is nonhomogeneous in all directions and changes very quickly. Under such conditions, the thyristor is vulnerable to destruction by fast-growing voltage V_{AC}. For this reason, the device has to be protected against external current or voltage disturbances either by internal protection structures (break-over diode) or by external recovery protection circuitry, which can be part of the gate unit [9]. Dangerous voltages are either limited or the thyristor is safely turned ON (= protective firing). The control of power is limited, but the thyristor survives such a fault event and is ready to control power in the subsequent period.

From an application viewpoint, t_q is the period of time without the possibility to control power. As a result, it is an important parameter in system design when it comes to a smooth power control without commutation failures, which can cause a temporary cessation of power transfer and system overloading. Even though a well-designed thyristor control unit avoids such failures, the shorter t_q means a smoother, broader and cheaper power control.

From the device design point of view, a well-designed thyristor features the shortest t_q by means of an optimised design of the cathode shorts and the AG structure. If the V_D pulse is applied at time $t < tq$, the thyristor turns ON safely starting from the gate region and the electron-hole plasma (triggered region) propagates laterally towards the edge of the wafer (chip). The t_q time can be further lowered by the electron irradiation (lower Q_{rr}) and even more by the optimised proton irradiation (locally reduced concentration of electron and holes in the ON-state) [8].

3.7 Serial and parallel connections

Many applications require either series or parallel connections of thyristors to obtain the required full output voltage and output current. Typical examples are HVDC transmission and FACTS, where ten up to one hundred thyristors can be connected in series. For best utilisation of the system, the thyristors must be specially selected so that they each take approximately the same share of current and/or voltage [13].

3.7.1 Serial connection of thyristors

For series connection, the reverse recovery charge Q_{rr} is normally the most critical parameter. It has a large influence on the dynamic voltage sharing, while the leakage current at the operating voltage is important for the static voltage sharing. Thyristors with different turn-on delay times t_d will have different turn-on times; but in general, the differences in t_d for a given design and technology are small. The need for banding of delay times can, therefore, be eliminated by using an appropriate gate triggering pulse.

Figure 3.32(a) shows a simplified application of two thyristors in series. The resistors R1 are used for a better static voltage sharing and R2 are damping resistors in the R2C snubbers. In a real application, many other factors will influence the reverse recovery peak voltage. The external factors are the commutation inductance and voltage. The internal factor is the softness of the thyristor reflected in the magnitude of the reverse voltage overshoot V_{RM} such as the one on Figure 3.30.

If we neglect the influence of the parallel resistors, the voltage difference due to the difference in Q_{rr} of individual thyristors can be estimated at $\Delta V = \Delta Q_{rr}/C$. If we consider a 1μF capacitor and thyristor with the Q_{rr} band of 1,000 μAs for the commutation $di/dt = -2.0$ A/μs at a given temperature, we obtain the worst-case voltage difference between the two thyristors $\Delta V = 1{,}000$ μAs/1 μF = 1,000 V. The thyristor with the lower Q_{rr} will stop conducting current earlier. It will, therefore, start taking up voltage earlier and end up with 1,000 V higher commutation voltage overshoot than the second thyristor. This unbalances results in a lower voltage margin and a higher risk of failure due to overvoltage. To avoid this a larger capacitor C or a lower ΔQ_{rr} is needed to reduce the voltage difference ΔV. A reduction of ΔQ_{rr} is the preferred solution since a large capacitor C is generally more expensive and bulky. In practice, the reduction of ΔQ_{rr} down to 500 μAs is realistic. Of course, for an increased cost of manufacturing, which can be compensated by no need to change the circuit components. The reduced Q_{rr} band then reflects in the reduced number of redundant thyristors in series.

The principle of Q_{rr} banding is sketched in Figure 3.32(b), which shows the technology curve V_T–Q_{rr} achieved by electron irradiation. The electron irradiation reduces Q_{rr} (also reduces t_q and increases dV/dt) at the penalty of increased V_T resulting in higher ON-state losses. At the same time, the original spread of 'As processed' V_T–Q_{rr} values can be reduced by a proper choice of irradiation doses for individual thyristors. The irradiation can be even repeated if the dose is still low to move the devices to the required Q_{rr} band. If the chosen irradiation dose is too high and the Q_{rr} becomes much lower than the lower band limit, the only way back to the original band is by annealing at temperatures above 400 °C, for several hours. This can work well for the low irradiation doses. In the case of higher irradiation doses, there can come other effects into play such as inverse annealing. In such a case, the original deep levels (radiation point defects) can be converted to the new ones with different capture cross-sections and emission coefficients. This effect may depend on the specific technology of silicon pulling of individual producers of silicon wafers.

The thyristors with lower Q_{rr} limits show a lower t_q and a higher dV/dt capability with the penalty of a higher V_T – see the additional points below the Q_{rr} band. The Q_{rr} band is chosen by application designers as a compromise between the required ON-state and turn-off losses.

The electron irradiation generates point defects acting as generation-recombination centres also called deep (energetic) levels. They occur deeper in the energy band ($E_C - E_{level} >> kT$) compared to the shallow levels ($E_C - E_{level} <<$

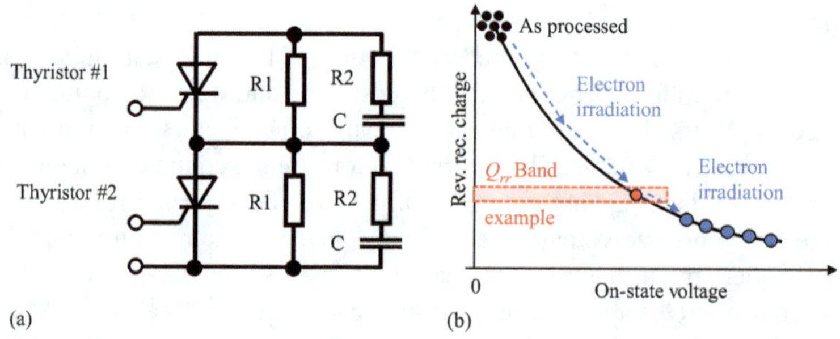

Figure 3.32 (a) Series connection of thyristors: R1 is the damping resistor, R2C is the snubber. (b) The principle of Q_{rr} banding: individually designed doses of electron irradiation for as processed thyristors with different Q_{rr} bring the Q_{rr} values closer to each other

kT), which are used as the donors and acceptors for the common N-type or P-type doping, respectively. The kT is the thermal energy of lattice (phonons), where k is the Boltzmann constant and T is the absolute temperature.

Figure 3.33(a) shows the two dominant point defects E1 and E4 in the silicon bandgap received after the post-irradiation annealing performed below $T = 300$ °C, which is typically used for the thyristors. The annealing is always performed to remove the point defects unstable in the whole range of device operation temperature, e.g., the vacancy phosphorus complex not shown here. After annealing, there is a plurality of deep levels as a result of reacting to the radiation defects (vacancies and interstitials) with the elements present in the original silicon crystal-like phosphorus, oxygen or carbon. However, only the electron deep levels E1 and E4 described further can significantly modify the electrical parameters of thyristors.

The deep acceptor level E4, which is the single negative charge state of the divacancy 0.42 eV below the conduction band, dominates the reverse bias. This means that this level is responsible for the magnitude of leakage current. As shows the Figure 3.33(a), this is because the probability of the two-step 'Generation' process from E_V to E4 and from E4 to E_C is higher than the one from E_V to E1 and from E1 to E_C (compare the product of multiplication of probabilities of individual transitions 1 and 2).

The deep acceptor level E1, which is the single negative charge state of the vacancy oxygen complex 0.16 eV below the conduction band, dominates the forward bias. The dominance means that it reduces the carrier lifetime at high carrier concentrations ($\geq 10^{16}$ cm^{-3}) much more than the level E4 as shown in Figure 3.34. Moreover, as there is a nonzero probability of de-trapping, the recombining electron back to the conduction band E_C (see the 'Recombination' in Figure 3.33(a)), the lifetime reduction caused by

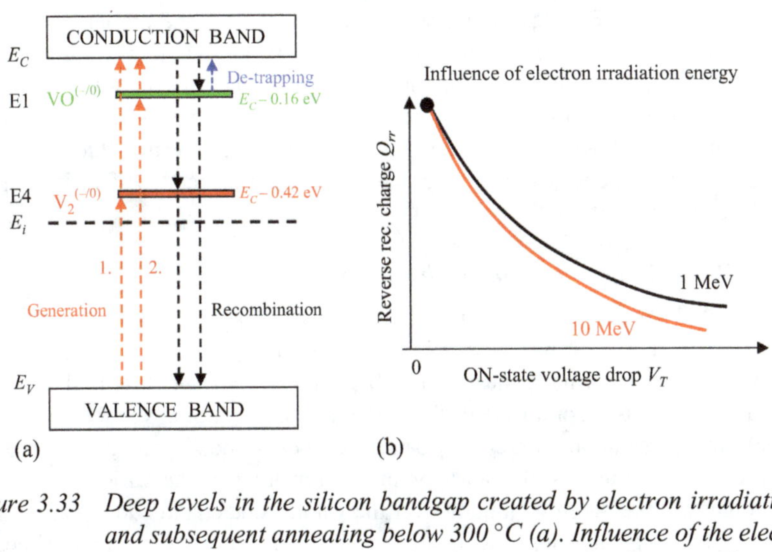

Figure 3.33 Deep levels in the silicon bandgap created by electron irradiation and subsequent annealing below 300 °C (a). Influence of the electron irradiation energy on the technology curve of a thyristor (b)

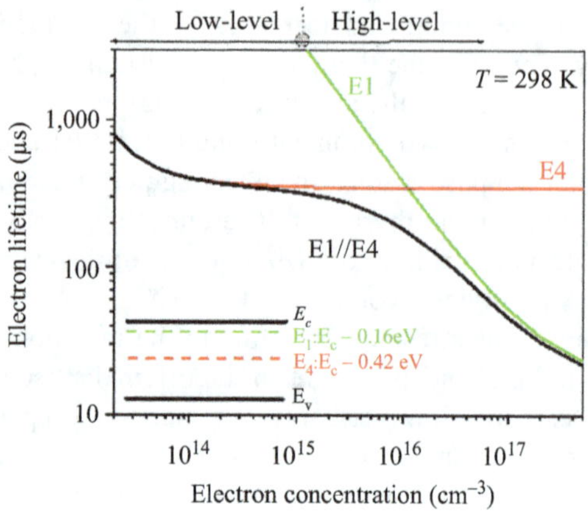

Figure 3.34 Dependence of electron lifetime on the excess carrier concentration after electron irradiation from Figure 3.33 calculated for room temperature and concentration ratio N_{E1}:N_{E4} = 4:1

the level E1 is strongly concentration and temperature-dependent. Consequently, the decrease of the ON-state voltage V_T due to the recombination through the level E1 can be lower and the carrier lifetime higher at elevated temperatures. As a result, a higher concentration of the level E1 results in a more negative temperature coefficient of V_T (*tempco* V_T), which is generally not welcome in power devices, especially when paralleling of chips or wafers is required.

The concentration ratio between E1 and E4 depends on the energy of the electron irradiation. With growing energy, the concentration of the levels E4 grows compared to the E1. Besides the more positive *tempco* V_T, the higher energies of electron irradiation result in a better technology curve as shown in Figure 3.33(b). Also, the penetration depth into silicon wafers at higher irradiation energies is higher. The electron irradiation energies above 5 MeV are therefore needed for devices with breakdown voltages $\geq$6.5 kV, which require wafer thickness over 1 mm as the consequence of the above-explained fact, that the thyristor is the non-punch through device.

3.7.2 *Parallel connection of thyristors*

For the parallel connection of thyristors, the ON-state voltage V_T is normally the most critical parameter since it has a large influence on the static current sharing. A V_T-band of 50/100 mV, normally measured at T_{jmax} and at a current close to the average rated current I_{TAV}, is recommended. Since this often does not cover the entire production spread, a solution with 2–5 V_T-bands may be the most economical approach. The necessary condition is also an appropriate mounting of the various devices to assure a uniform pressure distribution over the whole contact and cooling area.

For simple calculations, the nonlinear forward *I–V* curves can be linearly approximated by means of the differential resistance r_T and threshold voltage V_{T0}.

Then, the voltage drop V_T for a given current I_T reads

$$V_T = V_{T0} + r_T . I_T .$$

For two parallel-connected thyristors $T1$ and $T2$ with total current I giving the difference of V_T by ΔV_T between the first and second thyristors, we can write the following set of two equations:

$$I = I_{T1} + I_{T2} \tag{3.2}$$

$$V_{T1} + r_{T1} . I_{T1} = V_{T2} + r_{T2} . I_{T2} = V_{T1} + \Delta V_T + r_{T2} . I_{T2} . \tag{3.3}$$

Substituting $I_{T1} = I - I_{T2}$ from (3.2) into (3.3), we obtain

$$I_{T2} = r_{T1} . \frac{I}{r_{T1} + r_{T2}} - \frac{\Delta V_T}{r_{T1} + r_{T2}} .$$

For a hypothetical case of $r_{T1} \approx r_{T2}$, we obtain

$$I_{T2} = \frac{I}{2} - \frac{\Delta V_T}{2} . 2 . r_{T1} .$$

It confirms that the lower ΔV_T band gives a better sharing of currents between the two thyristors and, consequently, the lower temperature differences between devices. In practice, the differential resistances r_T are not equal and much bigger complication is that the I–V curves change with temperature in a way that they can change the polarity of the tempco V_T depending on current. For the paralleling, it is important that the crossing point of I–V curves at room and elevated temperature I_{Xing} appears well below the maximal rating current, optimally at a very low current (see Figure 3.35).

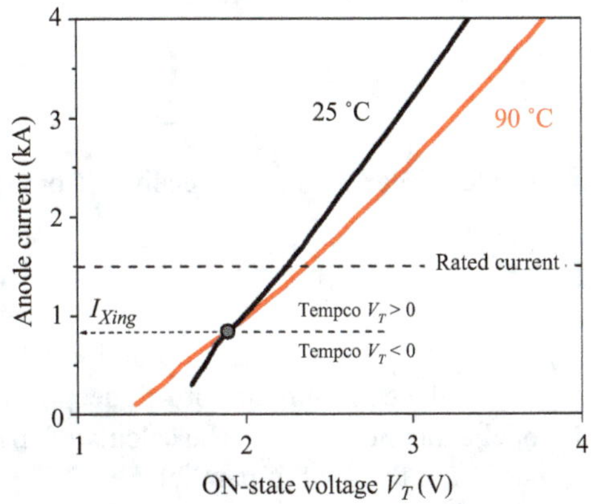

Figure 3.35 ON-state I–V curves of a thyristor at low and high operating temperature. The position of I_{Xing} is essential for the reliability of parallel-connected devices. For normal operation, the existence of I_{Xing} below rated current is sufficient

The positive *tempco* V_T assures that if a device gets hotter, its V_T increases and its current decreases. This negative feedback leads to the decrease of junction temperature and more uniform current sharing between devices in parallel.

The negative *tempco* V_T not only obstructs a reliable device paralleling. It also reduces the magnitude of surge current within a single device by bringing it to the negative differential resistance mode (NDR). The NDR creates electrical instabilities manifested by current filaments leading to device destruction at lower currents than in the devices with the positive *tempco* V_T.

The existence of the I_{Xing} is given by the competition between the following physical effects:

- temperature dependence of injection efficiency–intrinsic concentration $n_i = f(T)$,
- carrier mobility $\mu = f(T)$ and
- excess carrier lifetime $\tau = f(T)$.

An effect that increases the V_T with increasing temperature, e.g., the always decreasing function $\mu = f(T)$, helps in increasing the positive *tempco* V_T and supports the reliable paralleling of devices. The always increasing function $n_i = f(T)$ acts just the other way around and its final effect depends on the injection efficiencies of emitters (doping profile). The dependence $\tau = f(T)$ is more complex because it depends on the chosen irradiation technique (electron or ion irradiation) and parameters (energy, dose, annealing temperature) [14,15].

The impact of the two dominant deep levels from the electron irradiation on the function $\tau = f(T)$ is illustrated in Figure 3.34. The line E1 shows the dependence of excess carrier lifetime in the N-base of the 8.5-kV thyristor on the electron concentration in the ON-state (see Figure 3.20), if only the single level E1 was present. The line E4 shows the same dependence if only the level E4 was present and the full line E1//E4 shows the superimposed effect of these two levels [16,17].

The electron lifetime of *i*th level τ_{ni} in Figure 3.34 was calculated for a given concentration of excess electrons n using the equation

$$\tau_{ni} = \frac{n}{R_{ni}},$$

where the recombination rate of electrons R_{ni} is received from the equation

$$R_{ni} = N_{Ti} \cdot \frac{C_{ni} \cdot C_{pi} \cdot n \cdot p - E_{ni} \cdot E_{pi}}{C_{ni} \cdot n + C_{pi} \cdot p + E_{ni} + E_{pi}}.$$

Here, the C_{ni}, C_{pi}, E_{ni} and E_{pi} are the capture and emission coefficients of electrons and holes and N_{Ti} is the concentration of *i*-th deep level according to the Shockley–Read–Hall model explained in [16]. The calculated magnitudes of the τ_{ni} for the two levels E1 and E4 are then put in parallel

$$\tau_n = \frac{\tau_{nE1} \cdot \tau_{nE4}}{\tau_{nE1} + \tau_{nE4}}.$$

The lifetime of excess electrons τ_n is calculated for the N-type region, which is the N-base of thyristors. Analogous way, we could calculate the lifetime of excess holes τ_p for the P-base taking into account the parameters of deep donor type levels.

Except for the N_{Ti}, all the parameters are temperature dependent. The lifetime due to the deep level E4 is independent of electron concentration with the exception of the very low concentrations close to the N-base doping ($\approx 1.10^{13}$ cm^{-3}). On the other hand, the position of the level E1 close to the conduction band causes the strong dependence of lifetime on the carrier concentration that also reflects in the total lifetime E1//E4 at high carrier concentration. As the electron irradiation results in the concentration ratio of deep levels $N_{E1}/N_{E4} \approx 4{:}1$, the effect of the level E1 is pronounced. As this ratio decreases with increasing energy of the electron irradiation, the higher irradiation energy results in a lower crossing point current because the deeper level E4 is less temperature-dependent. The concentration ratio above is given by the fact that higher energies, as well as bigger projectiles (proton, alpha particles, …), generate more vacancies, which results in more divacancies (E4), while the concentration of oxygen ($\approx 10^{16}$ cm^{-3}) in the starting silicon is independent of irradiation parameters.

As shown in Figure 3.36, the lifetime grows with increasing device temperature in the whole range of carrier concentrations. The competition between the decreasing function $\mu = f(T)$ and increasing function $\tau = f(T)$ can be seen from the equation for the voltage drop

$$V_T \approx \frac{1}{\mu(T) \, . \, \tau(T)} \, .$$

A stronger effect of $\mu(T)$ over that of the $\tau(T)$ provides devices with a lower I_{Xing}. As we cannot influence the function $\mu = f(T)$, the only way to reduce the I_{Xing} is to choose the irradiation projectile, energy, dose and annealing temperature

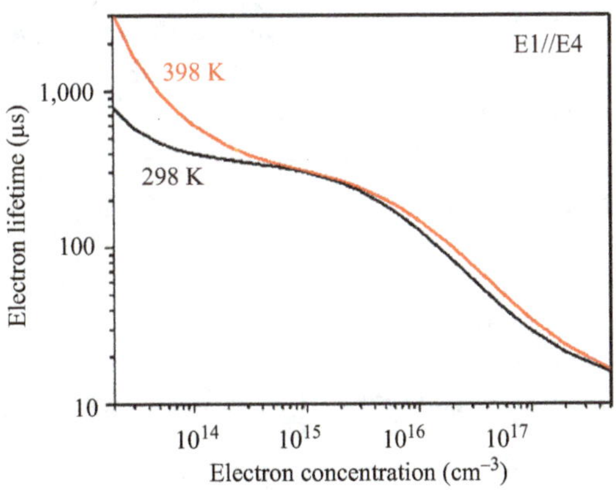

Figure 3.36 Dependence of electron lifetime on the excess carrier concentration after electron irradiation for room and elevated temperature. The curve for 298 K is from Figure 3.34

giving the least growing function $\tau = f(T)$, if possible. This translates into the requirement of having the concentration of the level E4 (with the weakest temperature dependence of the lifetime) close to that of the E1 as possible. We should be also aware that having only the level E4 would result in very high leakage current. Designers are, therefore, looking for an optimal concentration ratio N_{E1}/N_{E4} with respect to a given temperature dependence of injection efficiency given by the doping profile of emitters compared to that of the P- and N-base. However, the magnitude of I_{Xing} is not the primary design parameter. For example, the dose of electron irradiation is given by the required Q_{rr} band. The remaining parameters to choose is the irradiation dose and annealing temperature. As illustrated in Figure 3.32, the highest possible irradiation energy of 10 MeV (irradiation safety) is a good choice especially for the high-end thyristors, where the increased cost of high-energy irradiation is acceptable.

3.8 Summary

Thyristor has the lowest voltage drop for a given current density from all available semiconductor switches and, therefore, it can be found at the top of the power pyramid. State-of-the-art silicon thyristors provide forward and reverse blocking up to 10 kV. Last but not least, only a simple gate electrical or optical pulse is required for turn ON. Altogether with a relatively low production cost, all this keeps the thyristor business with the compound annual growth rate (CAGR) at about 6 per cent in spite of being the 65-year old device concept. Good knowledge of the basic four-layer p-n-p-n structure is also important in the cases where this appears as a parasitic one and must be suppressed by design.

Future research work will further increase the power density (not necessarily already very high power capability) provided by the silicon thyristors, their speed and robustness. There will appear new concepts of bidirectional thyristors (antiparallel thyristors at single wafer), fast thyristors with low turn-off time as well as increased *di*/*dt* capability. One cannot exclude the MOS control to appear again at least at the pulse power applications.

Note to the reader

This chapter is part of an edited book. To cite or use this chapter as a reference, please use the following format:

Vobecky J, "Thyristors." In: Iannuzzo F (ed.), *Modern Power Electronic Devices: Physics, Applications, and Reliability*. Stevenage UK: IET, 2020, pp. 49–89.

References

[1] J. L. Moll, M. Tanenbaum, J. M. Goldey and N. Holonyak, *"P-N-P-N transistor switches"*, Proc. IRE, vol. 44, pp. 1174–1182, 1956.

[2] N. Holonyak, *"The silicon p-n-p-n switch and controlled rectifier (thyristor)"*, IEEE Trans. Power Electron., vol. 16, pp. 8–16, 2001.

[3] M. S. Adler, K. W. Owyang, B. J. Baliga, R. A. Kokosa, *"The evolution of power device technology"*, IEEE Trans. Electron Devices, vol. ED-31, pp. 1570–1591, 1984.

[4] K. Lark-Horowitz, "*Nucleon-bombarded semiconductors"*, in Semiconductor Materials, Proc. Conf. Univ. Reading, Ed. H. K. Henish, London, U.K.: Butterworths, pp. 47–69, 1951.

[5] M. Tanenbaum, A. D. Mills, *"Preparation of uniform resistivity n-type silicon by nuclear transmutation"*, J. Electrochem. Society vol. 108, pp. 171–176, 1961.

[6] M. Schnoller, *"Breakdown behavior of rectifiers and thyristors made from striation-free silicon"*, IEEE Trans. Electron Devices, vol. 2, pp. 313–314, 1974.

[7] J. Vobecký, V. Botan. K. Stiegler, U. Meier, M. Bellini, "*A novel ultra-low loss four-inch thyristor for HVDC*", Proc. ISPSD 2015, Hong-Kong, pp. 413–416, 2015.

[8] J. Vobecký, V. Botan. U. Meier, K. Tugan, "*Local lifetime control for enhanced ruggedness of HVDC thyristors*", Proce. ISPSD 2018, Chicago, pp. 156–159, 2018.

[9] J. Vobecký, H.-J. Schulze, P. Streit, et al., *"Silicon thyristors for ultra high power (GW) applications"*, IEEE Trans. Electron Devices, vol. 64, pp. 760–767, 2017.

[10] R. A. Kokosa, E. D. Wolley, "*Design criteria for amplifying gates of triode thyristors"*, Proc. Int. Electron Device Meeting (IEDM), pp. 431–434, 1974.

[11] D. Silber, H. Maeder, M. Fuellmann, "*Light-activated thyristors"*, in Semiconductor Devices for Power Conditioning, Eds. R. Sittig and P. Roggwiller, NY: Plenum Press, pp. 49–82, 1981.

[12] B. E. Danielsson, *"HVDC valves with light-triggered thyristors"*, in Power Semiconductor Devices and Circuits, Ed. A.A. Jaecklin, Berlin: Springer, pp. 239–270, 1992.

[13] "*Parameter selection of high power semiconductor for series and parallel connection*", Application Note 5SYA 2091–01, ABB Switzerland Ltd. Semiconductors.

[14] J. Vobecký, P. Hazdra, O. Humble, N. Galster, *"Crossing point current of electron and proton irradiated power P-i-N diodes"*, Microelectron. Reliability, vol. 40, pp. 427–433, 2000.

[15] J. Vobecký, P. Hazdra, V. Zahlava, *"Impact of the electron, proton and helium irradiation on the forward I–V characteristics of high-power P–i–N diode"*, Microelectron. Reliability, vol. 43, pp. 537–544, 2003.

[16] D. K. Schroder, "*Semiconductor material and device characterisation*", Hoboken: John Wiley & Sons Inc., p. 391, 2006.

[17] H. Bleichner, P. Jonsson, N. Keskitalo, E. Nordlander, *"Temperature and injection dependence of the Shockley–Read–Hall lifetime in electron irradiated n-type silicon"*, J. Appl. Phys., vol. 79, no.12, pp. 9142–9148, 1996.

[18] M. Schenk, J. J. Przybilla, U. Kellner-Werdehausen, et al., "*State of the art of bipolar semiconductors for very high power applications*", Proc. PCIM Europe 2015, pp. 930–937, 2015.

Further reading

[1] P. D. Taylor, *Thyristors Design and Realisation*, Wiley, Chichester, 1986.
[2] S. Linder, *Power Semiconductors*, CRC Press, Boca Raton, FL, 2006.
[3] B. J. Baliga, *Fundamentals of Power Semiconductor Devices*, Springer Science, Berlin, 2008.
[4] J. Lutz, U. Scheuermann, R. De Donker, *Semiconductor Power Devices*, Springer, Berlin, 2011.

Chapter 4
Silicon MOSFETs

Gerald Deboy[1]

4.1 Introduction

The use of silicon for power switches dates back now for nearly four decades. While microelectronic circuits are fabricated laterally along the top surface of the chip, power devices use the silicon bulk material in their vertical dimension. Current flows from a source terminal on the top side of the device toward a drain terminal at the backside.

Power MOSFETs are unipolar devices. Only majority carriers contribute to the electric current flow. For n-type silicon, electrons do transport the current while holes carry the current for p-type devices. The current is controlled by field-effect through the formation of a strong inversion layer along with a silicon-oxide interface, the so-called MOS channel. Applying a positive voltage to a gate electrode being isolated from a p-type silicon body through a thin gate oxide creates a highly conducting inversion layer with a high electron concentration. Connecting this channel, on one side, with an electron-rich n^+ region and, on the other side, with a voltage sustaining lightly doped n-type region and a subsequent highly doped n^+ region forms a first power MOSFET. Source and gate electrodes are nested into each other along the top surface in a three-dimensional pattern. For more details, we are referring the interested reader to, e.g., Chapter 9.4 of the book "*Semiconductor power devices*" by Josef Lutz *et al.* [1].

Figure 4.1 shows a top view of a power MOSFET in a three-terminal transistor package (TO-220). The source electrode is connected to the right-hand package pin via a thick bond-wire, whereas the gate electrode is contacted by a relatively thin bond-wire. The chip is soldered to the lead frame of the package, thus forming an electric connection of the drain electrode to the center pin of the package.

As the gate electrode is arranged in a structured layer along the top surface, this concept is typically called a planar transistor. The power transistor consists of many paralleled individual cells, in large power switches up to several hundred thousand. The current flows from the source contact laterally through channel and accumulation layer and vertically through the drift zone toward the n^+-doped drain region. Figure 4.2 shows a cross-section of the structure.

[1]Infineon Technologies Austria AG, Villach, Austria

Figure 4.1 Power MOSFET in a three-pin transistor package

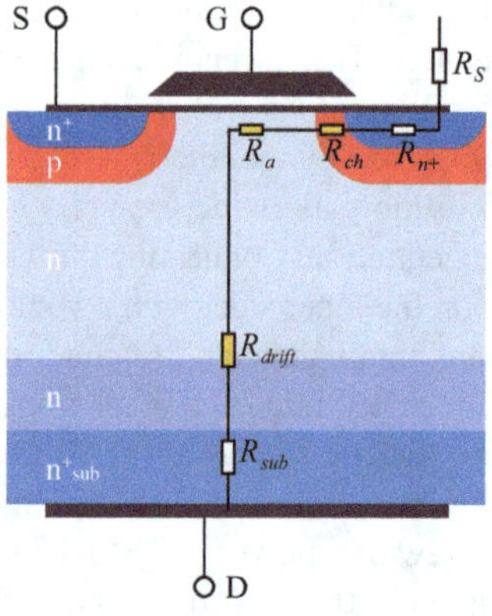

Figure 4.2 Cross-section of a planar power MOSFET

The resistance along this current path and the shares of the individual resistive contributions depend significantly on the intended blocking capability of the device. Table 4.1 shows simulation results for a 30-V-rated and a 600-V-rated power MOSFET, respectively.

For a low-voltage MOSFET channel, accumulation and drift layer contribute nearly evenly to the total resistance of the device, whereas for high-voltage MOSFETs, only the drift layer is of significance. The optimization strategy for high-voltage and low-voltage MOSFETs is hence very different, leading to two

Table 4.1 Contributions of individual MOSFET elements to the total on-state resistance

Power MOSFET $R_{DS(on)}$	
$V_{DS} = 30$ V	$V_{DS} = 600$ V
$R_s = 7\%$	$R_s = 0.5\%$
$R_{n+} = 6\%$	$R_{n+} = 0.5\%$
$R_{ch} = 28\%$	$R_{ch} = 1.5\%$
$R_a = 23\%$	$R_a = 0.5\%$
R_{drift}[1] $= 29\%$	R_{drift}[1] $= 96.5\%$
$R_{sub} = 7\%$	$R_{sub} = 0.5\%$

[1] R_{JEFT} is included in R_{drift}.

distinct concepts for modern power MOSFETs. For voltage classes above 400 V, the Superjunction principle has gained significant traction, while the shielded-gate or field-plate concept is used for voltage classes ranging from 20 to 300 V.

We will organize this chapter accordingly into sections for high-, medium- and low-voltage MOSFETs.

4.2 High-voltage MOSFETs

4.2.1 The silicon limit

The key challenge for high-voltage MOSFETs is the resistance of the drift layer. In the simplest model, we assume a pn-junction with a constantly doped n-region. The on-resistance of this n-layer can be calculated from Poisson's equation as follows:

The gradient of the electric field is proportional to the density of donor atoms N_D. With the constant doping, the breakdown voltage (BV) can be expressed as:

$$W = \frac{2BV}{E_{max}}$$

Substituting these equations into Boltzmann's transport equation $j = N_D \mu E$ with j as current density and μ the mobility of the majority carriers, we reach:

$$R_{on} = \frac{4BV^2}{\mu \varepsilon E^3{}_{max}}$$

This formula is of fundamental importance for all unipolar device concepts. It shows a *quadratic* increase of the on-state resistance R_{on} with blocking voltage. Classic power MOSFETs are hence at a strategic disadvantage in high-voltage applications. Consequently, bipolar devices, such as the Insulated-gate bipolar transistor (IGBT), have gained traction in high-voltage applications.

The formula further links the theoretical best on-state resistance to material properties, such as the maximum electric field strength E_{max} and the mobility of its majority carriers. Specifically, the inverse cubic proportionality to the maximum electric field strength has spurred research for wide-bandgap semiconductor materials such as SiC and GaN. With an inherently 10 times higher field strength, these materials promise a theoretically 1,000 times lower on-state resistance.

In silicon, the maximum electric field strength slightly increases with higher doping levels (or lower breakdown voltage), resulting in an even more pronounced dependency of $R_{on} \sim E_{max}^{2.5}$. With the material properties of silicon, the formula finally reads as:

$$R_{on} = 8.3 \times 10^{-9} BV^{2.5}\ \Omega\text{cm}$$

This is the so-called silicon limit, describing the theoretical best on-state resistance for silicon power MOSFETs at a given breakdown voltage.

Many efforts have been made to come technically as close as possible to the silicon limit: one of these concepts is to use a doping profile varying in concentration as a function of depth. Chenming Hu [2] already described, in 1979, the solution to this one-dimensional optimization problem. The ideal profile starts at a low doping concentration on the top side of the structure and increases with depth following an inverse square-root function. Such a profile results in a curved electric field profile with a higher breakdown voltage at a given thickness of the structure. The on-state resistance is 12 percent better than with a constant doping profile. Practical implementations use a two-step approach, where the upper two-third of the drift layer is relatively low doped and the lower one-third has roughly double the doping concentration.

More challenging is the topic of approaching an undistorted one-dimensional field profile. According to the theory, the highest electric field occurs at the top surface. Real devices have a deeply implanted p-body with a distinct curvature. Furthermore, the highest electric field should not be at the silicon-oxide interface. During the breakdown, electrons and holes are created within the region of the highest electric field through impact ionization; while the electrons flow vertically downward to the drain, holes flow laterally below the implanted n^+-region to the metallic source contact. This lateral current flow may forward bias the parasitic n-p-n bipolar transistor, being inherently present in every power MOSFET. Hence, the position of the highest electric field should be pinned to the bottom of the p-body to allow a vertical flow of holes from their origin to the contacts in case of avalanche breakdown. This design approach, motivated and mandatory from ruggedness requirements on the device, moves practical implementations of power MOSFETs away from the theoretical reachable limit. Furthermore, a buffer for production spread needs to be taken into account as well. The best-published results for power devices reach an $R_{DS(\text{on})}$ around 20 percent above the silicon limit [3].

4.2.2 The Superjunction principle

With the aforementioned arguments, even coming close to the silicon limit is challenging for commercial classic power MOSFETs. The massive reduction of the on-state resistance needs, hence, to take a radically novel approach. Rather than optimizing doping profile and thickness of the n-layer both toward $R_{DS(\text{on})}$ and breakdown voltage simultaneously, the Superjunction principle offers a new degree of freedom: the on-state resistance is optimized alone by a relatively high level of doping while the breakdown voltage requirement is satisfied by compensating the additional n-charges in the drift layer by adjacently arranged p-columns. The resulting electric field is now a superposition of a lateral field component based on the space charge created by p-and n-pillars and a vertical field component arising from the space charge of the net doping of the drift layer. As the local doping concentration in the p-pillar and n-pillar does not necessarily have to be equal, effective doping levels of p- or n- or a combination of both can be created [4].

The lateral field component has to stay below the maximum field strength of silicon, which ties the maximum doping concentration N_D to the width w of the pillars as:

$$N_D * w < 2 * 10^{12}/\text{cm}^2$$

Practical designs use a lateral charge density in the vicinity of 1 to $1.5 \times 10^{12}/\text{cm}^2$. The origin of the Superjunction principle goes back to several decades as well: Shirota and Kaneda [5] described a multi-pn-junction varactor already in 1978. The principle was then introduced into lateral power devices, where a reduction of the surface field (RESURF principle) could be observed if the current-conducting n-layer was put onto a p-substrate [6]. Dr. David Coe [7] from the Phillips research laboratories in the United Kingdom extended this idea into a multilayer RESURF structure and obtained a key patent on this idea in 1988. His patent shows, besides a number of lateral embodiments, a vertical power device structure, as shown in Figure 4.3. This embodiment is the foundation of modern Superjunction devices.

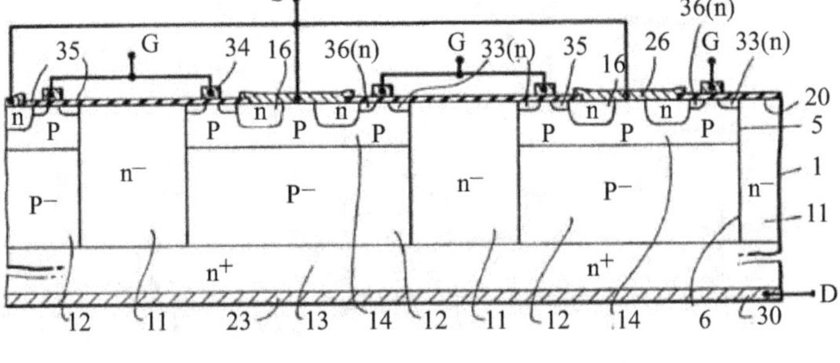

Figure 4.3 A vertical Superjunction device [7].

Dr. Tatsuhiko Fujihira [8] was the first to publish the modeling of this new type of power MOSFET in 1997. Assuming an ideal case of uniformly and equally doped p- and n-columns, he calculated a novel relationship between R_{on} and breakdown voltage (BV) with a linear dependency on the cell-pitch cp:

$$R_{on} = \frac{cpBV}{2\mathcal{E}\mu_n E_c^2}$$

In contrast to the silicon limit, the on-state resistance is now directly proportional to the breakdown voltage instead of the previous $R_{on} \sim E_{max}^{2.5}$ relationship. Furthermore, ever better on-state resistance can be reached by increasing the doping level of p- and n-pillars while reducing the cell-pitch. Figure 4.4 shows the structure and field profiles.

Prof. Florin Udrea [9] from Cambridge University reported the first ever application of the Superjunction principle on a 3D-Resurf diode in March 1998. Through simulations, he could show that the breakdown voltage of such a diode would increase from 145 to 780 V when adding a p-zone to the lateral $p^+/n^-/n^+$ structure.

The first commercial implementation [10] of the Superjunction principle was introduced by the author of this chapter, Dr. Deboy, and his team from Infineon Technologies in December 1998. For the first time, the silicon limit was broken within a silicon structure. The $R_{DS(\mathrm{on})}$ achieved was 50 percent lower than

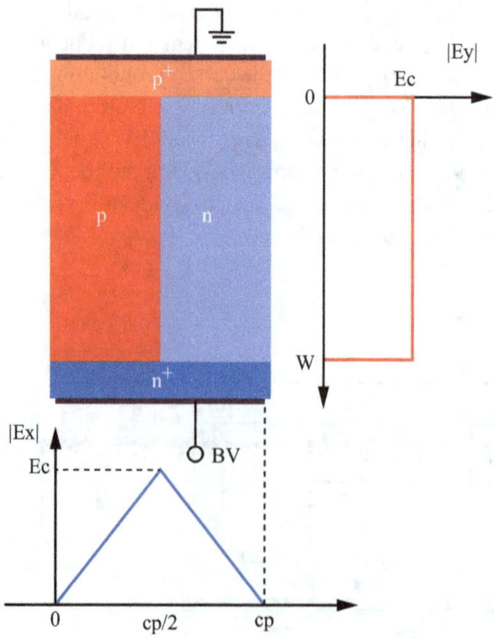

Figure 4.4 Structure and field profiles of a vertical Superjunction power device

calculated from the silicon limit formula (see in Section 4.2.1). Unlike suggested in the literature, doping profiles were tailored in a way to create a surplus of p-doping at the top surface of the transistor and a surplus of n-doping at the lower end of the drift zone [4], thus creating a net doping sequence of $p^+/p^-/n^-/n^+$. This implementation serves a double purpose: on the one hand, it pins the maximum of the electric field well in the center of the structure away from the oxide interface, which is beneficial both for reliability and stability of the edge termination; on the other hand, it provides avalanche capability to the device and enhances the usable process window in case of production tolerances. The avalanche capability can be designed by choosing the appropriate gradient of the electric field both in the net p^- and n^- regions of the drift layer. The steeper the gradient, the higher the capability of the device will be to support avalanche current. This feature, however, trades off with a lowering of breakdown voltage or, respectively, an increase of the thickness of the drift layer and, hence, an increase of the on-state resistance. The avalanche capability is consequently always associated with higher costs and does not come for free.

4.2.3 Electric characteristics of Superjunction devices

The key advantage created by the Superjunction principle is the lowering of the on-state resistance. Over the last two decades, the on-state resistance of 600-V-rated power devices has been lowered by more than one order of magnitude in comparison to the classic high-voltage MOSFET. Figure 4.5 shows the historical development of commercially released technologies versus the silicon limit and reported literature results [11].

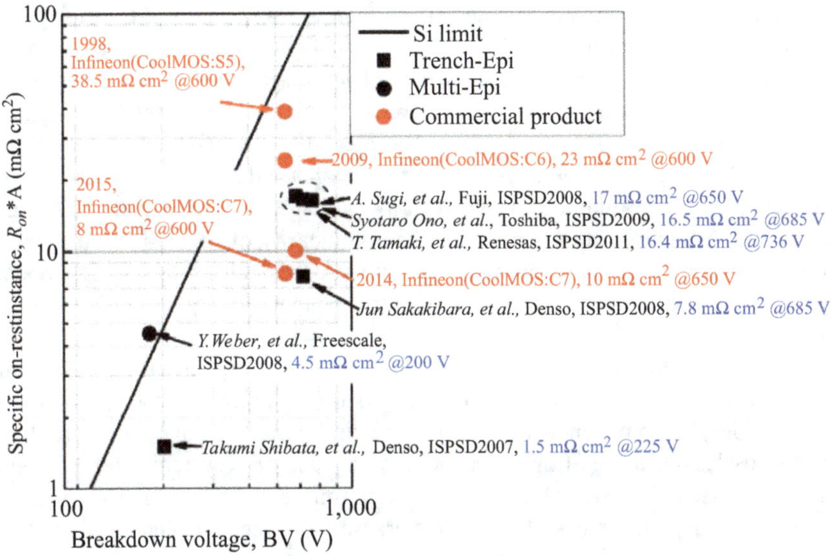

Figure 4.5 Specific on-resistance for commercially released Superjunction technologies versus the silicon limit, and literature references

The current technology node of Infineon's superjunction technology CoolMOS^{TM} C7 reaches in the 600-V class an area-specific on-resistance of 0.8 Ω mm^2. Further progress toward 0.5 Ω mm^2 can be expected from the next generation.

It is noteworthy that the on-resistance of any power MOSFET is temperature-dependent. Scattering of electrons with quantized lattice vibrations or phonons and with impurities such as doping atoms are the two major contributions for Superjunction devices. Between 25 and 125 °C, the on-state resistance is roughly doubling. On the other hand, the breakdown voltage will increase by more than 10 percent. Due to the scattering phenomena, electrons will need a higher electric field to reach the impact ionization threshold, which will initiate the avalanche breakdown of the device.

Definition 1: *CoolMOS^{TM} is the trademark for Infineon's proprietary high-voltage MOSFETs spanning the voltage range from 500 to 950 V. The technology uses the Superjunction principle and is now available in its 6th generation.*

The fine-structured, narrow-pitch sequence of p-and n-columns inside the drift layer creates a huge 3D area of the pn-junction. The term "Superjunction" refers exactly to this structural characteristic. When applying voltage to the device, both p- and n-columns are depleted. With an increasing doping concentration and lower cell pitch, the voltage at which the p-n-columns are fully depleted decreases. Beyond that voltage, a trapezoidal field is created with its gradient being defined by the remaining net space charge of the drift layer. The output capacitance in this voltage range is defined by the width of the space charge layer (typically, 50 μm, e.g., for 600-V devices) and the active area of the power device. At low voltages applied to drain, the output capacitance is defined by the 3D area of the undepleted pn-junction in contrast. As both the width of the space charge layer is very small in this voltage range and the effective surface of the pn-junction is large, the output capacitance at low voltage may be 3–4 orders of magnitudes higher than at high voltage. With further improvements in the area-specific on-state resistance, the output capacitance at high voltage will further decrease due to the reduction of the active area, while the output capacitance at low voltage will increase due to a larger surface of the Superjunction structure. The nonlinearity of the output capacitance becomes, hence, ever more pronounced with progress along technology nodes. Figure 4.6 shows a comparison of three subsequent Superjunction technology platforms.

It is noteworthy that during turn-off, both the electron current as well as the hole current is transported by majority carriers only. As long as the space charge layer builds up homogenously without creating stranded charges, turn-off is entirely lossless. Currents flow as capacitive displacement currents without creating Joule losses. In physical terms, both electrons and holes move as majority carriers in undepleted areas. This characteristic is further enabled by the strong nonlinearity of the output capacitance: as C_{oss} is at the beginning of the turn-off phase very large, the voltage across the device is nearly not changing for several ten

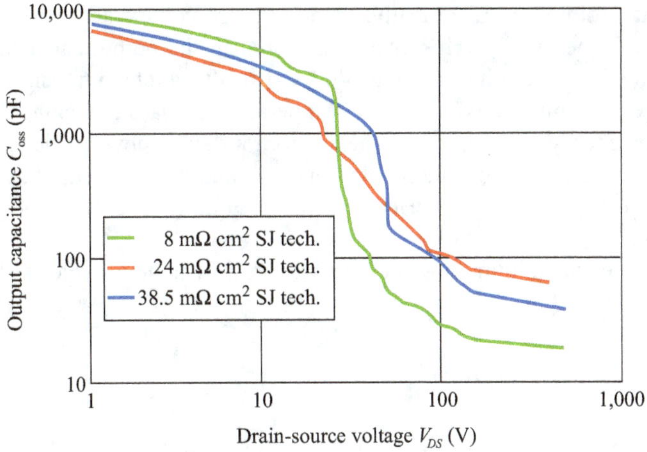

Figure 4.6 Double-logarithmic graph of the output capacitance of three consecutive Superjunction technology nodes

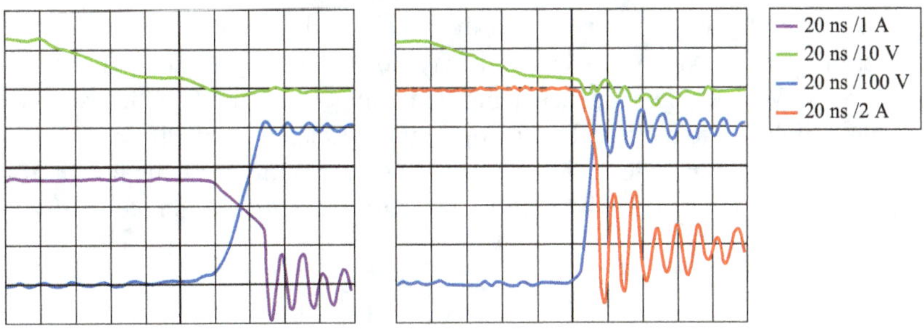

Figure 4.7 Turn-off waveform of a Superjunction transistor at 1/7 and 1/2 of the nominal current

up to hundred nanoseconds. This time allows for shutting down the MOS channel entirely. The load current I_l commutates into the output capacitance charging C_{oss}. The voltage slope is given by:

$$dv/dt = \frac{I_l}{C_{oss}}.$$

The load current and the shape of the output capacitance determine, hence, the voltage transition: in single-ended applications (a power MOSFET switching against a diode), the voltage shows a characteristic turn-off delay time followed by an ever-increasing voltage rise. Figure 4.7 shows characteristic waveforms for turn-off at 1/7 and 1/2 of the nominal current.

It is obvious that the transition is significantly faster at half of the load current. As long as the channel is completely turned off before the voltage rises, switching losses are at a minimum. Electrically, the overlap of voltage and load current creates a turn-off energy E_{off}. This energy is stored in the output capacitance being charged to the DC link voltage of, e.g., 400 V. In soft switching transitions, such as zero voltage switching, this energy is recuperated. In hard switching turn-on, in contrast, this energy is dissipated as heat. From a device physics viewpoint, electrons from the channel cross the depletion layer and start to neutralize the space charge region in the n-column. Holes from the p-body neutralize the p-column, respectively.

The theoretical minimum switching loss in a switching cycle is, hence, the energy E_{oss}, defined as:

$$E_{oss} = \int_0^U C_{oss}(U) \cdot U \cdot dU$$

Figure 4.8 shows a comparison of the measured turn-off energy as a function of the load current with the gate resistor as a parameter.

With increasing gate resistor, the turn-off of the channel is slowed down; consequently, a part of the electron current still flows through the channel and crosses the space charge layer building up with rising voltage. This electron contribution creates losses. Similarly, with the rising load current, the voltage transition becomes faster. Again, the channel may not be entirely turned off while the voltage rises. We see, hence, increasing switching losses both with rising load current and higher gate resistor. For the optimum efficiency, turn-off with very low gate resistors is recommended. This requires, however, a low-parasitic layout of the switching cell, which can handle both high *dv*/*dts* and *di*/*dts*. It is noteworthy that

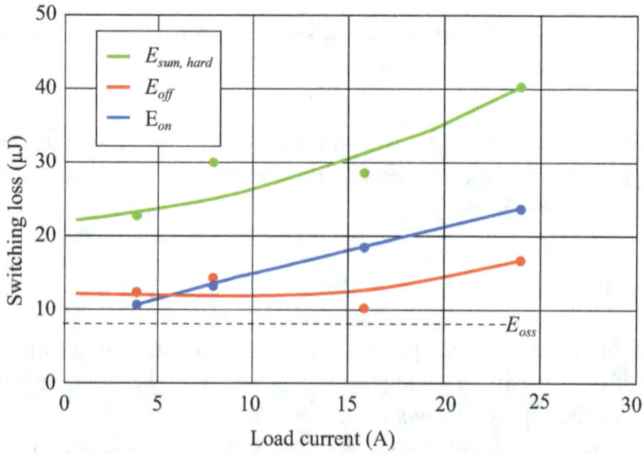

Figure 4.8 Turn-off losses as a function of the gate resistor and load current for the latest generation of Superjunction transistors of Infineon, CoolMOS™ C7

modern Superjunction devices can switch at far more than 100 V/ns and several thousand amperes per microsecond.

To further optimize efficiency in hard switching applications, the energy stored in the output capacitance needs to be minimized. The resulting figure-of-merit $R_{on} \cdot E_{oss}$ is known in the literature as the Nakajima limit of switching losses [12]. As the integral formula for E_{oss} puts more weight on the C_{oss} values at relatively high voltage, the strategy for optimizing E_{oss} relies on shrinking the die size of the transistor. This measure reduces the high-voltage end of the C_{oss} curve linearly and allows – with more dies per wafer – to lower production costs.

It has hence always been a research target to lower the area-specific on-state resistance of Superjunction transistors, as outlined in Figure 4.5. The beneficial effect on the energy stored in the output capacitance E_{oss} is shown in Figure 4.9. It is noteworthy that current Superjunction technologies, such as the CoolMOS^{TM} C7 and G7 from Infineon Technologies, are leading in this parameter being already on parity with the latest GaN high electron mobility transistors (GaN HEMTs) and will have even a better figure-of-merit in the next upcoming technology platform C8.

We can conclude that in hard switching single-ended applications, Superjunction devices are the first choice outperforming even significantly more expensive wide-bandgap counterparts.

In half bridge-based applications, however, E_{oss} (μJ) switching losses are proportional to

$$E_{sw} = (Q_{oss} + Q_{rr}) \cdot V_{DC}$$

Instead of E_{oss} with Q_{oss} and Q_{rr} corresponding to the charge stored in the output capacitance and the reverse recovery charge, respectively, wide-bandgap devices

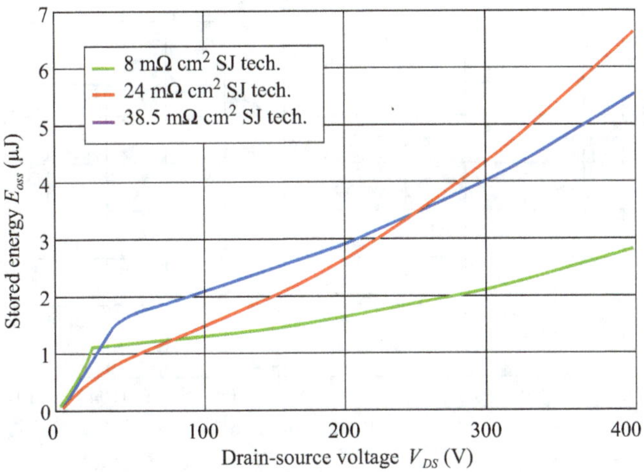

Figure 4.9 Comparison of the energy E_{oss} being stored in the output capacitance for three consecutive Superjunction technologies

such as GaN HEMTs and SiC MOSFETs are at least one order of magnitude better [13].

Q_{oss} can be understood in a simplified picture as the number of electrons needed for a given on-state resistance at a given breakdown voltage. With the relatively low breakdown field in silicon, the drift layer, e.g., for a 600-V MOSFET, has to be in the order of 50 μm whereas only one-tenth of this dimension would be needed in a wide-bandgap device. In addition, the reverse recovery charge as being injected into the drift layer during reverse operation of the power MOSFET may be another magnitude higher than the Q_{oss} charge. While doping levels of recent Superjunction MOSFETs are in the upper range of 10^{15} cm^{-3}, the concentration of an electron-hole plasma flooding the entire p- and n-column may be as high as 10^{16} cm^{-3} or even few 10^{17} cm^{-3}.

A hard commutation event, which means forcing a power device conducting in reverse direction to block and to take over blocking voltage, creates, hence, huge losses with Superjunction transistors. The current and voltage transitions are highly dynamic and show an abrupt discontinuation of the reverse recovery current and a very steep voltage rise. Figure 4.10 shows an example.

The survival strategy for Superjunction devices for hard commutation events comprises two key measures: first, slowing down the turn-on of the transistor to limit the reverse recovery peak by current saturation in the device turning on; second, providing recombination centers in the drift layer to reduce the reverse recovery charge. While the user of the device can apply the first measure, the latter-

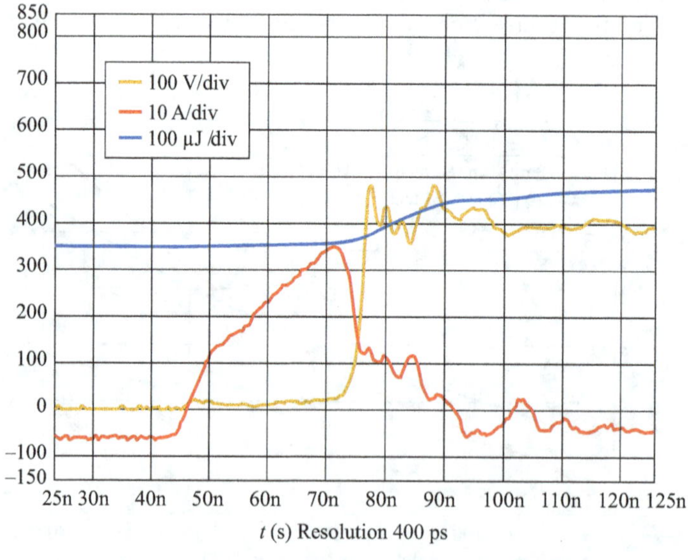

Figure 4.10 Hard commutation event of a Superjunction transistor showing extremely high di/dt and dv/dt values: current in red, voltage across the device in yellow and energy in blue

mentioned measure is a technology step creating a separate product family typically named "FD" for "fast diode." Reverse recovery charges can be typically lowered by one order of magnitude by Pt-doping or electron and He-ion irradiation, respectively.

4.3 Low- and medium-voltage MOSFETs

While the optimization of Superjunction transistors primarily focuses on the on-state resistance of the drift layer, the situation for low- and medium-voltage MOSFETs is more complex. With channel, accumulation and drift layer contributing equally strong to the overall resistance of the device, a different device concept is required. Furthermore, both switching figure-of-merits, such as $R_{on}*Q_g$ or $R_{on}*Q_{gd}$, as well as synchronous rectification figure-of-merits, such as $R_{on}*Q_{oss}$, need to be simultaneously optimized to create a MOSFET concept scalable from low voltage, such as 30 V, to medium voltages, such as 300 V.

4.3.1 The vertical trench MOSFET versus the shielded-gate MOSFET

Historically, low-voltage MOSFETs started from planar concepts similar to high-voltage MOSFETs. However, due to the large resistance contribution from both channel and accumulation region, the performance remained limited and could not be scaled by shrinking the cell pitch [14,15]. Transferring the channel into the vertical axis eliminates the accumulation resistance entirely and allows a roadmap toward ever lower channel resistance by shrinking the cells and using literally millions of paralleled cells per power MOSFET. This bold step required a number of novel production technologies, such as creating stable gate oxides on the sidewall of trenches as well as subsequent trench filling and planarization steps. Against all these odds, the vertical trench MOSFET emerged as the leading technology and is today ubiquitous in numerous applications [16].

Even though the vertical trench MOSFET creates a high-performance device on the low-voltage end of the spectrum, it does not address the resistance of the drift layer. The on-state resistance of the device, therefore, scales with the classic silicon limit formula given in Section 4.2.1. Furthermore, the increase of the trench density leads to a massive coupling capacitance between the gate electrode located in the trench and the drain electrode. This gate-to-drain capacitance slows down switching transients and conflicts with the optimization of the technology as an active switching element such as needed, e.g., for the control FET in a buck stage. The issue can principally be addressed with a thick bottom oxide; however, this concept creates a more complex process flow and may cause thermomechanical expansion stress. The third issue of the vertical trench MOSFET is the reliability of the gate oxide in the trench: as the channel needs to connect to the drift layer to form an uninterrupted electron path, the trench penetrates into the n-layer. The corners of the trench are, hence, exhibited to very high electric field peaks, which may lead to the creation of "hot" electrons. These electrons being accelerated in the electric field may gain that much energy that they can overcome or tunnel through the barrier between silicon and oxide, creating traps within the oxide. These traps lead to a shift in threshold voltage and may eventually destroy the device once a percolation path has been created through the oxide. Therefore, it is important to protect the gate oxide [17,18].

All the arguments as mentioned above – resistance of the drift layer, reduction of gate-drain capacitance and reliability of the gate oxide – have led to the concept of the shielded-gate or field-plate vertical trench MOSFET.

Figure 4.11(a) and (b) shows the two concepts. In addition to the vertical gate electrode, a second electrode is embedded within an even deeper trench. This second electrode is coupled to the source electrode. It shields the gate electrode and helps, therefore, to reduce significantly the overlap of the gate electrode with the drain electrode. The peak of the electric field can be shifted to the blocking pn-junction into the center of the mesa region – the silicon area between the trenches – thus, eliminating the reliability risk of hot electrons close to the gate oxide. Last but not least, the second electrode acts as a field plate providing counter-charges and creating a horizontal component of the electric field very similar to the Superjunction principle. The mesa region can hence be higher doped to lower the on-resistance of the drift layer [19,20]. This concept consequently also breaks the limit line of silicon and enables an optimization path with ever lower mesa width and higher doping levels.

The shielded-gate transistor concept allows a scalable device technology spanning voltage classes from 30 to 300 V [21]. The upper limit of the technology toward even higher voltages is given by the blocking capability of the oxide around the field plate. As the field plate is coupled to the source electrode, this oxide needs to block the full-rated voltage of the transistor. The required thickness, hence, creates a lower limit to the minimum width of the trenches and, hence, indirectly to the achievable pitch of the cells. Nevertheless, higher breakdown voltages can be

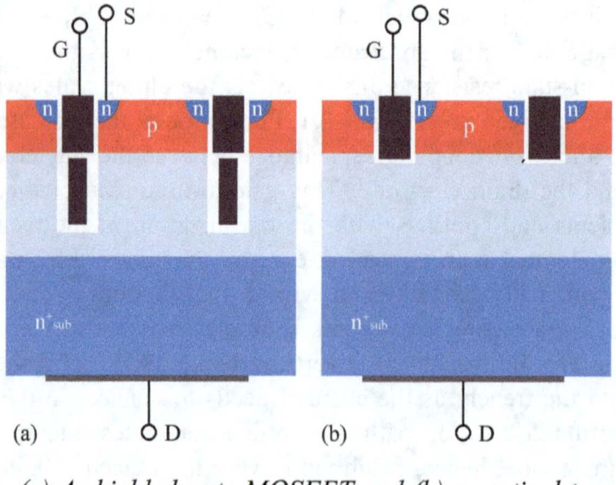

Figure 4.11 (a) A shielded-gate MOSFET and (b) a vertical trench MOSFET

realized by combining the shielded-gate concept with an epitaxial n^--layer below the trench [22].

4.3.2 *Electric characteristic*

The shielded-gate concept allows a continuous and dramatic improvement of both switching figure-of-merits, such as $R_{on}*Q_{gd}$, as well as the on-state resistance, as shown in Figure 4.12. The on-state resistance of Infineon's 30-V MOSFETs was reduced through seven generations by factor 50 showing an average reduction of 17 percent year-on-year, while the product of gate-drain charge and on-state resistance was improved by a factor 30 in the same time frame. With this simultaneous improvement of key figures-of-merit, the technology creates with every generation faster-switching devices while increasing the number of dies per wafer and, hence, reducing cost per die.

Both the threshold voltage of the MOS channel as well as the on-resistance are dependent on temperature. While the threshold voltage decreases with increasing temperature, the on-resistance will increase due to phonon and impurity scattering mechanisms. Figure 4.13(a) and (b) shows the temperature dependence of threshold voltage and on-resistance for a 60-V shielded-gate device and a 600-V Superjunction device, respectively.

The 60-V device shows a lower temperature dependence of the on-resistance than the 600-V-rated Superjunction device. As discussed earlier, low voltage (LV) and middle voltage (MV) transistors have a more balanced share of MOS channel and drift layer contributions to the on-state resistance in comparison to HV MOSFETs. As the on-resistance contribution of the MOS channel improves with temperature (due to lowering of the threshold voltage), the increase of the on-state resistance of the mesa region and drift layer is partially compensated leading to an overall lower temperature dependence.

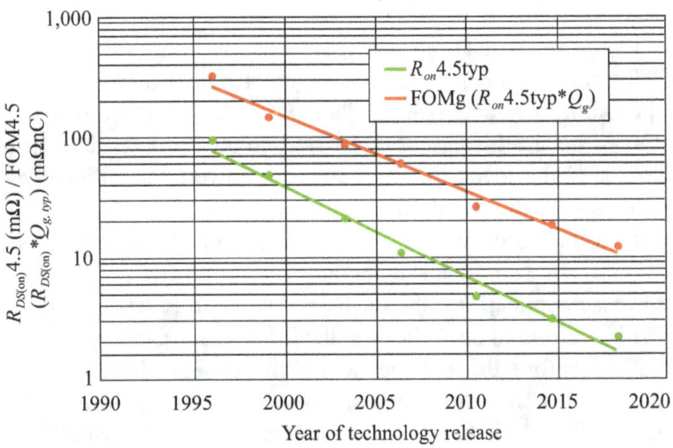

Figure 4.12 Figures-of-merit for 30-V vertical MOSFETs

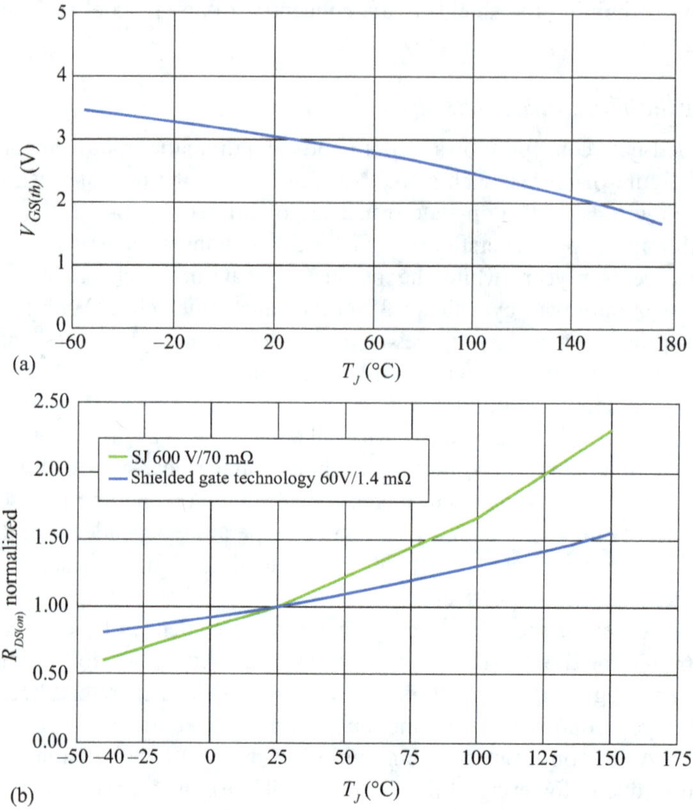

Figure 4.13 Temperature dependence of the threshold voltage (a) and on-state resistance (b) of a 60-V shielded-gate transistor and a 600-V Superjunction transistor

Besides its use in applications switching actively at high frequencies, LV and MV MOSFETs have major use cases as the replacement of diodes [23]. The key benefit is the ohmic characteristic of the switch in reverse direction when the channel is turned on. In this case, the forward voltage drop is defined by the on-state resistance and the load current and can be lowered significantly below the threshold voltage of Schottky diodes, thus increasing the efficiency of the converter. Examples include synchronous rectification on the secondary side of isolated DC/DC converters and the synchronous buck converter.

In all of these applications, the device conducts in reverse direction with the MOS channel being turned on. When the device needs to block, the channel needs to be turned off before the voltage across the device changes direction. The respective phases of such a hard commutation transition are shown in Figure 4.14.

The device conducts initially in reverse direction; the MOSFET is turned on. At the beginning of phase 1, the MOS channel is turned off; the device still conducts in reverse direction. The negative forward voltage drop across the device

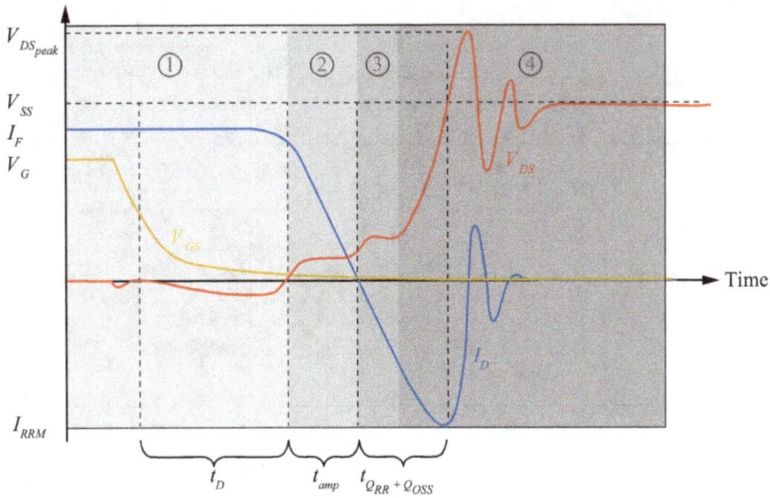

Figure 4.14 Hard commutation event showing current and voltage waveforms throughout distinct phases

increases to the level of the body diode threshold, typically up to –0.8 V. Phase 2 begins when the device is forced to take overvoltage, e.g., by turning on a device at the opposite position in a half-bridge circuit. The current starts to fall. The forward voltage drop is now the negative threshold of the body diode plus the voltage drop across the parasitic inductance of the device created by the negative *di*/*dt* transient. At phase 3, the current changes its direction, now removing the stored charge from the device. This charge consists of both of the bipolar reverse recovery charge Q_{rr} as well as the charge stored in the output capacitance Q_{oss}. The body diode no longer conducts, the forward voltage drop is now given by the package inductance alone. At the end of phase 3, the Q_{rr} charge is entirely removed; the negative current is now depleting the active region of the device building up the space charge layer. The device consequently starts to take overvoltage, which initiates phase 4. The voltage within the commutation circuit is now split between both devices: the device initially conducting on its body diode and the device turning on actively. This increase of voltage slows down the reverse recovery current until the di_{rr}/dt approaches zero when the device supports the full DC link voltage. The discontinuation of the reverse recovery current with its now positive *di*/*dt* may create oscillations with the parasitic inductance of the commutation loop leading to some voltage overshoot.

The development target for LV and MV transistors for this type of application is the continuous reduction of the charge stored in the output capacitance Q_{oss} as well as the bipolar charge Q_{rr}. The reverse recovery current waveform should show a soft discontinuation phase with relatively low voltage overshoot. The output capacitance should be as linear as possible to create a constant *dv*/*dt* at the switching node during the change of the current direction. Figure 4.15 shows such a

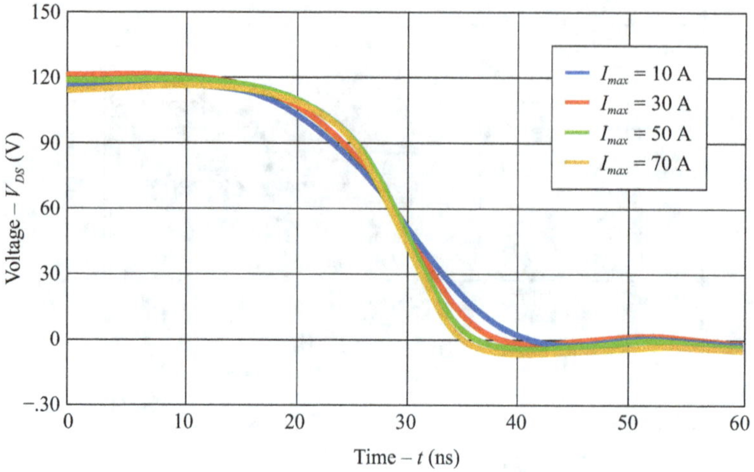

Figure 4.15 Voltage transition in a half-bridge circuit measured on an upcoming generation of 200-V MOSFETs

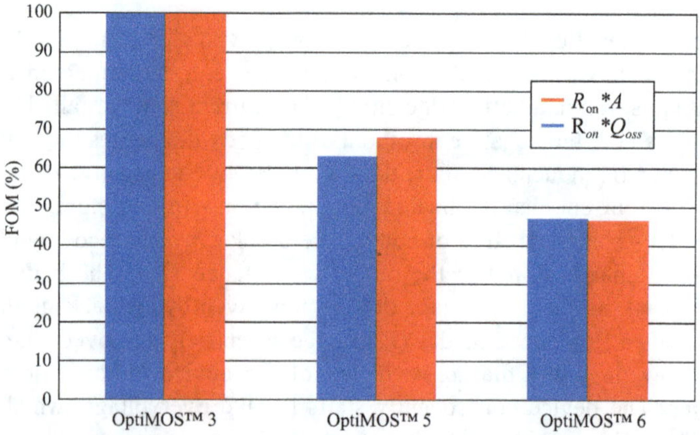

*Figure 4.16 Performance improvements of the figure-of-merit R_{on}*A and R_{on}*Q_{oss}, respectively, across three generations of 100-V transistors (projection for OptiMOS™ 6)*

transition for an upcoming generation of 200-V MOSFETs by Infineon Technologies.

Definition 2: *OptiMOS™ is the trademark for Infineon's proprietary low- and medium-voltage power MOSFETs spanning the voltage range from 25 to 300 V. The technology uses the shielded-gate concept and is now available in its 6th generation.*

The charge stored in the output capacitance can be lowered through optimization of the field-plate structure. Figure 4.16 shows performance improvements across three generations of Infineon's MV transistors.

The reverse recovery charge created through injection of electron-hole plasma into the mesa region during reverse conduction of the device can be optimized, on the one hand, by built-in recombination centers such as Platinum, and, on the other hand, again through field plate and mesa width optimization. Especially, the latter approach limits the volume physically for the reverse recovery charges and allows an overall optimization of the technology to near-perfect commutation ruggedness.

4.4 Summary

Using both, the Superjunction principle and the shielded-gate concept, silicon MOSFETs have been able to break the limit line of one-dimensional pn-junctions. Continuous progress along technology nodes has created tremendous advantages both in terms of area-specific on-resistance as well as in terms of switching figure-of-merits, such as $R_{on}*E_{oss}$, $R_{on}*Q_{oss}$ and $R_{on}*Q_{gd}$.

Future research work will further enhance the capabilities of silicon MOSFETs and will secure a significant portion of the market for the technology. Wide-bandgap devices based on SiC or GaN will complement and enhance the technology portfolio of silicon MOSFETs.

Note to the reader

This chapter is part of an edited book. To cite or use this chapter as a reference, please use the following format:

Deboy G. "Silicon MOSFETs." In: Iannuzzo F (ed.), *Modern Power Electronic Devices: Physics, Applications, and Reliability*. Stevenage, UK: IET; 2020. pp. 91–111.

References

[1] J. Lutz, H. Schlangenotto, U. Scheuermann, and R. de Doncker, "Semiconductor power devices," Springer Verlag, ISBN 978–3-642–11124–2.

[2] C. Hu, "Optimum doping profile for minimum ohmic resistance and high-breakdown voltage," *IEEE Trans. Electron Devices*, vol. 26, no. 3, pp. 243–244, 1979.

[3] T. Kobayashi, H. Abe, Y. Nimura, *et al.*, "High-voltage power MOSFETs reached almost to the Si limit," Proc. ISPSD, 2001, pp. 99–102, 2001.

[4] G. Deboy, D. Ahlers, H. Strack, M. Rueb, and H. Weber, "High-voltage semiconductor component," *U.S. Patent,* 6,630,698, 1998.

[5] S. Shirota and S. Kaneda, "New type of varactor diode consisting of multilayer p-n junctions," *J Appl. Phys.*, pp. 6012–6019, 1978.
[6] A. W. Ludikhuize, "A review of RESURF technology," *Proc.12th Int. Symp. Power Semicond. Device ICs*, pp. 12–18, 2000.
[7] D. J. Coe, "High voltage semiconductor device," *U.S. Patent*, 4,754,310, 1988.
[8] T. Fujihira, "Theory of semiconductor superjunction devices," *Jpn. J. Appl. Phys.*, vol. 36, pp. 6254–6262, 1997.
[9] F. Udrea, A. Popescu, and W. Milne, "The 3D RESURF double gate MOSFET: A revolutionary power device concept," *IEEE Electron. Lett.*, vol. 34, no. 8, pp. 808–809, 1998.
[10] G. Deboy, M. März, J. P. Stengl, H. Strack, J. Tihanyi, and H. Weber, "A new generation of high voltage MOSFETs breaks the limit line of silicon," *Int. Electron Devices Meet., 1998. Tech. Dig.*, pp. 683–685, 1998.
[11] F. Udrea, G. Deboy, and T. Fujihira, "Superjunction power devices, history, development and future prospects," *Trans. Electron Devices*, vol. 64, no. 3, pp. 713–727, 2017.
[12] A. Nakajima, M. Shimizu, and H. Ohashi, "Power loss limit in unipolar switching devices: Comparison between silicon superjunction devices and wide bandgap devices," *IEEE Trans. Electron Devices*, vol. 56, no. 11, pp. 2652–2656, 2009.
[13] G. Deboy, O. Haeberlen, and M. Treu, "Perspective of loss mechanisms for silicon and wide bandgap power devices," *CPSS Trans. Power Electron. Applicat.*, vol. 2, no. 2, pp. 89–100, 2017.
[14] D. Ueda, H. Takagi, and G. Kano, "A new vertical power MOSFET structure with extremely reduced on-resistance," *IEEE Trans. Electron Devices*, vol. ED-32, no. 1, pp. 2–6, 1985.
[15] R. K. Williams, M. N. Darwish, R. A. Blanchard, R. Siemieniec, P. Rutter, and Y. Kawaguchi, "The trench power MOSFET: Part I—history, technology, and prospects," Trans. Electron Devices, vol. 64, no. 3, pp. 674–692, 2017.
[16] H.-R. Chang, R. D. Black, V. A. K. Temple, W. Tantraporn, and B. J. Baliga, "Self-aligned UMOSFET's with a specific on-resistance of 1 mcm^2," *IEEE Trans. Electron Devices*, vol. ED-34, no. 11, pp. 2329–2334, 1987.
[17] F. Pfirsch and C. Schöffer, "Field-effect transistor configuration with a trench-shaped gate electrode and an additional highly doped layer in the body region," *U.S. Patent,* 6,541,818, June 2001.
[18] F. Hirler, M. Kotek, J. Larik, and F. Pfirsch, "Trench MOS transistor," *U.S. Patent,* 6,720,616, December 2001.
[19] M. Zundel and F. Hirler, "MOS field plate trench transistor device," *U.S. Patent,* 7,372,103, March 2006.
[20] K. Kobayashi, T. Nishiguchi, S. Katoh, T. Kawano, and Y. Kawaguchi, "100 V class multiple stepped oxide field plate trench MOSFET (MSO-FP-MOSFET) aimed to ultimate structure realization," *Proc. ISPSD,* 2015, Hong Kong, 2015.

[21] A. Schlögl, F. Hirler, J. Ropohl, *et al.*, "A new robust power MOSFET family in the voltage range 80V-150V with superior low-*RDS*(on), excellent switching properties and improved body diode," Proc. Eur. Conf. Power Electron. (EPE), Dresden, Germany, 2005.

[22] R. Siemieniec, C. Braz, and O. Blank, "Design considerations for charge-compensated fast-switching power MOSFET in the medium-voltage range," *IET Power Electron.*, vol. 11, no. 4, pp. 638–645, 2018.

[23] R. Siemieniec, C. Mößlacher, O. Blank, M. Rösch, M. Frank, and M. Hutzler, "A new power MOSFET generation designed for synchronous rectification," *Proc. EPE,* 2011, Birmingham, 2011.

Chapter 5
Silicon IGBTs

Munaf Rahimo[1] *and Paula Díaz Reigosa*[2]

5.1 Introduction

Close to four decades ago, a seemingly simple variant of the silicon power MOSFET [1–4] set on to change the power electronic landscape, namely, the insulated gate bipolar transistor (IGBT) [5]. The device presents characteristics combining both MOSFET and bipolar junction transistor (BJT) structures [6]. As a result of this combination, a blend of the physics of both devices would represent the IGBT principle of operation with very desirable performance and gate-driving features for power system designers – mainly low conduction and switching losses, high input impedance allowing comparatively small voltage-controlled gate drivers, robust turn-off performance and short circuit withstand capability. The basic functional IGBT designs were researched and developed during the 1980s; while the first commercially available IGBTs did not exceed blocking voltages above 1,200 V [7,8] and currents of few tens of amperes, and the development effort continued aimed solely at increasing the power-handling capability of this device concept [9]. During the 1990s, more advanced IGBT designs were realised and widely implemented in many mainstream power electronics applications with higher voltage and current ratings. High-voltage IGBTs and their antiparallel freewheeling diodes with ratings up to 6.5 kV [10] were successfully manufactured in high-current IGBT modules employing many chips connected in parallel. Today, such components with current ratings in the order of thousands of amperes are employed in many power electronics applications such as in rail and automotive traction systems, grid and renewable converters, and industrial drives and applications. Furthermore, the availability of such a wide range of IGBTs with different power ratings and electrical characteristics has enabled new power electronic applications to emerge, such as for solid-state circuit breakers and solid-state transformers with different emphasis on the performance requirements for each type of application.

[1]MTAL GmbH, Unterer Schafmattweg 39A, Gänsbrunnen, Switzerland
[2]Institute of Electric Power Systems, University of Applied Sciences and Arts Northwestern Switzerland, Klosterzelgstrasse 2, Windisch, Switzerland

5.2 The IGBT structure, equivalent circuit and operation

In principle, the IGBT structure is based on the vertical power MOSFET with one exception by adding a highly doped P+ collector layer (also referred to sometimes as the anode) to replace the MOSFET highly doped N+ drain region. For comparison, Figure 5.1 shows both the N-channel MOSFET and IGBT vertical power structures based on a planar MOS cell platform. The MOSFET source and IGBT emitter cell design concepts are the same as they consist of a deep P-base region to form the main blocking junction and MOS inversion channels along with highly doped N++ regions. In addition, highly doped P++ regions are required at the source/emitter contact opening. The MOS gate is isolated from the main silicon bulk by a thin gate silicon dioxide (SiO_2) layer. The behaviour of the IGBT differs from that of a unipolar MOSFET in that the IGBT is a bipolar device during conduction since the introduction of the P+ collector region provides minority carrier (i.e., holes) injection and, hence, conductivity modulation in the device low doped N– drift region. This feature will ultimately lead to a much lower on-state resistance for enabling high voltage-rated devices. It is important to note that Figure 5.1 shows a punch-through (PT) IGBT, which incorporates the N buffer or field-stop

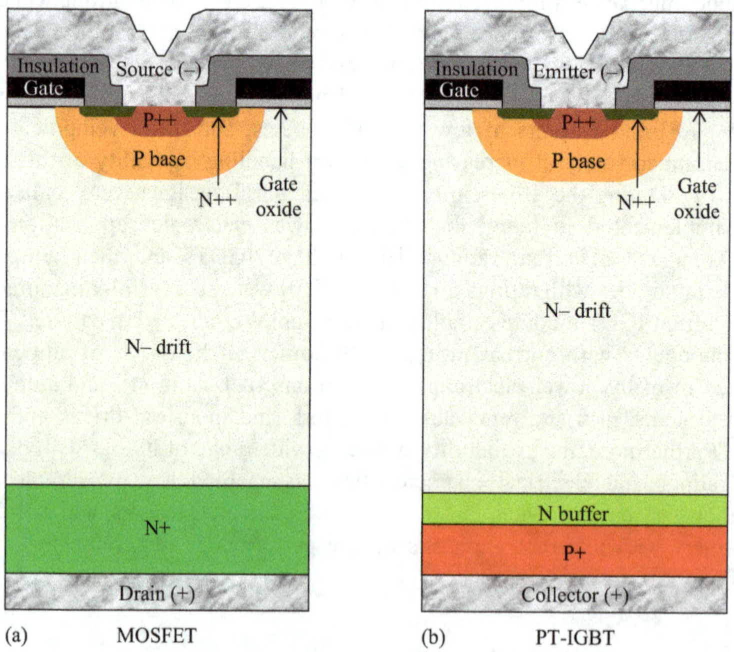

Figure 5.1 The N-channel MOSFET (a) and PT-IGBT (b) vertical power structures based on a planar MOS cell platform

(FS) region to prevent the space charge region (SCR) from reaching the P+ collector layer.

The simple IGBT conduction mode equivalent circuit is shown in Figure 5.2, which consists of an N-channel power MOSFET and a PNP bipolar transistor. Naturally, the MOSFET part of the IGBT also incorporates an internal NPN bipolar transistor as depicted in the same figure. The drain of the MOSFET is connected to the gate of the PNP transistor through a drift region resistance R_{DRIFT} (the N– drift region is the base of the PNP transistor). On the other hand, the source of the MOSFET is connected to the emitter of the PNP transistor through a base region resistance R_{BASE} which represents a lateral resistance of the P-base region (the P-base region is the base of the NPN transistor). Both the PNP and NPN transistors also form a parasitic thyristor element in the IGBT, which plays an important role in defining the device safe-operating-area (SOA) limits.

When a positive collector-emitter voltage V_{ce} is applied across the IGBT ($V_{ce} > 0$), it implies that the collector PN junction is forward biased and the base PN junction is reverse biased. If the applied gate-emitter voltage V_{ge} is below the threshold voltage V_{th} ($V_{ge} < V_{th}$), the MOS inversion channel in the P-base cannot be formed and no electrons will flow into the N– drift region. Under such conditions, the base PN junction remains in reverse bias as the device is in an (OFF) state where it can support voltages up to the specified voltage ratings. When a gate-emitter voltage V_{ge} is more than the device threshold voltage V_{th} ($V_{ge} > V_{th}$), a MOS inversion channel in the P-base region is formed between the N++ and N– drift regions. Subsequently, electrons will flow from the emitter into the N– drift region towards the forward-biased collector PN junction. Therefore, a given number of holes from the collector will be injected into the N– drift region depending on the injection efficiency of the P+ collector region. The electrons and holes will constitute a quasi-neutral plasma in the drift region (i.e., conductivity modulation) as the collector current i_c flows freely during conduction in an ON-state as shown in Figure 5.3.

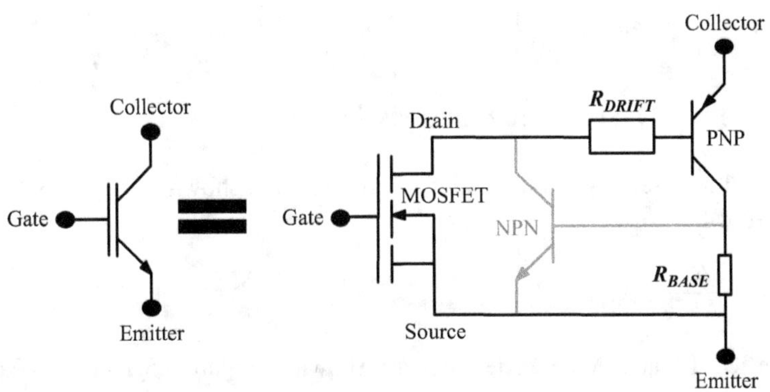

Figure 5.2 The IGBT circuit symbol and equivalent circuit

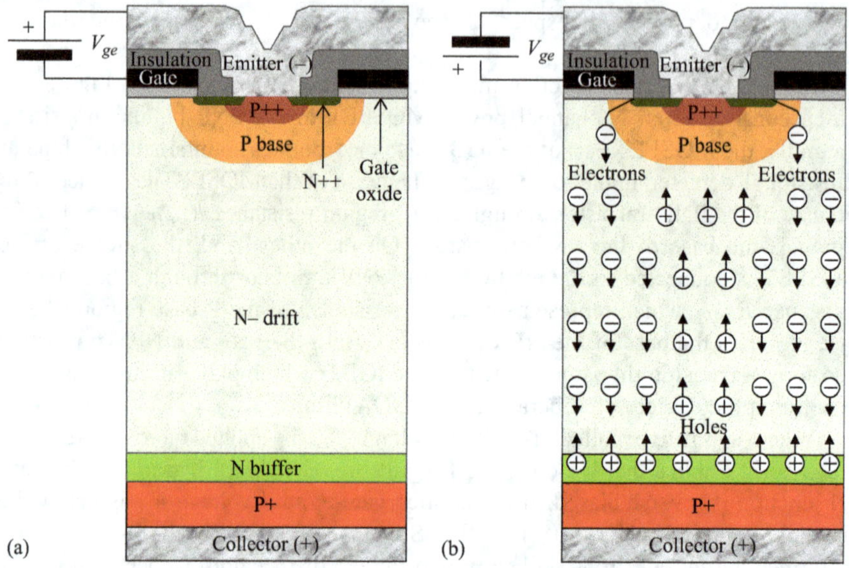

Figure 5.3 The IGBT in OFF state (a) and ON state (b)

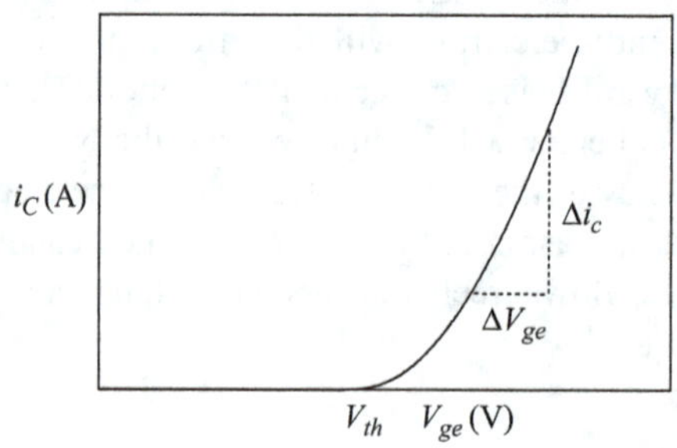

Figure 5.4 The IGBT IV transfer characteristics

5.3 The IGBT static characteristics

The typical transfer IV characteristics of an IGBT are shown in Figure 5.4, where the transconductance g_m is defined as:

$$g_m = \frac{\Delta i_c}{\Delta V_{ge}} \tag{5.1}$$

The IGBT static IV characteristics are shown in Figure 5.5. The IGBT operation during the static or steady-state conditions can be divided into four regions of operation:

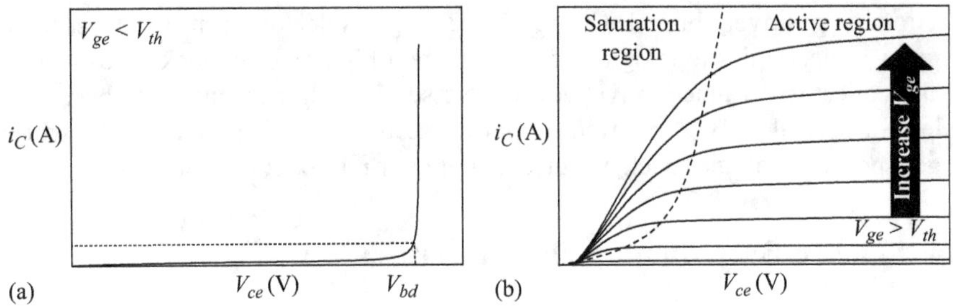

Figure 5.5 The IGBT IV static characteristics in OFF-state/blocking characteristics (a) and ON-state/output characteristics (b)

(a) Blocking or cut-off region (OFF) at ($V_{ge} < V_{th}$), ($V_{ce} > 0$) and ($i_c = i_{leakage}$).
(b) Active or linear region ($V_{ge} > V_{th}$), ($V_{ce} > (V_{ge} - V_{th})$) and ($i_c = i_{c(sat)} = g_m$ $(V_{ge} - V_{th})$).
(c) Saturation or Ohmic region (ON) at ($V_{ge} > V_{th}$), ($V_{ce} < (V_{ge} - V_{th})$) and ($i_c = i_{load}$).
(d) Reverse blocking region (reverse bias) ($V_{ce} < 0$), not shown in the figure.

The voltage blocking characteristic is shown in Figure 5.5(a) when $V_{ge} < V_{th}$ and the device is in OFF-state. The device voltage rating is defined by avalanche breakdown at the voltage level referred to as breakdown voltage V_{bd}. The IGBT leakage current is typically low but is strongly dependent on the IGBT PNP bipolar gain β_{PNP} and temperature. When $V_{ge} > V_{th}$ as shown in Figure 5.5(b), the device can perform in two operational regions. The first is the saturation or ohmic region where the IGBT is conducting current in the ON-state. IGBTs normally operate in this mode in most applications under standard operational conditions, because the low V_{ce} drop results in low conduction or on-state losses. The collector current level i_c is defined by the load and the IGBT behaviour is as described in Section 5.2. For a given collector current, the voltage drop across the IGBT is referred to as $V_{ce(sat)}$, which is an important parameter for defining the conduction losses of the IGBT at different current levels. The second region is referred to as the active or linear region where the collector saturation current $i_{c(sat)}$ is mainly determined by the applied gate-emitter voltage V_{ge} and in accordance with the following equation:

$$i_{c(sat)} = \frac{1}{1-\beta_{pnp}} \cdot \frac{\kappa}{2} \cdot \left(V_{ge} - V_{th}\right)^2 \tag{5.2}$$

where κ is the channel conductivity.

The above-described static characteristics can also be referred to as the forward bias characteristics since the collector electrode is positively biased with reference to the emitter electrode. In relation to this biasing arrangement, the opposite can be referred to as the reverse bias characteristics when the emitter electrode is positively biased with reference to the collector electrode. Most modern power electronics topologies do not require a high reverse blocking capability of the collector PN junction. However, reverse blocking capability has been provided for special IGBTs, referred to as reverse blocking (RB) IGBT [11], which target special topologies such as current source or matrix converters.

5.4 The IGBT switching characteristics

The unipolar MOSFET is inherently faster than all bipolar devices because there are no minority excess carriers that must be removed during the switching transients. Nevertheless, apart from the tailing of the collector current during turn-off, the IGBT has similar overall switching characteristics to those of the MOSFET due to the IGBT emitter MOS cell structure and voltage-controlled gate drive principle. For both devices, the switching characteristics are heavily influenced by the internal voltage-dependent capacitors, which are shown in Figure 5.6 and described in the following paragraph.

The gate-emitter capacitor C_{ge} is a combination of the electrostatic capacitance of the oxide layer in series with the capacitance of the depletion layer forming from the SiO_2–Si interface. The collector-emitter capacitor C_{ce} represents the depletion layer capacitance in the drift region and has, in principle, no serious effect on the switching characteristics of the IGBT as it is not a part of the gate circuit loop. The most significant capacitance affecting the switching characteristics is the gate-collector capacitor C_{gc}, which represents the extended drift region depletion layer

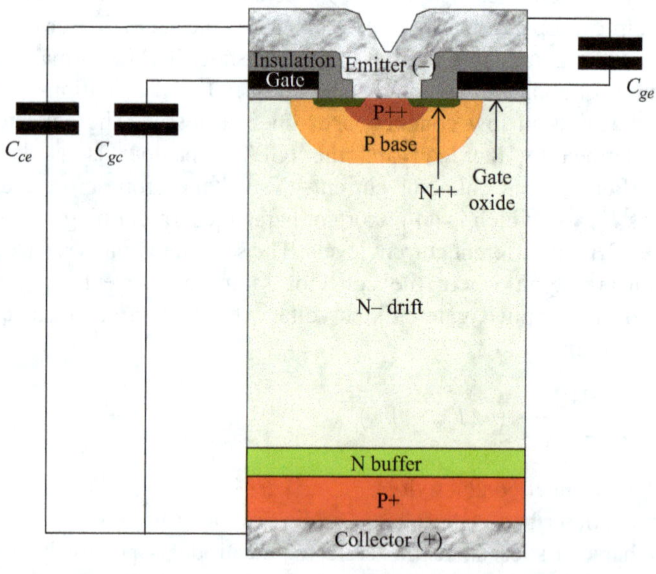

Figure 5.6 Origin of the internal capacitances that govern the switching characteristics of IGBTs

capacitance forming at the SiO_2–Si interface. The magnitudes of these three capacitors depend mainly on the geometry and topology used in the IGBT chip. Thus, higher current devices have larger capacitances.

In electrical terms, the capacitors described above are strongly dependent on the collector-emitter voltage V_{ce} and in general ($C_{ge} > C_{ce} > C_{gc}$). The switching characteristics are largely influenced by the change in values for these capacitors as a result of the change in V_{ce}. The voltage change across C_{gc} is much larger than that across C_{ge} and, usually, C_{ge} is assumed constant. The value of C_{gc} is usually approximated into two values C_{gc1} ($V_{ce} < V_{ge}$) and C_{gc2} ($V_{ce} > V_{ge}$), where ($C_{gc1} \gg C_{gc2}$).

The capacitive effects can also be given in terms of the input capacitance C_{iss}, the output capacitance C_{oss} and the transfer capacitance C_{rss}, where

$$C_{iss} = C_{ge} + C_{gc} \tag{5.3}$$

$$C_{oss} = C_{ce} + C_{gc} \tag{5.4}$$

$$C_{rss} = C_{gc} \tag{5.5}$$

To analyse the switching behaviour of the IGBT in a practical application, the chopper circuit, shown in Figure 5.7, is utilised and the waveforms are shown in Figure 5.8. A voltage V_{gg} is initially applied to the gate and the IGBT is ON, with a finite current i_{load} flowing through the circuit. When the gate voltage is removed, the IGBT turns off as the collector current i_c drops to zero and the IGBT supports the full voltage across the collector-emitter terminals. The voltage overshoot generated by the inductive load at turn-off will normally be high enough to reach close

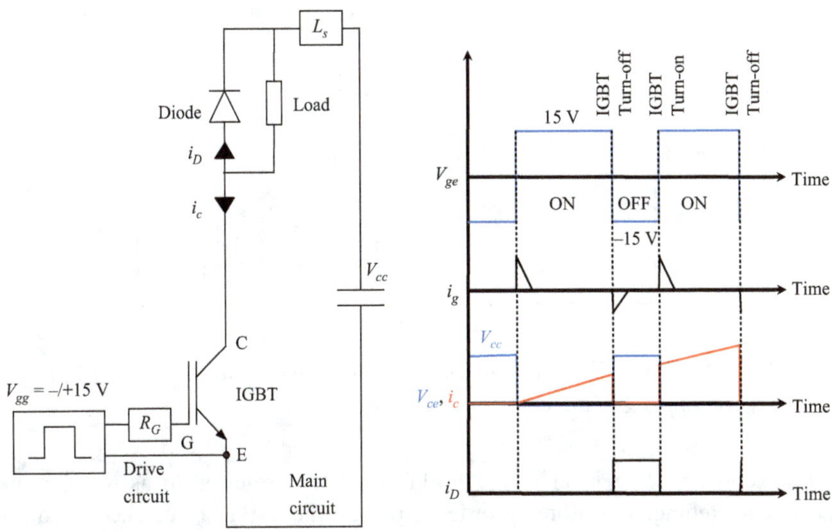

Figure 5.7 IGBT and freewheeling diode in a chopper circuit

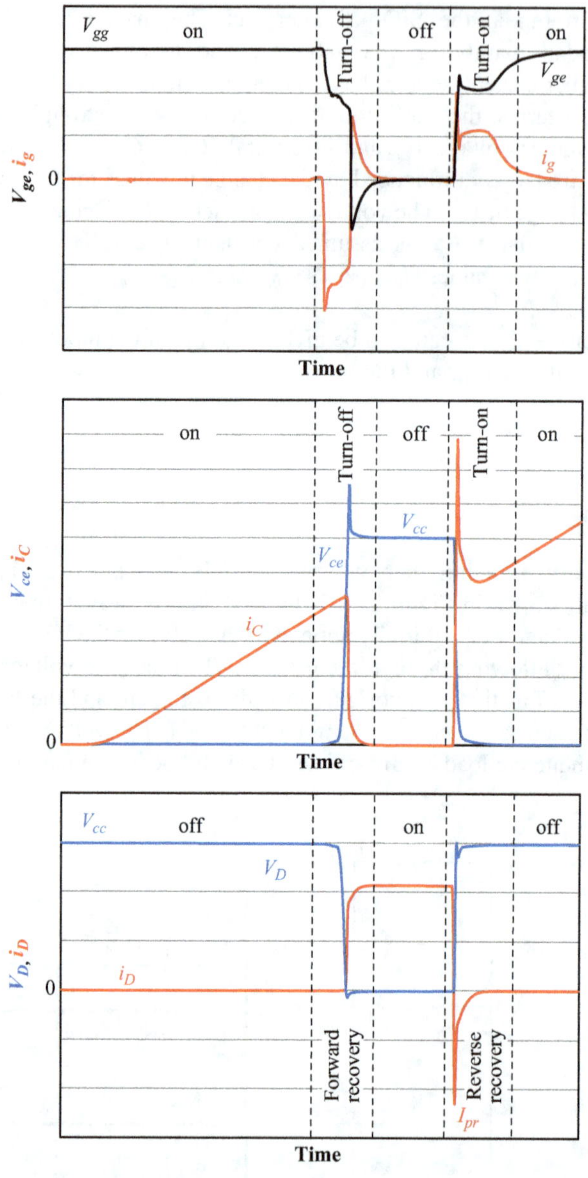

Figure 5.8 IGBT and freewheeling diode current and voltage waveforms in a chopper circuit

the breakdown voltage level of the IGBT in what is referred to as nonclamped inductive switching. Therefore, in order to protect the IGBT, a freewheeling diode is used to clamp the maximum device voltage to the supply voltage level V_{cc}. The stored load energy is dissipated after turn-off as a result of the current that flows in the freewheeling diode and load. Normally, the small resistive component of the freewheeling loop causes the inductive current to decay very slowly. When the positive gate voltage is reapplied, the IGBT current rises while the diode current falls at equal rates *di*/*dt*. The voltage across the IGBT also falls to the on-state value $V_{ce(sat)}$, whereas the diode supports the full supply voltage V_{cc}. Both the IGBT and diode suffer considerable power losses during the turn-off and turn-on switching periods.

The switching characteristic of the IGBT and a freewheeling diode in an inverter with an inductive load are strongly influenced by a number of different elements. During turn-off, the switching losses are mainly dependent on the drive circuit parameters, main circuit parameters and IGBT turn-off characteristics; while during turn-on, in addition to the above, the power losses depend greatly on the reverse recovery characteristics of the freewheeling diode. An important factor is that during the switching transient, the IGBT operates normally in the active region. Hence, the change in V_{ce} would result in a large increase in C_{gc} due to the Miller capacitance effect C_M. This large capacitance would dominate the switching characteristics of the IGBT and freewheeling diode during both the turn-on and turn-off periods [12], as discussed in the following paragraphs.

5.4.1 Turn-on transient

Figure 5.9 shows the voltage and current waveforms associated with the IGBT turn-on transient for the chopper circuit shown in Figure 5.7. The electric field and excess carrier concentration profiles in the IGBT during the turn-on transient are shown in Figure 5.10. The driving circuit elements include the gate resistance R_G and the applied gate voltage V_{gg}. The main circuit parameters include the supply voltage V_{cc} and the circuit stray inductance L_S.

It is important to note that ($C_{gc} = C_{gc2}$) in the analysis unless stated otherwise. The turn-on period can be divided into four phases depending on the IGBT modes of operation and the freewheeling diode recovery characteristics:

Delay phase (1): Initially, the IGBT is operating in the blocking or cut-off region and supports the full supply voltage, while the diode is conducting the load current. At t_1, the voltage V_{gg} is applied to the gate-emitter terminals of the IGBT. By injecting a constant current i_g into the gate terminal, the input capacitance C_{iss} begins to charge and V_{ge} increases. The initial gate current is given as:

$$i_g = \frac{V_{gg}}{R_G} \tag{5.6}$$

When V_{ge} reaches the threshold voltage V_{th}, the gate current drops at the end of this phase, while the gate-emitter voltage V_{ge} varies according to:

$$V_{ge} = V_{gg}\left(1 - e^{-t/R_G C_{iss}}\right) \tag{5.7}$$

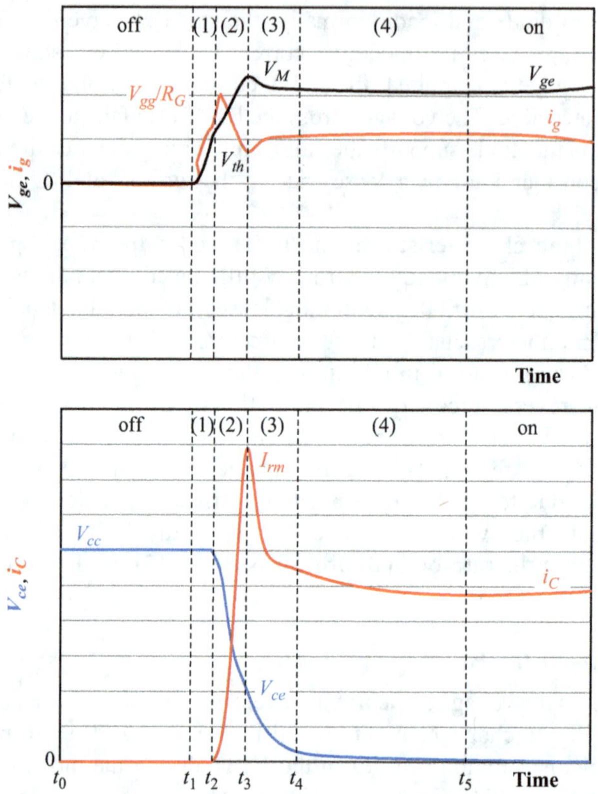

Figure 5.9 IGBT and diode interaction during IGBT turn-on

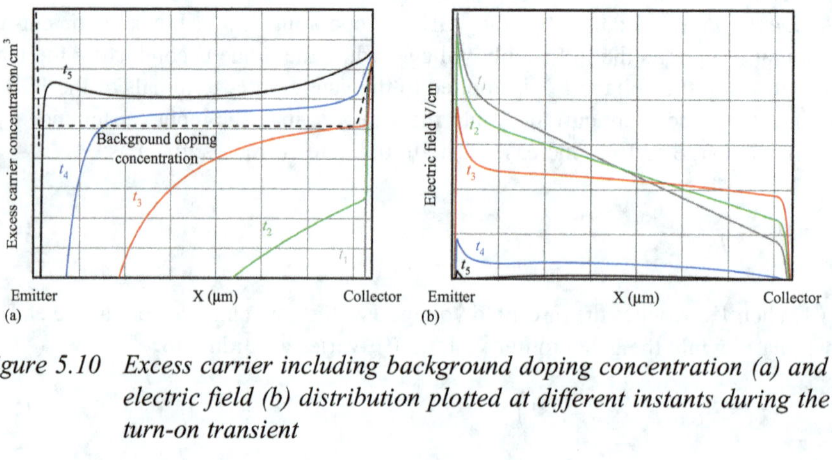

Figure 5.10 Excess carrier including background doping concentration (a) and electric field (b) distribution plotted at different instants during the turn-on transient

The time constant is $(\tau_{RC} = R_G(C_{ge} + C_{gc}))$ while the delay time is given as:

$$t_d = C_{iss}R_G ln\left(1 - \frac{V_{th}}{V_{gg}}\right) \tag{5.8}$$

Therefore, t_d is mainly dependent on V_{th}, R_G, V_{gg} and C_{iss}. The internal capacitances will have constant values during this phase as the voltage across them remains unchanged ($V_{ce} = V_{cc}$).

Commutating phase (2): This phase is initiated when ($V_{ge} = V_{th}$) and the collector current starts to increase as the IGBT operates in the active region. The commutating *di*/*dt* during this phase is the main parameter, which affects the switching speed and losses. The chopper circuit transients can be analysed by looking separately at both the input circuit (drive circuit) and the output circuit (main circuit):

$$V_{gg} = i_g(t)R_G + v_{ge}(t) \tag{5.9}$$

The collector current through the main circuit can be given in terms of the gate voltages

$$i_c(t) = g_m(v_{ge}(t) - V_{th}) \tag{5.10}$$

By differentiating, this equation would lead to:

$$\frac{di_c}{dt} = g_m \frac{dv_{ge}}{dt} \tag{5.11}$$

where

$$\frac{dv_{ge}}{dt} = \frac{i_g}{C_{iss}} \tag{5.12}$$

and, therefore,

$$\frac{di_c}{dt} = \frac{g_m(V_{gg} - v_{ge}(t))}{R_G C_{iss}} = \frac{V_{gg} - v_{ge}(t)}{R_G C_{iss}/g_m} \tag{5.13}$$

Ideally, (5.13) governs the rising rate of the collector current, stating that the commutating *di*/*dt* is only dependent on the drive circuit and IGBT internal capacitors. But this is only true if assuming that there is no stray inductance in the main circuit ($L_S = 0$); therefore, the output circuit equation during this phase is simply ($V_{ce} = V_{cc}$). For a given value of L_S, and assuming a very small main circuit resistance, V_{ce} will drop in value as a result of the inductive voltage $V_L = L_S \frac{di_c}{dt}$. Thus, the output circuit equation can be given as:

$$V_{cc} = L_S \frac{di_c}{dt} + v_{ce}(t) \tag{5.14}$$

therefore, *di*/*dt* can be written in terms of both the drive and main circuits as:

$$\frac{di_c}{dt} = \frac{V_{gg} - v_{ge}(t)}{R_G C_{iss}/g_m} = \frac{V_{cc} - v_{ce}(t)}{L_S} \tag{5.15}$$

Ideally, and assuming that C_{iss} remains unchanged, the change in ΔV_{ce} is equal to $L_S \frac{di_c}{dt}$, where *di*/*dt* is given in (5.13). Any changes in *di*/*dt* will be balanced by a change in ΔV_{ce}. This is not true because ΔV_{ce} would result in a large increase in the transfer capacitance C_{rss} or gate-collector capacitance C_{gc} due to the Miller effect. This would bring the influence of the main circuit parameters reducing the commutating *di*/*dt*, whereas the *di*/*dt* given in (5.13) only represents the maximum rate in the case when ($L_S = 0$). A large Miller capacitance would produce a feedback current via C_{gc} to the gate. This current is produced by V_{gg} and R_G given in (5.9); therefore,

$$\frac{dv_{gc}}{dt} = \frac{dv_{ce}}{dt} = \frac{i_g}{C_{gc}} \tag{5.16}$$

C_{gc} would increase due to the Miller effect to become

$$C_M = (1 - A_v)C_{gc} \tag{5.17}$$

where A_v is the voltage gain of the IGBT and is given as:

$$A_v = \frac{\Delta V_{ce}}{\Delta V_{ge}} = \frac{\Delta V_{ce}}{\Delta i_c}\frac{\Delta i_c}{\Delta V_{ge}} = -r_{ds} g_m \tag{5.18}$$

where r_{ds} is the IGBT on-state resistance and is given as

$$r_{ds} = \frac{\Delta V_{ce}}{\Delta i_c} \tag{5.19}$$

The gain can also be given as:

$$A_v = \frac{V_{ce} - V_{cc}}{V_{ge} - V_{th}} = \frac{-L_S\left(\frac{di_c}{dt}\right)}{V_{ge} - V_{th}} \tag{5.20}$$

Therefore, the input capacitance due to the Miller effect $C_{iss}{}^*$ can be written as

$$C_{iss}{}^* = C_{ge} + C_M = C_{ge} + C_{gc} + A_v C_{gc} \tag{5.21}$$

therefore,

$$C_{iss}{}^* = C_{iss} + A_v C_{gc}(v) \tag{5.22}$$

and the commutating *di*/*dt* is given as:

$$\frac{di_c}{dt} = \frac{V_{gg} - v_{ge}(t)}{R_G C_{iss}{}^*/g_m} = \frac{V_{cc} - v_{ce}(t)}{L_S} \tag{5.23}$$

and

$$\frac{dv_{ce}}{dt} = \frac{V_{gg} - v_{ge}(t)}{R_G C_{iss}^*} \tag{5.24}$$

This Miller effect will unbalance (5.24), forcing V_{ce} to drop to a certain level and at a certain rate *dv*/*dt* both of which are mainly dependent on the initial value of C_{gc}, the drive circuit and main circuit parameters. Variations in either the main or drive circuit fixture can vary *di*/*dt* independently while ensuring that both sides of (5.23) must equal each other. The main circuit is adjusted through V_{ce} while the driving circuit is adjusted through V_{ge}. It is important to emphasise that C_{gc} is a function of voltage. However, for high collector voltages, any change in the capacitance due to a change in the voltage would be negligible without the Miller effect; otherwise, any small change would be amplified and must be taken into account. This is important when considering the effect of the applied voltage on the commutating *di*/*dt*. Normally, the gain has a value >100, which would result in C_{iss}^* being 20 times larger than C_{iss}.

Charging the Miller capacitance depends mainly on the gate current. At later stages of the commutating phase, the Miller capacitance starts to decrease as the rate of change in V_{ce} is decreased. The Miller effect can also be seen clearly with the lower rate of change for V_{ge} during the commutating phase which is given as:

$$\frac{dv_{ge}}{dt} = \frac{i_g}{C_{iss}^*} \tag{5.25}$$

This phase ends as the freewheeling diode begins to block the supply voltage and the turn-on characteristics are dictated by the recovery characteristics of the freewheeling diode and C_{gc}. The collector current or recovery current reaches its peak value I_{pr} occurring when the inductance voltage is equal to zero or ($di/dt = 0$).

Fall phase (3): The IGBT remains in the active region and V_{ce} continues to fall, whereas the diode current enters into the recovery phase. Harder recovery characteristics would mean that V_{ce} would fall more rapidly. A higher *di*/*dt* during diode recovery would result in a higher dv_{ge}/dt providing an additional current to discharge C_{gc} in addition to the gate current. The Miller effect comes into account with a larger value than that during the commutating phase due to a lower collector voltage across C_{gc}. In addition, the diode voltage would reach its peak recovery value during this period. The phase ends when the diode recovery phase is over and the collector current reaches its stable load current value. The IGBT operation enters the saturation or ohmic region where $V_{ce} < V_{ge}$. Note that the diode might fully recover before the IGBT ends its operation in the active region. This would result in another phase with a lower dv_{ce}/dt as a result of the disappearance of the additional current provided by the diode recovery current.

Tail phase (4): During this phase, the collector-emitter voltage continues to fall at a lower rate corresponding to a large capacitance C_{gc1}, whereas V_{ge} remains constant at the Miller plateau level V_M. Once V_{ce} has dropped to its on-state value, V_{ge} becomes unclamped and continues to rise to V_{gg}, simultaneously the gate current decreases to zero where

$$\tau_{RC} = R_G(C_{ge} + C_{gc1}) \tag{5.26}$$

During turn-on, the switching losses depend mainly on *di/dt* that influences the peak current due to the diode recovery. The total switching power in the IGBT and diode is given as:

$$P_{Diode} = i_{diode}(t)v_{diode}(t) \tag{5.27}$$

$$P_{IGBT} = i_c(t)v_{ce}(t) \tag{5.28}$$

The peak value of the diode power losses occurs during the fall phase approximately when the reverse voltage reaches its peak value; while for the IGBT, it occurs at the end of the commutating phase or approximately when the collector current reaches its peak value.

5.4.2 Turn-off transient

Figure 5.11 shows the voltage and current waveforms associated with the IGBT turn-off transient. During this switching period, the gate charge curves are the mirror image of those during turn-on. The electric field and excess carrier concentration profiles in the IGBT during the turn-on transient are shown in Figure 5.12.

Delay phase (1): Initially, the IGBT is operating in the saturation region and conducts the load current, while the diode is supporting the full reverse voltage. At t_1, the gate voltage V_{gg} is removed and V_{ge} starts to decrease at a time constant $(\tau_{RC} = R_G(C_{ge} + C_{gc}))$. The delay phase ends when V_{ge} falls to the Miller voltage level V_M. During this phase, there is basically a very little variation of the device voltage V_{ce} and current i_c.

Miller phase (2): This phase is characterised in that the channel inversion layer is diminishing while V_{ge} is nearly constant in the Miller plateau V_M, where initially the collector-emitter voltage V_{ce} only rises. As the gate voltage approaches the threshold level, the IGBT enters the active region as the electron current ceases and the IGBT begins to block the full voltage. Therefore, V_{ce} increases rapidly whereas the collector current continues to conduct the load current. This phase ends when $(V_{ce} = V_{cc})$ and the collector current starts to fall.

Fall phase (3): During the current fall, two distinct phases can be seen. The fall phase is similar to the current fall of a power MOSFET and the parameters influencing the *di/dt* are the same parameters influencing the *di/dt* during turn-on but with a negative rate, which results in an inductive voltage overshoot as given in (5.14). The current fall rate is dependent on the transconductance g_m value.

Tail phase (4): After the fall phase, the IGBT presents a different behaviour from the MOSFET, which is related to a bipolar structure as the current in the IGBT decays with a tail, which is strongly dependent on the IGBT technology used. The tail phase occurs due to the minority carriers stored in the drift region. Thus, the tail amplitude depends on the collector injection efficiency, device thickness

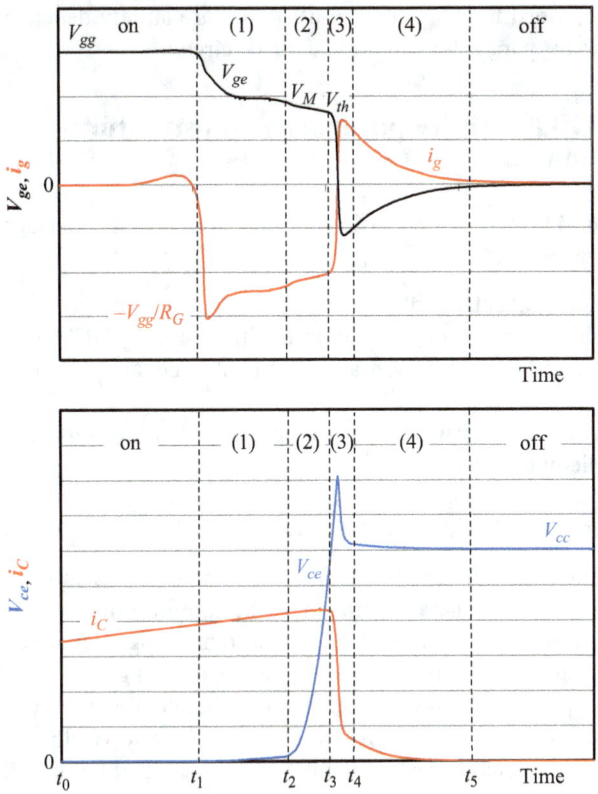

Figure 5.11 IGBT and diode interaction during IGBT turn-off

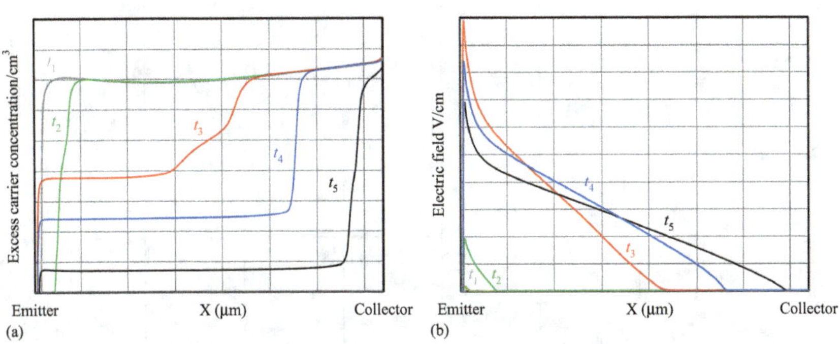

Figure 5.12 Excess carrier (a) and electric field (b) distribution plotted at different instants during the turn-off transient

and minority carrier lifetime. The turn-off losses are largely dependent on the tail current, the circuit parameters and on the drive circuit.

5.5 The IGBT main requirements and structural evolution

In order for IGBTs to function efficiently and reliably, it must demonstrate the following characteristics:

- low on-state/conduction losses,
- low switching losses and good controllability during IGBT turn-on,
- low switching losses with soft and controllable behaviour during IGBT turn-off,
- high turn-off current capability (safe-operating-area or SOA),
- controllable and robust short circuit characteristics,
- good current sharing between paralleled devices and
- stable blocking behaviour and low leakage current at elevated temperatures.

Since strong trade-offs exist to achieve the above targets simultaneously for a given design, the device characteristics must be carefully matched with the application requirements for a given voltage and current rating. Nevertheless, the main development trend has always been to focus on reducing the overall losses and especially the conduction losses. Hence, the IGBT structure evolved over the years by the continuous demand for higher power levels and lower losses while maintaining controllable and robust switching performance. Figure 5.13 shows the evolution of IGBTs since their first commercial products appeared around the early nineties [13].

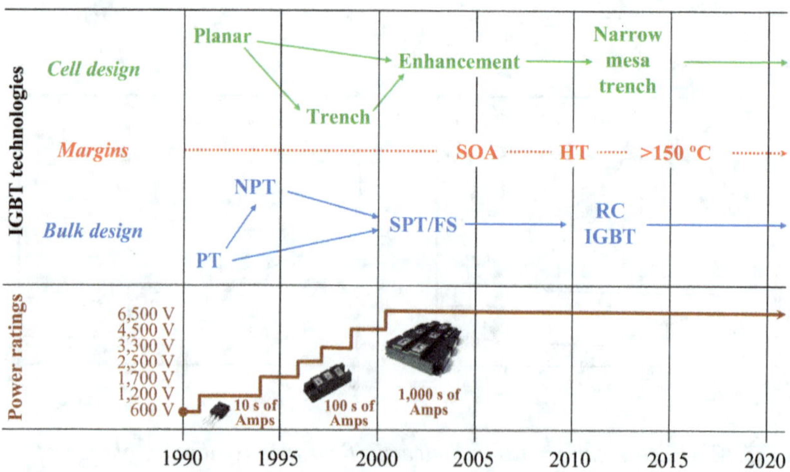

Figure 5.13 Evolution of IGBT technology

The structural development of the basic IGBT design can be divided into bulk (including drift, buffer and anode/collector) and MOS cell platform optimisation, where both target lower losses through drift region thickness reductions and excess carrier enhancements respectively. However, it is worth noting that during the 1990s, extensive development efforts were underway to replace the IGBT with lower loss, MOS-controlled thyristor-based structures. Such concepts never materialised into mainstream products due to their complex designs and associated processes and the lack of SOA capability during turn-off and short circuit modes of operation [14,15]. In parallel, the IGBT continued to develop from both the cell and bulk design platforms while also improving the SOA capability and enabling high-temperature (HT) operation beyond the standard 125 °C levels. Today, the resulting wide range of IGBT products continues to provide applications with optimum components, which have enabled with each improved generation a clear leap in power levels. In this section, the device evolution in the past will be covered while current and future development trends towards next-generation IGBTs will be included in Section 5.8.

5.5.1 Losses reductions due to bulk optimisation

Initially, IGBTs were available in the low-to-medium power range with voltages ranging from 600 up to 1,200 V. Such devices utilised the so-called punch-through structure PT-IGBT, employing epitaxial-starting material alongside strong and uniform minority carrier lifetime reductions due to the strong and thick P+ anode substrate having high injection efficiency as shown in Figure 5.14. The PT-IGBT suffered from major drawbacks such as a strong negative temperature coefficient during on-state and oscillatory turn-off switching behaviour due to the low lifetime in the drift region. Therefore, such devices were not suitable for heavy paralleling to target high power applications requiring high current ratings, while restricting their use only in discrete packages and low current modules. A solution for this problem followed with the introduction of the non-punch-through NPT-IGBT [8] that was manufactured on a thin-silicon substrate with a weak anode having low injection efficiency and, therefore, no lifetime engineering was necessary. The NPT-IGBT shown in Figure 5.14 was one of the first IGBTs to be employed successfully in high current modules with ratings up to 2,400 A. Although, due to the thicker drift region, the device suffered from higher on-state and turn-off losses, a strong positive temperature coefficient, soft turn-off behaviour and excellent short circuit ruggedness were the main reasons preferring the NPT-IGBT in such heavy paralleling configurations.

Following the NPT milestone, device designers resorted into an optimised version of the PT-IGBT by introducing a new generation, employing the so-called soft-punch-through 'SPT' or field-stop 'FS' design shown in Figure 5.14. The new IGBT concept [16,17] had thinner N– drift regions similar to the PT design but with an optimised low-doped N buffer region and a weak anode, which also required no lifetime engineering as in the NPT-IGBT. The main difference between the SPT and PT designs is that the first has a low-doped buffer optimised to 'stop the field penetration during the blocking state while still allowing conductivity modulation to occur during conduction', a principle already employed in fast-recovery diodes

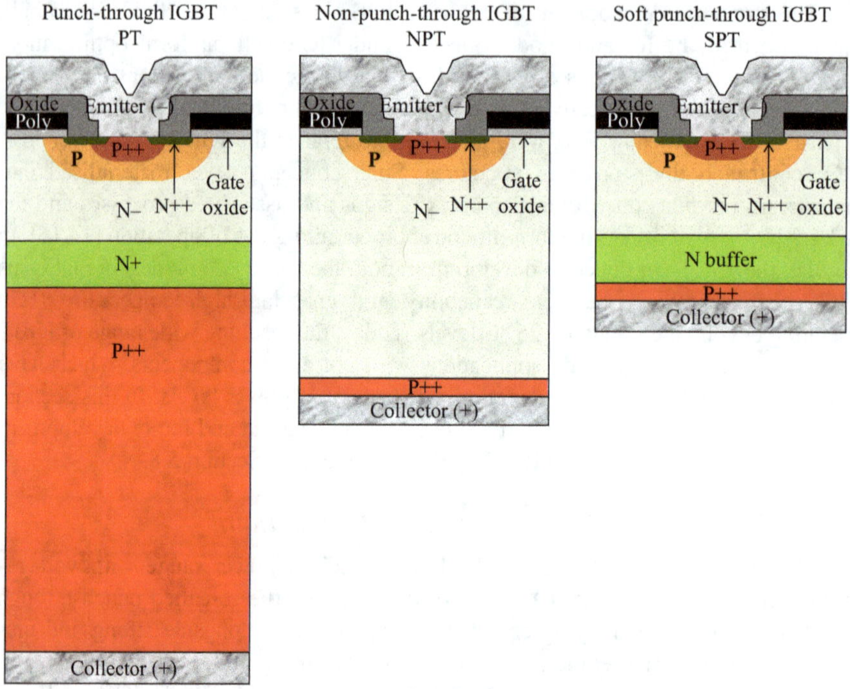

Figure 5.14 The IGBT bulk structural evolution

[18,19]. Therefore, the new SPT- or FS-IGBT combined the advantages of the low-loss PT-IGBT and the positive temperature coefficient and soft turn-off behaviour of the NPT-IGBT. For the IGBT, process challenges had to be met in terms of thin wafer processing, especially for the lower voltage-rated devices. However, the new approach allowed for the introduction of higher voltage-rated devices up to 6,500 V. Such SPT- or FS-IGBTs are now successfully implemented in high current modules exhibiting more than 20 per cent lower losses depending on the voltage class when compared to the NPT-IGBT. The doping profiles and electric field distribution are shown in Figure 5.15.

Despite all the benefits introduced with the SPT- or FS-IGBT, major design constrains surfaced since the reduction in device thickness has always been accompanied with a reduced doping concentration (i.e., higher resistivity) or the N– drift region to achieve the required blocking capability of the device and provide a low failure rate (FIT) due to cosmic rays. This trend will have the following impact on the device performance and becomes more critical when the device is operating with a higher stray inductance in high current modules due to

- reduced P-T voltage resulting in snappy/oscillatory and uncontrollable turn-off behaviour and
- large peak overshoot voltage 'stress' due to the faster *di/dt* commutation rates.

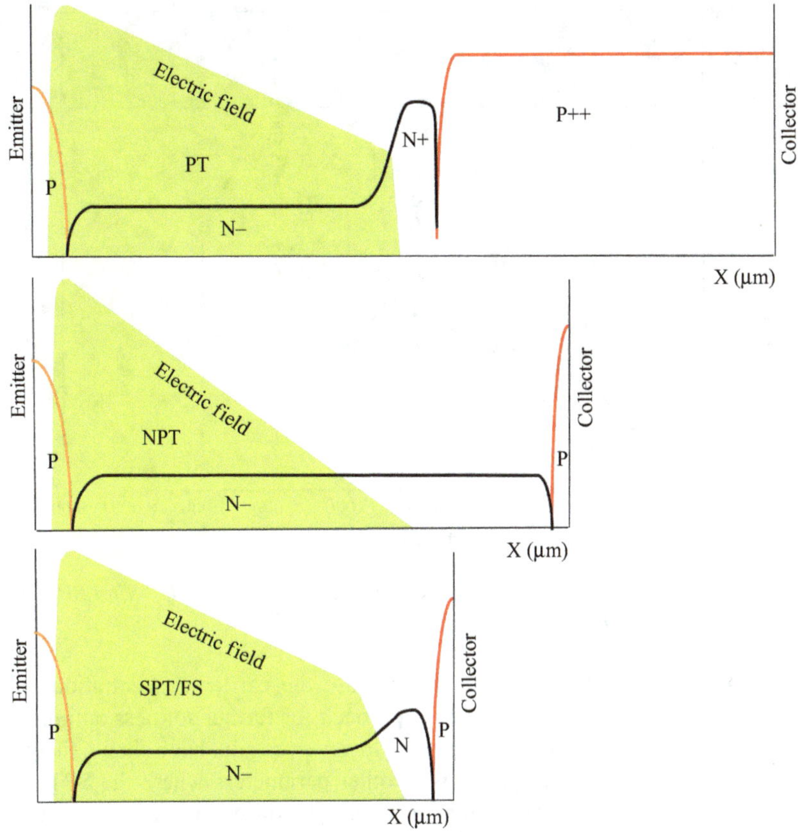

Figure 5.15 Doping profile and electric field distribution of different IGBT structures

The choice of the drift region thickness and doping concentration (resistivity) must be carefully reached to avoid or minimise the effects mentioned above. The higher the voltage rating of the device, the more critical this optimisation approach becomes due to the much lower doping concentration (higher resistivity) values and large circuit stray inductance accompanied with a high DC-link voltage. In general, the current state-of-the-art IGBTs is very close to their practical silicon design limits, and any further optimisation for lower losses will jeopardise the usefulness of these devices in many applications. The applied device thickness for modern SPT- or FS-IGBTs versus an equivalent NPT design over a whole range of voltage ratings is shown in Figure 5.16. The blocking capabilities are shown for the indicated voltage ratings per silicon design. The line indicated as the limit thickness refers to a practical design where the device required blocking capability margins and turn-off softness are not compromised.

The other important part of the bulk design is the choice of the buffer and anode doping profiles. Increasing the SPT buffer depth or employing a stronger

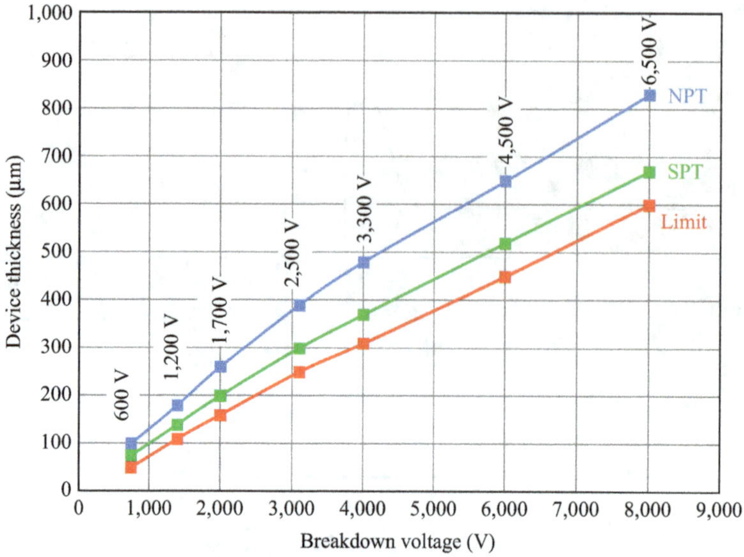

Figure 5.16 Blocking capability versus device thickness for NPT- and SPT-IGBT structures

anode with higher emitter efficiency to increase the carrier concentration in the device for softer behaviour has been one approach for further softness optimisation at the expense of increased turn-off losses. However, results have shown that there exists a strong trade-off relationship with other parameters where the SPT-IGBT behaviour under static, dynamic and short circuit conditions is strongly dependent on the PNP transistor gain under different operating conditions. These dependencies will be discussed in Section 5.7, in more detail.

5.5.2 Losses reductions due to MOS cell optimisation

For a given IGBT technology curve which represents the conduction losses versus the switching losses, optimising for lower on-state losses by adjusting the carrier concentration near the buffer/anode region will increase the turn-off losses, which is normally adjusted for the targeted frequency of operation. Therefore, an improved technology curve can only be achieved either by reducing the thickness of the N– drift region as described earlier or by lifting the excess carrier concentration near the emitter MOS cells, which will normally lead to a strong reduction of $V_{ce(sat)}$ without appreciably increasing turn-off losses.

The first IGBTs inherited planar MOS cell designs from the power MOSFET with relatively large scale dimensions and low excess carrier concentration near the emitter region. Further moderate improvements were made for the optimisation of the planar cell with finer dimensions but at the cost of lower SOA capability due to insufficient MOS cell protection by omitting p-well regions, which provided immunity against parasitic thyristor latch-up as shown in Figure 5.17.

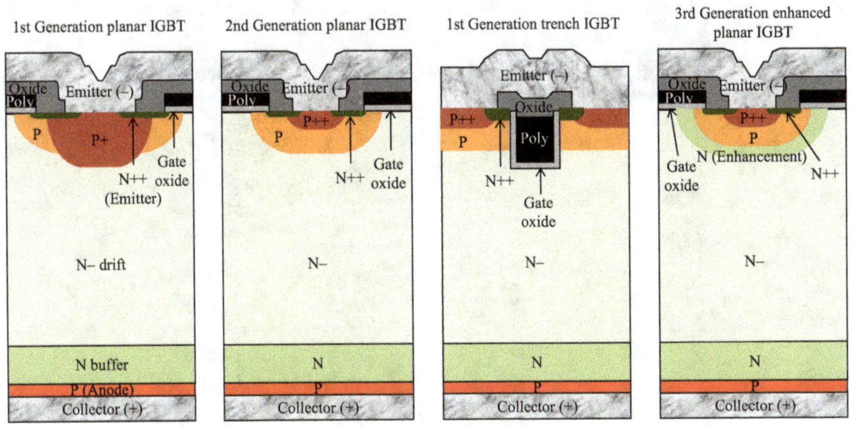

Figure 5.17 The IGBT MOS cell evolution

Nevertheless, planar designs maintained strong dependency on the following interacting effects for controlling the excess carrier concentration near the emitter:

- channel resistance, channel length and width (density),
- carrier spreading at the MOS channel,
- carrier drainage at the cells (PNP effect) and
- carrier accumulation between the cells (PNN effect).

The common aim of effective excess carrier concentration–enhancement techniques consists of decoupling the above effects as depicted in Figure 5.18. Therefore, lower losses were achieved by means of (a) fine pattern and (b) lower p-type doping concentrations in the cell for reducing the channel resistance and PNP hole drainage. On the other hand, wider cell to cell (pitch) dimensions were also employed for strengthening the PNN effect over the PNP effect [20,21]. Figure 5.19 shows a typical $V_{ce(sat)}$ optimisation curve for the planar MOS IGBT with respect to the cell to cell dimension. An optimum pitch dimension for achieving the lowest possible $V_{ce(sat)}$ level can be obtained for a given planar MOS cell design while ensuring a reasonable saturation current level for good short circuit capability. Improved IGBT technology, with lower losses due to carrier enhancement and or thickness reduction, is capable of lowering the whole curve but still maintaining the same pitch optimisation trend discussed above.

Nevertheless, the strong trade-off between the IGBT losses and SOA (during turn-off and short circuit) continued to limit the design window for further substantial reductions in the losses. Subsequently, two methods which were also originally developed for power MOSFETs were introduced for IGBTs, offering a better solution for improving the device losses.

To further reduce the on-state losses of the device through strong carrier enhancement, trench gate MOS cell emitter concepts were developed to replace the planar MOS cell [22,23]. The trench design benefits from its vertical MOS channel

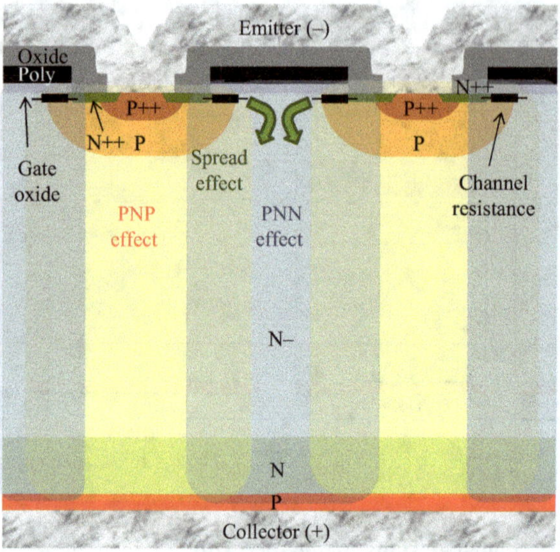

Figure 5.18 The IGBT planar MOS cell excess carrier effects

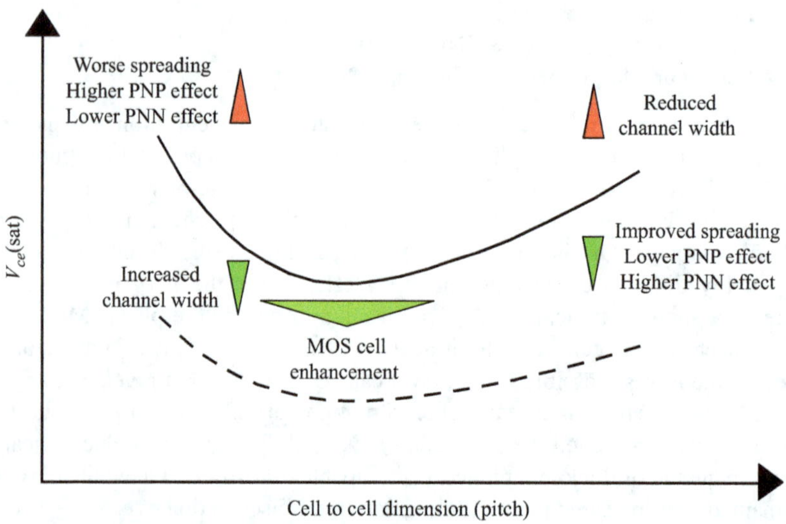

Figure 5.19 $V_{ce(sat)}$ optimisation curve for the planar MOS IGBT with respect to the cell-to-cell dimension (pitch)

for eliminating the current spreading effect and lowering the channel resistance by using highly packed narrow trench cells as in low-voltage MOSFETs. However, the IGBT trench architecture has inherently other design requirements. The desired excess carrier enhancement in the PNN region and the controlled short circuit

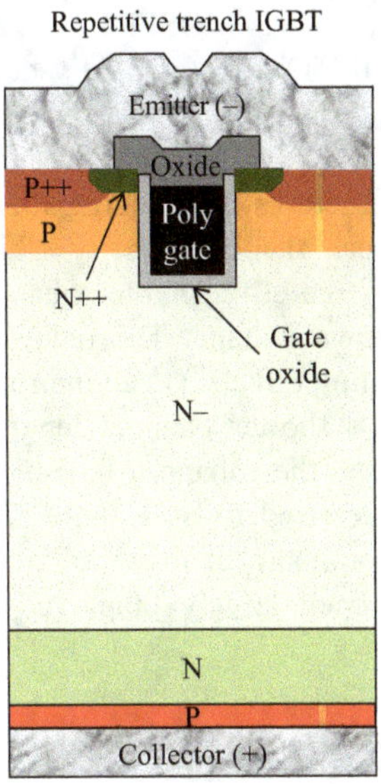

Figure 5.20 The IGBT trench MOS cell concepts

current level are not an outcome of using the classical repetitive, shallow and narrow trench designs, as shown in Figure 5.20. Therefore, wider and deeper trenches were investigated to obtain the strongest possible PNN effect despite the process challenges associated with such designs [24]. The trench MOS cell also prompted major changes in the device performance due to a large input capacitance of the trench cell compared to the planar cell [25] in addition to a number of undesirable characteristics, such as the high short circuit current levels due to the increased cell packing density. Modern trench IGBT layouts have evolved to nonrepetitive narrow trench cells separated by floating accumulation regions [26], which closely resembles the wide trench effect as shown in Figure 5.20. Trench designs with n-type enhancement layers were also proposed to further lower the on-state losses [27]. Furthermore, designers have also resorted to making many cells inactive (i.e., not contacted to the emitter) for enhancing the PNN effect while reducing the short circuit current density and input capacitance. However, both solutions no longer benefit from an effective electron accumulation layer adjacent to the MOS channel, which is very advantageous in terms of increasing the PNN effect. Therefore, when compared with the ideal wide and deep trench designs, a percentage of improvement in on-state losses had to be sacrificed. Nevertheless, today's trench IGBTs maintains a clear advantage over standard planar designs in terms of lower on-state losses but the challenge for providing a low input capacitance remains an important target for next-generation trench designs.

In parallel to trench developments, an enhanced planar IGBT concept was proposed [28] and realised to provide loss-reduction levels close to those of the state-of-the-art trench IGBTs [29]. For planar power MOSFET designs, in order to reduce the junction field-effect transistor (JFET) effect a lightly doped enhancement n-type layer (slightly higher than the N– drift doping) was placed between the cells or by encompassing the cell's P-base as shown in Figure 5.17. This method was proven to be also effective in IGBTs structures for bringing about the desired increase in the excess carrier density near the emitter. The higher the N-enhancement layer doping concentration is, the stronger the enhancement effect will be accompanied by a reduced breakdown voltage. The reduced conduction losses can be attributed to three main interacting effects: (1) an improved spreading of electrons from the MOS channel, (2) a shorter channel length and (3) an improved PNN effect between the cells due to the enhanced hole-accumulation at the periphery of the cell or what is also referred to as a 'hole barrier'. Up to recently, trench-based designs have been dominant for lower voltage-rated IGBT while the enhanced planar designs have been largely employed for the high voltage-rated devices.

5.6 Short circuit and related instabilities in IGBTs

During the operation of a power electronics converter, the IGBT is exposed to several stressful conditions which can lead to a component failure and subsequently a short circuit event. During the short circuit event, the IGBT is normally designed to withstand the overstress and successfully be turned off with the help of a desaturation protection function included in the gate driver [30]. However, in many cases, the IGBT may not withstand the short circuit event and, thus, fail. A typical IGBT current and voltage waveforms during the short circuit pulse are shown in Figure 5.21.

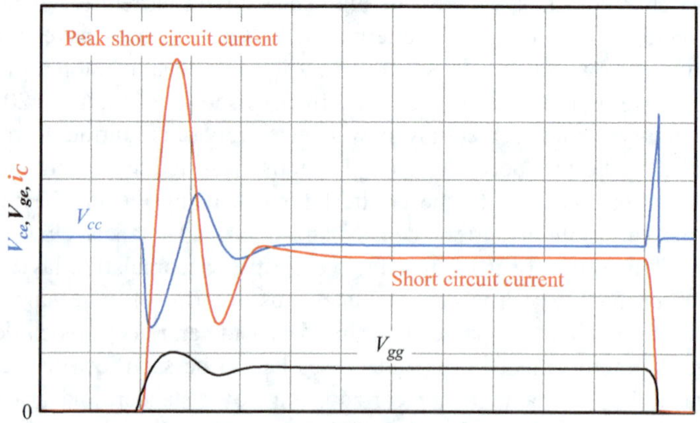

Figure 5.21 Typical IGBT short circuit pulse

In practice, the IGBT has to withstand the short circuit event under different states, which can be classified into two basic types:

- Short circuit type 1: the IGBT is turned on after a short circuit has occurred in the circuit.
- Short circuit type 2: the IGBT is conducting when a short circuit occurs.

5.6.1 Short-circuit turn-on transient

The short-circuit turn-on process of the IGBT differs from its normal turn-on process described previously in Section 5.4, as it will be highlighted in the following. Figure 5.22 shows the voltage and current waveforms associated with the IGBT turn-on process for the chopper circuit shown in Figure 5.7, when the load is short-circuited (i.e., $L_{Load} = 0$ nH). Additionally, Figure 5.23 illustrates the injection of the excess charge from the N– drift region of the IGBT and the electric field distribution across the device when the IGBT enters into a short circuit event. The turn-on period can be divided into three phases depending only on the IGBT modes

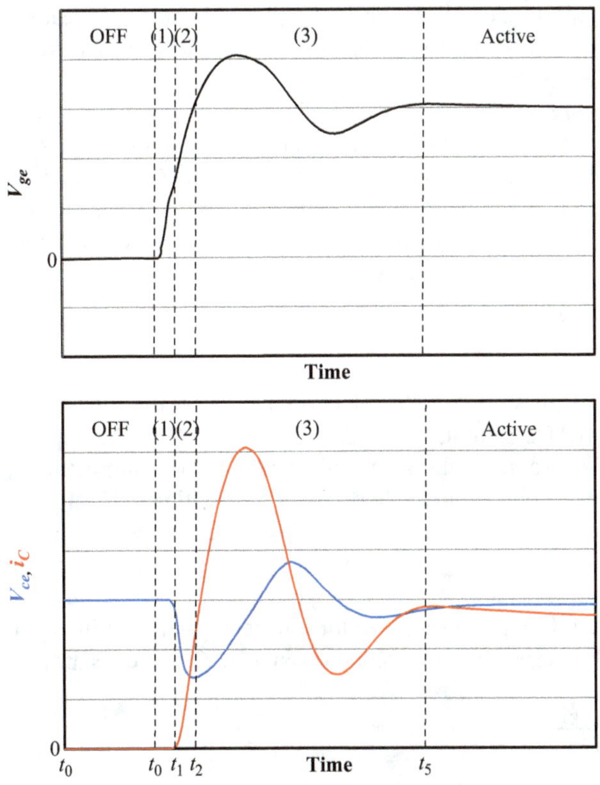

Figure 5.22 The IGBT turn-on process into a short circuit event

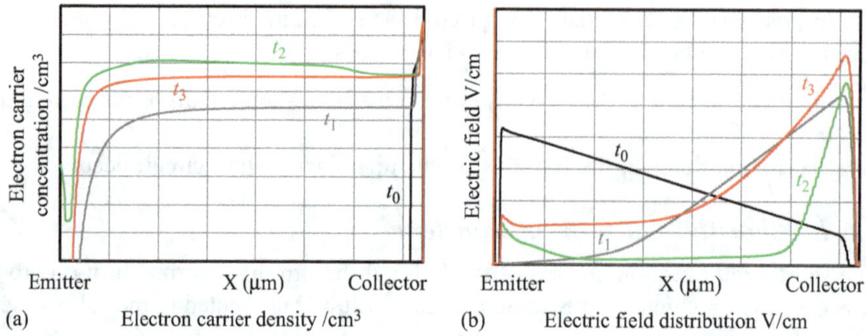

Figure 5.23 Electron carrier (a) and electric field (b) distribution plotted at different instants during the short-circuit turn-on

of operation, since the freewheeling diode does not play a role during the event of a short circuit.

Delay phase (1): Initially, the IGBT is operating in the blocking or cut-off region and supports the full supply voltage while there is already a fault in the circuit. This phase is equal to the normal turn-on transient, whose description can be found in Section 5.4.1

Commutating phase (2): This phase is initiated when ($V_{ge} = V_{th}$) and the collector current starts to increase with a higher di/dt than during normal operation, because of the low inductive path in the main circuit ($L_{load} = 0$). For a given stray inductance L_S and assuming a very small main circuit resistance, V_{ce} will drop in value as a result of the inductive voltage across the stray inductance. Thus, the output circuit equation can be given as:

$$V_{cc} = L_S \frac{di_c}{dt} + v_{ce}(t) \tag{5.29}$$

During this phase, the collector current through the IGBT rapidly increases towards the IGBT's saturation current, only limited by the applied gate-emitter voltage V_{ge}. The value of the stray inductance in the main circuit influences the collector-emitter voltage undershoot, whose variation ΔV_{ce} is typically small. Therefore, the value of the reverse transfer capacitance ($C_{rss} = C_{ge}$), which is collector voltage-dependent, does not become large. This means that the period where V_{ge} becomes clamped, which is typical for the normal turn-on process will not be observed during the short circuit transient. Therefore, the gate-emitter voltage rises with a nearly constant slope given as ($\Delta C_{iss} \approx$ constant):

$$\frac{dv_{ge}}{dt} = \frac{i_g}{{C_{iss}}^*} \tag{5.30}$$

This phase ends when the IGBT current reaches its saturation current and the voltage undershoot reaches a minimum value.

Rise phase (3): During this phase, the collector-emitter voltage starts to rise to the supply voltage level ($V_{ce} = V_{cc}$) when the IGBT current reaches an overshoot maximum value where the commutating *di*/*dt* becomes equal to zero. The collector-emitter voltage variation ΔV_{ce} due to the *di*/*dt* causes the transfer capacitance or C_{gc} to discharge causing a feedback current to flow into the gate circuit, given as:

$$\frac{dv_{gc}}{dt} = \frac{dv_{ce}}{dt} = \frac{i_g}{C_{gc}} \tag{5.31}$$

The feedback current results in an increased voltage drop V_M across the gate resistance caused by the negative feedback effect (the Miller effect):

$$V_M = i_g R_G \tag{5.32}$$

This phase ends when the IGBT reaches its final saturation current level. Since the collector-emitter voltage does not fall to the on-stage voltage level under a short circuit event, the additional phases observed during a normal turn-on transient are not here present. Following the turn-on period, the gate-emitter voltage V_{ge} remains constant at the applied gate-emitter voltage level ($V_{ge} = V_{gg}$). The IGBT operates in the active region, where the device has to withstand the saturation current and the supply voltage simultaneously.

5.6.2 *Short-circuit turn-off transient*

Figure 5.24 shows the voltage and current waveforms associated with the IGBT turn-off transient after a short circuit pulse or active duration. Additionally, Figure 5.25 illustrates the extraction of the excess charge from the N– drift of the IGBT and the electric field distribution across the device when the IGBT is turned off after a short circuit event.

Delay phase (1): Initially, the IGBT is operating in the active region and conducts the short circuit current while it blocks the supply voltage level V_{cc}. The gate voltage V_{gg} is driven negatively and V_{ge} starts to decrease at a time constant $\left(\tau_{RC} = R_G\left(C_{ge} + C_{gc}\right)\right)$. The saturation current decreases as the gate-emitter voltage falls, causing the collector-emitter voltage to rise, according to (5.14).

Fall phase (2): During the current fall, the gate-emitter voltage is still above the threshold voltage ($V_{ge} > V_{th}$). The parameters influencing the *di*/*dt* rate are the gate resistance R_G and the transconductance g_m value. The negative current rate results in an inductive voltage overshoot, whose peak is influenced by the stray inductance in the main circuit and the value of the turn-off gate resistance. The collector-emitter voltage variation ΔV_{ce} due to the *di*/*dt* causes the input capacitance to discharge.

Tail phase (3): After the fall phase, the IGBT presents a similar behaviour as its normal turn-off, since the current in the IGBT decays with a tail, which is strongly dependent on the IGBT technology used. During this phase, the gate-emitter voltage is below the threshold voltage value ($V_{ge} < V_{th}$), the collector

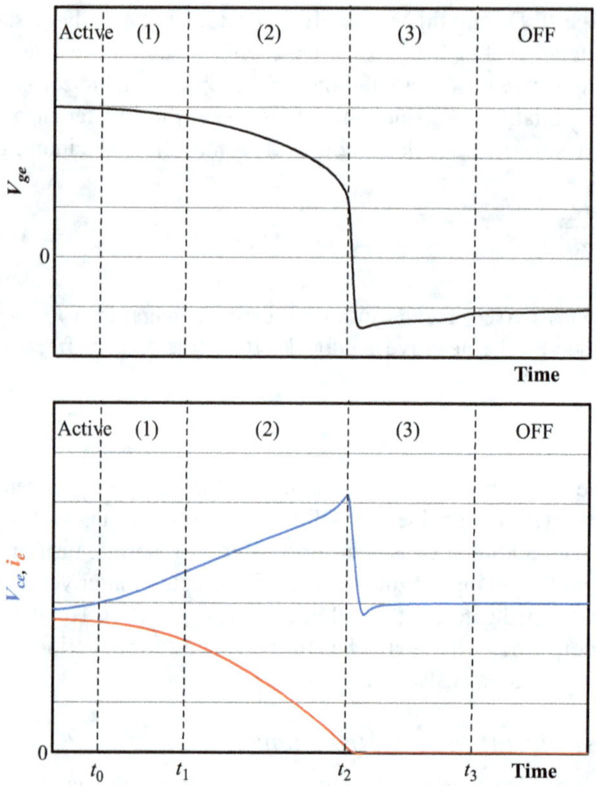

Figure 5.24 The IGBT turn-off process from a short circuit event

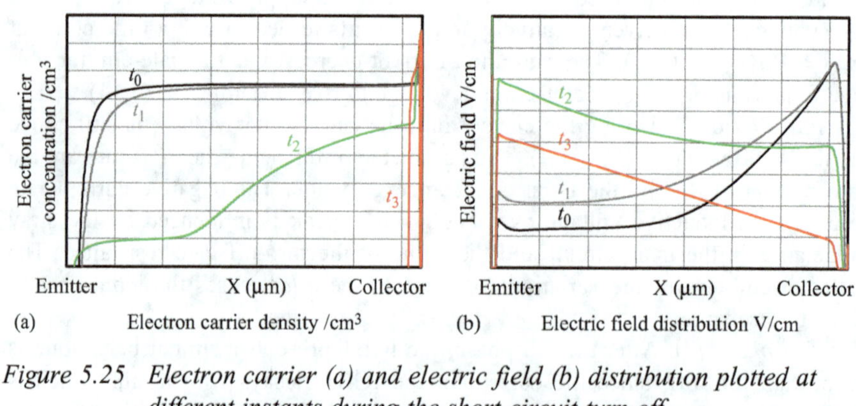

Figure 5.25 Electron carrier (a) and electric field (b) distribution plotted at different instants during the short-circuit turn-off

current falls to 0 and when the rate of *di/dt* becomes small, the collector-emitter voltage decreases to the supply voltage value ($V_{cc} = V_{ce}$).

5.6.3 Short circuit failure modes in IGBTs

In the event of a short circuit, most IGBTs have shown a good capability of withstanding high current and high voltage for a limited amount of time, which is referred to as the short circuit withstanding time t_{sc}. The high power released in the IGBT during this period leads to a rapid rise in temperature. This would lead to a significant decrease of the saturation collector current due to the reduction on the channel carrier mobility with increasing temperature. Note that the IGBT must safely operate for a maximum short circuit time t_{sc} of 10 μs, which is the standard desaturation protection time for modern gate drivers, and also represents the specified value on the device datasheet. During this t_{sc} period, the protection circuit must succeed in turning off the IGBT. Nevertheless, the IGBT could still fail due to a number of failure modes or could show unstable oscillatory behaviour before the gate driver protection is able to turn off the IGBT. Furthermore, lowering the short circuit current level is vital to avoid thermal failures after the short circuit pulse due to the high-dissipated energy in the limited volume of silicon [31]. Figure 5.26 shows the typical failure and unstable modes during and after the short circuit pulse. The first failure mode occurs during the turn-on transient, the second occurs during the short circuit pulse, the third during turn-off and the forth occurs after the turn-off transient.

- **Failure modes I and II**: When considering failure modes during or after the turn-on transient period (i.e., during the short circuit pulse), a current dependent failure mechanism, especially in SPT or field stop type IGBT structures, can be identified [32]. This type of failure occurs at a relatively well-defined,

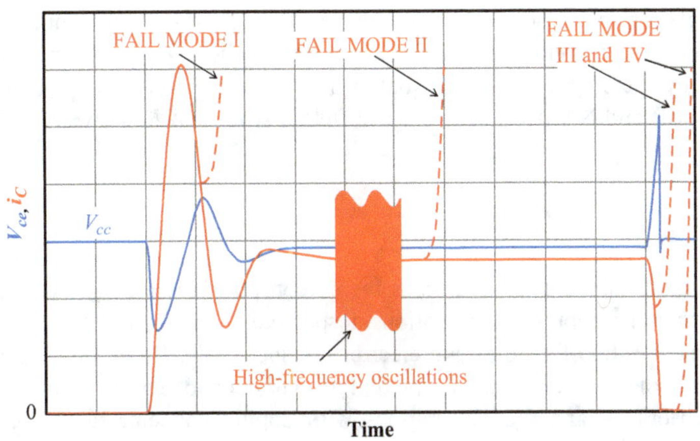

Figure 5.26 IGBT failure modes and instabilities during the short circuit pulse

design-specific current density. These failures will normally become more critical at low DC-link voltages close to the punch-through value when the space charge region is reaching to the N buffer. Therefore, they are not limited by the power levels applied during the short circuit pulse [33]. However, the higher the short circuit current, the less the short circuit withstand capability of the IGBT. Hence, for these two failure modes, lower temperatures and higher gate voltages become critical factors. Further details of the root cause of these two failure modes will be covered in Section 5.6.4.

- **Failure mode III**: This failure occurs during the short circuit pulse turn-off transient. During short circuit, the collector-emitter voltage V_{ce} is high and, therefore, the electric field will limit the presence of excess carriers in the drift region with limited dynamic avalanche occurring in the IGBT during turn-off. Hence, the IGBT can withstand higher turn-off current when compared to the normal operational turn-off transient as described in Section 5.7. Nevertheless, high turn-off overshoot voltages can still lead to device failures especially when a large stray inductance L_S is present in the main power circuit. The voltage overshoot can be reduced by using a larger gate resistance R_G during turn-off, which will reduce the di/dt rate and, therefore, the induced overvoltage.
- **Failure mode IV**: The failure type occurs after the IGBT is successfully turned off. The generated heat during the short circuit will cause a rapid temperature increase at the emitter side that will gradually dissipate towards the collector of the IGBT. Since the IGBT supports the DC-link voltage during and after the short circuit pulse, the much higher temperature levels will result in high leakage currents which could lead to thermal runaway failure many micro-seconds after the short circuit pulse. This failure mode is critical in low-voltage IGBTs due to their thin structures and, hence, small thermal capacitances.

5.6.4 Analysis of IGBT short circuit failure modes I and II

It is important to point out that power device behaviour during switching transients where both high current and voltage levels are applied simultaneously is mainly governed by the electric field distribution in the lowly n-type-doped bulk regions (e.g., N– drift and N buffer). The electric field distribution E is given by Poisson's equation:

$$\frac{dE}{dX} = \frac{q_e}{\varepsilon_s}(N_D - N_A + N_h - N_e) \tag{5.33}$$

where q_e is the electron charge, ε_s is the silicon permittivity, N_D and N_A are the fixed donor and acceptor concentrations, respectively, and N_h and N_e are the excess carrier concentration for holes and electrons, respectively. Based on the above, the electric field distribution in the n-type regions will depend heavily on the amount of excess carriers N_h and N_e. For regions with doping concentration levels below $1 \times 10^{16}/\text{cm}^3$, the excess carrier concentration at high currents can dominate the shape of the electric field. During the short circuit pulse, a high level of electrons

N_e will flow through the IGBT while will strongly compensate the background doping levels N_D, causing strong distortion of the electric field profiles, as shown in Sections 5.6.4.1. In order to compensate for the high electron charge, the short circuit will be dependent on the device design to supply enough holes with respect to the PNP transistor bipolar gain β_{PNP}, which is given as

$$\beta_{PNP} = \frac{n_{iB}^2}{n_{iC}^2} \cdot \frac{D_{pB}}{D_{nC}} \cdot \frac{L_{nC}}{\boldsymbol{W_B}} \cdot \frac{\boldsymbol{N_{AC}}}{\boldsymbol{N_{DB}}} \tag{5.34}$$

where W_B is the width of the space charge-free zone in the N buffer region, N_{DB} and N_{AC} represent the background doping of the space charge-free zone in the N buffer and the IGBT p-type collector, respectively. Also, D_{pB} and D_{nC} are the minority carrier diffusion coefficients in the space charge-free zone in the N buffer and collector, respectively. Finally, L_{nC} is the diffusion length of electrons in the collector. Hence, an optimum design of the device buffer and anode/collector regions, which provides a high bipolar gain for the compensation of the short circuit–generated electrons. If the bipolar gain β_{PNP} is low, the electron compensation during a short circuit will not be sufficient to maintain a normal electric field distribution and can result in unbalanced carrier concentrations in the N– drift region. Therefore, these carriers will modify the effective background doping as N_e governs the electric field distribution and subsequently leads a negative dE/dx slope and eventually a high electric field peak will be present near the N buffer. Such conditions will give rise to a negative differential resistance effect normally accompanied by high current filament formations that eventually can lead to the destruction of the device.

5.6.4.1 Effect of the voltage supply

To analyse the IGBT behaviour with respect to short circuit failure modes I and II, the effect of applying different supply voltages V_{cc} on the IGBT short circuit transient is shown in Figure 5.27. The IGBT is evaluated at different time instants plotted in Figure 5.28 to show the transition from blocking state into short circuit state, in terms of electric field and electron density distributions.

Before the short circuit transient, the IGBT is supporting the supply voltage V_{cc} and the electric field shows the typical triangular shape having its peak located at the emitter side. During the short circuit transient, the electric field flips as described earlier, leading to a low electric field at the emitter side and a large field at the collector side. Decreasing the supply voltage from 2 to 1 kV leads to lower electric field strength, heavily lower at the emitter side because the electric field flips during the short circuit transient.

5.6.4.2 Effect of gate voltage supply

The effect of applying different gate driving voltages V_{gg} on the IGBT short circuit transient is shown in Figure 5.29. The IGBT is evaluated at different time instants during short circuit to show the differences in terms of electric field and electron density distributions as shown in Figure 5.30.

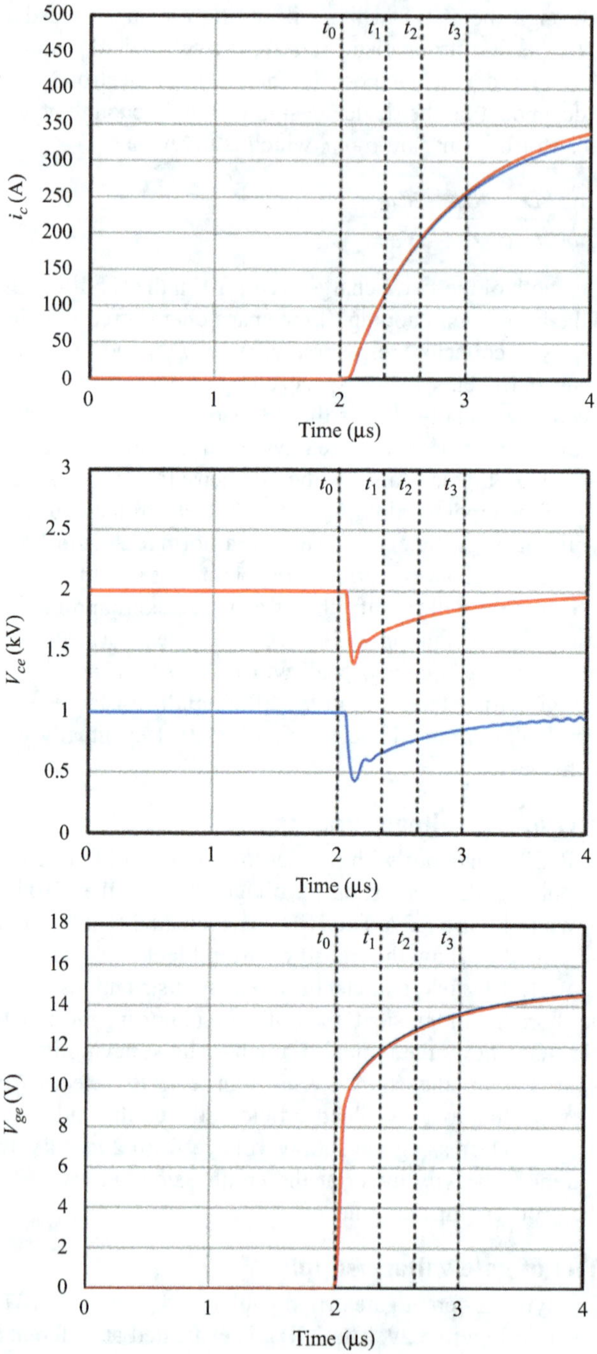

Figure 5.27 Short circuit simulations of a 3.3-kV planar SPT – IGBT at 1 and 2 kV supply voltage V_{cc}

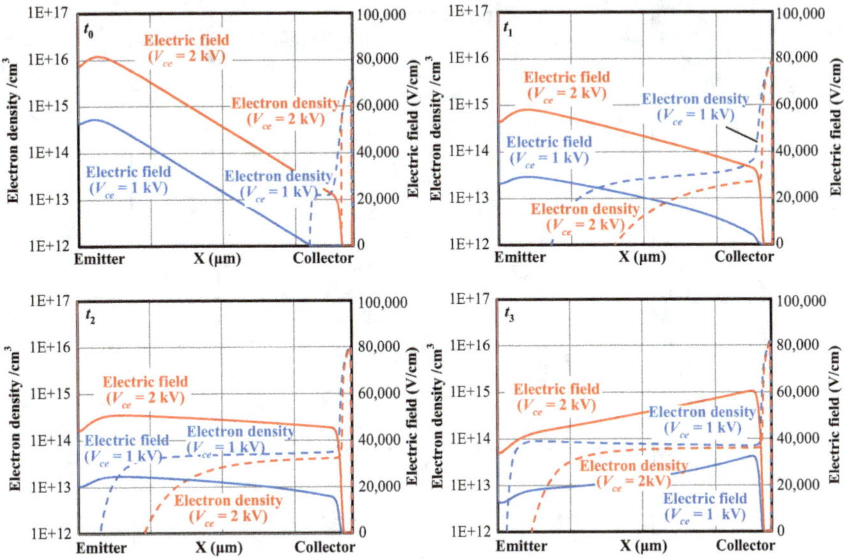

Figure 5.28 Simulated electric field (EF) and electron density concentration (e) at the time instants highlighted in Figure 5.26

Increasing the gate driving voltage from 15 to 17 V leads to a larger short circuit current and, therefore, to different current densities. The electron current injection level is stronger when driving the IGBT with $V_{GE} = 17$ V; therefore, the electric field strength is affected as can be observed in Figure 5.30. During short circuit, the low field region at the emitter due to the electric field flipping is even more lowered by the higher MOS injection current at $V_{GE} = 17$ V.

5.6.4.3 Effect of temperature

The short circuit current increases significantly at lower temperatures or as a result of gate voltage pumping during some failures experienced in the application (voltage increase during IGBT conduction leading to gate charge pumping of V_{ge} through the gate-collector capacitance). For that reason, it is necessary for IGBTs to withstand short circuit conditions under gate voltages exceeding the 15-V standard drive voltage. Increasing the short circuit capability by lowering the short circuit current of the IGBT has been the standard approach, which also aids thermally, but this is done at the expense of increased on-state losses and turn-on losses.

5.6.5 Short circuit oscillation phenomenon

One of the instabilities occurring during short circuit is the occurrence of high-frequency gate oscillations as illustrated previously in Figure 5.26. Many experimental tests show that under adverse circuit combinations (i.e., large circuit stray inductance and low gate resistance) and under certain operating conditions (i.e.,

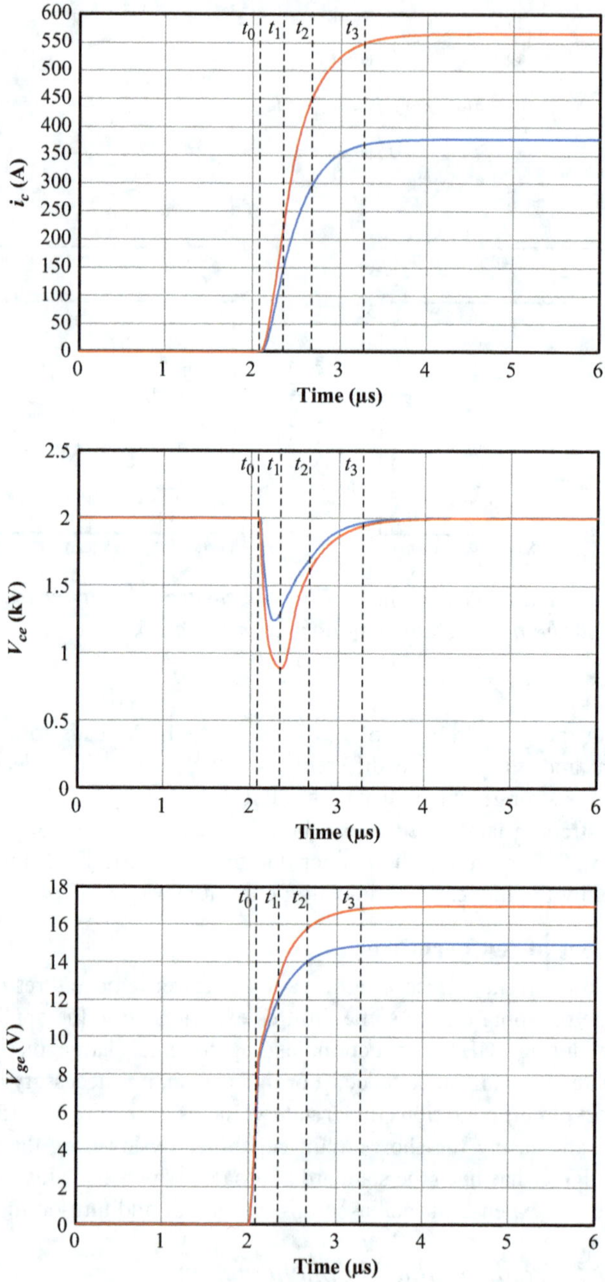

Figure 5.29 Short circuit simulations of a 3.3-kV planar SPT – IGBT at $V_{gg} = 15$ V and $V_{gg} = 17$ V

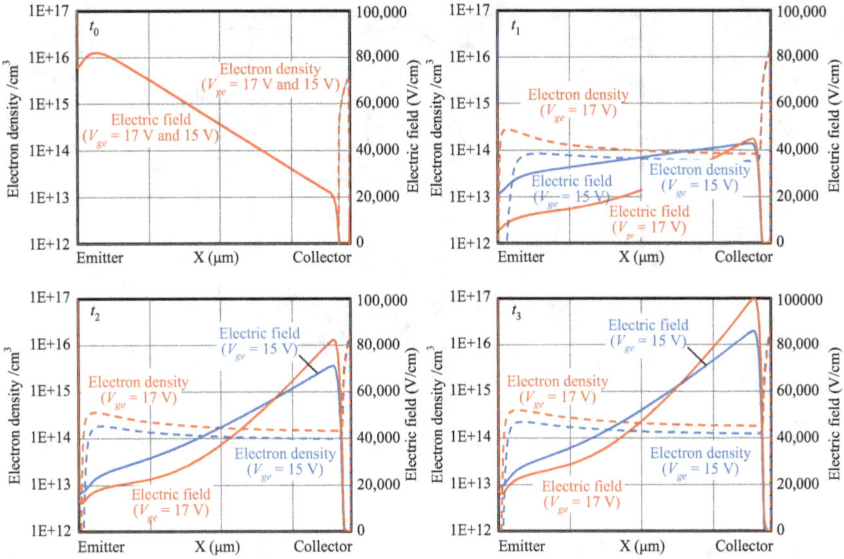

Figure 5.30 Simulated electric field and electron density concentration at the time instants highlighted in Figure 5.28

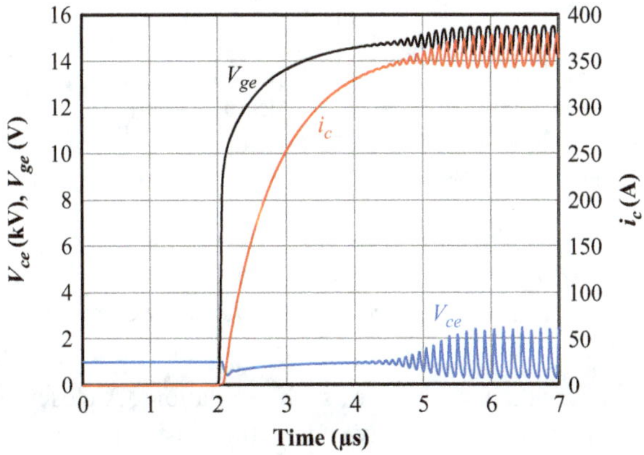

Figure 5.31 Short circuit simulation showing the oscillation phenomenon for 3.3-kV planar SPT – IGBT [34]

low DC-link voltage), IGBTs exhibit high-frequency current and/or voltage oscillations [34,35]. Such phenomena may result in high EMI levels or even cause the destruction of the gate-oxide since they could be mirrored in the gate-voltage signal resulting in high gate-overshoot voltages.

Figure 5.31 illustrates the oscillation phenomenon during a short circuit event, which serves as a basis to identify which parameters influence the short circuit

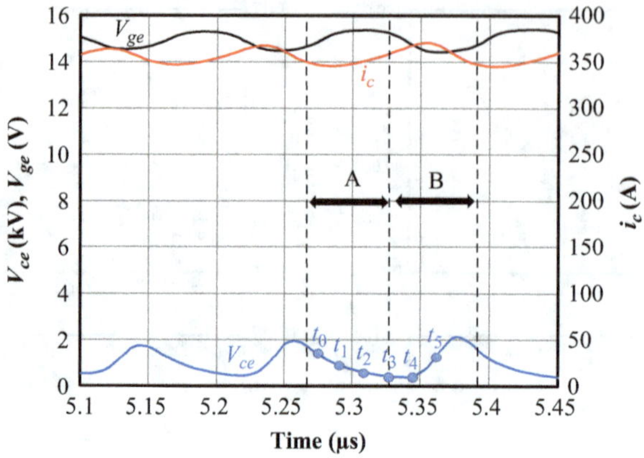

Figure 5.32 Short circuit simulation showing the oscillation phenomenon at a given time zone [34]

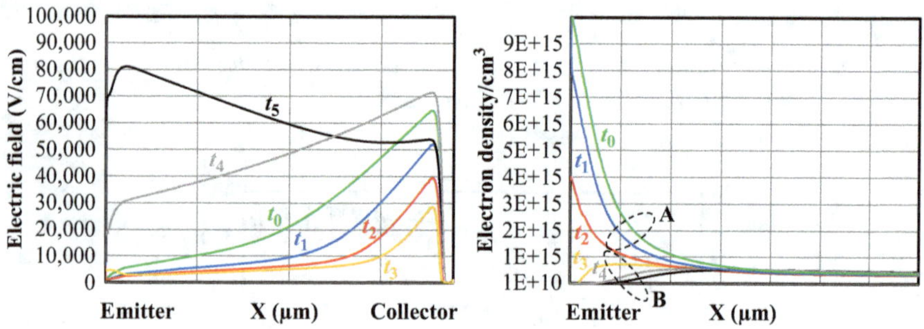

(a) Electric field distribution V/cm (b) Electron density distribution /cm³

Figure 5.33 Electric field variation (a) and electron density profile (zoom into 10 per cent of device thickness near emitter) (b) at the time instants illustrated in Figure 5.32, during the short circuit oscillations

process and later justify the root cause of such oscillations. A closer examination of Figure 5.31 is shown in Figure 5.32 between the time instants 5.1 and 5.45 μs, where the phase-shift relation of each electrical parameter is shown (i.e., V_{ce}, i_c and V_{ge}). The electric field and electron density concentration has been evaluated at different time instants during one oscillation cycle (from t_0 to t_5 in Figure 5.32) as shown in Figure 5.33. During one oscillation cycle, the electric field shows a large variation which leads to a periodic ripple of phase (A): charge storage at the time instants from t_0 to t_3, and phase (B): a charge removal effect at the time instants t_4 to t_5.

Phase A: (t_0 to t_3): A charge-storage effect at the emitter is observed which will coincide with an increase of the input capacitance, which is basically the Miller capacitance. This effect occurs when the collector-emitter voltage V_{ce} is low and, thus, the electric field is weaker, especially near the emitter MOS cell.

Phase B (t_4 to t_5): A charge-removal effect at the emitter is observed which will normally coincide with a decrease of the input capacitance (Miller). This effect occurs when the collector-emitter voltage builds up V_{ce} and, thus, the electric field is higher, especially near the emitter MOS cell.

The root cause of such oscillatory behaviour is a parametric oscillation involving the IGBT and the gate circuit. This instability is mainly dependent on the amount of excess carrier concentration and the strength of the electric field near the emitter, which in turn is dependent on the PNP bipolar gain (i.e., collector and buffer design), as described earlier where IGBTs with low bipolar gains show a low electric field at the emitter. However, in the case of oscillatory behaviour, the electric field starts to fluctuate and leads to a periodic charge storage and charge-removal effects as described above, provoking Miller capacitance variations. The time-varying Miller capacitance, combined with the stray inductance from the gate circuit, creates a parametric oscillation.

It is clear that to avoid failure modes I and II and oscillatory instabilities, IGBTs with high bipolar gain levels are preferred, but this comes at the expense of higher turn-off losses, higher leakage currents and potentially an increase in dynamic avalanche during normal IGBT turn-off as discussed in Section 5.7.

5.7 Safe operating area of IGBTs

Since the inception of the IGBT, one of the main performance challenges has been the 'SOA' capability during the IGBT turn-off especially under hard switching conditions [36]. A destruction point characterises the SOA limits whereas failures have been found to be largely related to the device design and/or process. In the past, in order to ensure that power devices and IGBTs, in particular, did not exceed their SOA limits, many restrictions were introduced for operating such devices where system designers resolved into setting many circuit and gate drive parameters accordingly. Such modifications include an increase in the gate resistance and the inclusion of protective active clamps or snubbers. This added complexity has had normally a negative impact on the performance, cost and size of power electronic systems. However, over the years, the SOA performance improved immensely with modern planar and trench MOS cell designs combined with SPT or FS buffer concepts. Furthermore, previous experience and literature have clearly pointed out that the SOA performance for higher voltage IGBTs degrades significantly when compared with the low-to-medium voltage class devices. This downtrend is due to physical constraints in high-voltage structures and the high stress-operating conditions they experience in high power applications. In addition, the trade-off between the optimisation of the overall losses and the SOA capability has imposed further restrictions in the design window of high-voltage IGBTs.

The consistent need for wide SOA limits is fuelled by many requirements in power electronics applications operating under hard switching conditions with no snubber circuit or clamps. The main aim is to provide higher power-handling capability the design optimisation for lower losses, higher operating temperatures and structural integration concepts as discussed in Section 5.7.1. At the same time, the increased SOA operational margins will also have a positive impact on the manufacturing yield, improved reliability and device controllability, ease of paralleling, reduced component count, improved system and gate-drive designs while employing optimised protection schemes aimed at reducing the total system losses.

5.7.1 Dynamic avalanche and IGBT failure mode during turn-off

Dynamic cell latch-up represents the main failure mode during the IGBT turn-off transient, especially under extreme dynamic avalanche conditions. These conditions include high current, high voltage, high temperature, large inductance and low gate-resistance values. As described in Section 5.6, the device behaviour is mainly governed by Poisson's equation (5.33). Contrary to the short circuit case where the conducting electron N_e flow dominates the electric field distribution by compensating the N– drift doping concentration, during turn-off the excess hole N_h flow towards the PN junction will result in a steeper dE/dx slope near the junction. Under such conditions, the critical electric field E_c is reached at lower voltages with high levels of electron-hole pairs being generated in what is referred to as dynamic avalanche. Figure 5.34 illustrates the carrier dynamics in the IGBT during turn-off under dynamic avalanche conditions.

As discussed previously, the IGBT consists of two bipolar internal elements, configured as an NPN and PNP transistors forming a PNPN parasitic thyristor configuration. Failures normally occur when the N+ emitter/P-base junction is forward biased due to the voltage drop generated by the high current flowing to the emitter contact. The high current consists mainly of holes generated from a

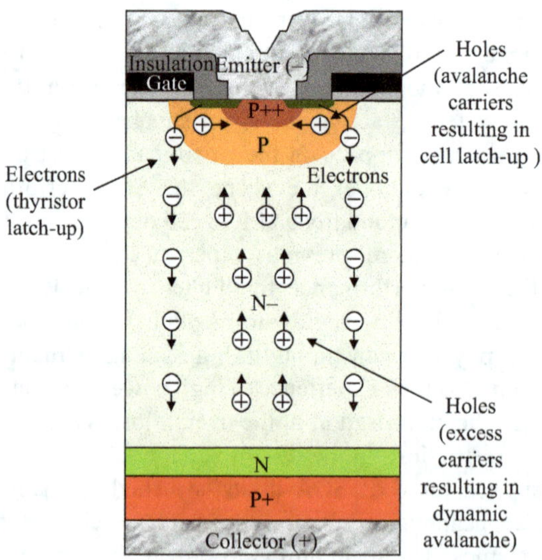

Figure 5.34 IGBT cross-section showing the parasitic thyristor triggering during turn-off

dynamic avalanche point in the PN junction. Therefore, the N+ emitter starts to inject enough electrons into the P-base region to cause uncontrolled parasitic PNPN thyristor triggering leading to device failure. Controlling the dynamic avalanche position in the main blocking PN junction and protection of the N+ emitter region by reducing the adjacent P-base resistance are the two key ingredients for increasing the cell latch-up immunity of IGBTs as shown in Figure 5.35. Figure 5.35(a) shows how a highly doped P++ protection region extends to the edge of the N+ emitter while remaining within the P-base region. Nevertheless, the avalanche point near the periphery of the P base will not provide the optimum position since the avalanche generated hole must still flow in the vicinity of the N+ emitter. The P+ well design in Figure 5.35(b) provides both protection for the N+ emitter and extends deeper than the P base for enabling a more optimum central position for a dynamic avalanche. Therefore, the majority of the avalanche-generated holes will flow directly to the emitter contact away from the critical N+ emitter unprotected regions. However, such as design will result in increased hole drainage levels and, subsequently, high on-state losses in the IGBT.

5.7.2 IGBT turn-off under SOA conditions

The early IGBT design suffered from low SOA capability, especially for high-voltage devices. The capability improved with the introduction of modern planar and trench cell designs. It was even demonstrated that under extreme test conditions [37] when the device is capable of sustaining the dynamic avalanche mode, a self-clamp avalanche mode during turn-off similar to that obtained in a standard

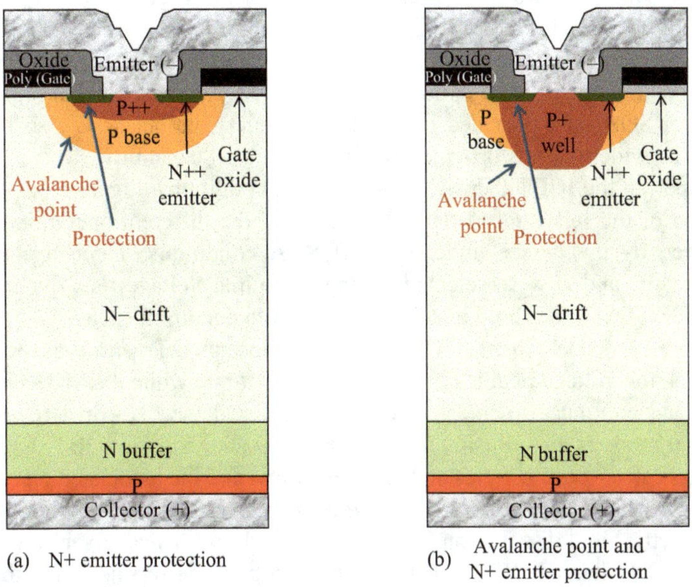

Figure 5.35 IGBT cross-sections showing the MOS cell design concepts for protection against parasitic thyristor triggering during turn-off

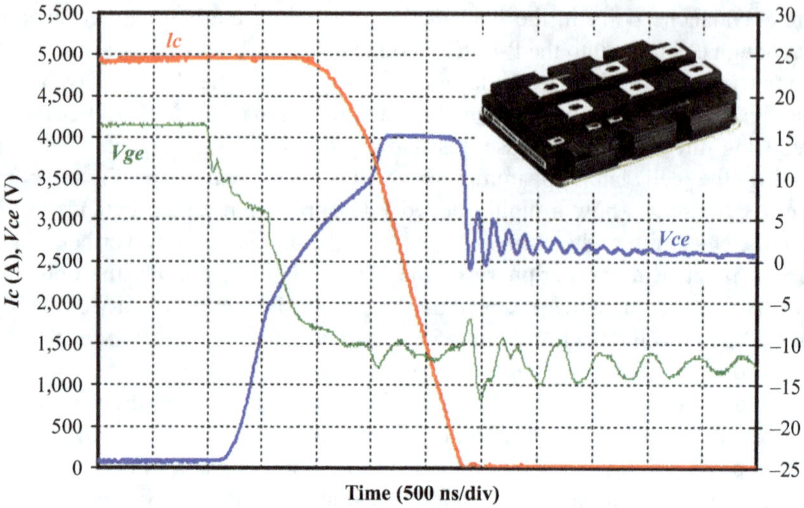

Figure 5.36 3.3 kV/1,200 A IGBT module turn-off SOA at 125 °C V_{cc} = 2,600 V, I_c = 5,000 A, R_{GOFF} = 1.5 Ω, L_s = 280 nH, energy = 20 J, peak power = 14 MW, V_{SSCM} = 4,000 V [37]

unclamped inductive test is observed [38]. We refer to this as the switching self-clamping mode (SSCM), where the device overshoot voltage reaches a value V_{SSCM} close to the static breakdown voltage of the device. Both dynamic avalanche and SSCM operational modes during turn-off are shown in Figure 5.36 for a 3,300-V IGBT module rated at 1,200 A and explained according to Figure 5.38.

This type of behaviour leads to an ultimate square SOA capability up to the device blocking voltage. The extracted square IV SOA can be seen in Figure 5.37. The typical SOA area provided in standard product data sheets is also indicated showing the wide margins provided by the device real capability.

To analyse the IGBT behaviour under these conditions, Figure 5.38 shows an illustration of the IGBT turn-off while indicating the different operational modes experienced by the device under turn-off SOA conditions. Three turn-off time instances t_1, t_2 and t_3 were selected during the initial voltage rise, dynamic avalanche and SSCM under both a stable and unstable condition each.

(Before t_1) MOS Channel ON: After the normal delay time as defined in Section 5.4, the voltage begins to rise at a high dv/dt rate while the MOS channel is still ON and continues to inject electrons into the device N– drift region. This operational mode is also illustrated in Figure 5.39 showing the IGBT doping profile, excess carrier profile, holes and electrons flow, and the associated electric field distribution. Under these conditions, the electrons and holes are fully compensated and the electric field distribution is governed mainly by N_D, as given in (5.33).

(t_2) MOS Channel OFF (dynamic avalanche): In the second stage, the IGBT goes into dynamic avalanche immediately after the MOS channel ceases to inject electrons into the N– drift region. Therefore, under such a condition, the lack of

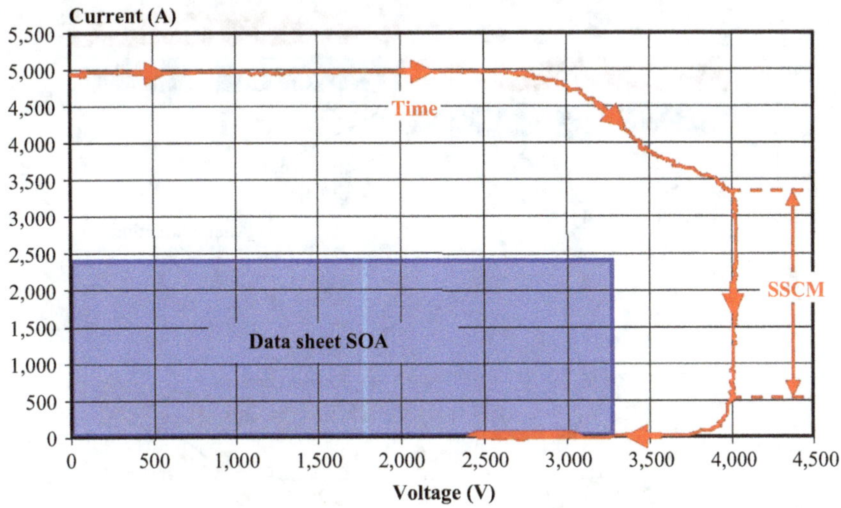

Figure 5.37 *3.3 kV/1,200 A IGBT module IV SOA curve at 125 °C $V_{cc} = 2{,}600$ V, $I_c = 5{,}000$ A, $R_{GOFF} = 1.5\ \Omega$, $L_s = 280$ nH, energy = 20 J, peak power = 14 MW, $V_{SSCM} = 4{,}000$ V [37]*

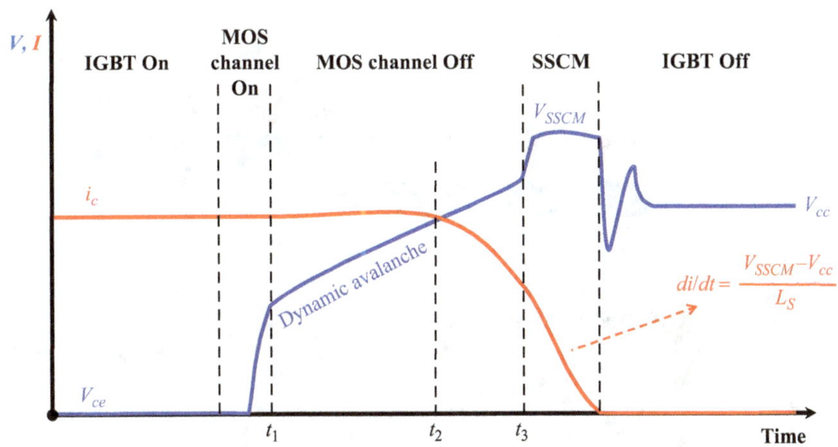

Figure 5.38 *IGBT turn-off modes under SOA conditions*

electron compensation for the recovering holes will modify the effective background doping and electric field distribution characterised by the lower *dv*/*dt* value during dynamic avalanche as shown in Figure 5.40.

The position of the dynamic avalanche source of holes is critical for the device SOA capability since these newly generated holes are mainly responsible for the parasitic thyristor latch-up. Unless a device failure occurs, the dynamic avalanche

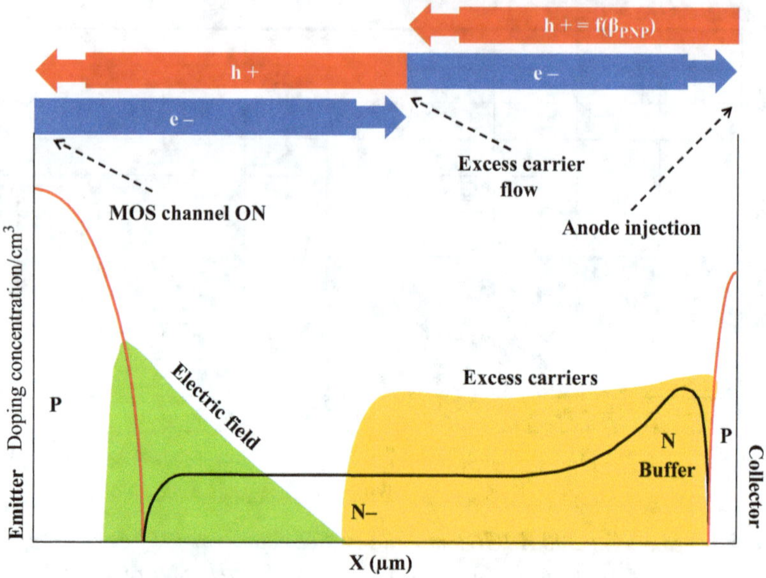

Figure 5.39 IGBT during MOS channel ON initial voltage rise phase

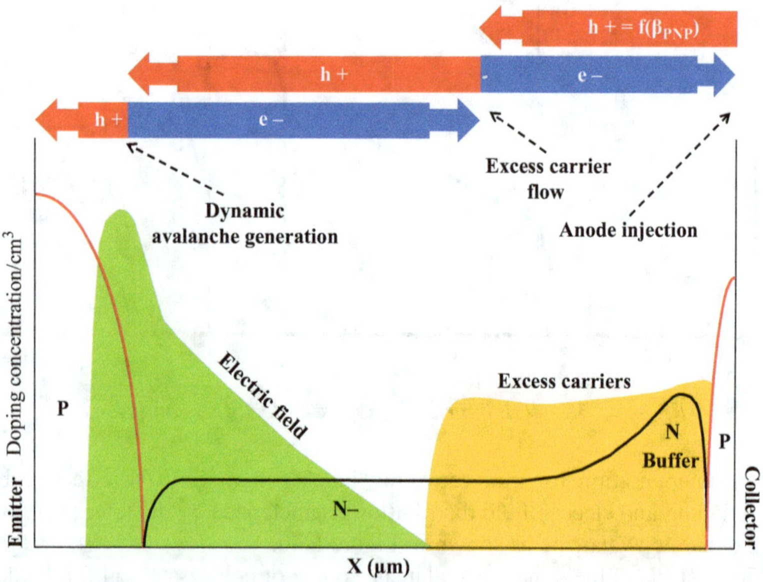

Figure 5.40 IGBT during MOS channel OFF dynamic avalanche phase

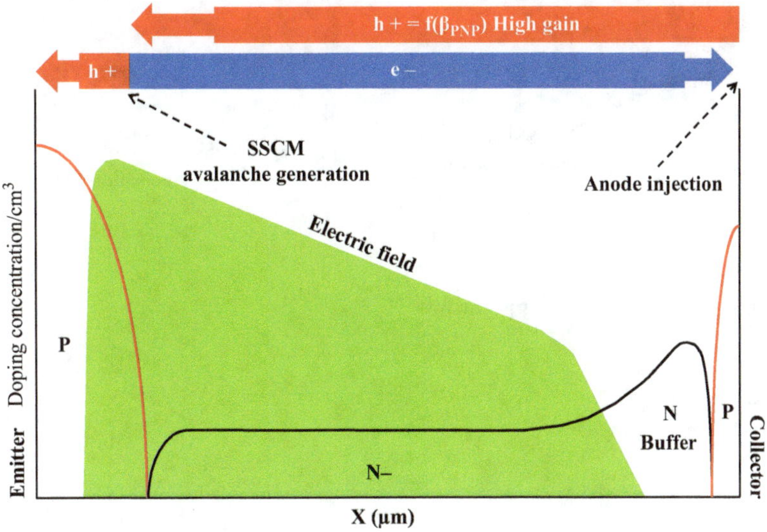

Figure 5.41 IGBT during stable switching self-clamping mode

phase continues until the remaining excess electrons and holes are used up and subsequently dynamic avalanche is suddenly eliminated.

MOS Channel OFF (SSCM): Because of the stray inductance in the commutation circuit, the voltage over the IGBT starts to rise and eventually reaches a quasi-breakdown voltage state of the PN junction. On the other hand, avalanche-generated carriers will carry the reverse current in the IGBT, as it enters into the SSCM mode by self-clamping the overshoot voltage successfully as shown in Figure 5.41. This represents a stable SSCM behaviour due to the compensation of the avalanche-generated electrons in the N– drift region. Therefore, sustaining SSCM mode of operation is fully dependent on the device design with respect to the PNP transistor bipolar gain β_{PNP}, which is given in (5.34). An optimum design of the device N buffer and anode/collector regions, which provide a high bipolar gain for the compensation of the avalanche-generated electrons, will enable the device to withstand SSCM conditions.

5.7.3 Switching self-clamp mode failure during turn-off

Very similar to the failure mode during short circuit as discussed in Section 5.6, Figure 5.42 shows the unstable SSCM mode. The only difference is that in the SSCM case, the supply of electrons is due to the SSCM avalanche-generated electrons; while during a short circuit, the electrons are supplied via the MOS channel.

During SSCM, if β_{PNP} is low, the hole N_h compensation of the SSCM avalanche-generated electrons N_e is not sufficient to maintain a normal electric field distribution shown in Figure 5.41 and will result in unbalanced carrier concentrations in the N– drift region. The electrons will modify the effective background doping as

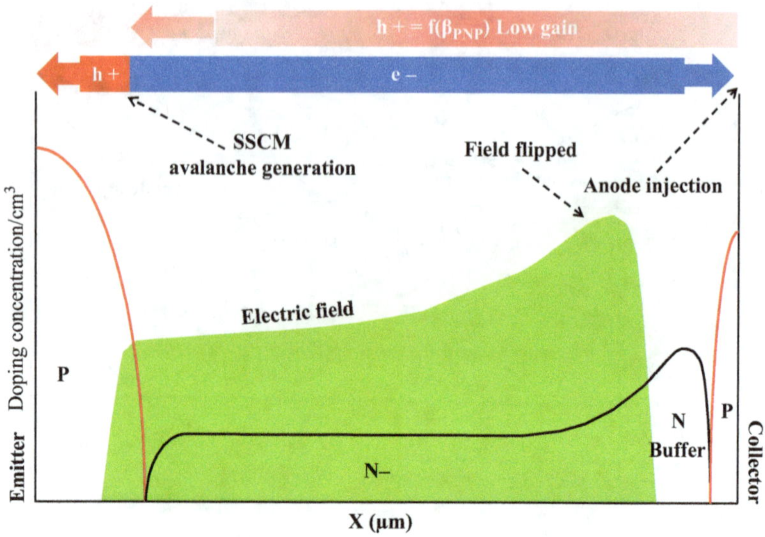

Figure 5.42 IGBT during unstable switching self-clamping mode

N_e governs the electric field distribution and, subsequently, leads a negative dE/dx slope and eventually a second and higher electric field peak will be present near the N buffer region, as shown in Figure 5.42. This type of behaviour gives rise to a negative differential resistance effect and the subsequently high current filament will form, which could eventually lead to the destruction of the IGBT [39].

5.8 IGBT development trends

For power devices in general, the main technology drivers for providing higher power was and is approached on two principle levels; first by providing higher absolute power levels by means of an increase in device area, paralleling multiple devices or by functional structure integration for a given package footprint and voltage rating. The second approach targets increased power densities for a given device-active area by means of losses reductions, higher operating maximum junction temperatures, improved thermal properties and increased SOA margins. Extensive process and design development efforts are required to enable such steps to take place. These technology drivers are illustrated in Figure 5.43; in the following paragraphs, we will define each of these trends on general terms. In addition, we will outline the current and future IGBT development trends with respect to the above-mentioned technology drivers.

5.8.1 Increase in absolute power

First, the trend towards increasing the absolute power by means of area increase, paralleling of devices or through the development of integration solutions.

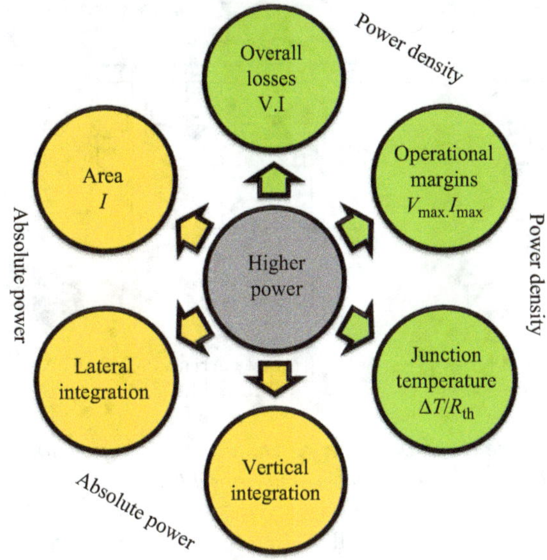

Figure 5.43 IGBT technology drivers

5.8.1.1 Area increase

MOS technologies, such as the MOSFET or IGBT, are based on producing small-size chips with active areas not exceeding 2 cm^2 in most cases. Therefore, increasing power levels by means of paralleling multiple chips in a single package or paralleling modules for a given inverter design is a standard approach for achieving the required power levels defined by the applications. Although at first glance, such an approach can appear to be less demanding from the technology point of view, it remains a fact that higher power levels by such means were only reliably possible with improved device, package and system designs. Critical parameters related to device current sharing, SOA, package layout and system parasitic elements were always in the focus to enable the IGBT to make this step at a practical level with well-defined derating rules [40]. For example, the stray inductance of the circuit is given by the converter setup. Thus, the stored inductive energy factor E_L dissipated per chip is given by

$$E_{L(Chip)} = L_S \frac{i_c^2}{2} \tag{5.35}$$

Equation (5.35) is applied normally for discrete and low-current components. However, for higher current modules with a number of paralleled chips (n), the equation for the effective inductive energy per chip is given as

$$E_{L(Chip)} = \frac{E_{L(Module)}}{n} = L_S \frac{n i_c^2}{2} \tag{5.36}$$

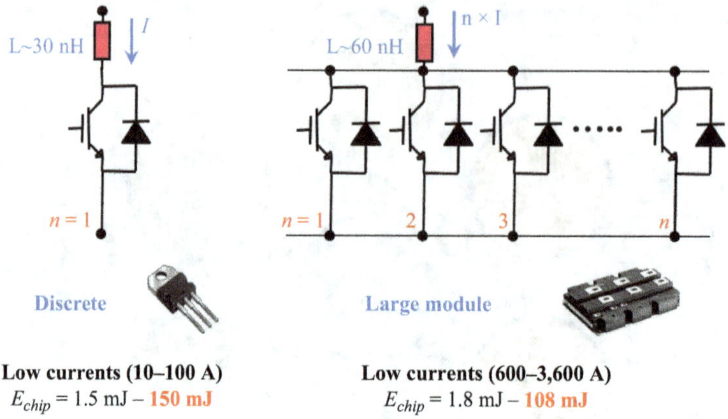

Figure 5.44 Discrete and module representation with respect to the effective stray inductance and inductive energies [40]

In addition to the increase in the effective inductance 'nL_S' value, the stray inductance in high current applications is also somewhat larger. Therefore, using typical known discrete and module current ratings, dissipated energy levels per chip represents an order or two of magnitude higher inductive stress for the IGBT chip when employed in large modules, as shown in Figure 5.44. Therefore, stress-related factors such as device controllability/softness and SOA are taken into account when heavy paralleling of chips or modules is required.

5.8.1.2 Integration solutions

A number of development trends fall into this category for enabling higher power capabilities. Integration solutions can be divided into (a) vertical integration and (b) lateral integration solutions.

(a) Vertical integration: Typical trends include
- Higher voltage ratings of IGBTs to replace or reduce series connection for lower losses and lower cost when viable. Also, for multilevel topologies, higher voltage ratings can reduce the number of devices employed in the system.
- Reverse blocking concepts (e.g., RB-IGBT) to replace an IGBT/Diode series configuration for providing lower losses and, hence, higher power in many applications based on current source converters or matrix converters.

(b) Lateral integration: Typical trends include
- Reverse conducting concepts (e.g., RC-IGBT) to integrate the IGBT and freewheeling diode into a single structure for higher power per package footprint.
- Improved narrower junction termination for maximising the active area of a given device total area.

The focus here will be on the recent development trend towards a hard switching single-chip RC-IGBT in order to eliminate the need for a separate fast recovery diode. Therefore, Section 5.8.1.3 will cover in more detail on this technology driver.

5.8.1.3 Reverse conducting RC-IGBT

The clear demand for increased power levels has led to the focus on an IGBT and diode integration solution or what has been normally referred to as the reverse conducting RC-IGBT. The practical realisation of a single-chip technology will provide an ideal solution for compact systems with higher power levels, which could prove to be beyond the capability of the standard two-chip approach regardless of the IGBT technology employed [41]. The basic RC-IGBT concept consists of an anode with N+ shorts regions to provide an internal antiparallel diode in the IGBT structure as shown in Figure 5.45 along with the RC-IGBT circuit symbol.

Conventionally, the realisation of such a device for mainstream hard switching applications has always been hindered by design and process issues resulting in a number of performance drawbacks and trade-offs, which are summarised as follows:

- Snap-back in the IGBT on-state IV characteristics (anode shorting effect),
- IGBT conduction versus diode switching losses trade-off (excess carrier shaping effect),
- IGBT versus diode softness trade-off (drift layer effect) and
- SOA (charge uniformity effect).

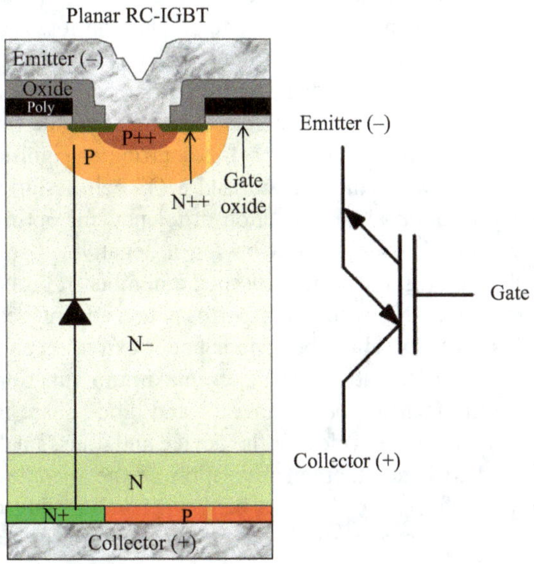

Figure 5.45 Cross-section and symbol of reverse conducting RC-IGBT

The early attempts for realising RC-IGBTs were focused on lower voltage–rated IGBTs targeting mainly soft switching applications where the diode is not optimised for fast and hard switching performance [41,42]. Nevertheless, development efforts aimed at tackling the above-mentioned issues have resulted in advanced RC-IGBT concepts with hard switching capability. High-voltage RC-IGBTs targeting mainstream high power applications have been demonstrated such as the bimode insulated gate transistor (BIGT) and the RC IGBT [43,44]. Today, the RC-IGBT technology is an important development trend for next-generation IGBT components and applications.

At the P-type anode/collector, alternating N+ doped regions called anode shorts are introduced, which then act as a cathode contact for the internal diode mode of operation. The area ratio between the IGBT anode (P+ regions) and the diode cathode (N+ regions) determines which part of the collector area is available in IGBT or diode modes, respectively. During the RC-IGBT conduction in diode mode, the P+ regions are inactive and do not directly influence the diode conduction performance. On the other hand, the N+ regions act as anode shorts in the IGBT mode of operation, strongly influencing the IGBT conduction mode. One of the implications of anode shorting is the voltage snap-back referred to previously, which is observed as a negative resistance region in the device IGBT mode IV characteristics. This effect will have a negative impact when devices are paralleled, especially at low-temperature conditions, although not proven conclusively in prior literature.

Some of the advanced RC-IGBT design concepts, such as the BIGT shown in Figure 5.46, include optimum anode short layouts [45], which constitute a pilot IGBT region and stripe-shaped short designs to reduce and even eliminate the snap-back effect. Furthermore, optimum lifetime reduction methods were developed to improve the trade-off between the IGBT mode conduction losses and diode mode reverse recovery losses. In addition, it is important to note that the available technology platforms, such as the SPT- or FS-buffer, alongside miniaturised trench or planar cell concepts have been main enablers to bring forth the possibility of this integration. The buffer optimum doping profile contributes significantly for reducing the snap-back effect while the miniaturised MOS cell design plays an important role to reduce the diode–anode injection efficiency for optimising the diode conduction and switching losses without having a negative impact on the IGBT performance. Finally, the RC-IGBT has inherited a number of properties due to the anode shorts design, which has resulted in performance advantages in both operational modes, such as soft switching behaviour under extreme conditions and very low leakage currents for operating at higher maximum junction temperatures. Furthermore, due to the fact that both transistor and diode share the same silicon area, there are no more inactive cycles for the device and, thus, smaller temperature ripple. This will lead to better thermal utilisation of the module and eventually improved reliability performance due to a better thermal cycling capability. This feature will also enable a more reliable chip operation at higher junction temperatures.

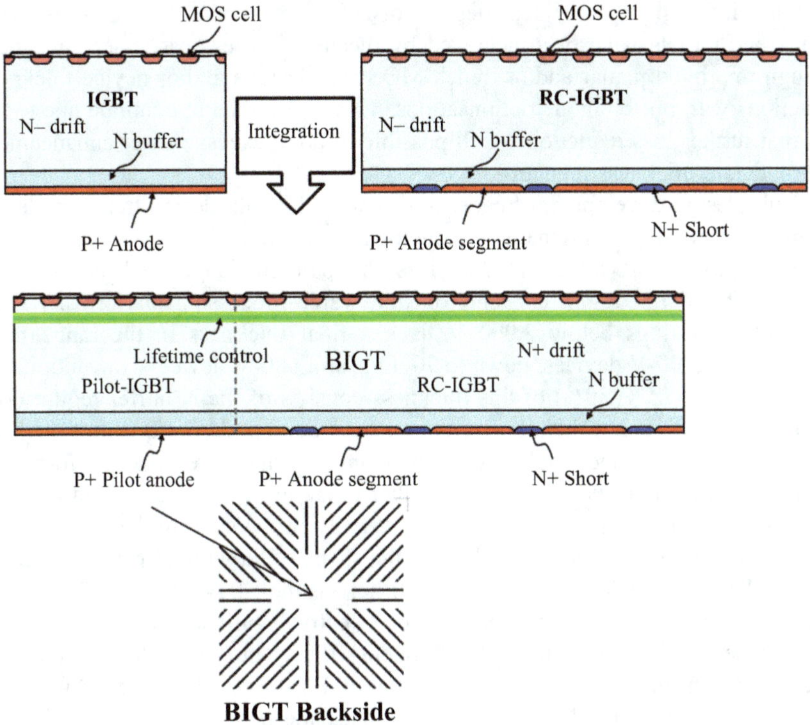

Figure 5.46 Advanced RC-IGBT concept referred to as BIGT [45]

5.8.2 Increase in power density

The second development trend focuses on increasing the device power density through the reduction of losses, increasing the SOA margins, and/or improving the thermal performance of the device and package. As a general rule, to enable higher power densities, first, it is important to identify the main limiting factor (losses, thermal or SOA) for a given device technology. In other words, if the SOA capability is limiting the power-handling capability of a given device technology as was the for the first generations of high-voltage IGBTs, then further reductions in losses could provide better efficiency. But it will not enable the device to operate at higher current densities and, hence, higher power levels. Once the SOA capability is improved and the losses become the limiting factor, a new generation of devices with lower losses become a viable development target for achieving another leap in power capability.

5.8.2.1 Losses reductions

Over the years, improved technology curves of successful IGBT generations through static and dynamic loss reductions have always been the traditional focus for achieving higher power densities. As discussed in Section 5.5, these development trend targets have been achieved by means of excess carrier enhancement through improved planar and/or trench MOS cell designs and/or device thickness reductions with buffer/anode optimisations. This trend is set to continue due to the fact that further loss reduction is still possible through excess carrier enhancement and moderate thickness reductions.

Bulk design development trends: In line with the bulk design trend, moderate optimisation possibilities remain by reducing the silicon thickness, as was shown in Figure 5.16. For low voltage-rated devices, the main challenge has been to realise modern SPT or FS design concepts on thin wafer–processing platforms. IGBTs with voltage ratings below 2,000 V have a final thickness in the range from 200 μm, for 1,700-V devices, down to 70 μm, for a 600-V device. Conventionally, a large portion (25%–30%) of this thickness consists of the N buffer region. The main purpose of such a thick buffer is to ensure good and controllable switching behaviour (softness) and stable reverse blocking. Furthermore, other performance parameters such as leakage current at higher operating temperatures and the ruggedness under short circuit conditions are heavily influenced by the bipolar transistor gain, which is largely dependent on the shape of the N buffer design and the anode/collector injection efficiency. Nevertheless, the SPT- or FS-buffer is both thick and relatively low-doped and is not effectively utilised for achieving the required device-blocking capability. Reducing the N buffer thickness gives the potential for further improvements for total IGBT losses. However, thin wafer processing requires that the majority of the backside processing is done after the front-side emitter is completed, including metallisation. This limits the temperature treatments after thinning to below 500 °C and, so, the means for forming both the buffer and anode regions have been very limited. In addition, the recent trend towards manufacturing IGBT on large diameter wafers (200–300 mm) increases further the challenges addressed above [46].

In spite of such obstacles, recent trends target thinner buffer regions even down to a few μm-range and depend mainly on single-sided high temperature and short pulse activation, using laser annealing process techniques to activate both the buffer and collector regions. Defect engineering at low temperatures after proton implantation to produce donor layers for the buffer are also applied in modern designs [47].

Furthermore, investigations have shown that the superjunction concept employed currently in MOSFETs could also bring forth lower losses for IGBTs [48]. The main feature is with regard to the lower turn-off losses achieved due to the lower excess carrier modulation and fast charge extraction in the N- and P-type bulk charge compensation regions. Nevertheless, when compared with the superjunction MOSFET design, finer dimensions and higher doping concentrations for those regions are needed for the IGBT for optimum performance. Hence, one can predict the advantage of such a concept for the lower voltage–rated device, with fast switching characteristics.

Cell design development trends: Despite the remarkable improvements in IGBT technologies over the years, the IGBT structure still suffers from inherently higher on-state losses when compared with thyristor-based devices, such as the integrated gate commutated thyristor or IGCT. The old quest from the 1990s for a thyristor-based MOS controlled device with a thyristor-like carrier distribution and IGBT-like overall performance is no longer pursued due to the limited SOA capabilities and the continuous improvements in IGBT designs. Therefore, MOS-controlled bipolar device evolution maintained an IGBT structure in principle, which remained intact since the basic properties to obtain good overall performance were kept.

The carrier-enhancement level in IGBTs is mainly influenced by the MOS channel resistance, the hole accumulation effect between cells (PNN effect) and the hole drainage effect in the cell (PNP effect). The electron spreading from the channel is an integrated part of the above three effects. These aspects of the IGBT design have been optimised with both planar and trench MOS cell designs with certain trade-offs in terms of blocking capability, controllability and SOA to achieve low losses as discussed previously, in Section 5.5. Looking forward, any further loss reductions can be still achieved by developing an improved trench cell design. The ultimate goal for IGBT loss reductions is to reach levels close to those of a thyristor or IGCT as shown in Figure 5.47.

For higher voltage devices, a second generation of enhanced or advanced trench designs have been developed while targeting higher current ratings, lower losses and controllable switching performance [49–51]. For improved controllability and sustainable short circuit current levels, the design must incorporate within the MOS cell inactive trench regions, such as floating or grounded trenches. For lower voltage–rated IGBTs, the next evolutionary step has already started with

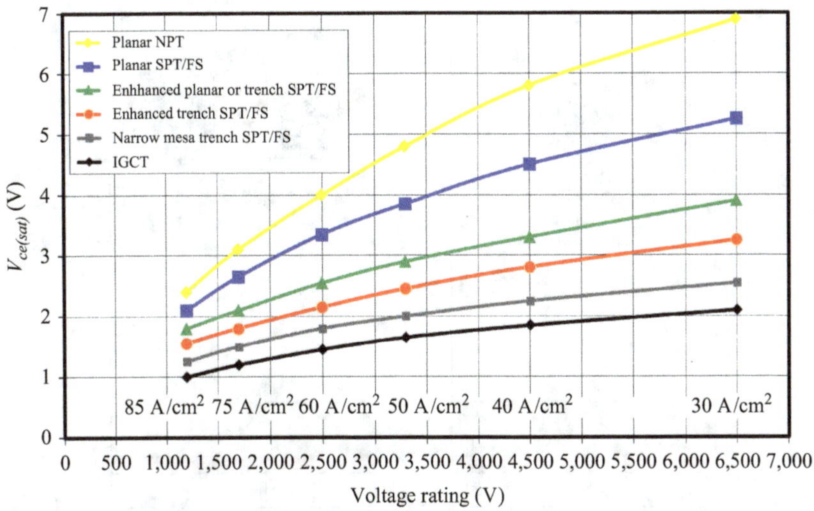

Figure 5.47 IGBT technologies on-state losses $V_{ce(sat)}$, same turn-off losses per voltage class at 125° [13]

optimised trench cell and layout designs. By adopting narrow trench cell mesa dimensions below 1 μm as shown in Figure 5.48, the hole drainage effect [52] can be strongly limited for providing higher excess carrier levels at the emitter compared to previous generations of trench based IGBTs [53–55].

5.8.2.2 Higher operating temperature

Increasing the maximum operating junction temperature of IGBT modules is an important trend to achieve higher power densities for next-generation systems. Device optimisation for realising this target has been focused on reducing the reverse bias leakage current at higher temperatures for stable thermal performance. It is important to point out that this trend is strongly accompanied with high-temperature package developments in terms of improved filling materials and joining technologies with lower thermal resistances, while also maintaining high-reliability power cycling capabilities for the targeted elevated temperatures.

The target towards increasing the maximum operating junction temperature of IGBTs has shifted some of the traditional design focus towards improved blocking characteristics especially in terms of reduced leakage currents for stable high-temperature performance. The important design parameters for achieving this target can be listed as follows:

- Improved junction termination design and passivation layers with lower leakage currents and stable performance.
- Reduction of the IGBT bipolar gain impact on leakage current. By optimising the buffer maximum doping concentration and anode-doping levels, a lower bipolar gain is achieved with lower leakage current levels for the same overall

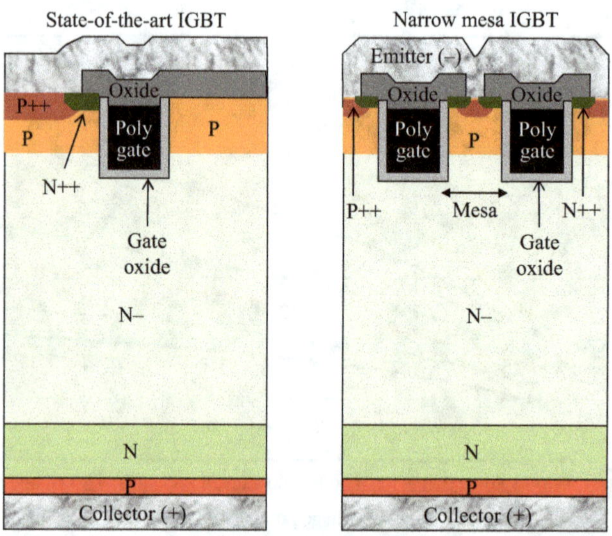

Figure 5.48 Trench-based IGBT technologies with wide mesa and narrow mesa design

static and dynamic losses. Furthermore, introducing lowly doped anode regions opposite the junction termination region has shown significant improvements towards this target.

In principle, it has been shown that purely from the electrical performance viewpoint, there is no clear device-related limits for enabling higher temperature operation up to 200 °C in terms of the dynamic and static performances [56]. Nevertheless, reliable performance at device and package level will remain in focus for the coming years as this trend continues to evolve. Today, 600 up to 1,700-V devices are commercially available with maximum junction temperature specifications up to 175 °C while 3,300 to 6,500-V IGBTs have reached 150 °C. There are developments to extend this trend towards higher temperatures.

5.8.2.3 Safe-operating-area and short circuit margins

As described in Section 5.7, modern IGBTs designs over a wide range of voltage ratings are capable of withstanding higher power switching transients while sustaining avalanche mode capability. Nevertheless, higher power densities must be always accompanied by the need for higher SOA margins and improved controllability for reliable operation. For next-generation IGBTs following the previously discussed development trends such as RC-IGBT integration, narrow mesa trench MOS cells, thinner buffer structures and higher operating temperatures, the need to maintain the required SOA margins will remain a challenge to overcome. The main aim for the IGBT will still focus on improving the parasitic thyristor latch-up capability in the MOS cell and the overshoot avalanche capability during device turn-off under extreme operational conditions. Therefore, the narrow mesa trench cell design, in particular, will present a challenge for targeting higher SOA capability especially for high voltage-rated IGBTs due to the limited hole drainage associated with such designs. Furthermore, preserving the short circuit protection capability for IGBTs is a must in most applications. The trend towards higher current densities with smaller chip dimensions, thinner devices with shallow buffer layers combined with higher operating temperature requirements is presenting a clear challenge for allowing the state-of-the-art 10 μs short circuit pulse requirement to be maintained in future device specifications.

5.9 Summary

Chapter 5 covers silicon-based IGBT devices with a focus on device structure, operation, performance requirements and development trends. First, a brief historical background is presented covering the conception and evolution of the IGBT structure from the basic principles of the unipolar power MOSFET and bipolar power transistor. This is followed in Section 5.2 with a detailed explanation of the basic IGBT structure and principle of operation from both the device physics and equivalent circuit model viewpoints. Sections 5.3 and 5.4 cover the basic static and dynamic characteristics of the IGBT. The static characteristics discuss the forward and reverse bias modes of operation while also including the transfer characteristics for the IGBT MOS cell structure. The dynamic performance is explained for both the turn-on and turn-off periods with the help of the internal IGBT capacitive effects and excess carrier charge dynamics during the switching transients. Section 5.5 is focused on the main requirements of the IGBT with respect to mainstream applications and how such demands have influenced the device structural evolution over the years. A number of critical design and performance trade-off relationships are presented with respect to the structural changes made at both MOS cell (planar, trench) and bulk design (PT, NPT) and their impact on the total device losses, switching control and robustness. Sections 5.6 and 5.7 present the IGBT requirements and design considerations for increasing the device SOA under both short circuit fault mode and switching transients. The different failure modes occurring under such conditions and their causes are discussed and analysed while highlighting the device design impact for providing wider margins and more robust behaviour. Section 5.8 provides the reader with an overview of the IGBT current and future development trends. The main target is to provide continuous increments in power handling capability through a number of independent and often combined technology paths. This includes the traditional trend with a focus on reduced losses and increased margins and new trends with respect to device integration, such as for reverse conducting IGBTs and allowing operation at increased junction temperatures.

Note to the reader

This chapter is part of an edited book. To cite or use this chapter as a reference, please use the following format:

Rahimo M. and Diaz Reigosa P. "Silicon IGBTs." In: Iannuzzo F (ed.), *Modern Power Electronic Devices: Physics, Applications, and Reliability*. Stevenage, UK: IET; 2020. pp. 113–170.

References

[1] B. J. Baliga, M. S. Adler, P. V. Gray, R. P. Love, N. Zommer, "The insulated gate rectifier (IGR): A new power switching device", IEEE IEDM Tech. Dig., pp. 264–267, 1982.

[2] J. P. Russell, A. M. Goodman, L. A. Goodman, J. M. Neilson, "The COMFET—A new high conductance MOS-gated device", IEEE Electron Device Lett., vol. 4, no. 3, pp. 63–65, 1983.

[3] A. Nakagawa, H. Ohashi, M. Kurata, H. Yamaguchi, K. Watanabe, "Non-latch-up 1200V 75A bipolar-mode MOSFET with large ASO", IEEE IEDM Tech. Dig., pp. 860–861, 1984.

[4] M. F. Chang, G. C. Pifer, H. Yilmaz, *et al.*, "Comparison of n and p channel IGTs", IEEE IEDM Tech. Dig., pp. 279–281, 1984.

[5] N. Iwamuro, T. Laska, "IGBT history, state-of-the-art, and future prospects", IEEE Trans. Electron Devices, vol. 64, no. 3, pp. 741–752, 2017.

[6] S. Linder, "Power semiconductors". EPFL Press (1st edition, 2006, ISBN 0–8247–2569–7).

[7] A. Nakagawa, H. Ohashi, "600- and 1200-V bipolar-mode MOSFET's with high current capability", IEEE Electron Device Lett., vol. 6, no. 7, pp. 378–380, 1985.

[8] G. Miller, J. Sack, "A new concept for a non-punch-through IGBT with MOSFET like switching characteristics", Proc. IEEE PESC Record, vol. 1, pp. 21–25, June 1989.

[9] M. Otsuki, S. Momota, A. Nishiura, K. Sakurai, "The 3rd generation IGBT toward a limitation of IGBT performance", Proc. Int. Sym. Power Semiconductors and ICs, Monterey, USA, pp. 24–29, May 1993.

[10] J. Bauer, F. Auerbach, A. Prost, R. Roth, H. Ruething, O. Schilling, "6.5kV-modules using IGBTs with Field Stop technology", Proc. Int. Sym. on Power Semiconductor Devices and IC's ISPSD'01, Osaka, Japan, pp. 121–124, June 2001.

[11] M. Takei, Y. Harada, K. Ueno, "600 V-IGBT with reverse blocking capability", Proc. Int. Sym. on Power Semiconductor Devices and ICs, Osaka, Japan, pp. 413–416, June 2001.

[12] M. T. Rahimo, D. J. Chamund, N. Y. A. Shammas, "Analysis of IGBT/freewheeling diode switching behaviour during Turn-on in hard switching applications", 27th International Conference on Power Electronics and Variable-Speed Drives, PEVD'98, London, UK, pp. 381–386, Sept. 1998.

[13] M. T. Rahimo, "Future trends in high-power bipolar metal-oxide-semiconductor controlled power semiconductors", IET Circuits Devices Syst., vol. 8, no. 3, pp. 155–167, 2014.

[14] J. Baliga, "The MOS-gated emitter switched thyristor", IEEE Electron Device Lett., vol. 11, no. 2, pp. 75–77, 1990.

[15] V. A. K. Temple, "MOS controlled thyristors (MCT's)", IEEE IEDM Tech. Dig., pp. 282–285, 1984.

[16] T. Laska, M. Munzer, F. Pfirsch, C. Scaeffer, T. Schmidt, "The field stop IGBT (FS IGBT). A new power device concept with a great improvement potential", Proc. Int. Sym. on Power Semiconductor Devices and ICs, Toulouse, France, pp. 355–358, May 2000.

[17] S. Dewar, S. Linder, C. V. Arx, A. Mukhitinov, G. Debled, "Soft punch through (SPT)-setting new standards in 1200V IGBT", Proc. PCIM Europe Conference, Nurnberg, Germany, p. 593, May 2000.

[18] M. T. Rahimo, W. J. Findlay, L. Coulbeck, "An improved design for ultra soft – Fast recovery diodes suitable for (600–1200) V IGBT application", Proc. PCIM Europe Conference, Nurnberg, Germany, pp. 409–417, May 1998.

[19] J. Lutz, "Advantage of the new controlled axial lifetime diode", Proc. PCIM Europe Conference, Nurnberg, Germany, p. 163, 1994.

[20] M. Kitagawa, I. Omura, S. Hasegawa, T. Inoue, A. Nakagawa, "A 4500 V injection enhanced insulated gate bipolar transistor (IEGT) operating in a mode similar to a thyristor", IEEE IEDM Tech. Dig., pp. 679–682, 1993.

[21] I. Omura, T. Ogura, K. Sugiyama, H. Ohashi, "Carrier injection enhancement effect of high voltage MOS devices-device physics and design concept", Proc. Int. Sym. on Power Semiconductor Devices and ICs, Weimar, Germany, pp. 217–220, May 1997.

[22] M. Harada, T. Minato, H. Takahashi, H. Nishimura, K. Inoue, I. Takata, "600V trench IGBT in comparison with planar IGBT", Proc. Int. Sym. on Power Semiconductor Devices and ICs, Davos, Switzerland, pp. 411–416, 1994.

[23] T. Laska, F. Pfirsch, F. Hirler, J. Niedermeyr, C. Schaffer, T. Schmidt, "1200 V-trench-IGBT study with square short circuit SOA", Proc. Int. Sym. on Power Semiconductor Devices and ICs, Kyoto, Japan, pp. 433–436, June 1998.

[24] F. Udrea, G. Amaratunga, "Theoretical and numerical comparison between DMOS and trench technologies for insulated gate bipolar transistors", IEEE Trans. Electron Devices, vol. 42, no. 7, pp. 1356–1366, 1995.

[25] I. Omura, H. Ohashi, W. Fichtner, "IGBT negative gate capacitance and related instability effects", IEEE Electron Device Lett., vol. 18, no. 12, pp. 622–624, 1997.

[26] A. Pfaffenlehner, J. Biermann, C. Schaeffer, H. Schulze, "New 3300V chip generation with a trench IGBT and an optimized field stop concept with a smooth switching behaviour", Proc. Int. Sym. on Power Semiconductor Devices and ICs, Kitakyushu, Japan, pp. 107–110, June, 2004.

[27] H. Takahashi, H. Haruguchi, H. Hagino, T. Yamada, "Carrier stored trench gate bipolar transistor (CSTBT) – A novel power device for high voltage applications", Proc. Int. Sym. on Power Semiconductor Devices and ICs, Maui, USA, pp. 349–352, 1996.

[28] M. Mori, Y. Uchino, J. Sakano, H. Kobayashi, "A novel high-conductivity IGBT (HiGT) with a short circuit capability", Proc. Int. Sym. on Power Semiconductor Devices and ICs, Kyoto, Japan, pp. 429–432, June 1998.

[29] M. T. Rahimo, A. Kopta, S. Linder, "Novel enhanced–planar IGBT technology rated up to 6.5kV for lower losses and higher SOA capability", Proc. Int. Sym. on Power Semiconductor Devices and ICs, Naples, Italy, pp. 33–36, 2006.

[30] N. Iwamuro, A. Okamoto, S. Tagami, H. Motoyama, "Numerical analysis of short-circuit safe operating area for p-channel and n-channel IGBTs", IEEE Trans. Electron Devices, vol. 38, no. 2, pp. 303–309, 1991.

[31] F. Hille, F. Umbach, T. Raker, R. Roth, "Failure mechanism and improvement potential of IGBT's short circuit operation", Proc. Int. Sym. on Power Semiconductor Devices and ICs, Hiroshima, Japan, pp. 33–37, June 2010.

[32] T. Laska, G. Miller, M. Pfaffenlehner, P. Türkes, D. Berger, "Short circuit properties of Trench/Field Stop IGBTs design aspects for a superior robustness", Proc. Int. Sym. on Power Semiconductor Devices and ICs, Cambridge, UK, pp. 152–155, 2003.

[33] A. Kopta, M. Rahimo, U. Schlapbach, N. Kaminski, D. Silber, "Limitation of the short-circuit ruggedness of high-voltage IGBTs", Proc. Int. Sym. on Power Semiconductor Devices and ICs, Barcelona, Spain, pp. 33–36, 2009.

[34] P. D. Reigosa, F. Iannuzzo, M. T. Rahimo, C. Corvasce, F. Blaabjerg, "Improving the short-circuit reliability in IGBTs: How to mitigate oscillations", IEEE Trans. on Power Electronics, vol. 33, no. 7, pp. 5603–5612, 2018.
[35] P. Diaz Reigosa, F. Iannuzzo, M. T. Rahimo, "TCAD analysis of short-circuit oscillations in IGBTs", Proc. Int. Sym. on Power Semiconductor Devices and ICs, Sapporo, Japan, pp. 151–154, 2017.
[36] B. J. Baliga, M. S. Adler, P. V. Gray, R. P. Love, "Suppressing latch-up in insulated gate transistors", IEEE Electron Device Letters, vol. 5, no. 8, pp. 323–325, 1984.
[37] M. T. Rahimo, A. Kopta, S. Eicher, U. Schlapbach, S. Linder, "Switching-self-clamping-mode 'SSCM', a breakthrough in SOA performance for high voltage IGBTs and diodes", Proc. Int. Sym. on Power Semiconductor Devices and ICs, Kitakyushu, Japan, pp. 437–440, 2004.
[38] M. Otsuki, Y. Onozawa, S. Yoshiwatari, Y. Seki, "1200V FS-IGBT module with enhanced dynamic clamping capability", Proc. Int. Sym. on Power Semiconductor Devices and ICs, Kitakyushu, Japan, pp. 339–342, 2004.
[39] M. T. Rahimo, A. Kopta, S. Eicher, U. Schlapbach, S. Linder, "A study of switching self-clamping-mode 'SSCM' as an over-protection feature in high voltage IGBTs", Proc. Int. Sym. on Power Semiconductor Devices and ICs, Santa Barbara, US, pp. 67–70, May 2005.
[40] M. T. Rahimo, U. Schlapbach, A. Kopta, S. Linder, "An assessment of modern IGBT and anti-parallel diode behaviour in hard-switching applications", European Conference on Power Electronics and Applications EPE 2005, Dresden, Germany, Sept. 2005.
[41] H. Takahashi, A. Yamamoto, S. Aono, T. Minato, "1200V reverse conducting IGBT", Proc. Int. Sym. on Power Semiconductor Devices and ICs, Kitakyushu, Japan, pp. 133–136, 2004.
[42] S. Voss, F-J. Niedernostheide, H-J. Schulze, "Anode design variations in 1200V trench field-stop RC-IGBTs", Proc. Int. Sym. on Power Semiconductor Devices and ICs, Orlando, USA, pp. 169–172, 2008.
[43] M. T. Rahimo, A. Kopta, U. Schlapbach, J. Vobecky, R. Schnell, S. Klaka, "The bi-mode insulated gate transistor (BIGT) a potential technology for higher power applications", Proc. Int. Sym. on Power Semiconductor Devices and ICs, Barcelona, Spain, pp. 283–286, 2009.
[44] D. Werber, F. Pfirsch, T. Gutt, *et al.*, "6.5kV RCDC, for increased power density in IGBT-modules", Proc. Int. Sym. on Power Semiconductor Devices and ICs, Waikoloa, USA, pp. 35–38, 2014.
[45] L. Storasta, A. Kopta, M. Bellini, M. T. Rahimo, U. Vemulapati, N. Kaminski, "The radial layout design concept for the bi-mode insulated gate transistor", Proc. Int. Sym. on Power Semiconductor Devices and ICs, San Diego, USA, pp. 56–59, 2011.
[46] H. J. Schulze, H. Öfner, F. J. Niedernostheide, *et al.*, "Use of 300 mm magnetic Czochralski wafers for the fabrication of IGBTs", Proc. Int. Sym. on Power Semiconductor Devices and ICs, Prague, Czech Republic, pp. 355–358, June 2016.

[47] J. Vobecky, M. T. Rahimo, A. Kopta, S. Linder, "Exploring the silicon design limits of thin wafer IGBT technology: The controlled punch through (CPT) IGBT", Proc. Int. Sym. on Power Semiconductor Devices and ICs, Orlando, USA, pp. 76–79, 2008.

[48] M. Antoniou, F. Udrea, F. Bauer, "Optimisation of super-junction bipolar transistor for ultra-fast switching applications", Proc. Int. Sym. on Power Semiconductor Devices and ICs, Jeju Island, Korea, pp. 101–104, 2007.

[49] K. Nakamura, K. Sadamatsu, D. Oya, H. Shigeoka, K. Hatade, "Wide cell pitch LPT(II)-CSTBT(III) technology rating up to 6500 V for low loss", Proc. Int. Sym. on Power Semiconductor Devices and ICs, Hiroshima, Japan, pp. 387–390, 2010.

[50] Y. Toyota, S. Watanabe, T. Arai, *et al.*, "Novel 3.3kV advanced trench HiGT with low loss and low *dv*/*dt* noise", Proc. Int. Sym. on Power Semiconductor Devices and ICs, Kanazawa, Japan, pp. 29–32, 2013.

[51] M. Andenna, Y. Otani, S. Matthias, *et al.*, "The next generation high voltage IGBT modules utilizing enhanced-trench ET-IGBTs and field charge extraction FCE-diodes", European Conference on Power Electronics and Applications (EPE'14-ECCE), Lappeenranta, Finland, 2014.

[52] A. Nakagawa, "Theoretical investigation of silicon limit characteristics of IGBT", Proc. Int. Sym. on Power Semiconductor Devices and ICs, Naples, Italy, pp. 5–8, 2006.

[53] M. Sumitomo, J. Asai, H. Sakane, K. Arakawa, Y. Higuchi, M. Matsui, "Low loss IGBT with partially narrow mesa structure (PNM-IGBT)", Proc. Int. Sym. on Power Semiconductor Devices and ICs, Bruges, Belgium, pp. 17–20, 2012.

[54] K. Eikyu, A. Sakai, H. Matsuura, *et al.*, "On the scaling limit of the Si-IGBTs with very narrow mesa structure", Proc Proc. Int. Sym. on Power Semiconductor Devices and ICs, Prague, Czech Republic, pp. 211–214, 2016.

[55] F. Wolter, W. Roesner, M. Cotorogea, T. Geinzer, M. Seider-Schmidt, "Multi-dimensional trade-off considerations of the 750V micro pattern trench IGBT for electric drive train applications", Proc. Int. Sym. on Power Semiconductor Devices and ICs, Hong Kong, China, pp. 105–108, May 2015.

[56] U. Schlapbach, M. T. Rahimo, C. von Arx, A. Mukhitdinov, S. Linder, "1200V IGBTs operating at 200 °C? An investigation on the potentials and the design constraints", Proc. Int. Sym. on Power Semiconductor Devices and ICs, Jeju Island, Korea, pp. 9–12, 2007.

Chapter 6

IGCTs

Eric Carroll[1]

6.1 Introduction

The integrated gate-commutated thyristors (IGCTs) are the latest of the high-power turn-off devices (ToDs). It is a further development of the gate turn-off (GTO) thyristor, which, as the name indicates, is a thyristor capable of being turned-off by gate-control, as will be discussed later (Section 6.4). Thyristors (see Chapter 3) are the most 'powerful' of semiconductors, as measured in terms of rated current multiplied by rated voltage but they lack the ability to be turned off while in full conduction, as can, for instance, transistors.

This chapter will attempt to cover every important aspect of IGCTs and as most aspects are imbricated, the reader will be referred back and forth to the various sections to avoid repeating earlier explanations, in the hope and belief that this will impart clarity more than frustration.

6.2 History

The IGCT was invented at ABB Corporate Research, Switzerland, in 1992. It was the result of a project to enable the cost-effective, series-connection of GTOs in order to build very high-power self-commutated converters with ratings in the order of 100 MW [1,2]. At the time, the GTO was the ToD with the highest power capability but its switching speeds, in general, were low, especially the turn-off times. This gave rise to large tailing currents at the end of each current extinction with correspondingly large differences between devices in the same series-connection. Such differences are usually 'absorbed' by parallel-connecting large 'snubber' capacitors across each semiconductor in order to dynamically 'soak' up the current differences during the turn-off process, thus limiting the resulting voltage differences across the series-connected devices. However, in view of the high currents and voltages (kiloamps and kilovolts), the energy dissipated at each capacitor-discharge was deemed excessive; although such approaches were common practice at the time, it was felt that a means of 'speeding up' the GTO should be found.

[1]EIC Consultancy, Montferrier sur Lez, France

Initial attempts focused on increasing the gate-drive voltage by a factor 10 (from 20 to 200 V), which had the desired effect but resulted in very large and highly dissipative gate-drive units. This device was initially called the 'Hard-Driven GTO' or 'HD-GTO'.

It was realised that reducing gate-circuit impedance by a factor 10 would have the same effect and allow the gate-drive voltage to remain at its usual 20-V level for a GTO but this required a very significant redesign of the semiconductor housing and an equally drastic redesign, both electrically and mechanically, of the gate-drive unit.

The required 10-fold increase of gate-current speed (*di/dt*) was finally obtained by eliminating gate-connection cables and physically 'integrating' the device into its gate unit [3].

ABB's targeted application at the time was that of 'Interties', those large converters which allow power grids of different frequencies and phase-count to be connected and to transfer power to each other in either direction. In some European countries (e.g. Germany and Switzerland), the electrical traction network is powered by a single-phase, 16.7-Hz overhead-line whereas the utility grid is typically three-phase, 50 Hz – so these grids must be 'tied' together by intertie converters in order to support each other.

Mitsubishi Semiconductors was the first manufacturer to accept fabrication of these new devices and they chose to call them 'Gate-Commutated Thyristors' since, as we will see later (Section 6.5), the principle of operation calls for the entire cathode current to be commutated out via the gate. As the potential applications for this fast and powerful device started to grow, ABB Semiconductors also agreed to produce them, initially still known as 'HD-GTOs'. It soon became apparent that because the gate unit was physically device-specific, ABB would sell the semiconductor integrated into its gate-unit, which gave rise to the new name of 'IGCT' which became commercially available in 1997 and as such, it is the most recent of the high-power ToDs. An ABB GCT is shown in Figure 6.1 and complete IGCTs in Figures 6.2 and 6.3.

6.3 Device types

Thyristor devices offer the greatest flexibility in terms of characteristics in that they are available as asymmetric, reverse-blocking and reverse-conducting types and devices are therefore available for both voltage-source and current-source converters (VSIs and CSIs) (Figure 6.3). These inverter-types will be discussed later

Figure 6.1 An ABB 4″ GCT (Courtesy ABB, Switzerland Ltd)

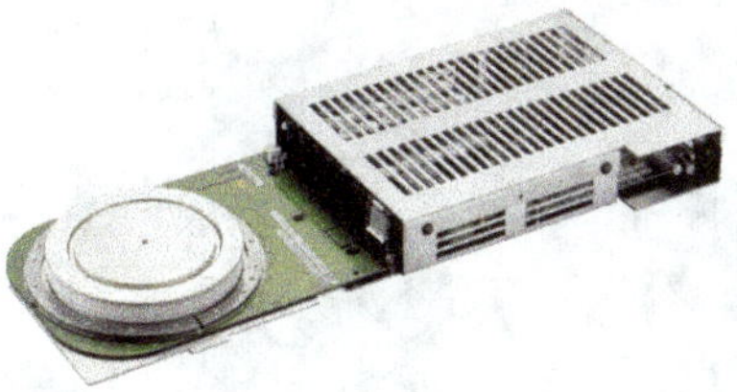

Figure 6.2 An ABB IGCT showing a 4″ GCT integrated into its gate-drive unit (Courtesy ABB, Switzerland Ltd)

Figure 6.3 A CRRC 6″ IGCT showing a 6″ GCT integrated into its gate-drive unit (Courtesy CRRC)

(Section 6.13.1) but at this point, it is sufficient to say that they require semiconductors which can conduct current in the reverse direction (VSIs) or block current in the reverse direction (CSIs). Such functions can be fulfilled by devices having no reverse conducting or blocking capabilities in association with a separate diode as shown in Figure 6.4, where an asymmetric device is associated with a

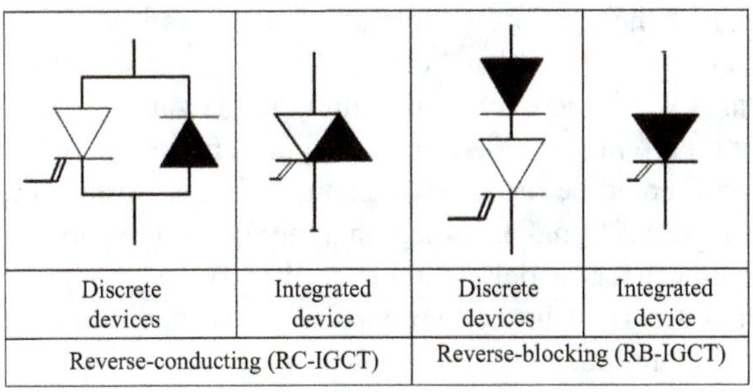

Figure 6.4 Asymmetric, RC and RB-IGCTs and their topologies based on discrete devices

Figure 6.5 A Mitsubishi 6.5-kV SGCT, as used in current source inverters (Courtesy Mitsubishi)

diode to form either a reverse-conducting or a reverse-blocking function. Reverse-blocking devices are also known as symmetric (I)GCTs or SGCTs. A 6.5-kV SGCT from Mitsubishi can be seen below in Figure 6.5.

As we will see later in Section 6.4, the turn-off operation of the IGCT requires unity-gain extinction which aims to sweep all the cathode current out of the gate as quickly as possible (ideally, instantly).

Other devices, such as the MTO (MOS turn-off thyristor) [4] and the ETO (emitter-switched turn-off thyristor) [5], also rely on this mode of operation but have not achieved the same performance, as will be discussed later (Section 6.7).

6.4 Gate turn-off thyristors (GTOs)

To appreciate the operation of the IGCT, it is important to understand the GTO from which it is derived.

The GTO, first commercialised in 1979, achieved gate-controlled turn-off, which, until then, had not been possible with thyristors.

As explained in Section 3.3, once a thyristor is turned on, it cannot be turned off except by external means. This can be intuitively understood since the gate structure covers only a small percentage of the wafer area; so, reversing the gate current would not be 'noticed' by remote parts of the wafer.

However, contrary to what has been said so far, a very small thyristor, all of whose cathode area would be in close proximity to the gate, could be turned off by reversing the gate current, provided the anode current were not too large and the reverse gate current could be made large enough. This is because the condition for latching, described in Section 3.3, namely that the sum of the pnp and npn transistor gains in the two-transistor model should exceed unity ($\alpha_{pnp} + \alpha_{npn} > 1$), depends on the instantaneous current in the two transistors, as shown in the current dependence of gain in Figure 6.6.

The thyristor was originally conceived to turn on easily, quickly and latch reliably. Designing a GTO meant compromising some of these attractive features, in order to make it turn-off without too great an effort. One change that was required

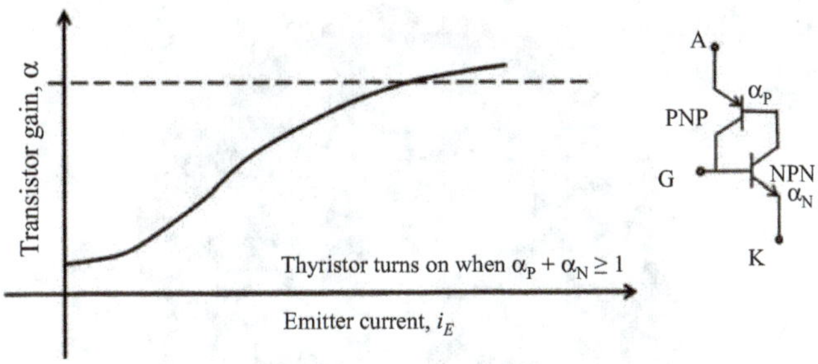

Figure 6.6 Transistor gain as a function of emitter current

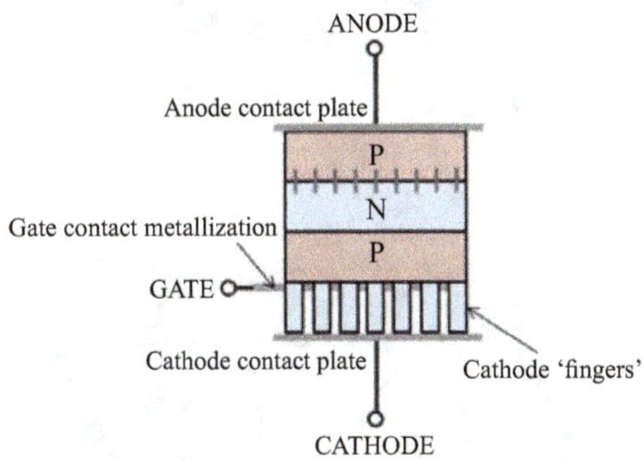

Figure 6.7 Sectional representation of a GTO

was to reduce the gain of the transistors; primarily that of the pnp transistor (Section 3.3) and one technique for doing this was to distribute shorting points across the anode p-n junction to reduce its efficiency. The thyristor's cathode shorts (Figure 3.22(a)), initially introduced to enhance the thyristor's *dv/dt* capability, had to be removed as they rendered the turn-off current even less effective and the amplifying gate structure (see Figure 3.22) also had to be eliminated. The result was a device that needed about ten times more current to turn on but that could be turned off by reversing the gate current to about 20 per cent of the load current, i.e. about 800 A were needed in the gate circuit to turn off 4,000 A in the anode. These changes lead to the GTO structure whose sectional representation is shown in Figure 6.7.

A real GTO wafer is shown in Figure 6.8, in which about 4,000 individual cathode islands can be roughly discerned, arranged in ten concentric circles.

A 4-kA GTO, as shown in Figure 6.8 (in wafer form), is typically driven by a 20-V gate unit capable of providing a turn-on gate-current, I_G of about 50 A and a turn-off current, I_{GQ}, of about 800 A or more and of biasing the gate negatively at about $V_{GR} = 15$–V in the off-state. This ensures that the GTO enjoys two stable states: that of a latched conducting thyristor and that of a blocking pnp-*transistor*, as illustrated by Figure 6.10. In effect, in the off-state, the cathode-side PN-junction is reverse-biased and effectively 'out of action' leaving a pnp transistor with an open n-base. Here, the

Figure 6.8 A 4,000-A GTO wafer (Courtesy Dynex Semiconductor Ltd)

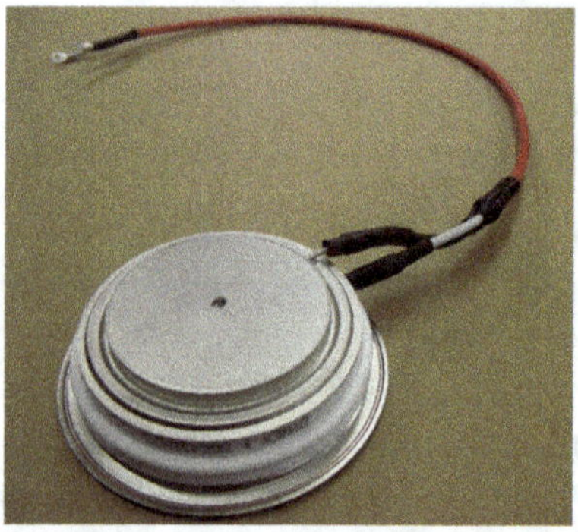

Figure 6.9 An encapsulated GTO (Courtesy Dynex Semiconductor Ltd)

GTO differs considerably from the original thyristor as its cathode-shorts, described in Chapter 3, would continuously draw current from the gate unit in the off-state. This ability, indeed, *need*, to be reverse biased when 'off', makes the GTO virtually insensitive to leakage currents, capacitive currents (due to *dv/dt*) or gate-noise, unlike the thyristor which needs to be protected from such electromagnetic interferences (EMI); the GTO, *without* its V_{GR} reverse-bias from a low-impedance source, is practically unable to block full voltage at rated temperature.

The wafer of Figure 6.8 is typically encapsulated in a press-pack ceramic housing, such as shown in Figure 6.9.

However, the 'down-side' to these rugged and stable *static* states is the dynamic transition from conduction to blocking where the GTO is neither a fully conducting thyristor nor a fully blocking transistor and undergoes some stressful and potentially destructive conditions, as depicted in Figure 6.11.

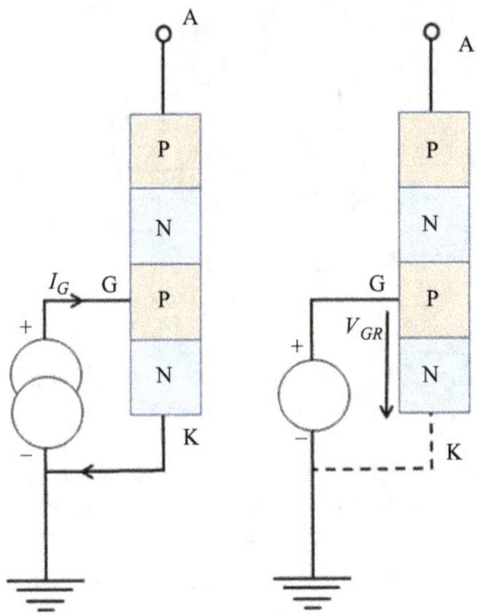

Figure 6.10 Conducting GTO (left) and blocking GTO (right)

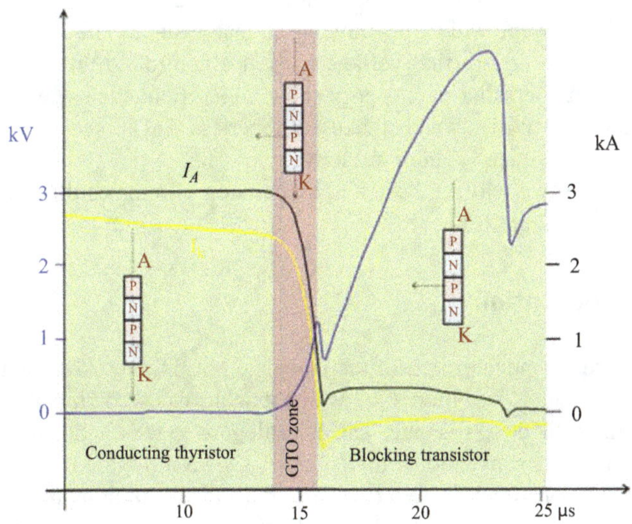

Figure 6.11 The three phases of GTO conduction with the stable zones, shown in green, and the stressful transition zone, shown in red

Turning on a GTO is relatively unproblematic: provided the gate-unit supplies a 'healthy' turn-on pulse of several tens of amps, it turns on quickly and uniformly, like a thyristor. Turning *off*, however, is a difficult process because the negative gate drive tries to break the regenerative transistor action by reducing the loop gain and as it does so, voltage rises across the device producing a *dv/dt* which reinjects

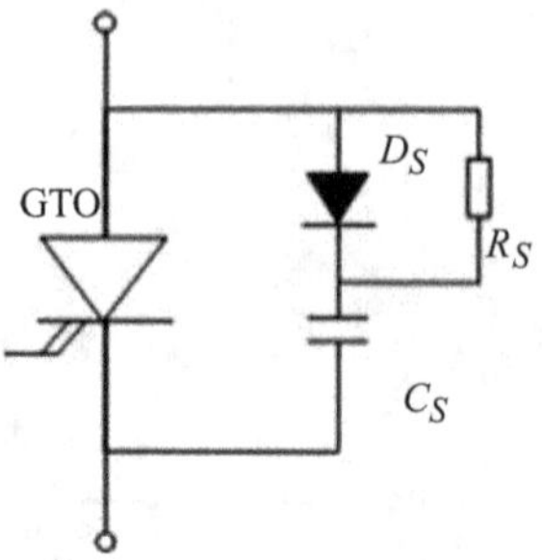

Figure 6.12 GTO with its turn-off snubber

capacitive current, raising the loop gain once more and tending to trigger the GTO back on. To avoid this, the GTO requires a large snubber of several microfarads connected across it, as shown in Figure 6.12.

During the 'dangerous' transition period, illustrated in Figure 6.11 and highlighted as the red 'GTO zone' during which both cathode and gate current are flowing, current distribution is nonuniform and the device could latch back on, locally – and fail – because the anode current would rise once again but not necessarily through the full device since some parts would have been turned off and no positive gate current would be flowing at that instant. The large snubber is needed, therefore, to ensure that voltage rises at a controlled rate (<1,000 V/μs). But these large snubbers dissipate energy at the higher frequencies needed for pulse width modulation (PWM), which makes snubberless ToDs, such as IGBTs (see Section 5.4.2), economically more attractive.

In Section 6.5, we will see how IGCT operation eliminates the dangerous red transition zone of Figure 6.11.

6.5 IGCT operation

The GCT (the semiconductor part of the complete IGCT) aims to minimise, indeed *eliminate*, the 'red zone' of Figure 6.12 which it achieves by commutating the entire current from the cathode to the gate. This is analogous to the anode current commutation used to turn off conventional thyristors, as shown in Figure 6.13, by which an external commutation circuit is used to reverse the anode current by forcing it to zero or even driving it to a negative value, limited by the reverse recovery current I_{RR} of the thyristor (see Section 3.6). In this arrangement, the current in the main thyristor is driven negatively by discharging the commutation capacitor C_{COM} via the auxiliary thyristor $T_{AUX.}$ Here, the commutation circuit is designed to provide a commutation pulse, which is significantly higher than the main thyristor's current, I_O; and the 'surplus' commutation current returns via the fly-back diode. This anode commutation is achieved by a negative current of amplitude $I_O + I_{RR}$.

Anode commutation is very effective: it treats the thyristor as a diode, by forcing reverse recovery but it requires a large number of auxiliary components for

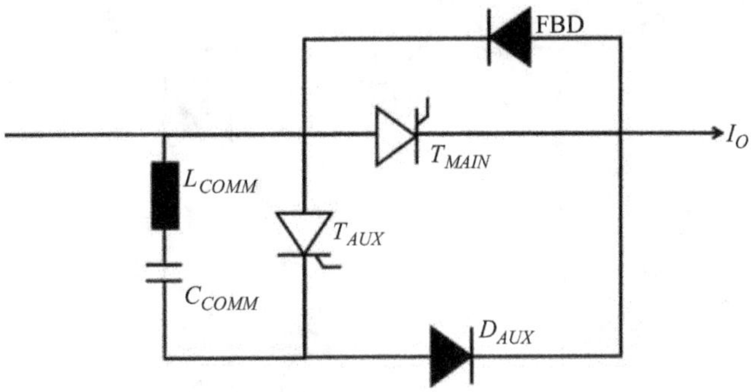

Figure 6.13 A thyristor commutation circuit

the circuit's main switches as well as large currents to oscillate throughout the system. Not only are the anode currents large but they operate at high (anode) voltage such that the auxiliary thyristor is of similar rating to the main thyristor and the passive components must handle large reactive powers.

If the same result could be achieved by commutating the cathode current out of the *gate* instead of the anode, the process would occur at gate voltage (20 V) rather than at full anode voltage (several kV) and would require much smaller passive components. This is what is achieved by the eponymous *gate-commutated thyristor*.

As described in Section 3.3, the PNP transistor of the thyristor allows high reverse-blocking voltage (in symmetric devices) at J1 (Figure 3.5) and the NPN allows high forward blocking at J2 (in all devices) but the gate reverse voltage capability at J3 is about 25 V due to the thin N-layer of the cathode emitter. This thin layer gives the gate-cathode junction, J3, the characteristics of a low-voltage diode, which means that the charge stored in this region is low compared to that stored at the HV junctions, J1 and J2. Thus, a large current from cathode-to-gate, equal to $I_O + I_{RRg}$ where I_{RRg} is the reverse recovery current at this junction, quickly blocks J3 and effectively puts the NPN transistor out of action so that the GCT now appears as a PNP transistor with an open base per Figure 6.14. This thin N-layer is quickly cleared of its stored charge and J3 blocks rapidly (<1 μs) like a fast diode, while the charge stored in the thick N base sustains current flow for several microseconds. In other words, the thyristor structure degenerates 'instantly' to that of an open-base PNP transistor and the resultant waveform resembles that of Figure 6.15.

The first consequence of this rapid shut-off of the NPN transistor is that the 'red zone' of Figure 6.11 is not only reduced but also completely eliminated so that no snubber is required – there is no need for *dv/dt* limitation and the resultant *dv/dt* is simply that which results from the reestablishment of the depletion layer within the device, i.e. the *dv/dt* is a *characteristic*, not a rating (see Section 6.10).

On the gate-voltage waveform of Figure 6.15, measured at the terminals of the ceramic GCT housing, three points of interest can be noticed:

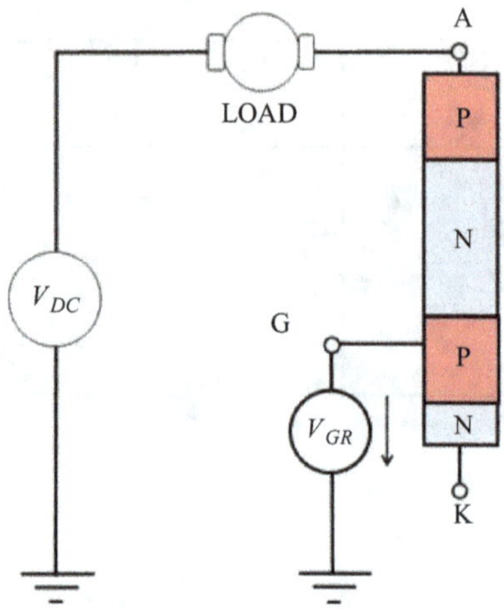

Figure 6.14 Blocking GCT in its circuit

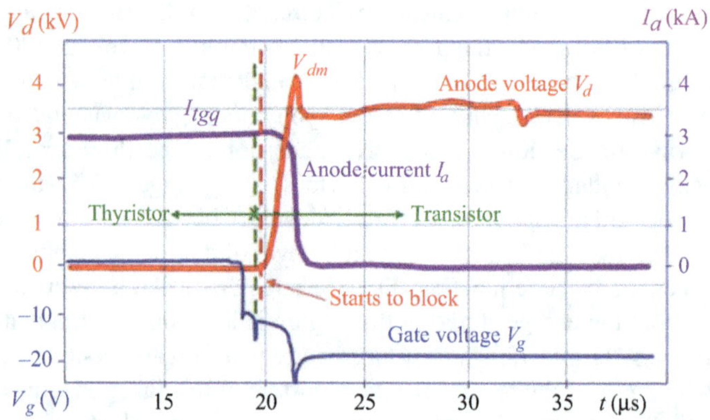

Figure 6.15 GCT turn-off waveforms at 2.8-kV$_{DC}$ and 90 °C (Courtesy ABB, Switzerland Ltd)

1. A V_{GK} step-rise from 0 V to about a 10 V plateau, corresponding to the application of the 20-V gate-drive source-voltage attenuated by the inductive potential divider of the housing inductance and the total circuit inductance (housing plus gate-drive circuit inductance),
2. Within less than 1 μs, a short peak occurs corresponding to the cathode current having been brought to zero or slightly less, as the gate-cathode junction begins recovering; a vertical dotted green line marks this point as the instant of conversion from thyristor to transistor; from this point on, anode current flows out of the gate, anode voltage rises and anode current falls.

3. The fast *di/dt* of the anode-gate current multiplied by the (albeit, small) housing gate-inductance, causes a voltage spike which is clamped to the gate-cathode avalanche voltage of about 25 V from which it recovers within about 2 μs, marking the end of gate-voltage movements and the start of the anode-current tail (of about 100 A and 7 μs, in this example).

From the above, we can state that the condition to achieve 'hard drive' or gate-commutation is that the NPN transistor be fully turned off before the PNP transistor begins to block (or 'knows what hit it'!).

A corollary of this is that the device be turned off at less than unity gain, as described earlier (i.e. $-I_G = I_O + I_{RRg}$).

The anode voltage in Figure 6.15 shows oscillation even after the end of the tail current and this is unrelated to the semiconductor but is due to the turn-on circuit, which will be discussed later under Section 6.12 'clamp circuit'. It should be noticed here that the turn-off anode and voltage waveforms from start to end of the tail-current are virtually indistinguishable from those of a high-voltage IGBT, which is understandable, as both devices are open-base PNP transistors at turn-off.

6.6 Silicon design

The operation of the GCT differs from that of the GTO as we have just seen but this difference in operation also leads to differences – in fact, substantial improvements – in the design of the wafer.

The 'red zone' of Figure 6.12 was partly due to the fact that the current tended to redistribute itself across the wafer during turn-off, leading to high current densities at certain points on the wafer during this critical turn-off phase and the thinner the wafer was, the worse this problem became. As a result, low-voltage devices needed just as much snubber capacitance as high-voltage devices and symmetric-blocking GTOs had better turn-off capability than their asymmetric counterparts. Having eliminated the 'red zone', the GCT wafer could be optimised in terms of thickness which led to the *buffered, transparent emitter* wafer design which is the direct counterpart of the buffer-layer transparent emitter IGBT. It is characterised by an additional n-layer buffer at the anode, which is itself weakly doped. This p-i-n structure stops the field at the anode, allowing a nearly *rectangular*, as opposed to *triangular*, field distribution (Figure 6.16). Since the sustained voltage is the integral of the field over distance (wafer thickness), the field, being almost constant, the required wafer thickness can be considerably reduced for the same forward-blocking voltage, accompanied by remarkable reductions in on-state voltage and turn-off energy, as shown in Table 6.1, which compares a standard GTO with its 'hard-driven' transparent emitter equivalent which gave rise to the current designs of GCTs.

The weakly doped p-i-n, field-stop structure resulted in an inefficient anode *emission* of the pnp transistor, thus, limiting its gain without recourse to the anode shorts described earlier for GTOs (Figure 6.7), which allows gate sensitivity to

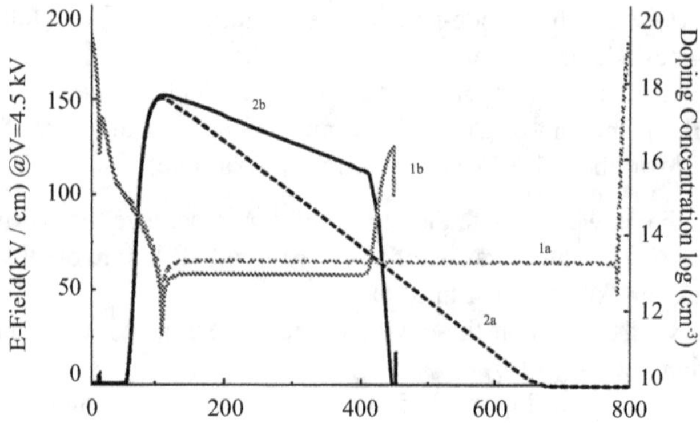

Figure 6.16 Doping profiles (1) and field distributions (2) in GTOs (a) and asymmetric GCTs (b) [3] (Courtesy ABB, Switzerland Ltd)

Table 6.1 Comparison of conventional GTO with a GCT ('HD-TGTO') [3]

	3 kA/4.5 kV 85-mm (T)GTO for 2.8-kV DC link	
Parameter	**GTO** **5SGA 30J4502 standard drive**	**HD-TGTO** **5SGT 30J4502 hard drive**
V_{TM} (on-state-voltage at 3 kA, 125 °C)	3.2 V	1.9 V
E_{OFF} at 3 kA, 125 °C,	10 Ws at 6 μF	5 Ws at 3 μF, 10 Ws at 0 μF
I_{GT} at 25 °C	3 A	0.2 A

increase by a factor of at least 10 [3]. The design was thus inherently that of an asymmetric device of negligible reverse-blocking capability, as required for VSIs.

The axial or vertical doping profile used in transparent emitter GCTs is of the type suited for high-performance fast-recovery diodes, allowing the design of high-performance reverse-conducting GCTs where the antiparallel 'fly-back' or 'free-wheel' diode can be efficiently diffused onto the same GCT wafer.

The GCT development of the 1990s, thus, represented a rare synergy of circuit techniques and semiconductor physics, which eliminated the design compromises of the GTO and heralded the advent of simple, efficient and reliable converters for very high-power operation, as will be described in Section 6.17.

6.7 Similar devices

From the foregoing, we have seen that the IGCT combines the robust static characteristics of thyristors and the robust dynamic turn-off properties of the transistor (turn-*on*, is relatively uncomplicated and will be discussed separately under Section 6.8).

The realisation that a GTO might be turned off like a transistor by driving it off with unity gain is not recent and was already exploited in the 1980s as 'Cathode-Switched GTOs' or 'Emitter-Switched Thyristors'. The idea behind these devices was to insert a switch in series with the cathode emitter of a thyristor-type device and achieve turn-off by blocking the emitter switch and diverting the full anode current out of the gate, according to the circuit of Figure 6.17, where the main device is a GTO, thus creating an emitter turn-off device or ETO.

The most successful of these devices consisted of parallel-connected MOSFET chips in series with a GTO chip either in a plastic module or as shown in Figure 6.18, with the MOSFETs mounted on the control board.

The advantage of the ETO [5] was that it required very little turn-off gate power in that current was simply diverted from the cathode to the gate by turning on Q_1 and turning off Q_2 (Figure 6.17), although a conventional high-current pulse was still needed at G_0 at turn-on (though a further variant would have allowed a high-voltage IGBT to be connected from the anode to G_0).

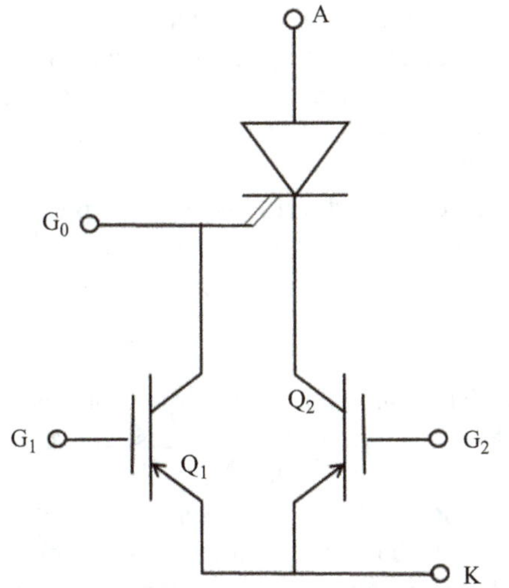

Figure 6.17 Basic circuit of the ETO

Figure 6.18 4,000 A/4,500 V ETO (Courtesy Virginia Tech)

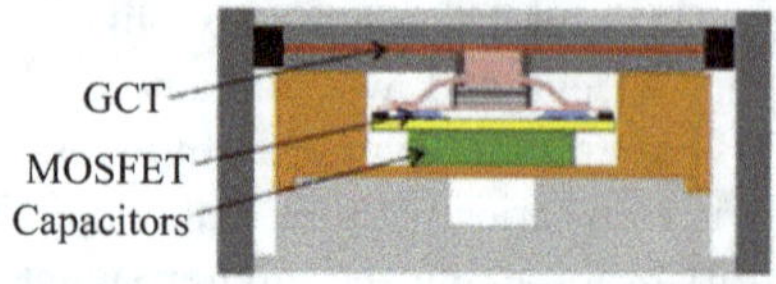

Figure 6.19 Cross-section of an ICT (Courtesy RWTH ISEA)

The obvious disadvantage, however, is the permanent presence of MOSFETs in the main current path, which increases the on-state voltage and cooling requirements.

An earlier variant of this kind of device was the MTO [4] or *MOS Turn-Off* device, which had no series-connected MOSFETs (no Q_2 in Figure 6.17) and simply required short circuiting the GTO's gate terminal with a battery of MOSFETs (Q_1).

The problem here was the difficulty of fulfilling the condition of unity-gain turn-off within 1 μs, as outlined earlier in Section 6.5 and the device, thus, required a snubber capacitor in order to survive the GTO's 'red zone' (Figure 6.12).

The ICT or internally commutated thyristor was essentially an IGCT with the gate-unit located within the GCT housing. This high degree of integration further reduced gate-circuit inductance allowing unity-gain turn-off at less than 7 V, which in turn reduced the size of the gate circuit components and allowed them to be located effectively within a machined recess in the device's cathode pole-piece, as illustrated in Figure 6.19.

The ICT has remained at the development stage, presumably due to concerns over manufacturability and thermal management.

6.8 Turn-on

Unlike a transistor (which is an amplifier), a thyristor cannot control the speed at which it turns on. This is not a problem for the GCT, because unlike the thyristor described in Chapter 3, a GCT is composed of many hundreds (even thousands) of small thyristors on the same wafer, as illustrated by Figure 6.8 and such a device, when driven by a positive gate pulse of tens or hundreds of amps, can turn on quickly and uniformly. However, in voltage source inverters, the turn-*on* of an active switch causes the turn-*off* of a diode and bipolar diodes are limited in the rate at which the current can be commutated off because a high reverse *di/dt* causes a high reverse current (I_{RR} or 'I_M') and a correspondingly high reverse-voltage peak (V_{RR} or 'V_M') the product of which (P_{RR} or 'P_M') could exceed the safe operating area (see Section 2.3.4).

The equivalent turn-on circuits for transistors and thyristors are shown in Figure 6.20.

Whereas a transistor can be turned on progressively, turning on a thyristor results in an uncontrolled rate of rise of current and, thus, a thyristor requires a *di/dt* limiting choke, *L*, if the allowable turn-off *di/dt* of the freewheel diode is not to be exceeded.

Of course, an inductance stores energy which must be dissipated before the cycle can be repeated, so a more practical *di/dt* limiting circuit for an IGCT is that of Figure 6.21.

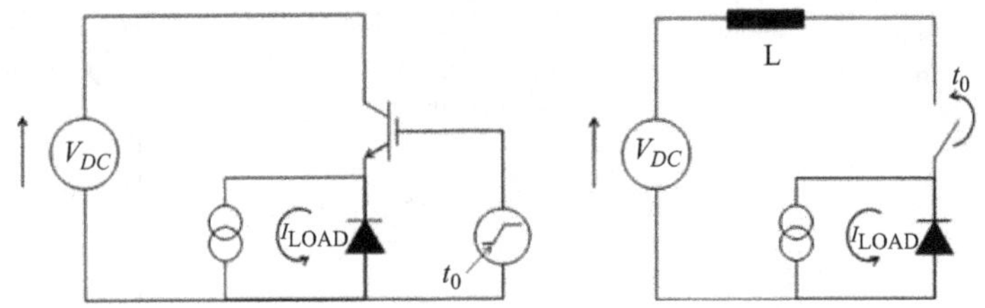

Figure 6.20 Turn-on of active devices: amplifier/transistor (left) and switch/thyristor (right)

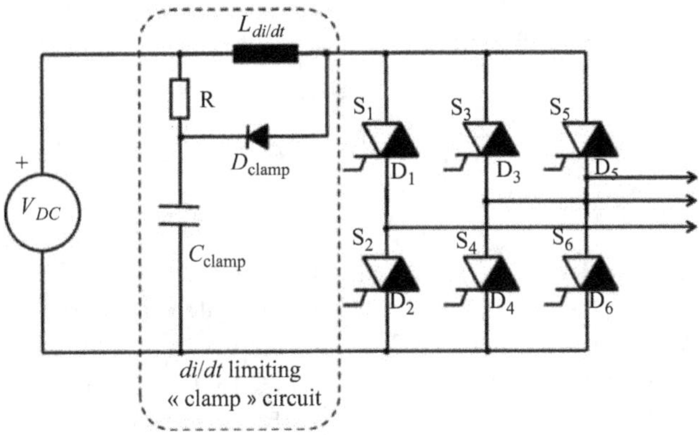

Figure 6.21 A di/dt limiting circuit inserted between a voltage source and a VSI

In the circuit (Figure 6.21), the energy stored in the choke is dissipated in the resistance during the off-time of the GCTs. The clamp capacitor initially absorbs the inductive energy, which then transfers to the resistance and 'clamps' the voltage, thus reducing transient peaks. The clamp circuit and its variants will be described in greater detail in Section 6.12.

If current rises too rapidly in a thyristor (or GTO or GCT), the anode voltage may be seen to rise again after its initial fall. However, with GTOs and GCTs which have the finely distributed gate structures of Figure 6.9, this fall can be avoided if sufficient gate current is supplied. This is illustrated by two measurements on an ABB GTO at two different gate-current levels designated 'hard' and 'soft' in Figure 6.22 which shows that, beyond a certain level of turn-on gate current, the GTO produces virtually no turn-on losses. The same result is obtained with GCTs using ten times less gate-current (see Table 6.1).

The presence of the inductance means that it is the inductance which sustains the source voltage during the transient turn-on phase and not the semiconductor although, as the current rises in the device, a large dynamic on-state voltage can be observed for the 500-A gate pulse which represents a turn-on loss in the semiconductor. This loss, however, is eliminated in the case of the 1,000-A gate pulse.

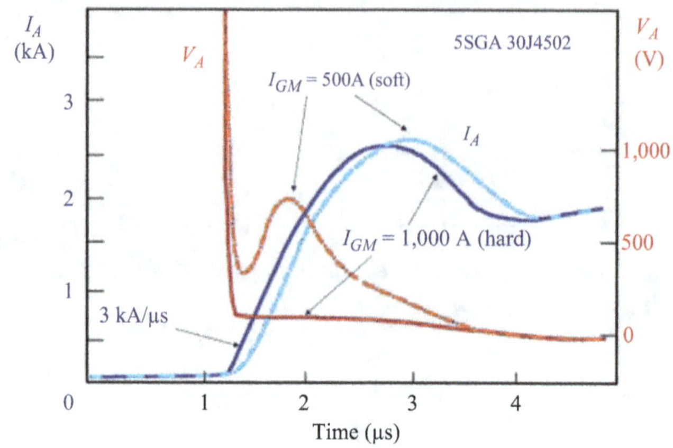

Figure 6.22 A 3-kA GTO turning on with a 500-A gate pulse and a 1,000-A gate pulse (Courtesy ABB, Switzerland Ltd)

IGCTs are supplied with their own gate-drivers, designed for their built-in GCT, which, as we have seen, have much higher gate sensitivity than GTOs. A 4-kA IGCT will typically generate a 200-A I_{GM} pulse, achieving the same turn-on waveform as the GTO of Figure 6.22 at $I_{GM} = 1{,}000$ A. In practice therefore, IGCTs have virtually no turn-on loss *in the semiconductor itself*, although, as mentioned earlier, a turn-on loss *is* generated but it is confined to the external (clamp) circuit and will be discussed later in more detail.

The high *di/dt* capability of GCTs has also been exploited for pulsed-power applications, achieving turn-on pulses of up to 150 kA at 30 kA/μs with 4″ devices [6].

6.9 Turn-off

Arguably the most important parameter of a ToD, the turn-off mechanism, was discussed under Section 6.5, 'IGCT operation'; but here, we will look at turn-off in greater detail.

Figure 6.23 shows the turn-off waveform of a 6.5-kV IGCT, which is characterised by an initially linear rising voltage which tends to flatten as the voltage peaks; the voltage continues to rise after the peak; the current fall is fairly linear down to 1,000 A and then falls at a lower rate at the start of a pronounced tail current.

Our first observation is that this wafer differs somewhat from that of Figure 6.15 which is for a 4.5-kV device at 90 °C of 1996 vintage, whereas the waveform below is for a much later generation (ca. 2010) at 130 °C but of virtually the same physical size (that of Figure 6.2). The 6.5-kV device, operating at twice the DC voltage, is seen to switch more than twice the current (>2 × 3 kA).

As a result of this increase in *safe operating area* (SOA – the product of instantaneous current and voltage), the semiconductor is severely stressed and the voltage rises faster than the space-charge region can be established, leading to high

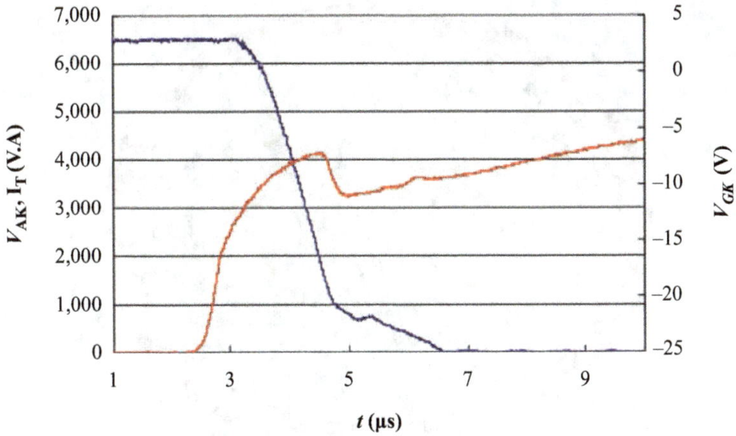

Figure 6.23 Turn-off waveforms for 91 mm 6.5-kV IGCT at $V_{DC} = 6$ kV, $I_T = 2$ kA and $T_j = 130$ °C (Courtesy ABB, Switzerland Ltd)

instantaneous fields and *dynamic avalanche*, causing charge to be reinjected as the device tries to remove it. The ability of the device to survive this condition, also found in modern IGBTs [7], is an indication of the uniformity of the turn-off process achieved by today's ToDs and some fast-recovery diodes.

In fact, by increasing the stray inductance in the circuit at turn-off, some designs allow the device to undergo a 'quasi-static' avalanche after the dynamic avalanche, further adding to device stress as shown in Figure 6.24. This is not a normal operating condition for an IGCT (the dynamic losses are considerable), but it illustrates the device's ruggedness, homogeneity and ability to withstand certain

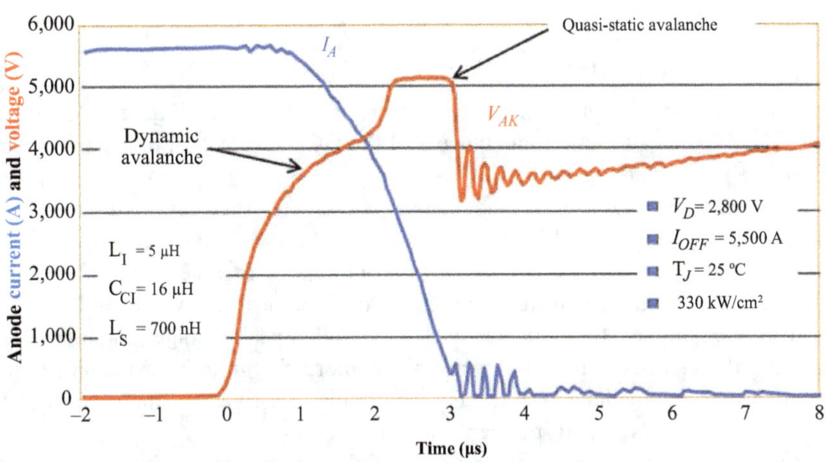

Figure 6.24 Example of extreme SOA testing on a 4.5-kV 4″ IGCT at 25 °C (Courtesy ABB, Switzerland Ltd)

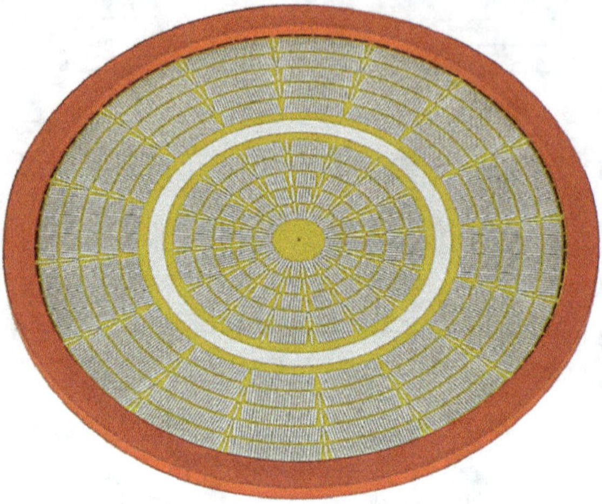

Figure 6.25 Fan-shaped cathode-finger layout of CRRC IGCT to improve cathode area utilisation and gate-current access to the individual cathodes (Courtesy CRRC)

fault conditions – and is a tribute to the manufacturer's R&D efforts to obtain such uniform turn-off. This mode of operation is known as *switching self-clamping mode* (SSCM), as described in detail in the literature [7].

The turn-off waveforms depicted so far show a variety of voltage and current curves due to numerous factors of which the most important are device design, junction temperature and stray inductance.

Both the vertical [8] and the horizontal designs of the GCT wafers continue to undergo continuous improvement as the novel finger layout of Figure 6.25 shows (see Section 6.9.2).

6.9.1 Stray inductance

The clamp circuit of Figure 6.21 should be refined to that of Figure 6.26, which shows the DC-link parasitic inductance and the GCT connection inductance, relevant to the present discussion.

Connection inductance L_{S2}, not being part of the freewheel paths of D_C or FWD, contains energy which will lead to additional losses in the GCT as well as to overvoltages (V_{DSP}) across it and this energy will also affect the shape of the turn-off waveforms, as illustrated in Figure 6.27, for three different values of L_{S2}.

In general, higher stray inductances lead to higher peak voltages and longer fall times and the fact that longer fall times allow more charge to be extracted during the fall results in smaller tail currents. Nevertheless, a five-fold increase in stray inductance still leads to a 30 per cent increase in E_{OFF}.

Again, the device's ability to withstand high stray inductance is not to say that efforts should not be made to minimise L_{S2} – low stray should always be a design objective – but once again, the robustness of these devices is illustrated.

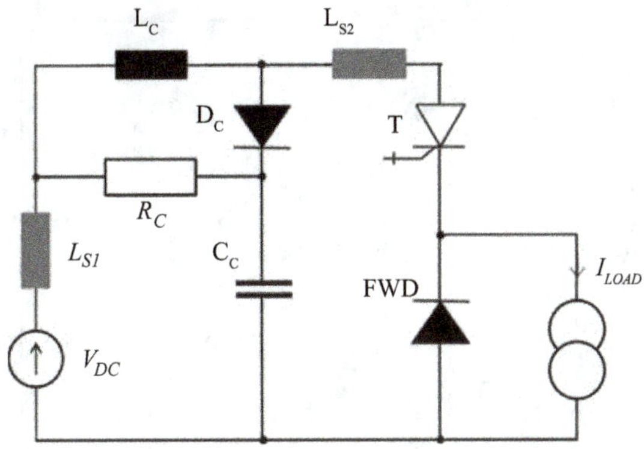

Figure 6.26 Complete equivalent circuit of an IGCT commutation cell showing the stray inductances in the DC link and in the GCT connection

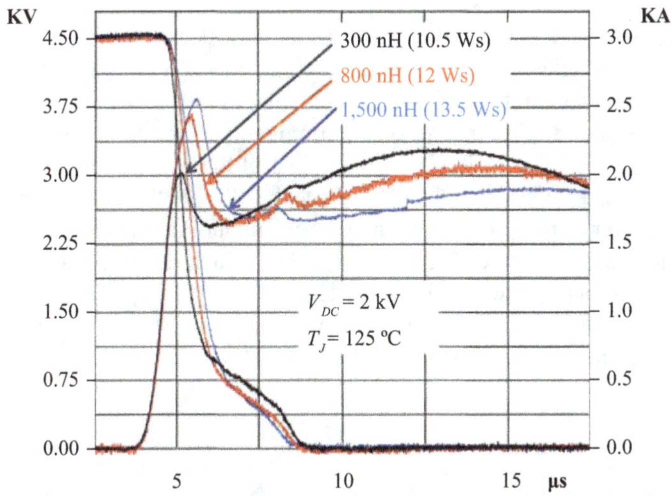

Figure 6.27 Effect of L_{S2} on V_{DSP} and E_{OFF} on a 4.5-kV 4" GCT (Courtesy ABB, Switzerland Ltd)

6.9.2 Device design

As with all bipolar semiconductors, there is a direct trade-off between conduction losses (e.g. on-state voltage, V_{TM}) and turn-off losses (or turn-off time or stored charge etc., depending on device type – see Figure 3.32(b)). This trade-off is described by a *technology curve* which, for GCTs, is usually a curve of turn-off energy, E_{OFF}, versus on-state voltage, V_{TM} (or *specific* E_{OFF} vs. V_{TM} – see Figure 6.28) and is usually achieved by some kind of lifetime control or anode efficiency design or combinations thereof, as mentioned in Section 5.8.2.1.

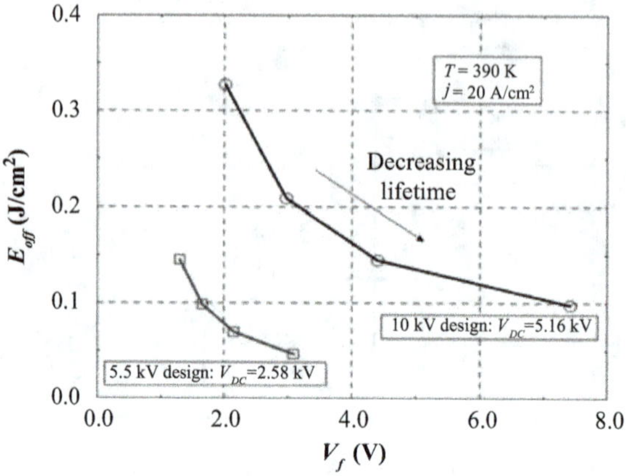

Figure 6.28 Technology curves for 5.5 and 10-kV GCTs (Courtesy ABB, Switzerland Ltd)

This means that a device with low conduction losses will have high switching losses and be suitable for low-frequency applications. Thus, for a given voltage class, some manufacturers may offer different devices, optimised for different applications such as circuit breakers (infrequent switching) and motor drive inverters (high-frequency switching). The effect of the chosen compromise can be seen in Figure 6.29 for 4.5-kV/4″ GCTs where the 'Type 12' device has a very slow voltage rise, a very slow current fall and a large tail current, indicative of very large switching losses as compared, say, to 'Type 11', which is based on a design with short carrier lifetime, allowing faster switching.

The three types of GCT measured dynamically in Figure 6.29 have the main characteristics and ratings, shown in Table 6.2.

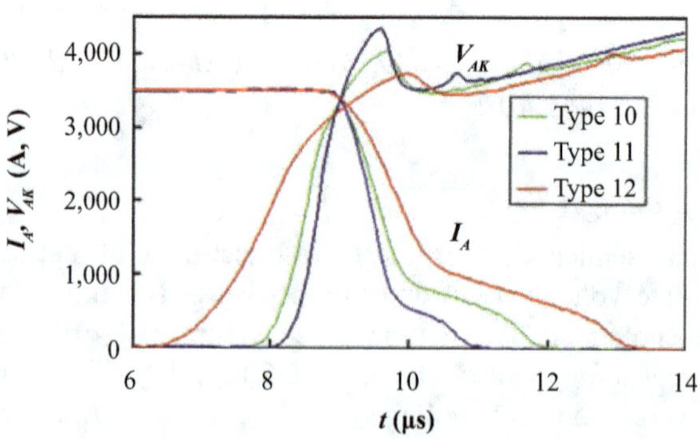

Figure 6.29 Effect of carrier lifetime engineering on turn-off (Courtesy ABB, Switzerland Ltd)

Table 6.2 Comparison of three different levels of lifetime control for 4.5-kV/4" ABB IGCTs

IGCT type (91 mm/4.5 kV)	Type 12	Type 10	Type 11
V_{TM} @ 4 kA/125 °C	2 V	2.7 V	3.5 V
E_{OFF} @ 4 kA/2.8 kV/125 °C	37 Ws	22 Ws	17 Ws
I_{RMS} @ $T_C = 85$ °C	3,300 A	2,700 A	2,200 A
Lifetime	High	Medium	Low
Typical operating frequency	Low	Medium	High

6.9.3 Temperature

The heat-loss P_{TH} that can be extracted from a semiconductor determines the amount of power that the device can control and, hence, the design rating of its converter; it is given by the relationship:

$$P_{TH} = [T_J - T_s]/\left[R_{TH(J-C)} + R_{TH(C-S)}\right]$$

where T_J and T_S are junction and heatsink temperatures, respectively, and $R_{TH(J\text{-}C)}$ and $R_{TH(C\text{-}S)}$ are the thermal resistances from junction to device-case and from case to heat-sink, respectively. Since the thermal resistances are fixed device characteristics, then the greater the difference between T_J and T_S, the greater the system power capability.

Thus, users demand as high a junction temperature rating as possible, as it determines their system cost and therefore, device manufacturers must juggle with the compromises, described in this section, in order to satisfy that demand.

Unlike a thyristor, the GCT is permanently controlled by its GU: in the on-state, a forward gate current flows gate-to-cathode; in the off-state, a voltage is applied cathode-to-gate. Thus, as already explained in connection with GTOs (Section 6.4), any dynamic or static current flowing from the anode is diverted to the GU and cannot flow over the gate-cathode junction and provoke spurious triggering, as described in Section 3.5.1.

However, spurious triggering is not the only nefarious effect of excessive leakage current, since a small current flowing across a high voltage will nevertheless produce a significant dissipation, all the more so that leakage currents are not necessarily uniformly distributed (they tend, in fact, to be greater at the edge of a pressure-assembled semiconductor since the junction termination is not contacted by the pole-pieces, i.e. not well cooled). Local heating produces more local leakage and yet, more heating, culminating in 'thermal runaway' and destruction of the wafer.

Roughly speaking, leakage current doubles for every 11 °C junction temperature increase as given by:

$$P_{TH} = V_0 \cdot I_0 \cdot 2^{\frac{(T_j - T_0)}{\Delta T}}$$

where ΔT represents the 11 °C mentioned above, I_0 is the leakage current at the applied voltage V_0 and the reference temperature is T_0 [9].

Another problem associated with high temperature is that the greater number of available charge carriers results in higher turn-off losses (partly mitigated by lower conduction losses due to those same carriers).

Increasing junction temperature results in more charge carriers being available, which has the same effect as increasing carrier lifetime, as shown in Figure 6.29 for three different lifetimes. Thus, device designers must strive to further minimise turn-off losses for high-temperature operation or the additional losses would negate any benefits of higher T_J.

Allowing the peak temperature to reach higher limits, automatically increases the amplitude of a power cycle, which, as we will see later under Section 6.17 'Reliability', is what determines a device's end-of-life. Therefore, measures must be adopted to mitigate against excessive 'scrubbing' action, which deforms the metallic interfaces within a press-pack device.

Figure 6.30 illustrates the results of just such a development by which the flattening and widening of the cathode fingers has been reduced from the usual 36 per cent to just 8.5 per cent after 100 temperature excursions of 100 °C. These are 'deep temperature cycles' used for accelerated cycling tests.

Most wear-out, however, is caused by *power* cycles whereby the semiconductor is heated internally by its own dissipation. The number of cycles that can be applied before the failure rate starts to increase will depend on the size and complexity of the device. In the case of GCTs, complexity is *low* (no bonds, solder joints glues or organic materials), however, size is *high* (monolithic chips of up to 150 mm in diameter). Thus, in Figure 6.30, a cathode finger from the inner ring/row of a 91-mm GCT is compared with a finger from the outer row. Both have been subjected to the same number of cycles (they are on the same wafer) but the outer fingers have necessarily undergone greater 'scrubbing' action.

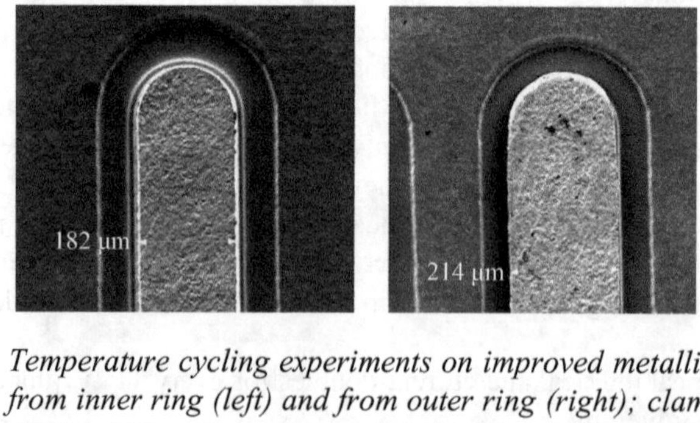

Figure 6.30 Temperature cycling experiments on improved metallisation finger from inner ring (left) and from outer ring (right); clamping pressure 40 kN; 103 cycles, –40 to 140 °C; reduced degradation of the cathode metallisation under thermomechanical stress 8.5 per cent widening compared to 36 per cent for conventional metallisation (Courtesy ABB Semiconductors Ltd) [8]

A final complication in this connection is that material hardness reduces with increasing temperature. So, a device will suffer more wear by cycling from 60° to 140° than from, say 40° to 120°, despite identical temperature excursions.

Having said all this, 140 °C IGCTs have indeed been developed [8] though the standard junction temperature rating remains 125 °C for most devices.

6.10 Data-sheet parameters

There are principally two kinds on device parameters: *ratings*, which are constraints which a device is guaranteed to withstand but beyond which it may fail and *characteristics*, which are measurable or 'virtually measurable' physical parameters.

6.10.1 Ratings

Ratings are the consequence of various device characteristics and they determine electrical, thermal and mechanical limit values which, if exceeded, may result in device failure. The most important ratings are listed below.

6.10.1.1 I_{FAV}, I_{TAV} – Maximum allowable average forward current

This is the maximum allowable forward current for a given case temperature T_C. It depends on current waveform and frequency but is usually quoted in data-sheets for 50/60 Hz sine-waves of 180° conduction (50 per cent duty cycle). This current rating is just one point on a set of curves for a variety of conduction angles and case-temperatures. This rating depends on the internal thermal resistance (see characteristics) and on another rating: the virtual junction temperature T_J. [The subscript 'F' applies to the reverse-conducting part of a GCT and 'T' to the thyristor part.]

6.10.1.2 I_{FRMS}, I_{TRMS} – Maximum root mean square current

This rating specifies the absolute maximum current for the internal connections of a semiconductor package for any current wave-shape or case temperature. It is particularly meaningful in the case of complex packages where heat generation due to I^2R heating of the leads and connectors is not negligible. It is of debatable importance in the case of press-pack semiconductors where the overwhelming bulk of the losses are generated by the nonohmic characteristics of the semiconductor junction (see Figure 6.31). [The subscript 'F' applies to the reverse-conducting part of a GCT and 'T' to the thyristor part.]

6.10.1.3 T_{vj} – Virtual junction temperature

The 'virtual junction' temperature is an equivalent or 'effective' temperature, which may be ascribed to a junction by calculation. Temperature affects almost all the characteristics and ratings of semiconductors but most particularly those associated with voltage because the depletion region of a p-n junction is highly temperature-sensitive. The semiconductor chip may be only a fraction of a

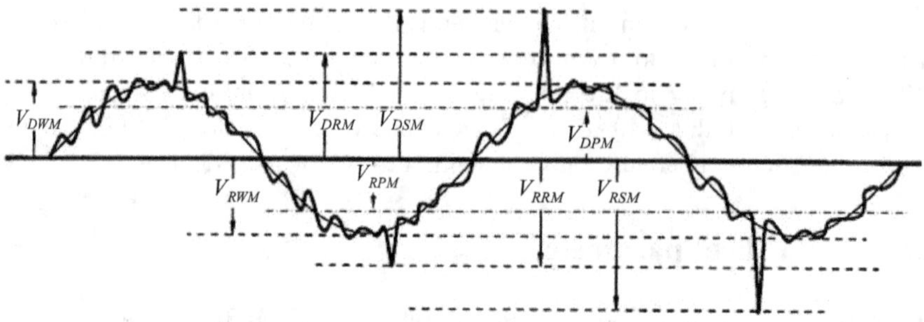

Figure 6.31 Definition of various voltage ratings

millimetre thick (about 1 mm at 10 kV) but contains one or more p-n interfaces. Each p-n junction is only a few tens of microns deep but may be tens of millimetres wide. Most of the heat may indeed be generated at the p-n interface ('built-in voltage') but the silicon 'bulk' also offers ohmic resistance as do also the internal device contacts or solder joints. Furthermore, the vertical and lateral current and thermal paths are not identical, resulting in vertical and lateral temperature gradients. It is, therefore, 'virtually' impossible to know where the junction is hottest or how hot it might really be.

Through single-shot factory tests in which the entire device is heated to an accurately known temperature (say, 125 °C), the behaviour of the junction can be measured. Under dynamic operating conditions whereby heat is generated internally and temperature gradients will exist across the silicon, the same behaviour can be guaranteed if the 'virtual junction temperature' (derived from the known power dissipation and the (virtually) known effective thermal resistance) is the same as the specified value. Maximum virtual junction temperature is, thus, the calculated value which must not be exceeded in normal operation.

6.10.1.4 V_{DRM} – Maximum direct repetitive voltage

This is the rated maximum voltage which may be applied to thyristors in the forward direction. It is also known as the 'off-state voltage'. This rating is usually established for a sinusoidal voltage of line frequency (see Figure 6.31). For IGCTs, this rating is given with the gate unit energised (negative gate voltage applied to GCT).

6.10.1.5 V_{RRM} – Maximum reverse repetitive voltage

This is the rated maximum voltage, which can be applied repetitively in the reverse direction of an SGCT (see Figure 6.31).

6.10.1.6 V_{LTDS} or V_{DC} – Long-term DC stability or permanent voltage

Also known as V_{DPM} for direct permanent voltage, this rating highlights the difference between the DC- and AC-blocking capabilities of high-voltage devices at elevated temperature.

In the case of the maximum direct repetitive-blocking rating, V_{DRM}, the peak voltage is applied cyclically and at a low duty cycle. If the same voltage were applied 'permanently', thermal runaway would probably occur [9]. Furthermore, the charge accumulation at the junction surface might cause 'ionic drift' on some devices resulting in long-term leakage current instability.

Certain applications nevertheless require that a device be continuously subjected to a DC voltage and type or reliability testing may require that the long-term stability of a device under continuous DC bias be assessed. In both DC and AC applications, devices are generally operated at voltages well below the repetitive rating (V_{DRM}). Where such devices are furthermore series-connected, close leakage current banding is required and this banding must be known to be stable over time.

Furthermore, cosmic radiation can cause catastrophic ionising events within the silicon wafer and the probability of this happening is strongly dependent on voltage [10].

6.10.1.7 V_{RPM} – Maximum reverse permanent voltage

This is the same rating as V_{DPM} above for devices with reverse-blocking capability.

6.10.1.8 V_{DSM} – Maximum direct surge voltage

This is the rated maximum forward voltage, which may be applied in a fault condition. The duration of the allowable surge is specified and is usually of the order of 1 to 5 ms. V_{DSM} is a nonrepetitive rating derived from the specifications of a thyristor, which is exposed to line-borne transient overvoltages. It is not common to GCT specifications since these are normally operated from DC-links, which are generally decoupled from the grid (see Figure 6.31).

6.10.1.9 V_{RSM} – Maximum reverse surge voltage

As for V_{DSM} above but for the reverse direction.

6.10.1.10 V_{DWM} – Maximum direct working voltage

This is the rated maximum forward (or 'off-state') voltage, which may be applied repetitively at any frequency or pulse width (see Figure 6.31). It is generally used as a recommended or allowable voltage in an AC line-operated environment. Its use presupposes knowledge of the transient voltage levels in a given installation, which depend on the electrical environment and surge suppression systems (if any). It is not usually specified for GCTs.

6.10.1.11 V_{RWM} – Maximum reverse working voltage

This is the same as V_{DWM} above but for the reverse direction.

6.10.1.12 I_{TSM}, I_{FSM} – Maximum forward surge current

This is the maximum nonrepetitive single-cycle surge-current, which may occur 'infrequently' in the life of the device. As with all nonrepetitive ratings, the device may have a 'memory' of the event.

The surge current is expressed as a single half sinewave of specified duration. Though this is typically 50 or 60 Hz for line-frequency thyristors, inverter devices such as GCTs operate off DC links of variable design, hence the pulse width t_P is specified for such devices (typically 3–30 ms). During a surge, the junction temperature may exceed 400 °C thereby subjecting the device to an extreme temperature cycle. At these temperatures, the device has little or no voltage-blocking capability. If full reverse blocking must be assured immediately (e.g. for SGCTs) after the pulse, it may be necessary to take steps to reduce the surge current amplitude. As a rough guide, any current surges which are to occur more than 100 times in the life of the device should not exceed 70 per cent of the rated surge current. (The subscript 'F' applies to the reverse-conducting part of a GCT and 'T' to the thyristor part.)

6.10.1.13 I^2t – Amp-squared-seconds for fusing

This limiting load integral is proportional to the maximum energy, which a protecting fuse may let through allowing a semiconductor to be matched to its fuse. Since a fuse is a resistive element, the energy required to melt it is given by:

$$E = R_{FUSE} \times I_{TSM^2}t.$$

The specified values are given for the same pulse widths as I_{TSM} and relate to the surge current by:

$$I^2t = I_{TSM^2} \times t_P/2.$$

6.10.1.14 *di/dt* – Maximum rate-of-rise of on-state current

Sometimes defined a *di/dt* CRIT for 'critical rate-of-rise of anode current', this parameter is of more importance for the FWD turning off than for the GCT turning on. In Section 6.8, we saw that the GCT has very little need for *di/dt* limitation so *di/dt* is determined by the FWD characteristics and the clamp circuit (see Section 6.12), which also determines the GCT's turn-off rating so the 'allowable' rate-of-rise of current is indirectly set by parameters other than the device's inherent capability.

6.10.1.15 *dv/dt* – Maximum (or critical) rate-of-rise of off-state voltage

IGCTs do not have a *dv/dt* limitation when the gate-unit is energised and it is not allowed to apply voltage to a GCT without a negative gate bias, in an application.

GCTs sold separately (i.e. for users who design their own gate-drive circuit) may have a *dv/dt* rating specified and the conditions should include the gate-cathode voltage and source impedance.

6.10.1.16 P_{GAV} – Average gate power

This is the maximal average allowable gate power dissipation in the forward direction for any wave-shape or duty cycle for a GCT.

IGCTs do not have such a specification; instead, the maximal input power to the gate-unit is specified.

6.10.1.17 P_{GM}, I_{FGM}, V_{FGM} – Maximum gate power, current, voltage

These are the maximal peak allowable gate parameters in the forward direction for a GCT. They are normally valid for a given pulse duration but this is generally deemed uncritical since the gate area is very large, turn-on pulses are very short and thus, the ability of the GCT to absorb a large power pulse is considerable and well beyond the ability of a gate driver to produce it. IGCTs do not have such a specification as the gate unit is designed to provide the necessary power.

6.10.1.18 V_{RGM} – Maximum reverse gate voltage

This is the maximal allowed negative gate-cathode voltage for a GCT and is not specified for an IGCT for which an input power-supply voltage range is given.

6.10.2 Characteristics

Characteristic parameters are one of two kinds: *characteristics* needed for system design and *control parameters* which are intended for production control, i.e. they describe the uniformity and conformity of device batches and may serve as a figure-of-merit for certain selection criteria.

6.10.2.1 V_{FM}, V_{TM} – Maximum on-state voltage

This is a control parameter as it expresses the on-states voltage at one current level only and frequently at 25 °C. It is, therefore, of little value in assessing the prospective conduction losses which are best derived from the on-state curve at rated junction temperature or from the V_{TO} and R_D values, which are used to approximate the curve as shown in Figure 6.32 for the 125 °C curve.

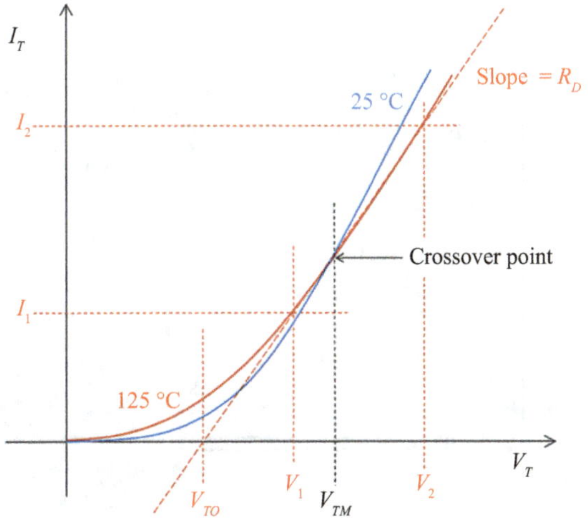

Figure 6.32 The on-state characteristics of GCTs

V_{TM} is frequently specified at or near 'the crossover point'. This is the current level at which the 25 °C on-state characteristic intersects the characteristic at rated operating temperature (usually at 125 °C). At this level, the on-state voltage is independent of temperature and the measurement is less likely to be affected by current wave-shape, duration or repetition rate.

The on-state curve can be approximated by the relation:

$$V_T(i) = V_{TO} + R_D \times I$$

and the conduction loss can be calculated from:

$$P_T = V_{TO} \times I_{AV} + R_D \times I_{RMS^2}$$

where I_{AV} and I_{RMS} are the average and RMS values, respectively, of anode current.

6.10.2.2 I_{DRM} – Maximum direct repetitive leakage current

This is the maximum characteristic leakage current at the rated maximum forward repetitive voltage (see 'V_{DRM}'). It is measured at rated temperature (on which it is highly dependent). Since V_{DRM} is specified as the peak of a sinusoidal voltage (50/60 Hz), I_{DRM} corresponds to the peak of this voltage and is itself the peak of the nonsinusoidal forward leakage current.

Most manufacturers test I_{DRM} with a repetitive 50 or 60 Hz voltage of amplitude V_{DRM} with the device externally heated to rated temperature ($T_C = T_J$). The product of leakage current and applied voltage, however, can cause the device's junction temperature to increase cumulatively at each successive line-frequency pulse leading to thermal runaway. For this reason, some manufacturers test their devices on a 'pulsed' basis in which a single line-frequency sinewave is applied at the reduced rate of, say, 1 Hz. An alternative approach, used for very high-voltage devices, as in high voltage direct current (HVDC), uses a special 'Prussian helmet' waveform as shown in Figure 6.33, which more closely

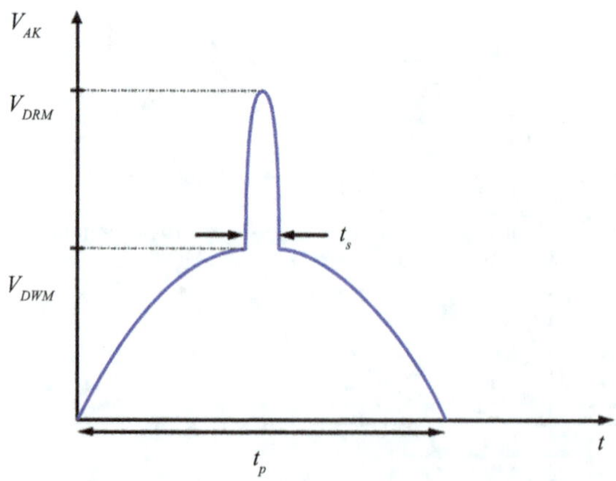

Figure 6.33 A practical test wave for measuring I_{DRM} at 50 Hz on HV devices (Courtesy ABB, Switzerland Ltd)

matches real applications as defined in Figure 6.31. The narrow pulse, t_S, is typically of 0.25 ms duration.

6.10.2.3 I_{RRM} – Maximum reverse repetitive leakage current

This is the same as I_{DRM} but in the reverse direction.

6.10.2.4 I_{DPM}, I_{RPM} – Maximum direct/reverse permanent leakage current

These are the maximal characteristic forward and reverse leakage currents at the rated maximal direct and reverse permanent voltages (see '$V_{DPM/RPM}$'). Since $V_{DPM/RPM}$ is less than $V_{DRM/RRM}$, $I_{DPM/RPM}$ will be very much lower than $I_{DRM/RRM}$ and will allow the optimised design of series sharing resistor networks where required. It is not a widely used parameter and rarely appears in device data-sheets. As with all leakage currents, it is very dependent on temperature.

6.10.2.5 V_{GT} – Gate-voltage to trigger

This is the gate voltage level above which a GCT is guaranteed to trigger in a given set of conditions which are typically:

- $T_J = 25$ °C,
- $V_A = 24$ V and
- $R_A = 0.1$ Ω (anode resistance).

From these 'unrealistic' test conditions, it can be recognised that V_{GT} is a control parameter. Under no circumstances should a GCT be operated with a gate voltage, which only *just* triggers the device as failure would result. Since bipolar semiconductors, in general, are current, not voltage, controlled devices, V_{GT} is a parameter of little importance except in ascertaining that the internal connections to the gate region have low contact resistance (see also 'I_{GT}').

V_{GT} is not specified for IGCTs.

6.10.2.6 I_{GT} – Gate current to trigger

This is the gate current level above which the device is guaranteed to trigger in a given set of conditions which are typically those of V_{GT}. I_{GT} is also a control parameter and cannot be used to turn on a GCT in an application.

6.10.2.7 I_{GM} – Maximum initial gate current

I_{GM} is a recommended turn-on gate current level for GCTs and it is usually 100 to 500 times greater than I_{GT}. The recommended I_{GM} must be used if the rated *di/dt* of the device is to be achieved. It is not normally necessary to maintain I_{GM} for more than a few microseconds (until anode current has reached its peak value following turn-on). After this period, the gate current may be reduced to typically six times I_{GT}. Figure 6.34 illustrates a typical gate pulse and defines the relevant quantities. Not only is I_{GM} important but also the rate at which it is established (di_{GM}/dt), which is typically a fifth of the anode *di/dt*.

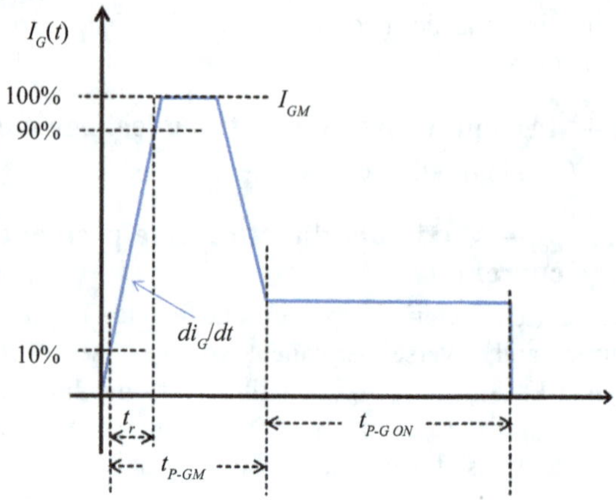

Figure 6.34 Typical gate pulse

6.10.2.8 E_{ON} – Turn-on energy

All semiconductors produce an energy loss when they switch from one state to another and this is associated with injecting carriers when changing from the blocking to the conducting state or extracting them when turning off (see E_{OFF}, below).

The turn-on energy of a GCT is mainly associated with the spreading of the conducting zone as the current ramps up from the instant of turn-on to the instant at which the full device area is in conduction. Since a GCT is composed of hundreds or thousands of very small 'thyristors', typically only 200 μm wide (see Figure 6.30), distributed across the wafer and triggered by a high gate current of very short duration, this process is very fast and practically lossless, as illustrated by Figure 6.22.

6.10.2.9 E_{OFF} – Turn-off energy

This is the energy dissipated by any ToD at turn-off as illustrated in Figure 6.23. The concomitance of high voltage and current during the short turn-off event leads to a significant energy loss measured in Joules (see also E_{ON} above).

6.10.2.10 P_T – Power loss

The total semiconductor losses are composed of conduction, turn-on and turn-off losses with leakage and gating losses generally neglected as being insignificant.

Thus, IGCT losses may be approximated to:

$$P_T = F \times (E_{ON} + E_{RR}) + V_{TO} \times I_{AV} + R_D \times I_{RMS^2}$$

6.10.2.11 $R_{TH(C\text{-}S)}$ – Thermal resistance case-to-sink

The contact between the case of a semiconductor (or the baseplate of a module) and the heat-sink on which it is mounted offers a thermal resistance to heat flow from the semiconductor. This resistance is measured in '°C per Watt'. The case temperature T_C of a semiconductor can be determined by the relationship:

$$T_C = T_{HS} + R_{TH(C-S)} \times P_T$$

where T_{HS} is the heat-sink temperature and P_T is the semiconductor power dissipation.

Case-to-sink thermal resistance depends on the following factors:

- pressure,
- surface finish,
- flatness and
- type of thermal interface compound used (if any).

A thermal compound (usually, a silicone grease sometimes containing thermal conduction-enhancing particles and oxide inhibitors) is used to fill the voids, which exist between the imperfect faces of the semiconductor case and heat-sink. Excessive use of thermal compounds will actually insulate those interface areas which could have made a direct contact.

In the case of modules, the case-to-sink interface is generally a thermal contact only; whereas in the case of press-pack and stud semiconductors, the interface is also electrical. Since thermal greases are poor electrical conductors, they should, again, be used sparingly.

As a general guide, a light silicone grease-containing oxide inhibitors should be used.

6.10.2.12 $R_{TH(J\text{-}C)}$ – Thermal resistance junction-to-case

This is the internal thermal resistance between the 'virtual junction' in the silicon chip and a point on the case. Because the internal construction of a semiconductor contains several thermally massive components, the response of the junction to a step-increase in power is gradual, the steady-state temperature being reached in a few seconds. Knowledge of this transient response allows the designer to optimise the use of his components if his current waveform contains narrow pulses of high amplitude. The figures quoted for $R_{TH(J\text{-}C)}$ are usually the steady-state values $R_{TH(J\text{-}C)\text{-}DC}$ and curves are given showing the 'transient thermal impedance' $Z_{TH(J\text{-}C)} = f(t)$. For short pulse durations, values of $Z_{TH(J\text{-}C)}$ are lower than the steady-state value. Using these values results in lower peak junction temperatures being calculated because the 'transient value' allows for thermal capacitance, which takes a certain time to 'charge up'.

6.10.2.13 T_C – Case temperature

The case temperature is neither a rating nor a characteristic but simply an operating condition for defining certain ratings. It is of particular importance in defining the allowable currents for different wave-shapes and frequencies. Since the thermal resistance junction-to-case is fixed (and specified), the allowable power-dissipation, P_T, is given by:

$$P_T = (T_J - T_C)/R_{TH(J-C)}$$

(allowance being made for junction temperature ripple – see $R_{TH(J\text{-}C)}$).

Since the current rating of a semiconductor depends on its cooling conditions, a large power dissipation (and, hence, a large current rating) can be allowed if either T_J is increased or T_C is lowered (provided I_{RMS} is not exceeded).

6.10.2.14 t_{gt} – Maximum turn-on time

The maximum turn-on time of an SCR is defined as the time-lapse between 10 per cent of the forward gate-current and 10 per cent of the anode voltage when switching into a resistive load, per Figure 6.35.

The turn-on time includes the delay time and the current rise-time such that:

$$t_{gt} = t_d + t_r$$

On a purely resistive load, such as that used for the test, the current rise-time and the voltage fall-time would be the same; but in practice, load currents are, at least partially, inductive so that the 'rise time' is defined on the falling voltage waveform.

Turn-on time is affected by:

- gate current,
- anode load resistance and
- junction temperature.

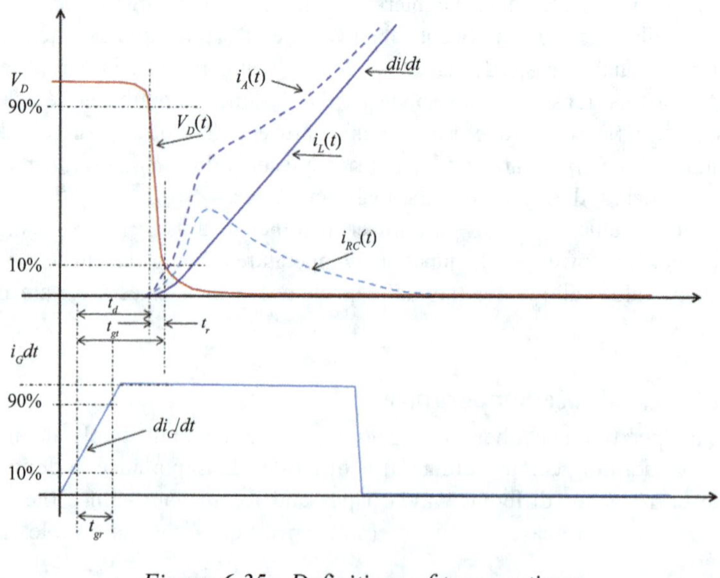

Figure 6.35 Definitions of turn-on times

Figure 6.35 shows the load current rise by which *di/dt* is generally defined. Additionally, if there is a snubber (see Figure 6.12), this also discharges its current, $i_{RC}(t)$, into the device with high (and undefined) *di/dt* and the resultant anode current, $i_A(t)$, is, therefore, the sum of these two currents.

6.10.2.15 $T_{ON\text{-}MIN}$

Although the IGCT turns on uniformly with its high I_{GM} gate pulse, of the order of hundreds of amps, a certain time is required for the hole/electron pairs to reach equilibrium, as can be deduced from Figure 6.22 and thus, during a few microseconds, the on-state voltage can be seen to settle towards its steady-state value. Turning the device off before the specified on-time might result in the unity gain turn-off condition not being fully met. In Section 6.5, the importance of turning off the NPN transistor *before* the PNP transistor was described and if the PNP transistor were not fully saturated, it *might* turn-off faster than expected and the full turn-off rating of the deice might, therefore, not be attained. Of course, turning the device off prematurely would result in lower switching losses which might be desirable for some applications, but operation at reduced on-time should only be envisaged in consultation with the IGCT manufacturer.

6.10.2.16 $T_{OFF\text{-}MIN}$

When the IGCT turns off, a tail current can be observed due to decay of the remaining charge in the PNP transistor. Towards the end of the tail, the current distribution in the device may not be perfectly homogeneous so that turning the device on might result in large currents suddenly flowing in small areas of the wafer, which would contradict the previous assertion that IGCTs turn on perfectly uniformly. To avoid this device-specific condition, the minimum off-time should be respected but a further constraint is circuit-specific and relates to the clamp-circuit design, which is discussed in Section 6.12.

6.11 Gate drive

IGCTs are generally supplied complete with their gate drivers (hence, the name *integrated*) but an understanding of the operation of the gate-unit (GU) is nevertheless important for the correct application of IGCTs.

Figure 6.36 shows an ABB GCT gate driver into which a GCT can be 'screwed' thanks to the open end of the printed circuit board (PCB), which can be observed in the lower left-hand corner of the picture. A matching slot can be seen in the upper left-hand corner of the GCT of Figure 6.1, allowing the GCT to be inserted with its double gate-flange into an otherwise rigid, multilayer PCB. This establishes a very low-impedance connection between the driver and the GCT [11] with minimal hardware.

The power consumption of a GCT GU is a function of the switching frequency and the commutated current (I_{TGQ}) as it will be recalled that the total peak device current is commutated out of the GU at turn-off (Section 6.9).

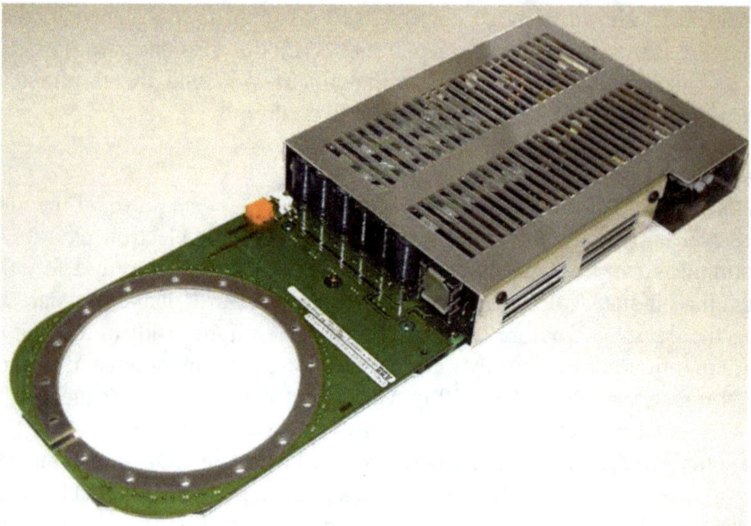

Figure 6.36 An ABB IGCT gate-drive (Courtesy ABB, Switzerland Ltd)

Because the anode current is commutated out of the gate at turn-off, once the gate-cathode junction is blocked (which occurs very rapidly), the gate unit appears in series with the DC-link voltage and the load, thus adding its 20 V to the (for instance) 3-kV supply. Current, at that point, flows through the GU and the load and this part of the GU power is thus not being dissipated in the GU itself but in the load, hence the caption to Figure 6.37 *'dissipated plus transferred'*. Thus, P_{GIN} represents what the power-source must supply to the GU and not the GU's dissipation.

The equivalent series-connection of gate and load circuits at turn-off is shown in Figure 6.38 and the corresponding waveforms can be seen in Figure 6.39, which reprises Figure 6.15 in greater detail.

Figure 6.38 will be considered again later when we consider the parallel operation of IGCTs (Section 6.13).

Referring to Figure 6.39, Phase I represents the steady on-state where only a small forward gate-current flows. At Phase II, the gate current reverses and ramps up to its peak reverse value equal to the initial forward anode current. At the start of Phase III, the gate-cathode junction blocks and from this point on, the gate unit is in series with the load and the pnp transistor of the GCT starts to desaturate enabling the device to sustain voltage within a microsecond. Current continues to flow in the GU until the anode current ceases to tail at the end of Phase IV. Phase V represents the steady-blocking state.

At 500 Hz, 1,000 A, 50 per cent duty cycle, chopper mode:

Power required per Figure 6.37 = 80 W, of which about 20 W are dissipated in the IGCT (about 5 W in the GCT and about 15 W in the GU).

Thus, some 60 W are transferred to the load.

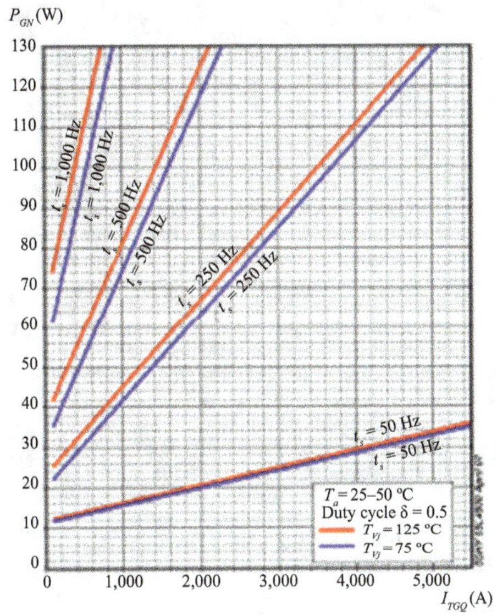

Figure 6.37 Maximum gate unit input power (dissipated plus transferred) in chopper mode, example of a 4" ABB IGCT (Courtesy ABB, Switzerland Ltd)

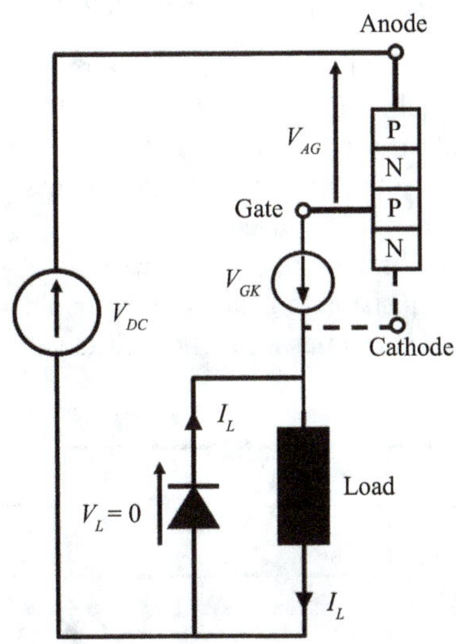

Figure 6.38 Equivalent circuit at the moment when the gate-cathode junction blocks due to current reversal in the GU

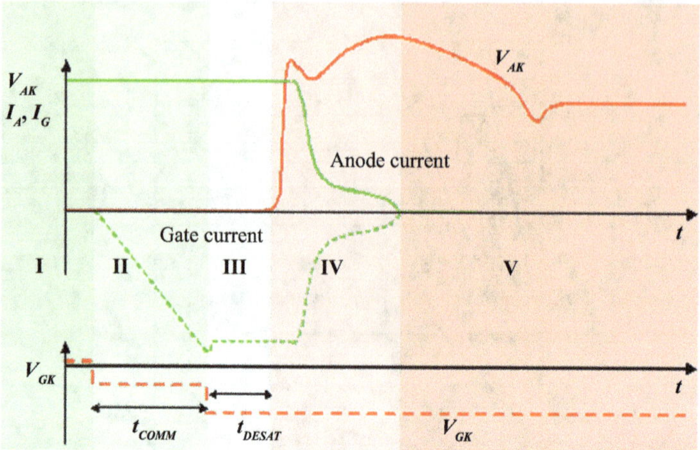

Figure 6.39 Five commutation phases at GCT turn-off

6.12 The clamp circuit

The clamp circuit shown in Figure 6.21 serves several functions:

- as a *di/dt* limiting *snubber* for the FWD,
- as a voltage-limiting *clamp* for FWD and IGCT,
- as a surge-limiting *impedance* and
- as a turn-on loss-eliminating *snubber* for the IGCT.

The clamp equivalent circuit, reprised in Figure 6.40, has the IGCT represented as a mechanical switch. The logic here is that at turn-on, the DC-link voltage appears across the inductance *L*, so S_1, if it is 'fast', could just as well be a mechanical switch. As a result, no voltage appears across the 'switch' and, hence, no power is dissipated by its rising current.

In the absence of *di/dt* limitation, such a mechanical switch would cause the diode to fail, so only a transistor (e.g. an IGBT) could be used, as this would modulate the rate of rise of current as a function of its gate-drive. In this case, V_{DC}

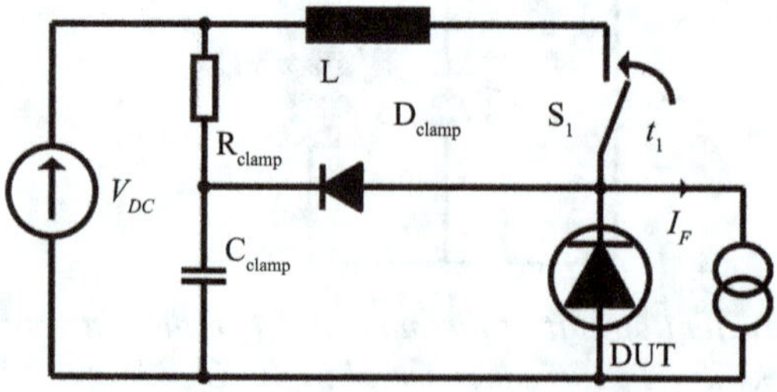

Figure 6.40 Equivalent clamp circuit for one commutation cell

would not be across L but across the IGBT ($V_L = V_{IGBT} = V_{DC}$) but the waveforms remain essentially the same as the allowable rate-of-rise of current is set by the FWD's rating.

Thus, the energy dissipated or stored at turn-on is given by:

$$E_{ON\text{-}CIRCUIT} = t_{ON} \cdot V_{DC} \cdot (I_{LOAD} + I_{RR})/2$$

where $E_{ON\text{-}CIRCUIT}$ is the energy determined by the circuit (i.e. by the FWD's rating). An additional small energy is dissipated in the device ($E_{ON\text{-}DEVICE}$) and this is the E_{ON} specified in IGCT data-sheets but is, in practice, little more than a measurement error!

Thus, whether an IGBT is used or an IGCT with clamp, $E_{ON\text{-}CIRCUIT}$ remains the same for the same type of FWD, however with a clamp, it is not dissipated in the semiconductor switch but stored in the choke and then dissipated in the resistor (or optionally returned to the DC-link using a more sophisticated clamp circuit).

The reverse recovery current of the diode (I_{RR} in Figure 6.41) increases with *di/dt* but t_{ON} *decreases*. Intuitively, one would expect there to be a minimal value of $E_{ON\text{-}CIRCUIT}$ and this is indeed the case, occurring where $I_{LOAD} = I_{RR}$ (Figure 6.42).

The turn-off waveform, already illustrated in previous oscillograms, is highly dependent on the clamp circuit parameters and the GCT's characteristics, which affects the tail-current waveform and the device's turn-off loss. These are difficult to calculate in detail but a good approximation can be obtained using simulation software with behavioural semiconductor models.

Below is an example of a simulation using an IGCT with a nonideal turn-off characteristic, approximated to an exponential current fall.

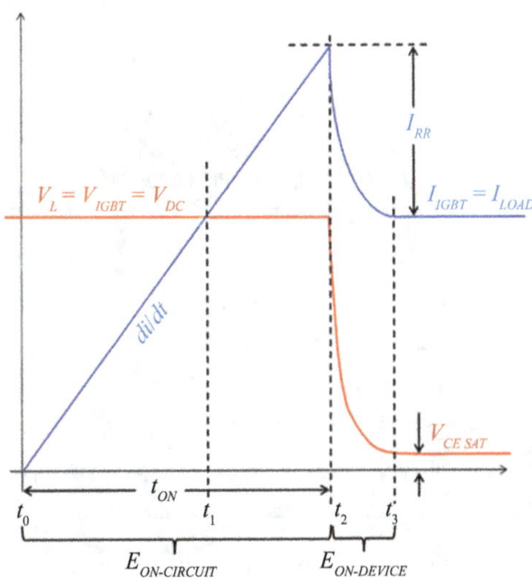

Figure 6.41 Turn-on current in IGCT (blue) and voltage across L (red)

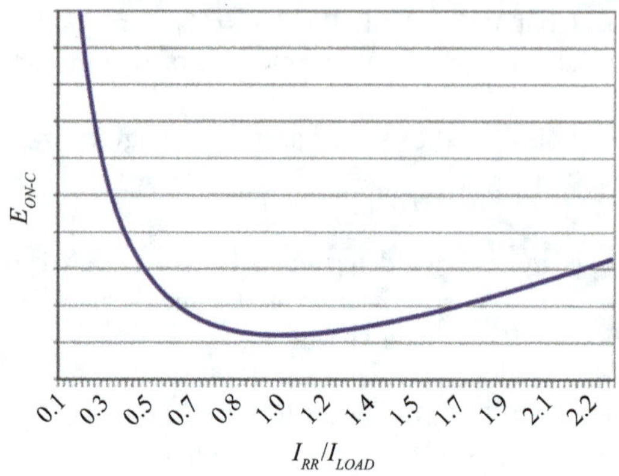

Figure 6.42 The minimum value of $E_{ON\text{-}CIRCUIT}$ occurs when $I_{RR} = I_{LOAD}$

Varying the different clamp-circuit parameters and stray inductances in Figure 6.43 allows an assessment of their impact on the commutation waveforms.

Of particular importance is the clamp-diode current duration of Figure 6.44, as this affects the minimum off-time during which no IGCT (of any phase) should be allowed to turn on as only the stray inductance L_4 (Figure 6.43) would limit the clamp-diode's reverse recovery. This is the 'circuit-specific' component of $T_{OFF\text{-}MIN}$ referred to in Section 6.10.2.16.

As mentioned earlier, the clamp circuit keeps the turn-on loss out of the semiconductor, thus improving its 'thermal budget' but this energy is lost nevertheless and is dissipated mainly in the clamp resistor with a small amount contributing to the GCT's E_{OFF} (roughly 10 per cent). Several circuits have been proposed to recover this energy, of which one is shown in Figure 6.45.

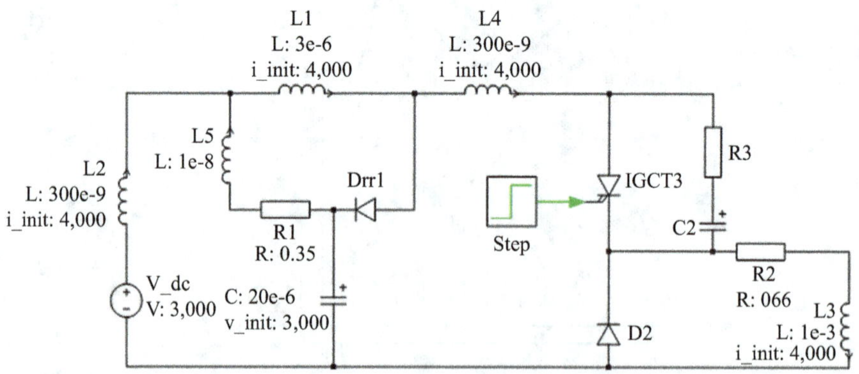

Figure 6.43 PLECS circuit for commutation cell simulation [12] NB: the small RC snubber across the ToD is a simulation artefact used to allow the simulation to start

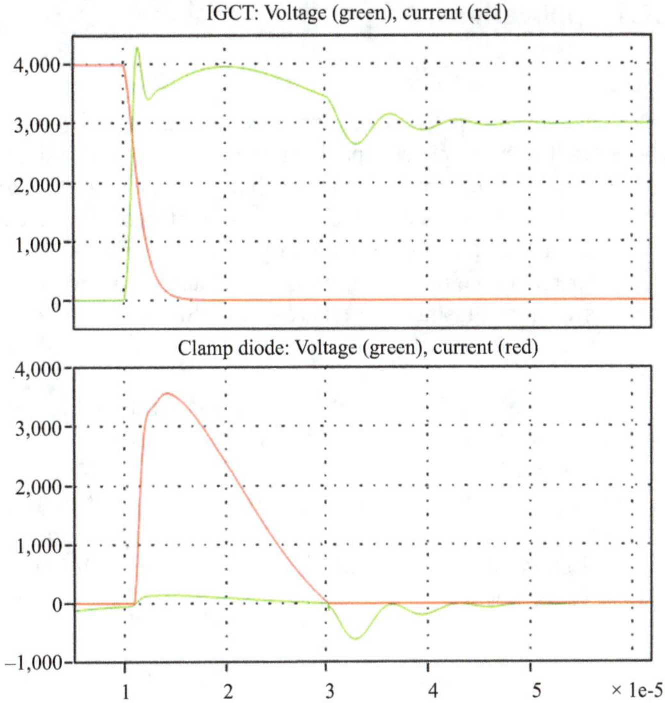

Figure 6.44 Figure 6.42 simulated turn-off waveform using PLECS

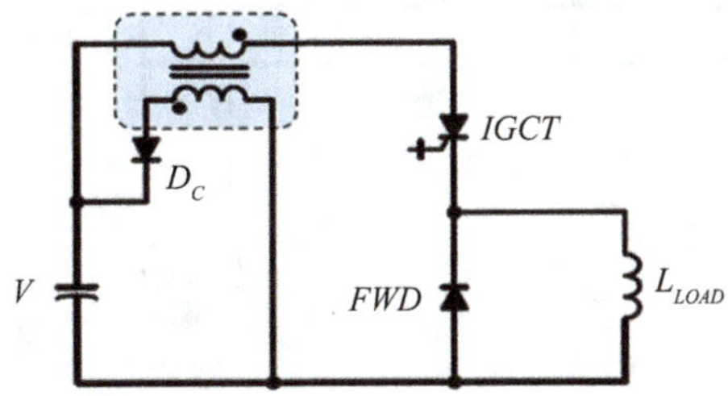

Figure 6.45 Fly-back clamp circuit [13]

As previously shown in Figure 6.41, the energy stored in the choke (to be later dissipated or recovered) is given by:

$$E_{ON\text{-}CIRCUIT} = t_{ON} \cdot V_{DC} \cdot (I_{LOAD} + I_{RR})/2$$

but it can be seen from Figure 6.41 that this occurs in two phases: from I_{RR} to I_{LOAD} and from I_{LOAD} to zero current. More accurate calculation methods, therefore, treat the two events separately to allow for independent variation of I_{RR} with respect to I_{LOAD} [14].

6.13 IGCT applications

6.13.1 IGCT VSIs and CSIs

Voltage, current and power control are required in various industrial applications. DC motors can be controlled by simple chopper circuits and AC motors by single or three-phase variable voltage, variable frequency (VVVF) drives which turn DC to AC (and sometimes vice versa). This is done on the basis of one of two converter principles: conversion of a DC current source to an AC current source or of a DC voltage source to an AC voltage source. In their simplest forms, this requires a matrix of six switches for a three-phase system as illustrated in Figure 6.46 (VSI) and Figure 6.47 (CSI).

6.13.1.1 VSIs

VSIs are the most common inverter types. At LV, it is virtually the only inverter topology used and is shown in Figure 6.48 in its basic two-level form. Most VSIs are based on this configuration.

Its popularity can be understood by the fact that the cheapest turn-off devices, (ToDs) or switches are IGBTs which generally block voltage in the forward direction and lend themselves well to the topology of Figure 6.48, which has the following salient properties:

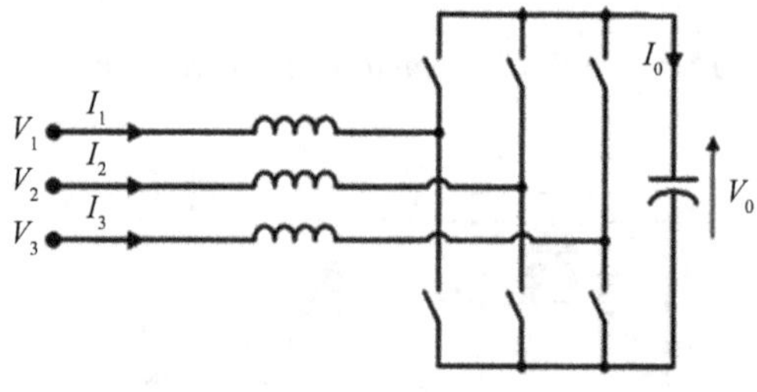

Figure 6.46 Basic VSI circuit

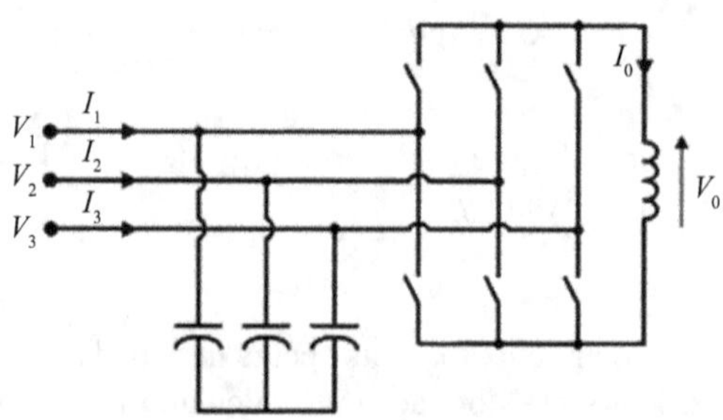

Figure 6.47 Basic CSI circuit

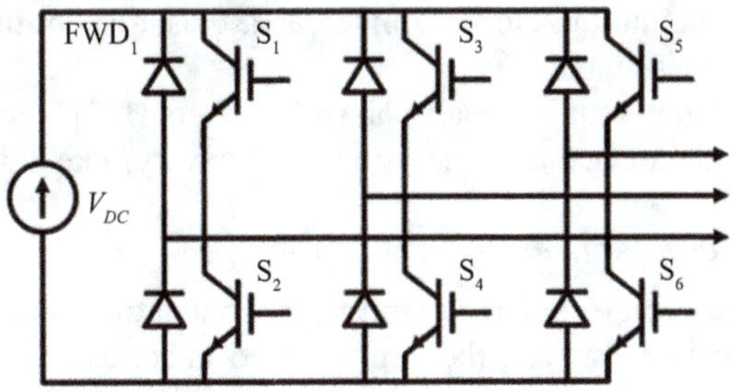

Figure 6.48 Basic two-level IGBT VSI

- unidirectional voltage,
- bidirectional current (needs asymmetric devices plus diodes or RC devices),
- shoot-through may cause explosive surge currents (no fault-current limitation between switch matrix and voltage source), unless a clamp-circuit is used (see Figure 6.21) and
- the semiconductors do not normally require voltage snubbers.

6.13.1.2 CSIs

The current source converter is the 'dual' of the VSI. Here, the DC-link consists of inductors providing a 'constant' current rather than a capacitor providing a 'constant' voltage. In the VSC, the voltage, by definition, cannot reverse and power reversal is achieved, therefore, by *current* reversal. In the CSC, this is the obverse: current cannot reverse, so power reversal is achieved by *voltage* reversal. These fundamental differences between the two topologies impose fundamental differences in the types of semiconductor switches they require: the VSC uses devices which allow *current* reversal and the CSC requires devices allowing *voltage* reversal, i.e. symmetric devices or SGCTs (Figure 6.49).

The CSI topology has the following salient features:

- essentially used at high power needing high voltage, high current devices (e.g. IGCTs),
- requires symmetric-blocking devices and IGCTs are available as such (SGCTs),

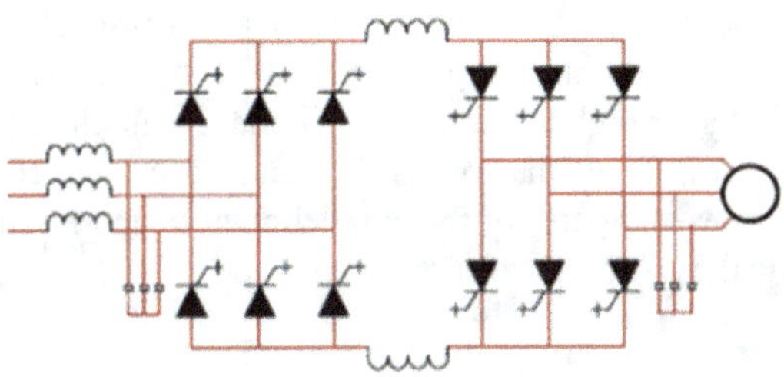

Figure 6.49 Basic two-level IGCT CSC

- the CSI is, by definition, current limited and, thus, a shoot-through is usually without consequence and
- the semiconductors require small voltage snubbers (RC), since when the voltage reverses, the device undergoes reverse recovery, like a diode.

6.13.2 Series connection

IGCTs can be series-connected as can most semiconductors, in order to operate at higher voltage levels. However, the number used in series is usually limited by the gate-drive power requirements (50–200 W), which means that each device must be supplied by a low-voltage supply in contrast to thyristors and IGBTs, which require very little gate power (e.g. 5 W), allowing the supply to be drawn from the snubber circuits used to ensure voltage-sharing, thus obviating the need for isolated power supplies which is problematic at very high voltages (e.g. hundreds of kV).

However, the need for the direct series connection of semiconductors for very high-voltage converters has become less important since the development of the modular, multilevel converter (MMC) [15], which series-connects inverter *cells* rather than semiconductors, thus eliminating the need for dynamic and static voltage-sharing components. Each level has a defined voltage source from which a low-voltage supply can be derived to power the gate units.

Figure 6.50 illustrates the principle but a practical MMC level with IGCTs would be that of Figure 6.51.

Because the MMC controls voltage by switching levels in and out, the output voltage can be made to closely approximate a sine-wave without switching at a high PWM frequency. This minimises switching losses and allows the use of devices with very low conduction losses (but correspondingly high turn-off losses) making the use of IGCTs in the MMC suitable for high-current and high-voltage applications such as HVDC transmission.

6.13.3 Parallel connection

IGCTs are available to high currents and it is unusual to need to connect them in parallel. There are, however, exceptional applications which may require parallel connections, such as for circuit breakers [16]. IGCTs turn on homogeneously and provided their on-state characteristics are well matched (V_{TO} and R_D – see Section 6.10.2.1), good static current-sharing is achievable.

Dynamically, the currents also balance well at turn-off because of the gate-unit, which finds itself connected in series with the GCT, as illustrated in Figure 6.52. An initially higher current in Anode 2 causes the gate voltage of IGCT2 to fall with respect to that of IGCT1 causing a compensating current to circulate via the anodes and gates of the parallel devices and reestablishing current balance at the critical SOA point of turn-off.

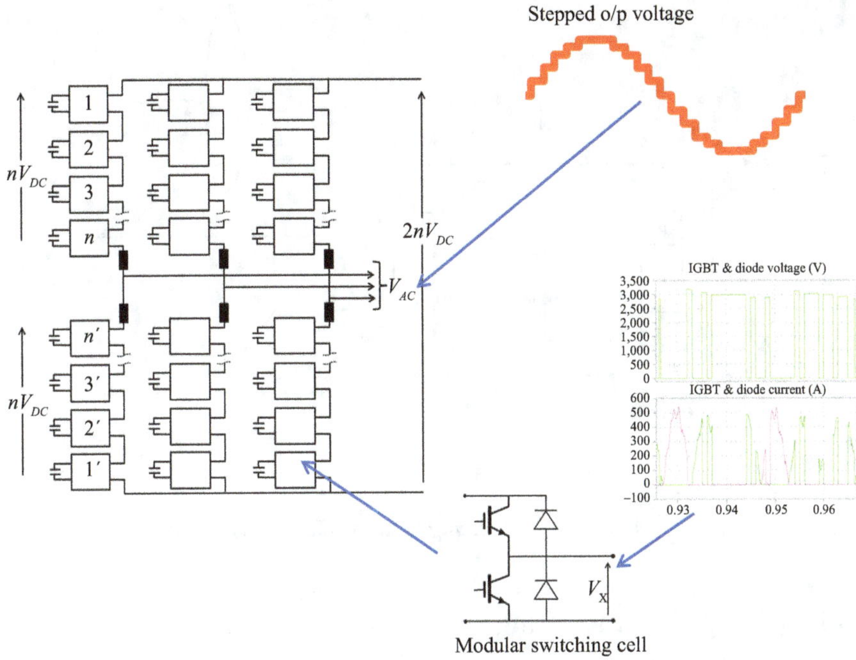

Figure 6.50 The MMC showing seriesed cells with the insert representing of a single cell

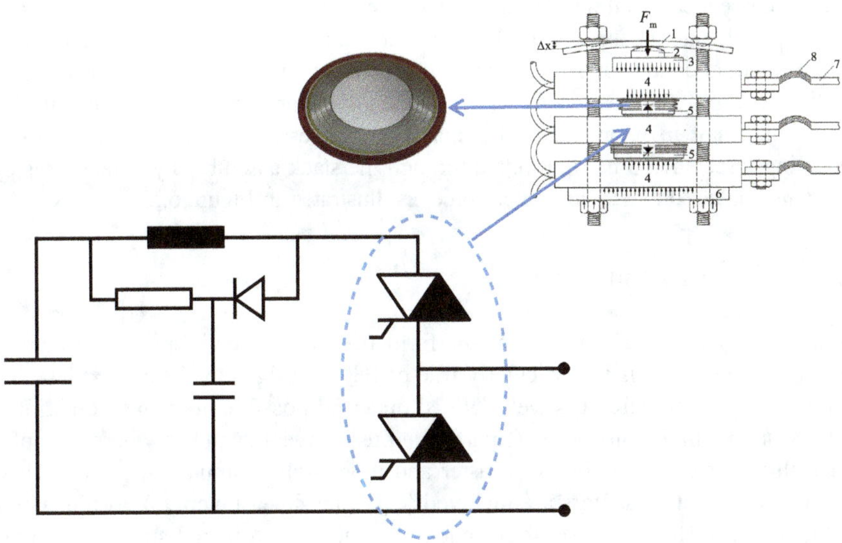

Figure 6.51 An RC-IGCT MMC cell with clamp circuit

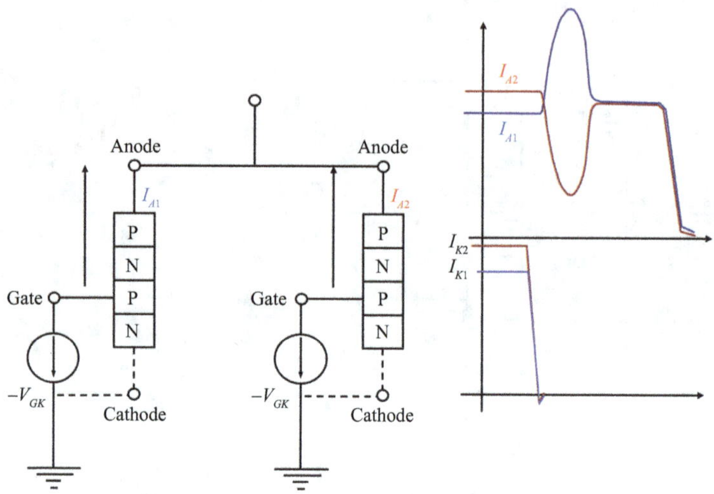

Figure 6.52 IGCT dynamic self-balancing of turn-off currents (Courtesy TU, Berlin)

6.14 Mechanical mounting

IGCTs are produced exclusively as monolithic, hermetically sealed press-pack components, which means that the metallic surfaces serve as both electrical and thermal contacts, implying, therefore, that these surfaces must be of a given flatness and the mounting force must respect the manufacturer's specification. Detailed requirements are either given in the device specifications or more usually in application notes [17]. The recommended metal interfaces are given along with suitable contact grease types (generally silicone grease without fillers as the contact must be a good *electrical* as well as thermal conductor). The mounting force must be within certain limits and be applied uniformly: excessive force will cause premature wear-out and insufficient force will result in poor electrical contacts and excessive thermal resistance. The mounting force should be transmitted through the stack assembly by components of the same contact diameter and rated force, as illustrated in Figure 6.53.

6.15 Circuit simulation

Simulation of electrical and thermal performance is an essential part of designing power converters. This is particularly true of IGCT designs to ensure that interaction with the clamp circuit is well defined under all possible operating conditions. Such system simulations are best performed using ideal switches where the software allows this, as this leads to faster and more stable simulations, particularly where many semiconductors are involved (e.g. three-level inverters), as illustrated in Figure 6.54. Here, the use of behavioural semiconductors and stray connection

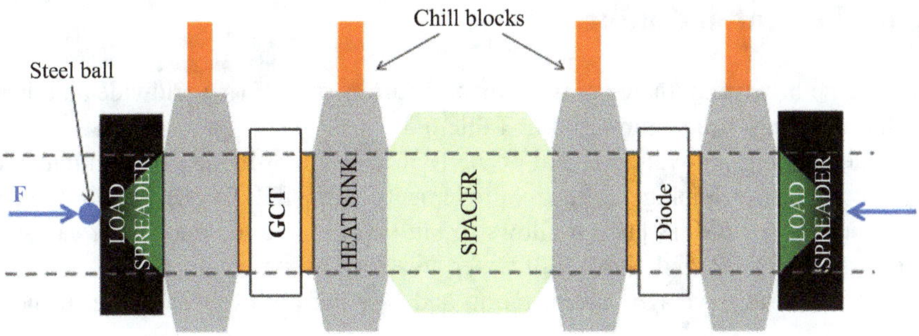

Figure 6.53 Example of a correctly designed stack assembly

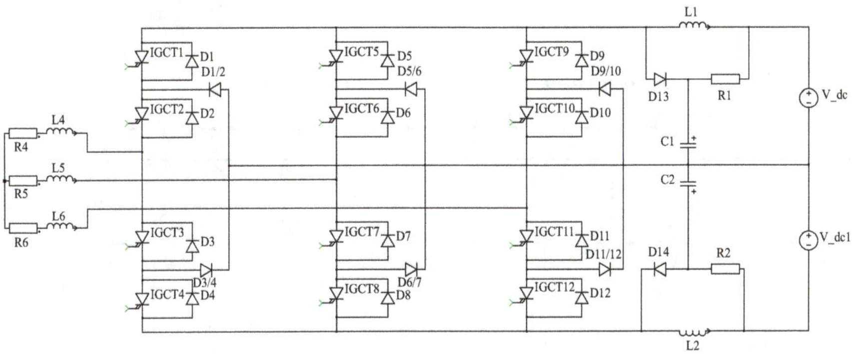

Figure 6.54 Three-level inverter: best simulated using ideal switches

inductances, shown in Figure 6.43 would be inappropriate and are only of interest for the evaluation of a single commutation.

Thermal simulation is also extremely valuable and again, should not be executed with behavioural models. Certain simulation applications allow semiconductor losses as well as clamp-circuit losses to be read from look-up tables, generated from the device manufacturer's data-sheets [12,14]. Data-sheet losses take into account parameter variations over many device production lots and are the sole 'legally' valid values for temperature calculations.

Finally, *real-time simulation* or *hardware-in-the-loop* (HIL) testing allows verification of the controllers prior to connection to the power electronics and comparisons with the earlier off-line simulations in order to verify operating conditions, all of which may not be feasible in the laboratory. Such simulations are essential when designing high-power systems using components costing hundreds or thousands of dollars in costly power laboratories! [12,18].

6.16 Present and future

At the time of writing, there are three manufacturers of IGCTs worldwide and their product offerings are summarised in Table 6.3.

One pending innovation is that of the BGCT or the bi-mode gate commutated thyristor. This technology, already implemented for IGBTs (BIGT), has been demonstrated for IGCTs [19]; it allows the full wafer area to conduct forwards as an asymmetric GCT and allows full wafer area for reverse conduction, as a diode, virtually doubling average current rating and SOA of both 'devices' (i.e. of both functions).

Table 6.3 Overview of available and developmental IGCTs

	Asymmetric		Reverse conducting					Reverse blocking			
Si Ø (mm)	91	150	38	51	68	91/94	150	35	48	68/62	91
V_{DRM} (kV)											
2.5											● (ABB)
3.3											
4.5	✓ (ABB) ✓ (CRRC)	● (CRRC)	✓ (ABB)	✓ (ABB)	✓ (ABB)	✓ (ABB) ✓ (CRRC)	● (ABB) ● (CRRC)				
5.5	✓ (ABB)		✓ (ABB)	✓ (ABB)	✓ (ABB)	✓ (ABB)					
6.0	✓ (CRRC)						✓ (Mitsubishi)				
6.5	✓ (ABB)					✓ (ABB)	● (ABB)	✓ (Mitsubishi)	✓ (Mitsubishi)	✓ (Mitsubishi) ● (ABB)	
10	● (ABB)					● (ABB)					

Legend:

Mitsubishi ABB CRRC

✓ – Available • – Developmental

6.17 Reliability

The reliability of all components or equipment can be represented by the 'classical' bathtub curve shown in Figure 6.55.

It shows three distinct phases:

- *Infant mortality*, which is little related to time and is measured in defects per million (DPM),
- *Useful life* is expressed by *Failure Rate* λ in Failures-In-Time (FITs)
- *Wear-out*, which is typically measured in number of operating cycles (N° of events).

From the failure rate λ, we can derive the mean time to failure (MTTF) for a component as $1/\lambda$ in hours. For an equipment, the MTBF (mean time *between* failures) is the reciprocal of the sum of the individual component failure rates.

The main cause of failure for GCTs (apart from misapplication) is cosmic rays, which affect all semiconductors sustaining an electric field. Manufacturers of high-power, high-voltage devices, destined for critical applications, specify their devices' cosmic ray withstand capabilities in their data-sheets or in separate application notes [10], which specify the probability of failure (failure rate) in terms of voltage, temperature and altitude above sea level.

The GCT consists of a large monolithic wafer assembled using established techniques, which go back to the early days of thyristor press-packs for which we have more than half a century of data and field-experience. Even GTOs, which are mechanically similar to GCTs, have been in service for nearly 40 years.

Because most GCTs are supplied as *IGCTs*, the failure rates are quoted for the complete device, which includes the gate-unit. The GU has the same basic ingredients as any GU for GTOs or transistors: positive and negative power supplies, logic circuits, optocouplers and turn-on and turn-off channels composed of transistors (MOSFETs) and electrochemical capacitors. The main difference of IGCT

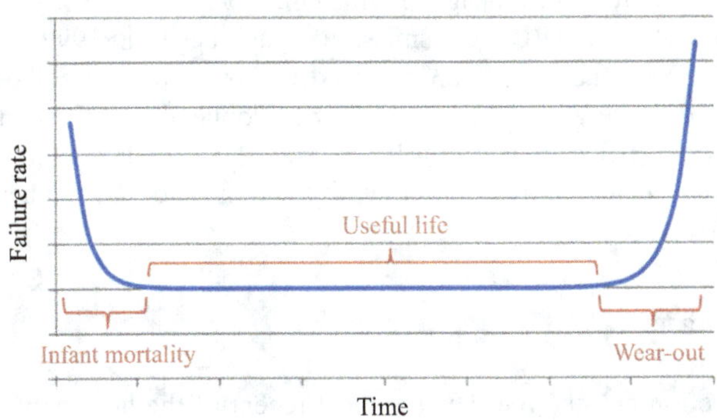

Figure 6.55 The reliability 'bathtub' curve

GU, with respect to IGBT GUs, is their large numbers of MOSFETs and capacitors in parallel, particularly in the turn-off channel, required to minimise impedance. While large numbers of components are conducive to higher failure rates, in this case, they lead to very low stresses per device which in turn has led to very low component-failure rates.

Failure rates can be calculated based on standard component reliability tables, which are typically established under rated operating conditions. A more accurate approach is obtained from *hindsight* when large numbers of similar devices have been in operation for many years. This is indeed the situation for the large projects (interties, MV-Drives, FACTS, etc.), which have seen over a quarter of a million (ABB) IGCTs in service since 1997 [20]. The low *observed* failure rates allow ABB to specify a λ of 100 FIT for their 4.5 kV/4 kA IGCT 'workhorse', which equates to 50 FIT for the GCT and 50 FIT for the gate-unit. These are among the lowest semiconductor and GU FIT-rates in the industry.

As mentioned above, MTTF applies to components which can be replaced but not repaired while MTBF applies to equipment which *can* be repaired. Generally, a repair entails 'down-time' and this can be minimised by adding 'redundancy'. This requires the use of supplementary (redundant) components in series for direct series-connections or of additional system levels where cascaded inverters are used (e.g. in the MMC). In either case, current must be allowed to flow in failed devices or levels in order to not interrupt operation – and this for considerable periods (e.g. years) – to allow for long service intervals.

This has led to the concept of *short-circuit failure mode* (SCFM) by which a semiconductor is designed to operate in the shorted mode for long and defined periods. Whereas this capability was inherent in the early massive series-connections of monolithic thyristors in HVDC, it was clearly not possible with semiconductor modules using bonded wires. Even press-packs, housing multiple chips, could not automatically ensure SCFM, as a chip failure would result in all the load current flowing through the shorted chip with nothing to induce the remaining good chips to conduct. The outcome of such situations is that the failed chip is rapidly electro-eroded, resulting in an open-circuit, internal arcing and eventual erosion of the housing and heatsink with overall system failure (coolant leakage, plasma emission, etc.).

The use of VSCs for HVDC transmission, which began in 1999 [21], required devices with SCFM capability [22,23] which drove press-pack IGBT development to achieve SCFM because the direct series-connection of semiconductors was simpler with IGBTs than with IGCTs (as explained in Section 6.13.2). However, IGCTs, being monolithic, like thyristors, also have inherent SCFM capability [24].

6.18 Summary

As stated in Section 6.2, the IGCT is the most recent of the high-power ToDs and has now been commercially available for nearly 25 years. It has always found its application, principally, in the highest power systems because the competing IGBT was generally able to grow in power rating to 'catch-up' with IGCTs and was generally deemed easier to use since a clamp-circuit was not required (chokes are not as easy to find in catalogues as capacitors!).

It was also true that some of the highest power applications also required high voltage and hence, series connection of devices. The rapid adoption of the MMC has obviated the need for semiconductor series-connection and, thus, lifted one of the difficulties of using IGCTs. Coincidentally, the need for SCFM, also in the MMC cell, along with the IGCT's surge capability, has rendered the IGCT *more* attractive in these configurations and its high-current capability makes parallel connections unnecessary. Finally, its simpler construction and ruggedness have been shown to increase reliability and reduce capital costs [25] and the imminent availability of 6″ reverse-conducting devices with blocking voltages up to 10 kV, further enhances the practicality of IGCTs in high-power conversion.

These concurrent evolutions have led to a flurry of design activities and interest in the use of IGCTs in STATCOMs and more recently, in HVDC [16,24–27] and the growing availability of these devices from three suppliers and in mechanically compatible forms; this trend is set to continue.

Note to the reader

This chapter is part of an edited book. To cite or use this chapter as a reference, please use the following format:

Carroll E. "IGCTs." In: Iannuzzo F (ed.), *Modern Power Electronic Devices: Physics, Applications, and Reliability*. Stevenage, UK: IET; 2020. pp. 171–223.

References

[1] P. K. Steimer, H. Grüning, J. Werninger, and D. Schröder, "State-of-the-Art Verification of the Hard-Driven GTO Inverter Development for a 100 MVA Intertie", Proceedings IEEE PESC, Baveno, 1996.

[2] P. K. Steimer, H. Grüning, J. Werninger, E. Carroll, S. Klaka, and S. Linder, "IGCT: A new emerging technology for high power, low cost inverters", IEEE Industry Applications Magazine 2(4):12–18, 1999.

[3] H. Grüning, B. Ødegård, J. Rees, A. Weber, E. Carroll, and S. Eicher, "High-Power Hard-Driven GTO Module for 4.5 kV/3 kA Snubberless Operation", Power Conversion May 1996 Proceedings, Nuremberg, Germany.

[4] D. E. Piccone, R. De Doncker, J. A. Barrow, and W. H. Tobin "The MTO thyristor–A New High Power Bipolar MOS Thyristor", IAS '96. Conference Record of the 1996 IEEE Industry Applications Conference Thirty-First IAS Annual Meeting, San Diego, Ca. USA.

[5] Y. Li, A. Huang, and F. Lee, "Introducing the Emitter Turn-Off Thyristor (ETO)", Conference Record – IAS Annual Meeting (IEEE Industry Applications Society) 2:860–864 vol. 2 November 1998, St Louis, USA.

[6] A. Welleman, E. Ramezani, and J. Waldmeyer, "Semiconductor Switches for Single-Pulse and Repetitive-Pulse Applications", 4th International AECV Conference, The Netherlands 24–26 Sept. 2001.
[7] U. Schlapbach, M. Rahimo, A. Baschnagel, A. Kopta, and E. Carroll, "Switching-Self-Clamping-Mode "SSCM" for Over-voltage Protection in High Voltage IGBT Applications", PCIM 2005, Nuremberg, Germany.
[8] M. Arnold, T. Wikstroem, Y. Otani, and Th. Stiasny, "High-Temperature Operation of HPT+ IGCTs", PCIM 2011, Nuremberg, Germany.
[9] R. Schnell, and N. Kaminski, "Thermal Runaway during Blocking" Application Note 5SYA 2045-01, ABB Switzerland Ltd, www.abb.com/semiconductors
[10] "Failure Rates of IGCTs Due to Cosmic Rays", Application Note 5SYA 2046-01, www.abb.com/semiconductors
[11] "Applying IGCT Gate Units", Doc. No. 5SYA 2031-03, Sept 07, www.abb.com/semiconductors
[12] www.plexim.com
[13] S. Shirmohammadi, and Y. Suh, "Low Dissipative Snubber Using Flyback Type Transformer for 10 kV IGCT in 7 MW Wind Turbine Systems", ICPE 2019. Busan, Korea.
[14] M. Mohan, E. Tsyplakov, Ch. Winter, and H. Marabathina1, "Converter Switching Based Generic Loss Evaluation in Semiconductor and Clamp Circuit of IGCT Based MMC Converter Cells for HVDC Applications", ICPE 2019. Busan, Korea.
[15] A. Lesnicar, and R. Marquardt, "An Innovative Modular Multilevel Converter Topology Suitable for a Wide Power Range", IEEE Power Tech Conference, Bologna, Italy, 23–26 June 2003.
[16] Q. Yi, Y. Wu, Y. Wu, F. Yang, Z. Zhang, and C. Gao, "Investigation of a Novel IGCT Module for DC Circuit Breaker", ICPE 2019. Busan, Korea.
[17] B. Backlund, and T. Schweizer, "Recommendations Regarding Mechanical Clamping of Press-Pack High Power Semiconductors", Doc. No. 5SYA2036-03 June 06, www.abb.com/semiconductors
[18] www.typhoon-hil.com
[19] T. Stiasny, M. Arnold, U. R. Vemulapati, *et al.*, "Experimental Results of a Large Area (91mm) 4.5kV 'Bi-mode Gate Commutated Thyristor' (BGCT)", PCIM 2016, Nuremberg.
[20] C. Winter, "Gotthard Tunnel's Power Semiconductors Give Insights to IGCT Reliability", White Paper, www.abb.com/semiconductors
[21] U. Axelsson, A. Holm, C. Liljegren, M. Aberg, K. Eriksson, and O. Tollerz, "The Gotland HVDC Light Project-Experiences from Trial and Commercial Operation", 16th International Conference and Exhibition on Electricity Distribution, Amsterdam, Netherlands, 2001.
[22] R. Chokhawala, B. Danielsson, and L. Ängquist, "Power Semiconductors in Transmission and Distribution Applications", Proceedings of the 13th International Symposium on Power Semiconductor Devices and ICs, 2001, ISPSD '01, Osaka Japan.

[23] R. Schlegel, E. Herr, and F. Richter, "Reliability of non-hermetic pressure contact IGBT modules", Microelectronics Reliability, 41(9–10): 1689–1694, 2001.

[24] F. Dijkhuizen, and S. Norrga, "Fault-Tolerant Operation of Power Converter with Cascaded Cells", EPE 2011, 30th August–1st September, Birmingham, UK.

[25] Ph. Ladoux, N. Serbia, and E. Carroll, "On the Potential of IGCTs in HVDC", DOI 10.1109/JESTPE.2015.2429116, IEEE Journal of Emerging and Selected Topics in Power Electronics.

[26] P. Steimer, D. Weiss, and B. Ødegård, "Robust, Low-Loss RCIGCT Technology and MV Applications", 10th International Conference on Power Electronics-ECCE Asia, May 27–30, 2019 / BEXCO, Busan, Korea.

[27] K. Kurachi, K. Taguchi, and G. Majumdar, "GCT Technologies and Their Applications", ICPE 2019. Busan, Korea.

Chapter 7
Silicon carbide diodes

Jens Peter Konrath[1]

7.1 Introduction

Since the introduction of the first commercial silicon carbide (SiC) Schottky barrier diodes (SBDs) by Infineon, and shortly thereafter by Cree, SiC diodes have found widespread applications in power electronics. From 650-V SiC SBDs to 1,700-V SiC merged-pn-Schottky (MPS) diodes, they can be found in switch-mode power supplies, solar inverters, battery chargers, and motor drives, just to mention a few. The widespread application is due to its favorable material properties, most notably the high critical electric field which is a direct consequence of the material's large bandgap. It allows for the design of unipolar diodes up to several kilovolts, greatly reducing switching losses as compared to bipolar silicon diodes of the same voltage and current class. Other material properties that enable the use of SiC diodes (and switches) in power electronics are its high thermal conductivity, a coefficient of thermal expansion that is closer to one of the commonly used DCB materials than silicon's, and its high sublimation point and chemical inertness [1]. Of course, there are some challenging properties one has to keep in mind when designing SiC power devices, such as its negligible diffusion constants for most of the dopant species [1] or its hardness of nine on the Moh's scale. These two features, among others, require some more refined methods of processing of SiC wafers.

A more detailed description of the physical properties of SiC is beyond the scope of this chapter and the reader is referred to the literature [1–4]. The aim of this chapter is to offer an insight into SiC power diodes and is organized as follows.

In Section 7.2, we present the most common SiC SBD device structures and discuss their properties important to applications. Section 7.3 is concerned with diode edge termination structures and their reliability and Section 7.4 covers electrical and thermal device properties. In Section 7.5, we present some example applications and demonstrate the advantages of SiC SBDs over their bipolar silicon counterparts. Finally, we conclude with some remarks about possible future developments in Section 7.6 (integration into switches).

[1]Infineon Technologies Austria AG, Siemensstraße 2, Villach, Austria

7.2 Review of silicon carbide SBD structures

Commercially available SiC diodes can be put in four groups, which are (from least to most bipolar conducting), junction-Schottky barrier diodes (JBD diodes), MPS diodes, and p-i-n diodes, the latter ones being “true” bipolar diodes.

Besides any edge termination structures, an SBD consists of a metal-semiconductor contact only. A sketch of its structure and the equivalent circuit is shown in Figure 7.1 [5]. In forward bias operation, there is a voltage drop U_{sc} across the Schottky contact, the drift layer U_{drift} and the substrate and backside contact U_{bs} that contributes to the total forward voltage drop U_F

$$U_F = U_{sc} + U_{drift} + U_{bs} \tag{7.1}$$

The voltage drop across the substrate and backside ohmic contact is given by the substrate doping and the ohmic contact resistance.

The voltage drop across the drift layer is given by

$$U_{drift} = w_{drift} \times R_{on,sp} \times J_F \tag{7.2}$$

where w_{drift} is the drift layer thickness, J_F the forward current density, and $R_{on,sp}$ the specific on-state resistance of the drift layer. The latter is given by [5],

$$R_{on,sp} = \frac{w_{drift}}{q\mu_n N_D} \tag{7.3}$$

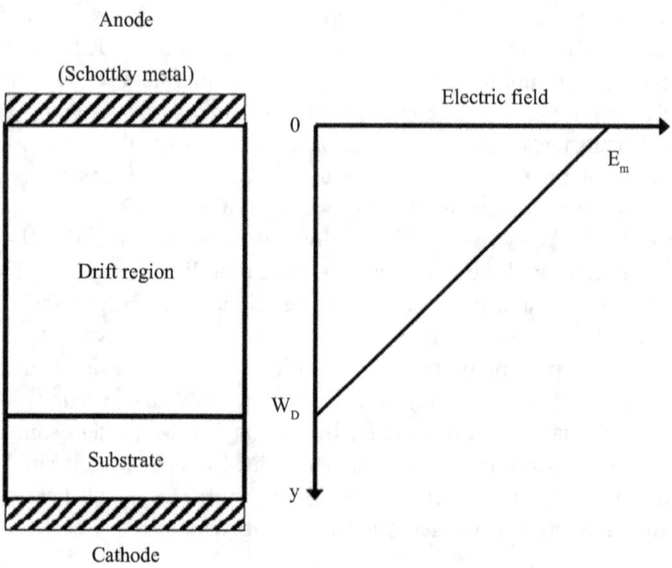

Figure 7.1 Structure of a Schottky barrier diode and the electric field within the drift layer

where q is the elementary charge, μ_n is the electron mobility, and N_D is the drift layer doping concentration. As all SiC diodes are punch-through diodes by design where the highly doped n^+ region is represented by the buffer layer in between the substrate and the drift region, the width of the latter can be derived as [6],

$$w_{drift} = \frac{\varepsilon E_C}{qN_D} - \sqrt{\left(\frac{\varepsilon E_C}{qN_D}\right)^2 + \left(\frac{2\varepsilon V_B}{qN_D}\right)} \tag{7.4}$$

where V_B is the desired breakdown voltage, ε is the permittivity of SiC, and E_C is the critical electric field given by [7],

$$E_C = \frac{2.49 \times 10^6}{1 - \frac{1}{4}\log(N_D/10^6)} \tag{7.5}$$

With the equations above, the optimization of the drift layer to achieve the lowest on-state resistance for the desired breakdown voltage can be performed by differentiation of (7.3) with respect to N_D after inserting (7.4) and (7.5).

The voltage drop across the Schottky contact depends on the work function of the metal used and the surface preparation before metal deposition [8]. Both determine the Schottky barrier height ϕ_B. With the effective Richardson constant A^{**}, the voltage drop can be expressed as [9],

$$U_{sc} = \frac{nk_BT}{q}\ln\left(\frac{J_F}{A^{**}T^2}\right) + \phi_B \tag{7.6}$$

for $U_{sc} \gg \frac{kT}{q}$ and J_F is the forward current density. The ideality factor n indicates the contribution of other mechanisms than thermionic emission, e.g. tunneling. For thermionic current transport only, $n = 1$ holds [9].

Reducing the Schottky barrier height by choosing a suitable contact metal and surface preparation process leads, on the one hand, to a reduction of the forward voltage drop and, on the other hand, to an increased reverse leakage current density, which is increased as well because of the thermionic and field emission through the barrier. The relationship between reverse leakage current density and the Schottky barrier height and surface electric field for thermionic emission reads,

$$J_R = A^{**}T^2\exp\left(-\frac{q\phi_B}{nk_BT}\right)\exp\left(\frac{q\sqrt{\frac{q\mathfrak{E}}{4\pi\epsilon}}}{kT}\right) \tag{7.7}$$

where the barrier lowering due to the Schottky effect has been taken into account [8]. The electric field strength at the contact $\mathfrak{E}$ is determined by the drift layer doping and reverse bias U as follows [9],

$$\mathfrak{E} = \sqrt{\frac{2qN_D}{\varepsilon}\left(U + \phi_B - \frac{kT}{q}\right)} \tag{7.8}$$

Here, $U > 0$.

As with the on-state voltage drop versus breakdown voltage, this trade-off can be utilized to optimize the diode for the intended use in applications. The leakage current caused by the space-charge generation current can be neglected because of the large band gap of 4H–SiC.

To improve beyond the leakage current versus Schottky barrier height trade-off of the simple SBD, the electric field strength at the contact has to be reduced. For this purpose, different shielding structures have been proposed that achieve the limitation of the electric field by the formation of a space-charge region below the Schottky contact. Dual metal trench Schottky (DMTS) diodes achieve the shielding by using a second Schottky metal with a considerably higher barrier that is filled in trenches dividing the Schottky contact area into small patches. Trench MOS barrier Schottky (TMBS) diodes form a MOS interface in said trenches, and pinch rectifiers or junction barrier Schottky (JBS) diodes use p-doped regions below the Schottky contact to form a space-charge region. The shielding structures introduced by these concepts form an n-doped channel in between them that carries the electron current injected by the Schottky contact. As an example, the JBS diode, as shown in Figure 7.2 [10] will be discussed in detail. Its main feature is the already mentioned p-doped regions for shielding the Schottky contact from high electric fields as follows.

With increasing reverse bias, the space-charge regions of the pn junctions merge and effectively shield the Schottky contact from the electric field. Early models [11] assumed the electric field to be kept constant, but a more recent analysis by Zhu and Chow [10] suggests it can be described by the relation,

$$\mathfrak{E} = \left(\alpha \frac{W^{\beta}}{X} \right) U^{\frac{1}{2}} \tag{7.9}$$

The parameters α and β can be determined by fits to simulation and experimental data. Zhu and Chow [10] report a good agreement between their

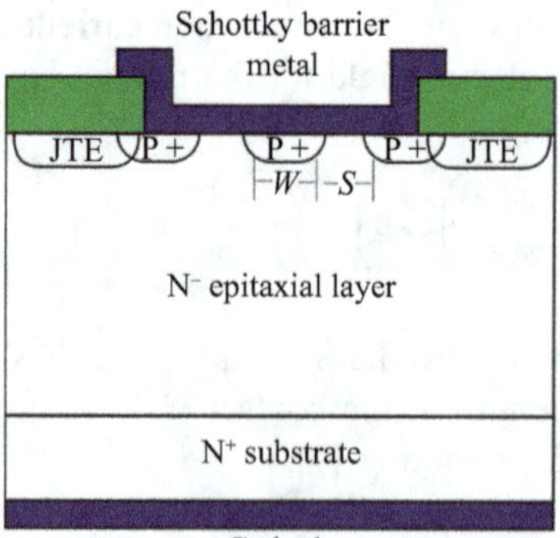

Figure 7.2 Structure of a junction barrier Schottky diode. The width and spacing of the p regions are denoted with w and s, respectively [10]

calculations and the experiment with $\alpha = 7 \times 10^3$ and $\beta = 0.7$ for Infineon's 3rd generation 4-A 600-V diode. While shielding the Schottky contact from high electric fields, the p-doped regions reduce the area of the Schottky contact. This leads to an increased forward voltage drop for a given current density as compared to the SBD. When calculating the Schottky area loss, the pn space-charge regions have to be taken into account as well, leading to a Schottky contact area A_{SC} per unit cell of [10],

$$A_{SC} = \frac{S - 2\sqrt{\frac{2\varepsilon}{qN_D}(U_{sc} - U_F)}}{P + S} \tag{7.10}$$

In (7.10), P and S are the p-doped area and masked area during p implantation of the unit cell as shown in Figure 7.3 [10], respectively. The second term in the numerator takes the Schottky area reduction by the space-charge region into account. Thus, the current density across the Schottky contact is enlarged by a factor $1/A_{SC}$. In addition, the voltage drop of the n channels U_{ch} in between the p-doped regions has to be taken into account as well as the voltage drop in the current spread region below the channels U_{sp} and (7.1) has to be changed to include these contributions as follows [10],

$$U_F = U_{sc} + U_{ch} + U_{sp} + U_{drift} + U_{sb} \tag{7.11}$$

Assuming a current spread angle of 45° as indicated in Figure 7.3, an electron mobility μ_n and a spread layer resistivity ρ_{sp}, the channel and spread layer

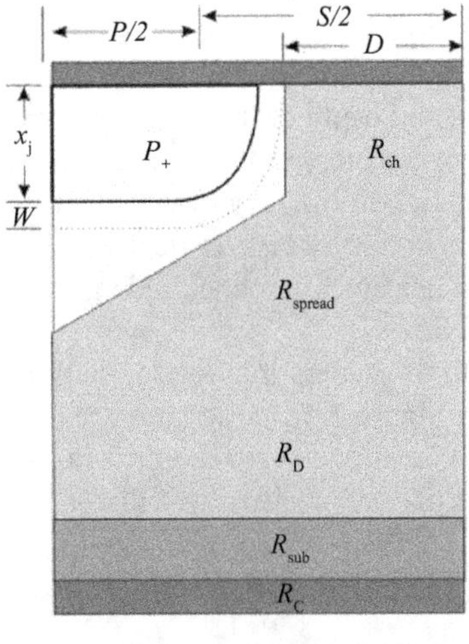

Figure 7.3 Different on-state resistance contributions of a junction barrier Schottky or merged PN-Schottky diode. The light gray area indicates the current flow

contributions to the voltage drop can be expressed as [10],

$$R_{ch} = \frac{x_j(P+S)}{2q\mu_n N_D D} \tag{7.12}$$

and [10]

$$R_{sp} = \frac{1}{2}\rho_{sp}(P+S)\ln\left(\frac{P+S}{2D}\right) \tag{7.13}$$

The current spreading layer resistance is depending on its doping in the usual way. The assumption of a 45° current spreading angle implies a current spreading layer thickness of [10],

$$d_{sp} = \frac{P+S}{2} - D - x_j \tag{7.14}$$

In these expressions, x_j is the depth of the pn junction.

$$x_j = \sqrt{\frac{2\varepsilon}{q}\left(\frac{N_A+N_D}{N_A N_D}\right)(U_{bi} - 2k_B T/q)} \tag{7.15}$$

These contributions, due to the channel and current spreading resistance, lead to an additional temperature dependence of the forward voltage drop, which is especially important for SiC power devices. Due to the nonnegligible ionization energy of donor atoms in 4H–SiC, not all of them are ionized at room temperature. This effect is known as incomplete ionization and leads to a reduction of the channel, current spread and drift resistivity with increased temperature until all donor atoms are ionized. An adverse effect is the reduced mobility at elevated temperatures [12], leading to an increase of the channel, current spread and drift layer resistivity.

By increasing the area and depth of the p-regions of the JBS diode, one arrives at the so-called merged PN Schottky (MPS) diode. This design not only provides the already discussed shielding to the Schottky contact but also decreases the drift layer resistivity by hole injection once a critical current density is reached. The dependence of this current density where the diode goes from unipolar into bipolar mode can be derived as follows.

Once the voltage drop of carriers moving laterally below the p-region and in the channel as depicted in Figure 7.4 [13] is larger than the built-in voltage U_{bi} of the pn-junction, holes will be emitted into the drift layer, an electron-hole plasma will form and, thus, the drift-layer resistivity will drop. This voltage drop can be calculated as [14],

$$U_{MPS} = \int_0^L \rho_s j_{pn} x dx \tag{7.16}$$

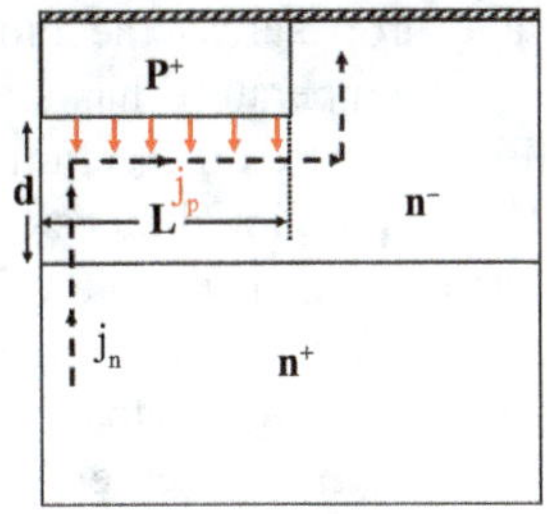

Figure 7.4 Current path beneath p+ region in JBS and MPS diodes [14]

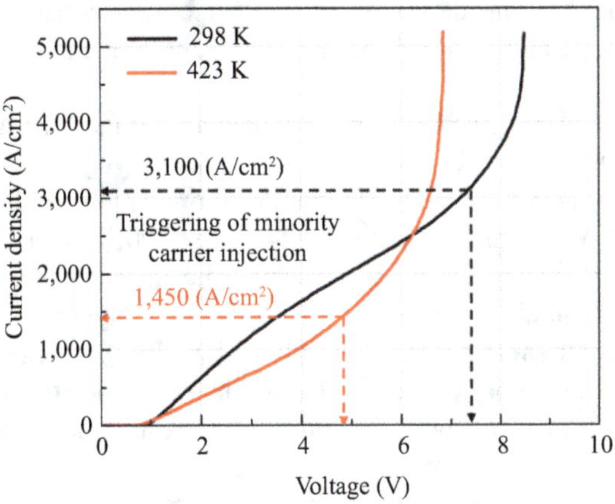

Figure 7.5 IV characteristics of 1,200-V SiC MPS diode with 10-A-rated current at 25 and 150 °C. Triggering is marked based on the pn injection onset equation

The integration can be easily carried out and, as a result, the onset of this so-called bipolar regime occurs at the current density [13],

$$j_{pn} = 2\frac{U_{bi} - U_{sc}}{RL^2} \tag{7.17}$$

From (7.17), it is clear that the introduction of a low-resistivity n-doped current spreading layer increases the turn-on of the bipolar operation. Indeed, achieving a low $R_{ds,on}$ and providing a low turn-on current density for bipolar turn-on are two conflicting design targets.

As the carrier mobility drops rapidly with increasing temperature, the drift layer resistance R beneath the p+ region increases strongly with temperature. This leads to a pronounced dependence of the bipolar onset current density on temperature as shown in Figure 7.5 [13]. As the forward voltage drop at j_{pn} is the sum of the voltage drop U_{sc} across the Schottky contact and the built-in pn junction voltage U_{bi}, it is clearly larger than U_{sc} at room and elevated temperatures. The behavior of the forward voltage drop across the diode and the current density needed for bipolar turn-on versus temperature has strong implications for surge-current robustness, which will be discussed in Section 7.4.

Of course, one can use different p region dimensions for shielding the Schottky contact and achieving hole injection. This is done for the CoolSiC Schottky diodes by Infineon, where small hexagons provide the Schottky contact shielding functionality, while larger hexagons are designed to trigger and sustain the bipolar mode during surge-current events [15]. However, enlarging the p regions at a given device active area reduces the area for the Schottky contact, therefore increasing the voltage drop at the rated current. An analytic model to find the optimum design point was developed by Huang *et al.* [16]. They achieve good agreement with 2D device simulations by considering an equivalent circuit consisting of a Schottky and pn diode in parallel and taking into account the channel and drift layer resistance.

For large-enough reverse blocking voltages and, thus, drift layer thicknesses, the voltage drop across the Schottky contact and drift layer will exceed the built-in voltage of the pn junction, rendering pn diodes more suitable at high voltages than Schottky diodes. While this blocking voltage is $\sim$ 300 V for silicon diodes, it is ~3.3 kV for 4H–SiC diodes. As for silicon ones, the structure of a SiC high-voltage diode is that of a p-i-n diode where the low-doped drift layer is always sandwiched between the p-doped emitter and the highly n-doped buffer region as depicted in Figure 7.6 [9]. As the theory of the pn junction is well understood and presented in many textbooks, e.g. [9], we will only present the results necessary to understand device properties.

In the case of pn junctions, we differentiate between low- and high-injection conditions. At high current densities, the injected minority carrier concentration is comparable with the majority carrier density and the device is operated under high-injection conditions. Most of the voltage drop occurs across the drift region, where the plasma density contributes significantly to the free carrier concentration, and at the ohmic contacts. With the ambipolar diffusion length L_a given by,

$$L_a = \sqrt{D_a \tau_a} \tag{7.18}$$

where D_a is the ambipolar diffusion constant and τ_a the ambipolar lifetime; the voltage drop across the “intrinsic” *i* region can be expressed as,

$$U_i = \frac{k_B T}{q} \frac{2b}{(1+b)^2} \left(\frac{w_{drift}}{L_a} \right)^2 \tag{7.19}$$

with the ratio of the electron and hole mobilities b and assuming a constant plasma density across the drift layer. The resistance of the drift layer as modulated by the electron-hole plasma

$$R_i = \frac{U_i}{I_F} = \frac{k_B T}{q} \frac{2b}{(1+b)^2} \left(\frac{w_{drift}}{L_a} \right)^2 \frac{1}{I_F} \tag{7.20}$$

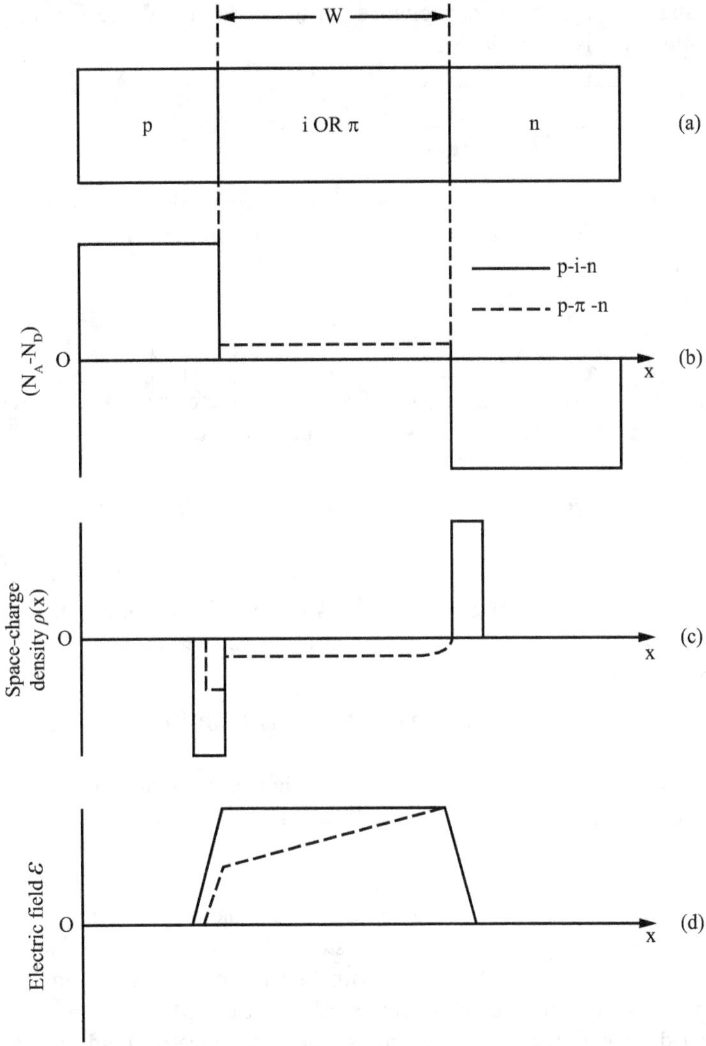

Figure 7.6 Schematic of the p-i-n structure, net doping concentration, space-charge density, and electric field (from top to bottom) [9]

can be deduced by plotting the diodes differential resistance at high-injection conditions versus the inverse of the load current. The intercept of the line fitted to the data points with the *y*-axis is then the contact resistance, while the slope of the line yields the voltage drop of the drift layer. To attain a low drift layer resistivity, it is therefore necessary to achieve a high carrier lifetime. As the defects induced by ion implantation lead to a decrease in carrier lifetime, either epitaxial *p* emitters have to be processed or a suitable post-implant annealing process has to be employed to bring carrier lifetimes to an acceptable level [17].

The space-charge region capacitance per unit area of both the Schottky and pn junction diode is given by [9,18],

$$\frac{C}{A} = \sqrt{\frac{\pm q\varepsilon\varepsilon_0(N_A - N_D)}{2(\pm U_{bi} \pm U - k_BT/q)}} \tag{7.21}$$

where the built-in voltage in the case of Schottky diodes is a function of the Schottky barrier height and a correction that depends on the number of states in the conduction band, N_c as

$$U_{bi} = \phi_B - \frac{k_BT}{d}\ln\frac{N_c}{N_D} \tag{7.22}$$

With the expression for the capacitance of the depleted space-charge region the energy stored at a given reverse bias U_R can be calculated using $dW = CUdU$ as

$$W_{cap} = A\sqrt{-q\varepsilon\varepsilon_0(N_A - N_D)}\left(\frac{1}{3}(U_R - 2(U_{bi} + k_BT/q))\sqrt{U_R - 2(U_{bi} + k_BT/q)}\right) \tag{7.23}$$

However, during turn-off of bipolar devices the energy to remove the plasma is considerably higher than the energy to build up the depletion region.

7.3 Edge termination and reverse bias reliability

All semiconductor devices have finite size, i.e. they are bounded by surfaces. Some of these surfaces serve as contacts while others have a gradient of the electric potential parallel to them. The most prominent of these surfaces are the device sidewalls created by some singulation process. These processes, e.g. sawing with diamond blades or laser-based sawing, create crystal damage that leads to increased surface roughness and surface density of states through dangling bonds. As with chip front- and backsides not covered with metal contacts, these dangling bonds arise from the surface atomic layers of the crystal lattice. Moreover, the singulation is performed through the drift layer that supports the electric field during reverse (blocking) operation creating a leakage current along the chip sidewalls and degrading the breakdown voltage and its long-time stability.

It is for the aforementioned reasons that special structures around the semiconductor edges have to be introduced. These edge terminations support the electric field during reverse operations leading to only little potential gradient along the chip sidewall and the chip surface between the edge termination and the chip edge. This can be achieved by nonplanar structures such as mesa edge terminations, etch terminations, or beveled chip sidewalls. Structures with a planar semiconductor surface include field plate terminations, floating field rings, and single and multiple zone junction termination extensions (JTE). These approaches can be combined, floating field rings and field plates being one prominent example. Directly

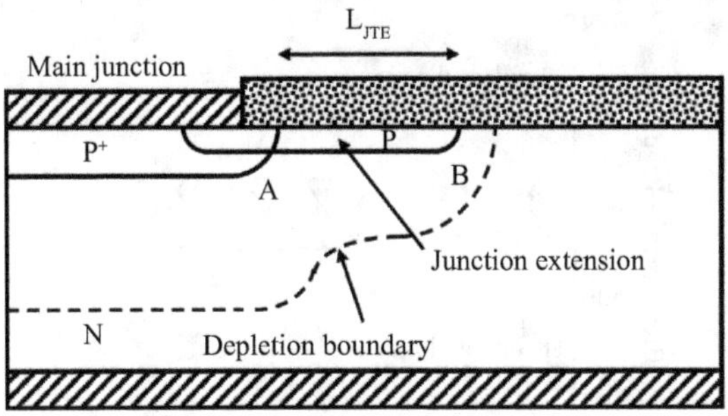

Figure 7.7 Junction termination extension (JTE)

transferring these edge termination techniques that were designed for silicon power devices to SiC power devices is difficult for two reasons. First, the processing techniques are different as, e.g. diffusion processes are not common in manufacturing SiC power devices as diffusion constants are magnitudes of order smaller for dopants as they are in silicon. Boron is a notable example to this rule. Second, the larger achievable electric breakdown fields call for larger oxide thicknesses when designing field plate terminations to reduce the high electric oxide fields. This leads to the fact that JTE structures are the most widely used edge termination for SiC power diodes, and we will constrain ourselves to the description of these terminations.

The JTE structure, as depicted in Figure 7.7 [6], was first described by V. Temple *et al.* as an attempt to overcome the disadvantages of the etch terminations created by etch, implantation and diffusion processes [19–21]. These process variations during dopant diffusion and the etch process lead to significant variations in breakdown voltage (VBR). The JTE method uses ion-implantation techniques to shape the space-charge region on both sides of the pn junction leading to better control of the electric field and, therefore, a larger process window. The desired charge within the JTE p region can be determined as follows. For a homogeneously doped region, the charge within the depleted portion of that region is related to the doping concentration and the magnitude of the maximum electric field at the surface by,

$$Q_0 = qN_A w_D = \varepsilon\varepsilon_0 E_m \tag{7.24}$$

The charge must be well controlled and matched to the breakdown voltage to be achieved. In case the charge is too small, it will have little effect on the electric field distribution at the semiconductor surface, and breakdown will occur at the p+/p− transition at point "A" in Figure 7.7. When the charge is too large, the breakdown will occur prematurely at the outer edge of the JTE, "B." It was found that the best results, i.e. a breakdown voltage closest to the theoretical optimum, are achieved with activated dopant concentrations of 60 to 80 percent of Q_0 [19]. However, the first experimental results obtained with the structure shown in Figure 7.7 fell short of the corresponding theoretical expectations as shown in Figure 7.8 [19]. The cause for this shortcoming is the implantation depth of the

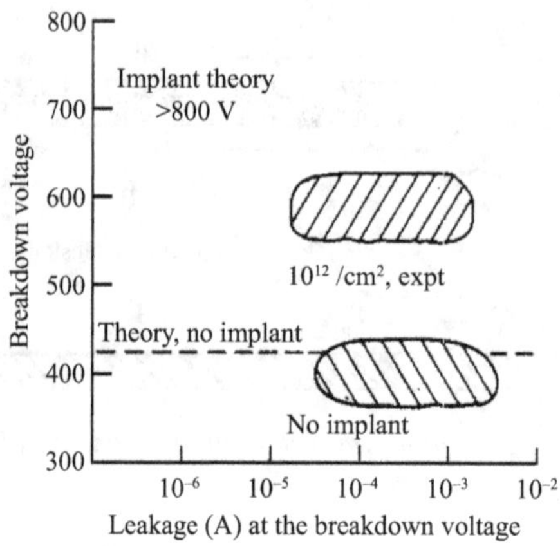

Figure 7.8 Comparison of JTE breakdown voltage for no implant and 1E12/cm² implant dose. The theoretical parallel plane breakdown voltage is >800 V

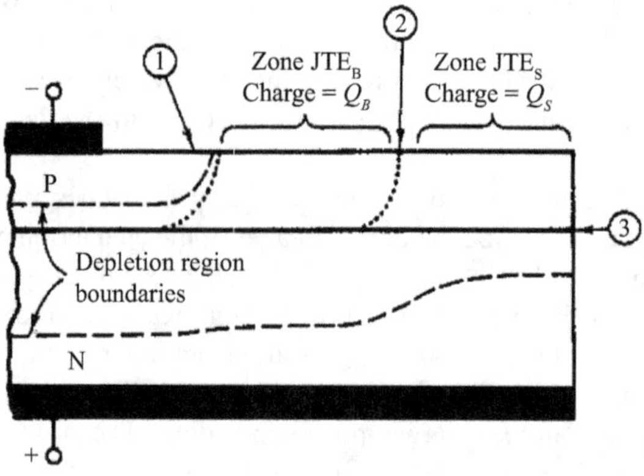

Figure 7.9 Two-zone JTE design as proposed by Temple in 1983

p-doped extension, which was shallow. Because the JTE is not implanted as deep as the p+ region, the radius of curvature of the JTE p region is smaller; thus, the electric field strength is higher than at the p+ edge "A." This causes a larger electric field strength and a premature breakdown as shown in Figure 7.8.

The maximum field at the points "A" and "B" reacts opposite to the change of the implanted ion concentration. The lower the concentration, the lower the surface electric field and the higher the field at point "A."

Using a two-zone JTE design as shown in Figure 7.9 [20], the electric field strength at both points can be adjusted separately [20,21]. For this purpose, the total charge in the zone "JTE_B" is chosen to achieve the best bulk breakdown, $Q_B \sim Q_0$ [20,21]. The surface field at point "3" is roughly proportional to the total charge Q_S

in implant zone "JTE$_S$." The proposed total charge for Q_S differs largely from 10 percent of Q_B in [22] to 30–70 percent of Q_0 in [20] and [21]. Moreover, the design window for the total JTE charge can be greatly enhanced by using multiple zone or ring JTE designs [20–24]. One notable development is the so-called field guard-ring assisted JTE, where the multiple guard rings are embedded within a JTE zone [22]. It further expands the total charge range in which the optimal breakdown voltage, i.e. the breakdown voltage closest to the theoretical maximum, can be achieved, see Figure 7.10 [22]. Here, the breakdown voltage versus the dose of the JTE is shown. For the double-zone JTE, the doping concentration of Q_B is plotted on the x-axis and the concentration of Q_S is assumed to be one tenth of Q_B. The doping concentration of the p rings for the guard-ring assisted JTE, "FGR+JTE," is assumed to be the same as for the p+ region, $2 \times 10^{19}/\text{cm}^3$. The combination of the guard-ring edge termination with a JTE termination eliminates the low-doped areas between the guard rings, which are sensitive to surface charges which form leakage paths on the low-doped zones. This reduces the influence of external fixed or mobile charges, e.g. fixed oxide charges, on the breakdown voltage as shown in Figure 7.11 [25]. The addition of positive interface charges leads to negative mirror charges within the edge termination, effectively reducing the dopant concentration and leading to a moderate reduction of VBR for charge densities up to $4 \times 10^{12}/\text{cm}^2$ for the guard-ring assisted JTE, as can be seen in Figure 7.11. The reduction of VBR for additional negative interface charges that induce positive mirror charges within the edge termination is more pronounced. This can already be deduced from Figure 7.10 and is clearly revealed in Figure 7.12 [25]. An additional interface charge larger than $4 \times 10^{12}/\text{cm}^2$ leads to a significant loss of reverse blocking capabilities. Thus, the presence of mobile or fixed charges close to the edge termination has to be prevented.

The discussion so far concerned the static reverse blocking properties of edge terminations. The dynamic behavior, i.e. the transition from forward conducting to

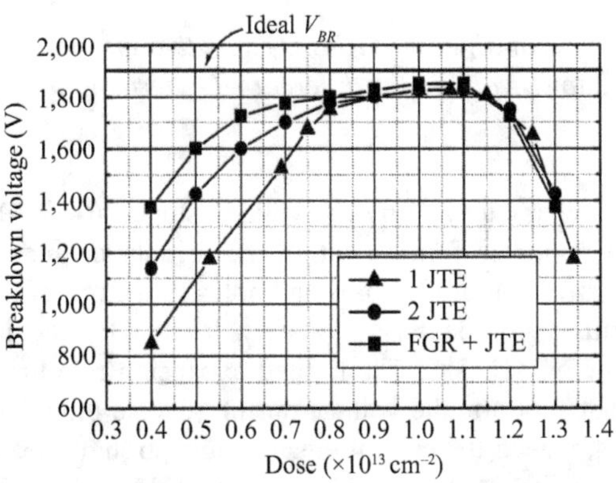

Figure 7.10 Breakdown voltage versus doping concentration for one- and two-zone JTE as well as guard-ring assisted JTE

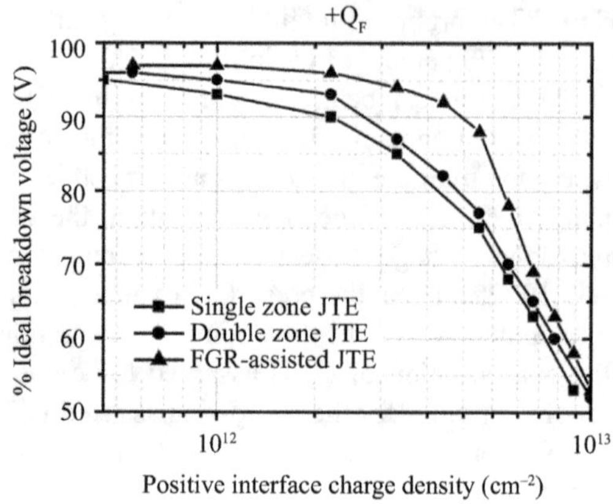

Figure 7.11 Influence of different SiC/SiO$_2$ positive interface charges on the breakdown voltage for different edge termination concepts

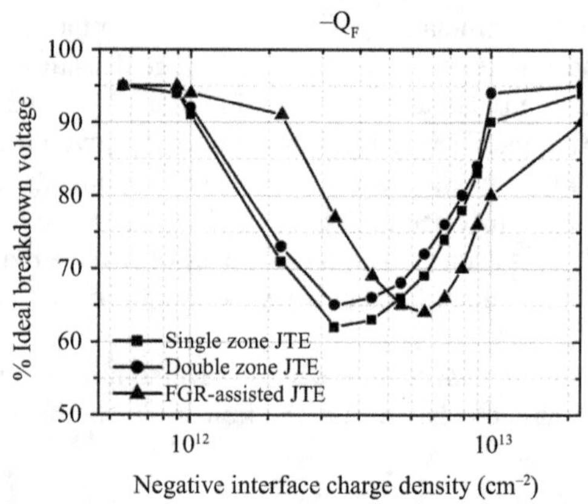

Figure 7.12 Influence of different SiC/SiO$_2$ negative interface charges on the breakdown voltage for different edge termination concepts

reverse blocking, is characterized by the voltage slope, *dU/dt*. During this transition, the depletion region is formed within the drift layer and the p-doped edge termination by the diffusion of electrons and holes to the respective sides of the pn junction. This diffusion process is ultimately limited by the ionization rate of acceptors and donors in 4H–SiC. The limited expansion rate limits the size of the depletion region that supports the voltage during reverse blocking and the electric field strength might exceed the critical electric field strength, leading to avalanche breakdown at lower voltages if *dU/dt* is high enough [26]. As the ionization rate is proportional to $\exp(E_A/k_BT)$, increasing temperatures lead to an increased *dU/dt* capability.

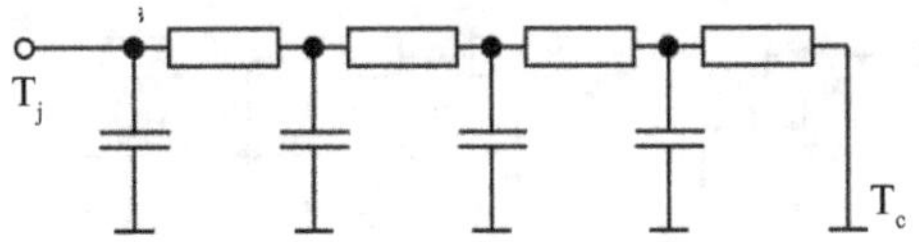

Figure 7.13 Foster model

Repetitive avalanche will be discussed in Section 7.4, as this device capability requires the avalanche breakdown to occur in the cell field due to the power densities involved.

7.4 Measurement of application relevant parameters

The most important parameters of a diode characterizing its performance are the forward voltage drop at nominal current and the leakage current at the rated reverse blocking voltage. Both can be easily measured during functional test at the end of either front-end or back-end manufacturing. For avalanche robust designs, a breakdown voltage at a given current may be specified; for surge-current capable designs, a value for i^2t in the datasheet is given, where the first two diode properties refer to standard operation, the latter often refer to some application-relevant failure mode, e.g. short circuit conditions or overvoltage. These properties are ensured and characterized by, and during, design and not measured during wafer or packaged device tests. Other notable properties of this kind are the thermal resistance and thermal impedance that describe the time-independent flow of heat from the device to the ambient and the time-dependent flow of heat due to, e.g. changes in load conditions or surge-current events, respectively.

For the measurement of thermal resistance of a packaged device, it is mounted on a heat sink whose temperature is monitored. The device under investigation is operated at its nominal current rating. After a constant heat sink temperature is reached, the load current is replaced by a small sense current and the voltage drop across the diode is measured. In this way, the junction temperature can be deduced using (7.6) and the thermal resistance can be derived. The extraction of thermal impedance requires some model for the thermal network under consideration, the most widely used being the Foster network shown in Figure 7.13 [27]. It models the packaged device and its mounting to a heatsink by a number of series connected parallel heat capacity and heat resistivity pairs. However, it is not possible to attribute a node in a Foster network to a geometric location within the chip-package arrangement. This is only possible when the physically correct Cauer network, shown in Figure 7.14 [27], is used. This implies that the capacitance and resistance values in this model can be calculated from material properties, which is not possible for the Foster network where the model parameters have to be determined experimentally as described in the next paragraph. Another consequence of the node – location correlation is the possibility that two Cauer models can be combined or one Cauer model can be subdivided, e.g. into a part modeling the thermal

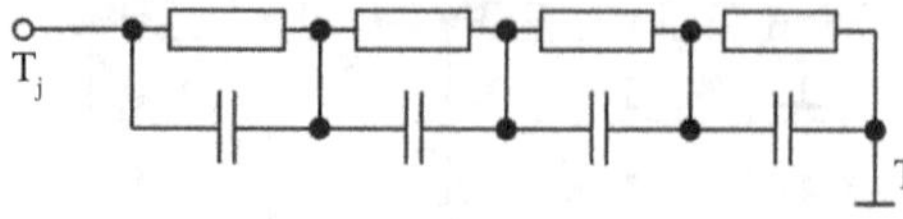

Figure 7.14 Cauer model

impedance of the chip-lead frame system and one part modeling the lead frame-heatsink system. The reason why the Foster model is still widely used is the simplicity with which the thermal impedance can be expressed through an analytical expression,

$$Z_{th} = \sum_{i=1}^{n} R'_i \left(1 - exp\left(-\frac{t}{R'_i C'_i}\right)\right) \tag{7.25}$$

To determine the Foster model parameters, the device under investigation is again mounted onto a heatsink wherein temperature is kept constant and it is operated at nominal current until a steady state, indicated by a constant junction temperature, is reached. The load current is then replaced by a sense current and the transient of the voltage drop across the diode is recorded, from which the temperature transient can be deduced. A fitting procedure yields the model parameters that can be employed to calculate the temperature response with respect to time for a given load transient. In most cases, e.g. in DC–DC converter or motor drive applications, the devices are not operated under a static load but with some duty cycle D for a given pulse length t_{pulse}. For such a series of constant pulses, the thermal impedance is given by,

$$Z_{th,\ dynamic} = \sum_{i=1}^{n} R'_i \left(\frac{1 - exp\left(-\frac{Dt_{pulse}}{R'_i C'_i}\right)}{1 - exp\left(-\frac{t_{pulse}}{R'_i C'_i}\right)}\right) \tag{7.26}$$

The maximum temperature that is reached after an infinite series of such constant pulses can be calculated as,

$$T_{max,equilibrium} = P_{on} \times Z_{th,dynmaic} \tag{7.27}$$

where P_{on} is the power dissipated during the on-state of the duty cycle.

Figure 7.15 shows a plot of the thermal impedance for different duty cycles for Infineon's 1,200-V/20-A diode in a TO-247 package. For the calculation, the thermal impedance was modeled by four terms according to (7.26). For short on-state pulses, i.e. small duty cycle and pulse length, the thermal impedance is small as the energy generated during the pulse is stored within the device and then dissipated to the much larger thermal capacity of the solder and lead frame. As the duty cycle approaches unity, i.e. DC operation, the dynamic thermal impedance approaches the steady-state thermal impedance.

Closely connected to the packaged device's response to load current transients is the overcurrent capability, which is often given as power integrated over one

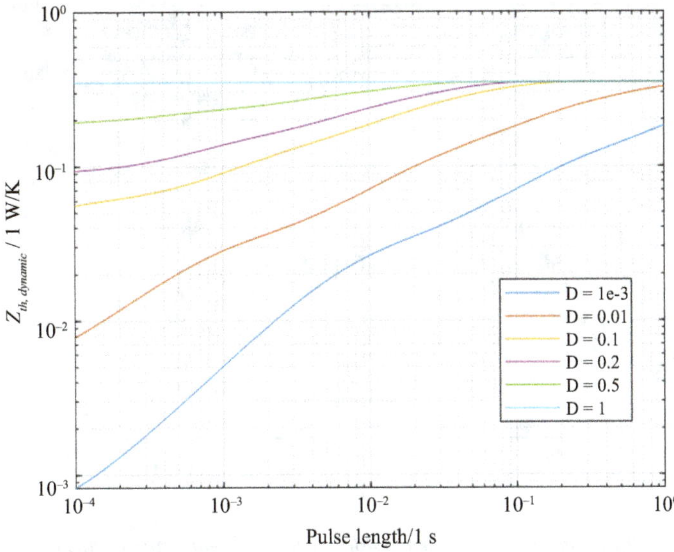

Figure 7.15 Thermal impedance of a 20-A/1,200-V Infineon SiC MPS diode

half-wave of the 50-Hz cycle or 10 ms, respectively. As the voltage drop is given by the *I(U)* characteristics of the device under investigation, this integral can be expressed as,

$$\int_0^{10\text{ ms}} i \times u(i)dt \propto \int_0^{10\text{ ms}} i^2 dt \tag{7.28}$$

This so-called "i^2t"-value is measured during device characterization and its typical value is given in the device datasheet. The voltage and current waveforms for a surge-current event of a 20-A/1,200-V MPS diode are shown in Figure 7.16 [28]. The pulse peak current was set at 220 A and the pulse duration was 10 ms. The current versus voltage characteristics during the pulse test is shown in Figure 7.17 [28]. The onset of bipolar conduction mode at ~150 A is accompanied by a negative differential resistance (NDR). This has to be considered when paralleling the MPS diodes, as the diode with the earliest onset of NDR will take over a major part of the load current. The transition from bipolar back to unipolar conduction mode occurs at a voltage drop of about 3.5 V. The device under test (DUT) in question was destroyed at a peak pulse current of 240 A, leading to a datasheet value for i^2t of 180 A at room temperature after taking the other devices in the test ensemble and some safety margin into account.

Of course, these tests are carried out at elevated temperatures during device characterization and the resulting "i^2t"-values are given in the datasheet as well. Closely related to the "i^2t"-value for a 10-ms sine half-wave waveform is the

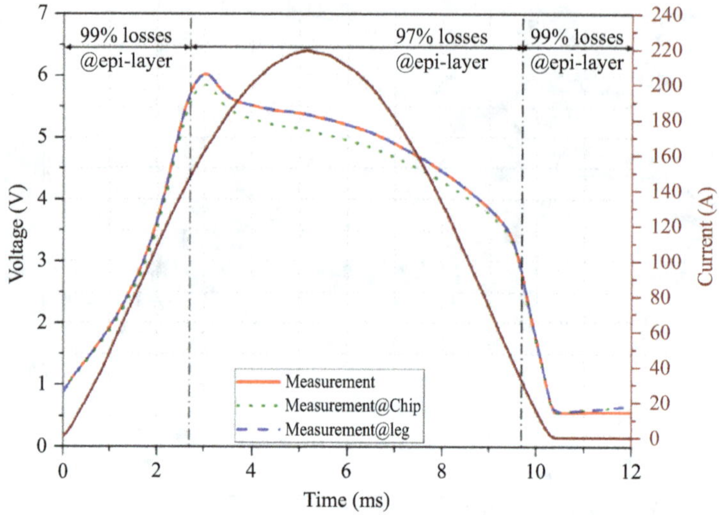

Figure 7.16 Voltage and current waveform for a 10-ms surge-current pulse. Simulation results are shown for comparison

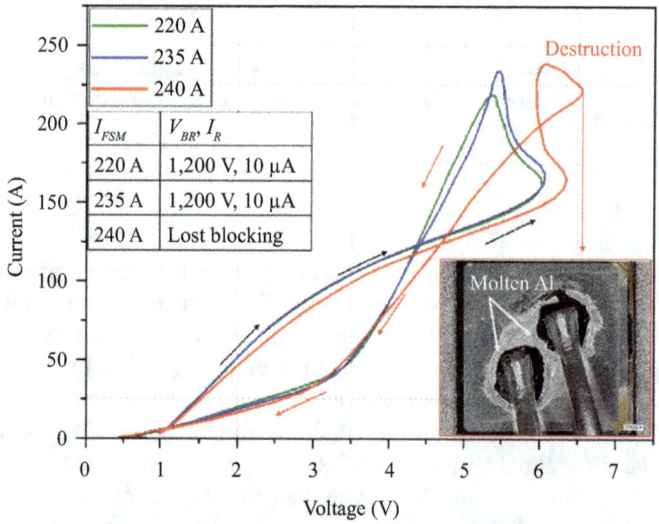

Figure 7.17 IV characteristics of single event surge-current limit under 10 ms half-sine pulse and destruction of the metal (molten Al)

repetitive sine half-wave surge current,

$$I_{F,RM} = \sqrt{\frac{i^2 t}{0.01s}} \tag{7.29}$$

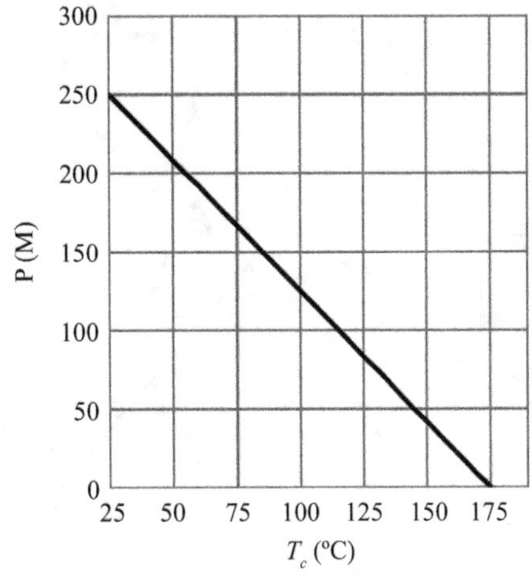

Figure 7.18 Maximum allowed power dissipation versus case temperature for a 20-A/1,200-V diode in TO-247 package with soft solder

These values have been successfully predicted by simulation for a TO-247 housing given the surge-current power losses and the thermal impedance by Basler *et al.* [28]. An electro-thermal simulation was used to simulate the temperature distribution, taking into account the Joule heating of the bond wires and the SiC substrate.

For static operations, the maximum allowed power dissipation versus ambient temperature can be derived with the help of the junction to casing thermal resistance, $R_{th(j-c)}$, and is given by

$$P_{max} = \frac{T_{max} - T_{case}}{R_{th(j-c)}} \tag{7.30}$$

For a maximum power dissipation of 250 W, a thermal resistance of 0.6 K/W and a maximum case temperature of 175 °C, the power dissipation versus case temperature diagram is given in Figure 7.18 [29].

Assuming the maximum casing temperature is limited by the mold compound and plastics used, the only way to increase the maximum allowed power dissipation is to reduce the thermal resistance, e.g. by using an advanced soldering method. This reduction which can be seen by comparing Figure 7.18 with Figure 7.19 exemplifies this by comparing the so-called soft solder die attach in TO-247 with "diffusion soldering" in a TO-220 casing [29]. By reducing the $R_{th(j-c)}$ from 0.6 to 0.46 K/W, the maximum power dissipation was increased from 250 to 330 W at 25 °C. Due to the thermal resistance being the steady-state limit of the thermal impedance, the maximum allowed forward current for duty cycles D with $D < 1$ is larger than the steady-state maximum forward current. The maximum steady-state forward current is determined by the allowed power dissipation and the voltage

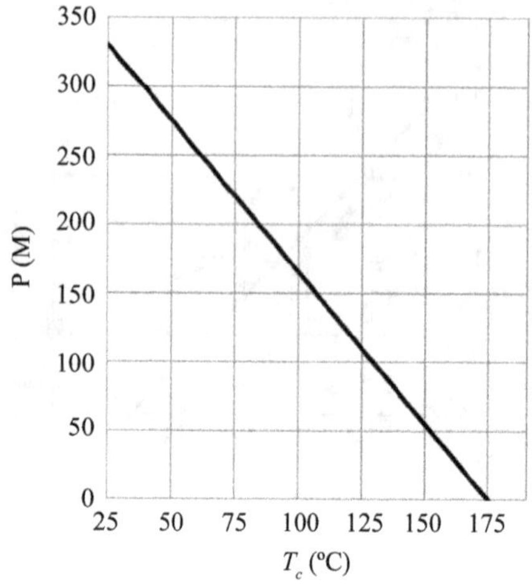

Figure 7.19 Maximum allowed power dissipation for a 20-A/1,200-V diode in TO-220 with diffusion solder

drop across the diode at the current applied. This voltage drop is one important parameter measured at wafer level during front end manufacturing as well as the packaged device level at room temperature. The forward voltage drop at different current levels and elevated temperatures is measured during device characterization and typical and maximum datasheet values are derived from the measured ensemble of devices.

The electric device properties discussed so far concerned the forward (conducting) operation. Reverse blocking parameters relevant for application are: (i) leakage currents at the rated blocking voltage at room and elevated temperatures, (ii) the reverse breakdown voltage in case the device can be brought into avalanche breakdown, and (iii) the space-charge region capacitance, which dominates the switching losses for unipolar devices. The leakage currents and breakdown voltage at room temperature are both measured at wafer level during frontend manufacturing and for packaged devices. For the front end of line, the leakage current level and the number of devices violating the leakage current limits are an indication of on-wafer defect density. In addition, these parameters are relevant for applications as they determine the power losses during blocking and the maximum allowed voltage overshoot when switching inductive loads in case the device must not be brought into avalanche breakdown. In case (repetitive) avalanche is allowed, the maximum energy released during such an event that can safely be handled is listed in the datasheet as measured by unclamped inductive switching (UIS) tests. These tests are performed with a circuit as shown in Figure 7.20 [15], where the waveforms of the voltage across the diode and current flowing through the diode are shown in Figure 7.21 [15]. By switching off the IGBT at a preset voltage, an overvoltage is induced across the switch and the diode is brought into avalanche.

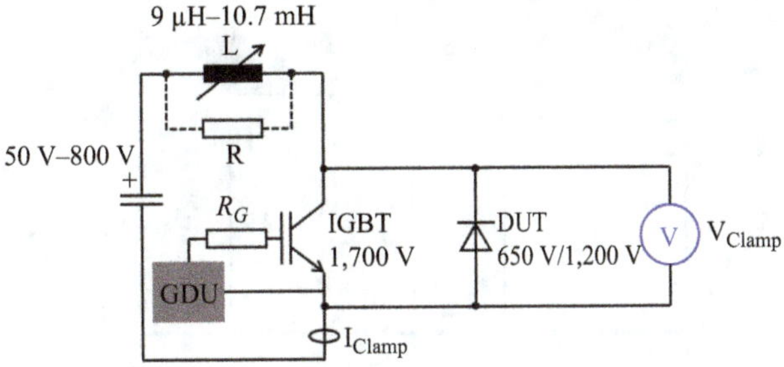

Figure 7.20 Test setup for unclamped inductive switching tests. GDU denotes the gate-drive unit

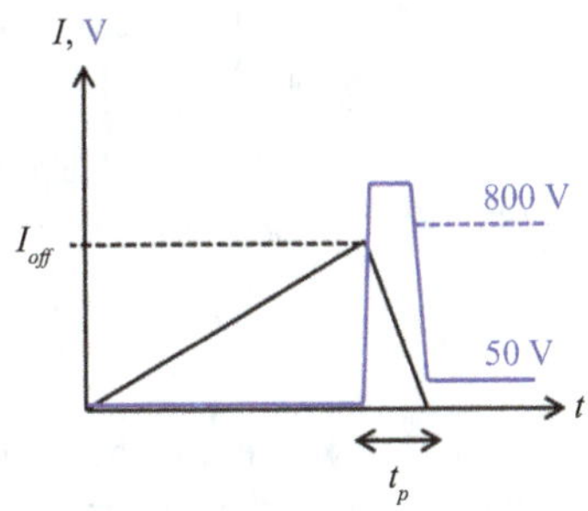

Figure 7.21 Voltage and current waveforms during an UIS test

The time span of the diode in avalanche, t_p, is given by [30],

$$t_p = \frac{L \times I_0}{V_{BR}} \tag{7.31}$$

By changing the inductance L, the so-called clamping time t_p and the avalanche energy E_{AV} can be varied as given below [15,30],

$$E_{AV} = \frac{1}{2} L I_0^2 \tag{7.32}$$

As for the surge-current characterization, the energy is increased until the DUT is destroyed and the last energy before destruction recorded. The mean value of some ensemble is then listed as the maximum avalanche energy, again taking some safety margin into account. The maximum allowed energy released during an avalanche event depends on the junction and casing temperature and can be calculated using (7.30–7.32):

$$P_{avalanche,max} = \frac{T_{max} - T_{case}}{Z_{th,dynamic}(t_p)} = \frac{E_{AV}}{t_p} = \frac{I_0 V_{BR}}{2} \tag{7.33}$$

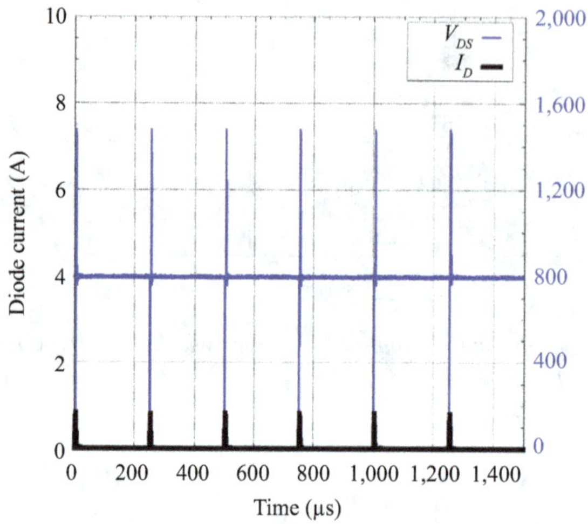

Figure 7.22 Repetitive UIS test of an 8-A/1,200-V MPS diode with a frequency of 4 kHz, $E_{AV} = 28$ mJ/cm^2, $V_{DC} = 800$ V, and L = 3 mH

To demonstrate the avalanche ruggedness of MPS diodes, Basler *et al.* performed 10,000 clamping events with an avalanche energy density of 2.85 J/cm^2 at different reverse currents [15]. The current and voltage waveforms during such a test are depicted in Figure 7.22 [15]. Up to 144 A for a 1,200-V diode rated at 8 A, no degradation could be observed. The clamping time during these pulses was 0.92 μs. The avalanche ruggedness of MPS 4H–SiC diodes may be used to protect a nonavalanche robust switch, e.g. in circuits with high (parasitic) inductances. In this case, the diode parallel to the switch has to clamp the avalanche energy with the switching frequency used. This use case was investigated by testing an 8-A/1,200-V MPS diode in a UIS configuration with a switching speed of 4 kHz until thermal equilibrium was reached [15]. During each event, an energy density of 28 mJ/cm^2 was dissipated, which did not lead to device destruction during the test. However, it must be noted that the use of an MPS diode for clamping protection leads to higher overall switching losses. Therefore, it should only be used at overload conditions.

As mentioned in Section 7.2, capacitive losses are the most important turn-off losses for unipolar diodes. The characterization of the depletion-region capacitance can be done statically by measuring the charge stored in the device at a given negative bias or with a dynamic measurement by applying a small sinusoidal voltage in addition to a negative DC bias and measuring the phase shift to the resulting current. To deduce the capacitance, an equivalent circuit has to be assumed ([18], Section 2.4.2). Using the static capacitance measurement method, the assumption of an equivalent circuit is not necessary, but now the leakage current leads to an additional capacitive contribution. In practice, both methods are used to characterize an ensemble of test devices yielding the datasheet values for the total stored capacitive charge and the total capacitance for different reverse biases.

Closely related to the total stored capacitive charge is the energy stored due to this capacity,

$$E_C = \int_0^{V_R} C(V)VdV$$

Figure 7.23 shows the stored energy versus the reverse blocking voltage for a 20-A/1,200-V MPS diode [31].

7.5 Operation in applications

Buck (step-down) and boost (step-up) converters are the basic forms of DC-DC converters. In these converters, energy is periodically stored within and released from an inductor's magnetic field to either decrease the output voltage and increase the output current or increase the output voltage and decrease the output current, respectively. Inverters that produce arbitrary output voltages can be created by combining the two converter types, limiting our discussion of DC-DC converters to buck and boost converters.

The concept of the buck converter, whose schematic is depicted in Figure 7.24, can be understood best in terms of the energy flow in and out of the inductance. Beginning with the switch S open, no current is flowing through the inductance and its field is zero. Once the switch is closed, the current flowing builds up a magnetic

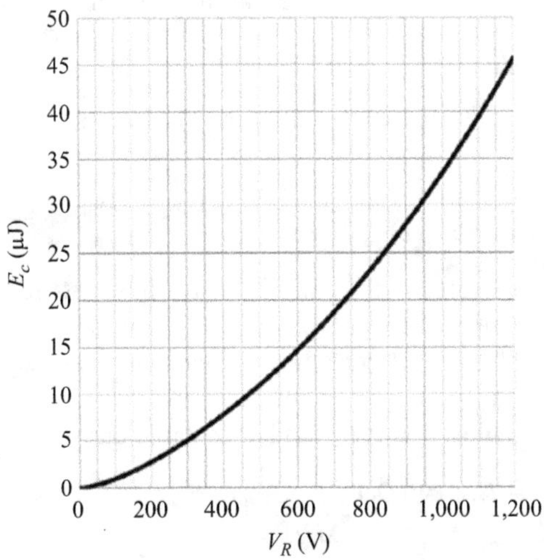

Figure 7.23 Capacitively stored charge versus reverse bias for a 20-A/1,200-V MPS diode

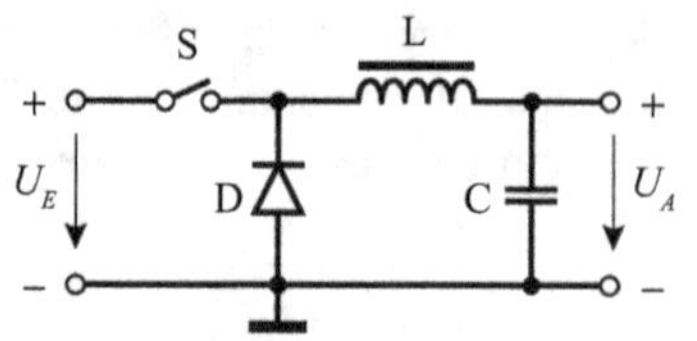

Figure 7.24 Buck converter schematic by Wdwd, licensed under CC BY-SA 3.0

field and causes an opposing voltage drop across the inductor by Lenz's rule. This voltage drop reduces the voltage across the connected load. As the rate of change of current decreases, the voltage across the inductor decreases and the voltage at the load will increase. When the switch is opened again, the voltage source is disconnected and the decreasing current will produce a voltage drop across the inductor that will support the current flow through the load. If, now, the switch is turned on before the magnetic field of the inductor has collapsed completely, the current of the voltage source and of the inductor will add up, raising the current through the load beyond the output current of the voltage source. At the same time, the output voltage is reduced below the voltage across the source. During the switch's off-state, the load current is flowing through the diode, which is going from reverse blocking to forward conducting mode at nonzero voltage and current. This mode of switching is called "hard" switching as opposed to "soft" switching during voltage crossing zero as, e.g. for rectifying diodes. The switching losses that occur during hard switching of unipolar Schottky diodes are mainly related to the capacity of the space-charge region that has to be built up or collapsed as there is no plasma that has to be injected or extracted during switching with frequency f_s,

$$P_{loss,cap} = W_{cap} f_s \tag{7.35}$$

The loss occurring during reverse and forward bias is given by,

$$P_{loss,reverse} = I_R U_R D \tag{7.36}$$

$$P_{loss,forward} = I_D U_F (1 - D) \tag{7.37}$$

where D is the duty cycle of the switch. The higher static losses of 4H–SiC Schottky diodes as compared to Si bipolar diodes are more than compensated by the reduced switching losses [32], enabling higher switching frequencies or reduced losses. As the size of the inductor L in the buck converter is inversely proportional to the product of switching frequency and acceptable current ripple ΔI_L [33],

$$L = \frac{U_{out}(U_{in} - U_{out})}{\Delta I_L f_s U_{in}} = \frac{D(U_{in} - U_{out})}{\Delta I_L f_s} \tag{7.38}$$

a high switching frequency is beneficial to reduce the size of the inductor. As the switching loss is significantly reduced by unipolar Schottky diodes, raising the switching frequency is possible.

Comparing power losses for the Schottky diode is now straightforward. Assuming an input voltage of $U_{in} = 800$ V, an output voltage U_{out} of 200 V, the forward voltage drop of U_F as listed in the datasheet for Infineon's IDH20G120C5, an operating temperature of 25 °C, and a switching frequency of 10 kHz, the duty cycle is given by $D = U_{out}/U_{in} = 0.25$ and finally the power loss can be calculated as

$$\begin{aligned} P_{loss,total} &= 23 \times 10^{-6}\text{ J} \times 10^{4}\text{ Hz} + 5 \times 10^{-8}\text{ A} \times 800\text{ V} \times 0.25 \\ &\quad +20\text{ A} \times 1.5\text{ V} \times 0.25 \\ P_{loss,total} &= 23 \times 10^{-2}\text{ W} + 1 \times 10^{-5}\text{ W} + 7.5\text{ W} \end{aligned}$$

At the same time, the total output power of the buck converter is $P_{out} = 200\text{ V} \times 20\text{ A} = 4\text{ kW}$. At elevated operating temperatures, e.g. 150 °C, the loss calculation is even more in favor of high switching frequencies as the depletion layer capacitance does not depend on the temperature, while U_F and I_R that are independent of the switching frequency increase. Our example calculation now reads

$$\begin{aligned} P_{loss,total} &= 23 \times 10^{-6}\text{ J} \times 10^{4}\text{ Hz} + 1.5 \times 10^{-6}\text{ A} \times 800\text{ V} \times 0.25 \\ &\quad +20\text{ A} \times 2\text{ V} \times 0.25 \\ P_{loss,total} &= 23 \times 10^{-2}\text{ W} + 3 \times 10^{-3}\text{ W} + 10\text{ W} \end{aligned}$$

Under these switching conditions, the thermal impedance is close to the thermal resistance of 0.46 K/W. The ambient temperature thus must not exceed

$$T_{amb,max} = 175°\text{C} - 10.3\text{ W} \times 0.46\text{ K/W} = 128°\text{C}$$

given an optimal connection of the DUT to the heatsink.

Aiming at a current ripple of 1%, i.e. 200 mA, this leads to an inductance of

$$L = \frac{0.25 \times (800\text{ V} - 200\text{ V})}{0.2\text{ A} \times 10^{4}\text{ Hz}} = \frac{150}{2 \times 10^{3}}\text{ H} = 75\text{ mH}$$

By increasing the switching frequency to 100 kHz, the switching losses of the diode increase to 2.3 W; however, the inductance can be cut down to 7.5 mH, with leads to reduced volume, weight, and resistive losses of the inductance.

Increasing the switching frequency does not only decrease the size of the inductor but also reduces the size of the output capacitor as well for a given voltage

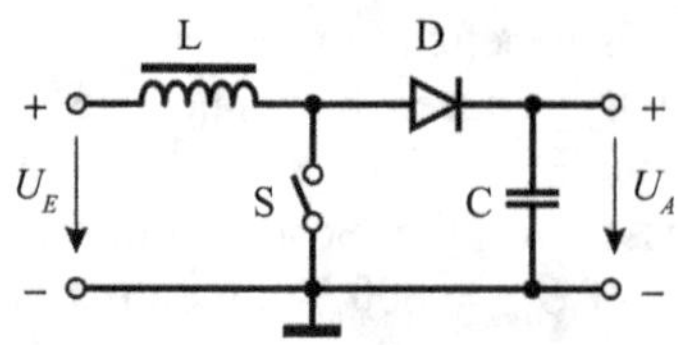

Figure 7.25 Boost converter schematic by Wdwd, licensed under CC BY-SA 3.0

ripple ΔU_{out} to be achieved,

$$C_{out,min} = \frac{\Delta I_L}{8 f_s \Delta U_{out}} \tag{7.39}$$

When aiming for an output voltage ripple of 1%, i.e. 4 V, the calculation for the output capacitor's value reads,

$$C_{out,min} = \frac{0.2\ \text{A}}{8 \times 10^4\ \text{kHz} \times 4\ \text{V}} = \frac{2 \times 10^{-1}}{3.2 \times 10^5}\ \text{F} = 6.25 \times 10^{-7}\ \text{F}$$

The output capacitance of 625 nF can be further reduced by increasing the switching frequency as for the inductance.

A complete investigation of buck converter losses is beyond the scope of this chapter, as an important contribution is due to the switch's losses, e.g. on-state and switching losses.

As for the buck converter, the key principle of operation of the boost converter is the reluctance of an inductor to changes in current by storing or releasing energy in to or from an electric field. Beginning with no current flowing and no energy stored in the inductance and output capacitance, closing the switch S will cause a current to flow through the inductance L, see Figure 7.25. This is building up a magnetic field and storing energy in the inductance. Once the switch is opened, the collapsing magnetic field will induce a current to maintain the current to the load which has a higher impedance than the switch. As a consequence, the voltage across the inductor U_L and the input voltage U_{in} add up, charging the output capacitor to that voltage. The voltage drop across the capacitor, $U_{out} = U_{in} + U_L$, is the output voltage of the boost converter. When the switch S is closed at the start of the next duty cycle (and before the magnetic field of the inductor drops to zero), the diode will prevent discharging the output capacity through the switch but support the output voltage, while the current driven by the voltage source through the inductance increases its magnetic field again. As the voltage across the load is increased over the output voltage of the source, the current through the load is smaller than the output current of the source. Once more, this type of converter benefits from increasing the switching frequency as the size of the inductance is given by [34]

$$L = \frac{U_{in}(U_{in} - U_{out})}{\Delta I_L f_s U_{out}} = \frac{(1 - D)(U_{out} - U_{in})}{\Delta I_L f_s} \tag{7.40}$$

Increasing the switching frequency while keeping the output current ripple constant allows the inductance to be reduced and the same is true for the output capacitance as,

$$C_{out,min} = \frac{I_{out,max} D}{f_s \Delta U_{out}} \tag{7.41}$$

holds [34]. Moreover, as the turn-off current tail of the Schottky diode is significantly smaller than the current tail of a pin diode, the overcurrent capability of the switch can be relaxed considerably. Again, we choose the 1,200-V/20-A MPS diode IDH20G120C5 diode to calculate the impact of the switching frequency on the inductor and output capacitor size. This time, the input voltage is assumed to be 240 V and the output voltage and current shall be 400 V and 20 A, respectively. The output voltage is chosen to be the voltage of electric (hybrid) vehicle batteries. The diode losses for this use case at 150 °C are

$$P_{loss,total} = 23 \times 10^{-6}\ \text{J} \times 10^4\ \text{Hz} + 6 \times 10^{-8}\ \text{A} \times 400\ \text{V} \times 0.6 + 20\ \text{A} \times 2\ \text{V} \times 0.4$$

The static forward losses are orders of magnitude larger than the switching losses, allowing for a significant increase of the switching frequency.

The calculation for the inductance reads

$$L = \frac{0.6 \times (400\ \text{V} - 240\ \text{V})}{10^4\ \text{kHz} \times 0.2\ \text{A}} = \frac{96}{2 \times 10^3}\ \text{H} = 48\ \text{mH}$$

As in the case of the buck converter, this value can be cut to 1/10 when the switching frequency is increased by an order of magnitude. The size of the output capacitance for the values given is

$$C_{out,min} = \frac{20\ \text{A} \times 0.4}{10^4\ \text{kHz} \times 4\ \text{V}} = \frac{8}{4 \times 10^4}\ \text{F} = 200\ \mu\text{F}$$

As before, increasing the switching frequency by a factor of 10 reduces the size of the output capacitance significantly.

A complete investigation of boost converter losses is beyond the scope of this chapter, as an important contribution is due to the switch's losses, e.g. on-state and switching losses.

Besides buck and boost converters, half bridges consisting of a low- and high-side switch with an antiparallel freewheeling diode are an important basic power electronic circuit. Starting from half bridges, H bridges, and three-phase inverters can be designed. Applications range from small to large motor drives, AC/AC and HVDC converters make use of H bridges. The topology of the half bridge can be derived starting with a buck converter by replacing the diode by a switch. As the diode was used during freewheeling, the switch as to be accompanied by an anti-parallel diode. For reversed current flow, i.e. current flowing from the half bridges' output to its input, a freewheeling diode has to be added to the high-side switch as

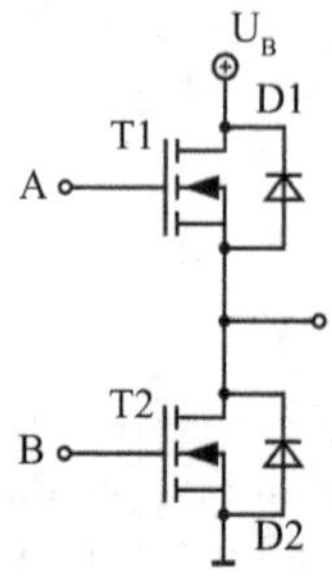

Figure 7.26 Half-bridge topology with two switches T1 and T2 and two antiparallel freewheeling diodes D1 and D2

well and we arrive at the topology shown in Figure 7.26. Half-bridges have two modes of operation called forward acceleration and forward deceleration, which indicate the use of these modes with a motor (inductive) load. During acceleration, T2 is always open and the switching frequency and duty cycle of T1 determine the current that is transferred to the load. When T1 is opened, D2 is acting as the freewheeling diode that takes the load current. When T1 is closed, it has to conduct the turn-off current of D2 in addition to the load current through the diode. As the diodes are Schottky diodes where the forward current is not supported by an electron-hole plasma, the turn off reverse current peak is negligible compared to a p-i-n diode. As a consequence, the turn-on losses of the switch T1 are greatly reduced. During deceleration, the switch T1 is always open and T2's duty cycle and switching frequency determine the current flowing from the motor to the power source. As in acceleration mode, the low turn-off current of the Schottky diode D2 reduces the turn-on power losses of the switch T2.

To summarize, the electrical properties of 4H–SiC Schottky diodes lead to a substantial reduction of passive elements in applications by increasing the switching frequency, leading to higher power densities of converters and motor drives. However, to quantify the power loss reductions possible in these applications is beyond the scope of this chapter as a detailed treatment of the switching process, especially the stored charge within the switch and the switching speed, i.e. dU/dt and dI/dt, is needed for this purpose.

7.6 Future developments

As 4H–SiC Schottky diodes were used in more fields of application, their blocking voltages increased from 300 to 1,700 V for commercially available products, while blocking voltages greater than 10 kV have been published [35–37]. As fast diodes are used in boost and buck converters, the commercialization of Schottky diodes with ever higher blocking voltages is mainly driven by DC/DC and DC/AC conversion, e.g. solar inverters. As the forward voltage drop within the drift zone scales with the square of the blocking voltage, it surpasses the on-state voltage drop of 4H–SiC p-i-n diodes at blocking voltages of about 3.3-kV. In applications that call

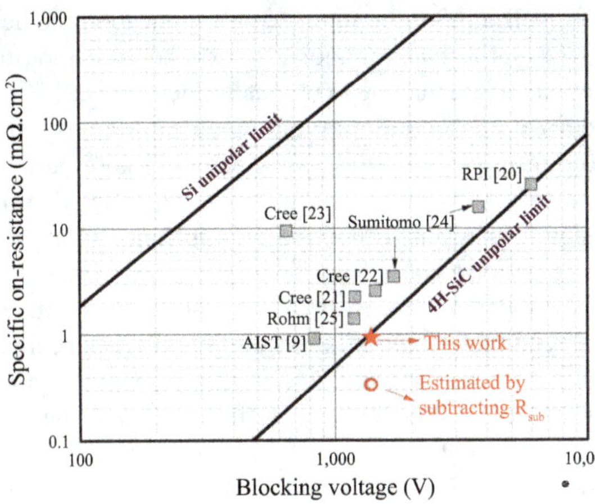

Figure 7.27 Performance of superjunction diode described in [37] with respect to the unipolar limit and work cited therein

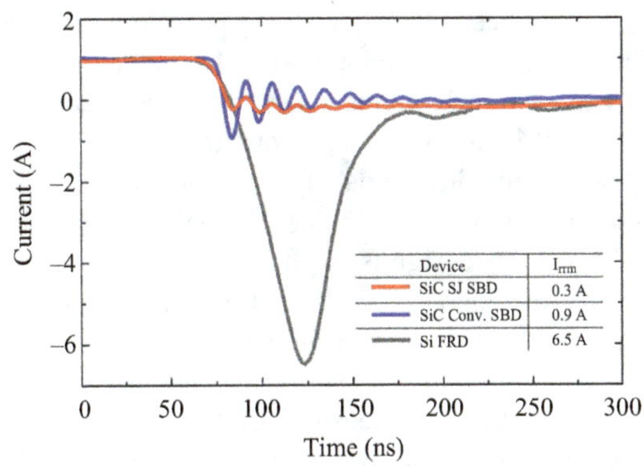

Figure 7.28 Reverse recovery waveforms of SiC superjunction SBD, conventional SiC SBD, and Si FRD. Load current was 1 A and DC-link voltage 400 V

for blocking voltages beyond 3.3 kV, the choice between Schottky and p-i-n diodes has to be made based not only on reverse blocking capabilities, but taking turn-off losses into account as well. One important development for going beyond 3.3-kV unipolar devices in general and Schottky diodes in particular is the realization of charge balanced devices, i.e. superjunction Schottky diodes [38–41]. However, the steep dependence of the superjunction drift layer capacitance on the applied reverse voltage leads to large *di*/*dt* during turn-off and large EMI noise and switching losses [42]. This was shown in dynamic tests of a 1,200-V superjunction SBD that has been compared to a conventional SiC SBD and a Si FRD [43]. The *di*/*dt* was found to be higher, and the turn off behavior more "snappy" as depicted in Figure 7.28. From this, it is clear that the optimum trade-off between static losses and switching performance has to be tuned to the application the device is intended for. Here, modern methods such as variational approaches as demonstrated in [44] can be used in conjunction with technology computer aided design (TCAD) simulations. As superjunction devices rely on the charge balance between n- and p-doped pillars, the imbalances from manufacturing process variations has to be taken into account carefully. This imbalance has to be taken into account as an additional factor when optimizing the design [39]. Even with a charge imbalance of up to 20 percent, the superjunction devices perform better than the unipolar limit (Figure 7.27).

While Schottky diodes are necessary for boost or buck converter topologies, modern power semiconductor switches integrate the freewheeling diode monolithically, e.g. planar or trench 4H–SiC MOSFETs or lateral channel JFETs. This raises the question of which diode concept to integrate into the MOSFET cell for least area consumption. Here, the planar MOSFET has an advantage over the trench cell design because a planar SBD can be integrated, using the surrounding p-doped channel regions for shielding of the electric field. The integration of a Schottky diode into the trench MOSFET turns out to be more challenging. Of course, one can add a planar Schottky contact by increasing the cell pitch but this is not favorable because of the high wafer cost of 4H–SiC wafers. New diode concepts or contact materials that have not yet been commercialized need to be introduced, e.g. trench Schottky diodes that use poly-Si as Schottky metal. The Schottky barrier height has to be determined for these trench contacts separately, as it depends on the crystallographic face the contact is on [45]. This is due to the polar nature of 4H–SiC.

Besides these two major directions of development, increasing the blocking voltage and integration into switches, the continuous improvement of the already broadly commercialized MPS design is the third major field of improving 4H–SiC Schottky diodes. Here, the increase of Schottky contact area and simultaneous control of the electric field at the contact is of foremost interest as it increases the current per area at a given forward voltage drop. This can be achieved by placing the Schottky contact in a trench, increasing the Schottky contact area per active chip area or by burying the shielding p-doped regions.

As the trend for continuously higher current and blocking voltage ratings goes on, the Schottky barrier diode stays in the focus of device and technology development for the next years to come.

7.7 Summary

The physics of 4H–SiC diodes and their consequences for practical device structures, characterization of electrical and thermal properties, as well as the impact of these diodes on applications have been discussed in this chapter. Starting with SBDs, we arrived by increasing the bipolar range of operation, first at junction barrier and merged pn-Schottky diodes and finally at p-i-n diodes. Although 2D or even 3D TCAD simulations are necessary to investigate device behavior in detail under all conditions of operation, valuable first order approximations for the forward voltage drop or reverse currents can be obtained by analytical models. Apart from the already cited paper by Zhu and Chow [10], the reader might be interested in the work by Benedetto about trench Schottky barrier diodes [46], and Latorre-Rey *et al.*, who derive a charge-based model of operation of JBS diodes [47].

Care has to be taken choosing the right edge termination as this part of the device is prone to environmental influences as moving ions that change the reverse bias behavior over time. The development of multizone and multiring JTE structures has mitigated the problem to a large extent.

The characterization of Schottky diodes can be divided into two areas, consisting of static electrical tests as forward voltage drops at different current levels or leakage currents at different reverse bias voltages on the one hand, and dynamic electric and thermal properties as surge-current behavior and thermal impedance on the other hand. One major aspect is the inhomogeneity of the Schottky barrier, which is rooted in the deposition process of the Schottky metal. The interested reader can find more on this topic in [48–50]. Further investigations about the ruggedness of JBS and MPS rectifiers are reported, e.g. in [51–54].

Finally, the negligible turn-off current and stored charge during forward conducting and reverse blocking modes of operation allow for higher switching frequencies in already existing and even more efficient new inverter topologies. To this extent, the impact of increased switching frequencies for buck and boost converters has been discussed. However, the reduction of losses due to the switch is beyond the scope of this chapter. More detailed calculations of and experimental results on losses can be found, e.g. in [55–58].

Since the commercialization of the first 4H–SiC Schottky diodes by Cree and Infineon, enormous efforts have been undertaken to improve the electrical and thermal performance of these devices. New structures, as well as the application of the well-known compensation principle, still leave room for improvement and new developments to come.

Note to the reader

This chapter is part of an edited book. To cite or use this chapter as a reference, please use the following format:

Konrath J. P. "Silicon carbide diodes." In: Iannuzzo F (ed.), *Modern Power Electronic Devices: Physics, Applications, and Reliabilit*. Stevenage, UK: IET; 2020. pp. 225–259.

References

[1] T. Kimoto and J. A. Cooper, *Fundamentals of silicon carbide technology*. John Wiley & Sons, 2014.

[2] M. Shur, S. L. Rumyantsev, and M. E. Levinshtein, *SiC materials and devices Vol. 1*, vol. 40. World Scientific, 2006.

[3] M. Shur, S. L. Rumyantsev, and M. E. Levinshtein, *SiC materials and devices Vol. 2*, vol. 43. World Scientific, 2006.

[4] Y. Hijikata, Ed., *Physics and technology of silicon carbide devices.* InTech, 2012.

[5] B. J. Baliga, *Silicon carbide power devices.* World Scientific, 2006.

[6] B. J. Baliga, *Fundamentals of power semiconductor devices*. Springer Science & Business Media, 2008.

[7] A. O. Konstantinov, Q. Wahab, N. Nordell, and U. Lindefelt, "Ionization rates and critical fields in 4H SiC junction devices," *Mater. Sci. Forum*, vol. 264–268, pp. 513–516, 1998, doi: 10.4028/www.scientific.net/MSF.264-268.513.

[8] N. Kwietniewski, M. Sochacki, J. Szmidt, M. Guziewicz, E. Kaminska, and A. Piotrowska, "Influence of surface cleaning effects on properties of Schottky diodes on 4H–SiC," *Appl. Surf. Sci.*, vol. 254, no. 24, pp. 8106–8110, 2008, doi: 10/dkdjq2.

[9] S. M. Sze, *Physics of semiconductor devices*, 2nd ed. John Wiley & Sons, 1981.

[10] L. Zhu and T. P. Chow, "Analytical modeling of high-voltage 4H-SiC junction barrier Schottky (JBS) rectifiers," *IEEE Trans. Electron Devices*, vol. 55, no. 8, pp. 1857–1863, 2008, doi: 10.1109/TED.2008.926638.

[11] B. J. Baliga, "Analysis of junction-barrier-controlled Schottky (JBS) rectifier characteristics," *Solid-State Electron.*, vol. 28, no. 11, pp. 1089–1093, 1985, doi: 10/b9tq3v.

[12] M. Roschke and F. Schwierz, "Electron mobility models for 4H, 6H, and 3C SiC [MESFETs]," *IEEE Trans. Electron Devices*, vol. 48, no. 7, pp. 1442–1447, 2001, doi:10/ck5h4q.

[13] S. Palanisamy, S. Fichtner, J. Lutz, T. Basler, and R. Rupp, "Various structures of 1200V SiC MPS diode models and their simulated surge current behavior in comparison to measurement," in *2016 28th International Symposium on Power Semiconductor Devices and ICs (ISPSD)*, Prague, Czech Republic, 2016, pp. 235–238, doi: 10/gf85m9.

[14] T. Ogura, H. Ninomiya, K. Sugiyama, and T. Inoue, "4.5-kV injection-enhanced gate transistors (IEGTs) with high turn-off ruggedness," *IEEE Trans. Electron Devices*, vol. 51, no. 4, pp. 636–641, 2004, doi: 10/c4wdww.

[15] T. Basler, R. Rupp, R. Gerlach, B. Zippelius, and M. Draghici, "Avalanche robustness of SiC MPS diodes," in *International Exhibition and Conference for Power Electronics, Intelligent Motion, Renewable Energy and Energy Management*, 2016.

[16] Y. Huang, T. Erlbacher, J. Buettner, and G. Wachutka, "A trade-off between nominal forward current density and surge current capability for 4.5kV SiC MPS diodes," in *2016 28th International Symposium on Power Semiconductor Devices and ICs (ISPSD)*, Prague, Czech Republic, 2016, pp. 63–66, doi: 10/gf85qh.

[17] T. Hiyoshi and T. Kimoto, "Elimination of the major deep levels in n- and p-type 4H-SiC by two-step thermal treatment," *Appl. Phys. Express*, vol. 2, no. 9, p. 091101, 2009, doi: 10/dgpgsf.

[18] *Semiconductor material and device characterization*, 3rd ed. Hoboken, NJ, USA: John Wiley & Sons, Inc., 2005.
[19] V. A. K. Temple, "Junction termination extension (JTE): A new technique for increasing avalanche breakdown voltage and controlling surface electric fields in P-N junctions," in *1977 International Electron Devices Meeting*, 1977, pp. 423–426, doi: 10.1109/IEDM.1977.189277.
[20] V. A. K. Temple, "Increased avalanche breakdown voltage and controlled surface electric fields using a junction termination extension (JTE) technique," *IEEE Trans. Electron Devices*, vol. 30, no. 8, pp. 954–957, 1983, doi: 10.1109/T-ED.1983.21243.
[21] V. A. K. Temple and W. Tantraporn, "Junction termination extension for near-ideal breakdown voltage in p-n junctions," *IEEE Trans. Electron Devices*, vol. 33, no. 10, pp. 1601–1608, 1986, doi: 10.1109/T-ED.1986.22713.
[22] R. Pérez, D. Tournier, A. Pérez-Tomás, P. Godignon, N. Mestres, and J. Millán, "Planar edge termination design and technology considerations for 1.7-kV 4H-SiC PiN diodes," *IEEE Trans. Electron Devices*, vol. 52, no. 10, pp. 2309–2316, 2005.
[23] Y. Huang, Y. Wang, X. Kuang, W. Wang, J. Tang, and Y. Sun, "Step-double-zone-JTE for SiC devices with increased tolerance to JTE dose and surface charges," *Micromachines*, vol. 9, no. 12, p. 610, 2018, doi: 10.3390/mi9120610.
[24] D. C. Sheridan, G. Niu, and J. D. Cressler, "Design of single and multiple zone junction termination extension structures for SiC power devices," *Solid-State Electron.*, vol. 45, no. 9, pp. 1659–1664, 2001, doi: 10.1016/S0038-1101(01)00052-1.
[25] R. P. Rodríguez, "Planar edge terminations and related manufacturing process technology for high power 4H-SiC diodes," Universitat Autònoma de Barcelona. Departament de Física, 2005.
[26] E. Van Brunt, G. Wang, J. Liu, *et al.*, "Operation of 4H-SiC Schottky diodes at dV/dt values over 700 kV/μs," in *2016 28th International Symposium on Power Semiconductor Devices and ICs (ISPSD)*, Prague, Czech Republic, 2016, pp. 67–70, doi: 10/ggbhdb.
[27] Infineon Technologies AG, "Transient thermal measurements and thermal equivalent circuit models," Munich, Germany, Application Note AN 2015-10, Oct. 2018.
[28] S. Palanisamy, G. Vishwamitra Yoganath, G. Zeng, J. Kowalsky, J. Lutz, and T. Basler, "Temperature determination of SiC MPS diodes during surge current event with measurement and simulation," 2019.
[29] "Datasheet IDWD20G120C5." Infineon Technologies AG, 30-Jan-2019.
[30] C. Di Marino and B. A. Hull, "Avalanche capabilities of commercial 1200 V 4H-SiC power MOSFETs," *Mater. Sci. Forum*, vol. 858, pp. 872–875, 2016, doi: 10/ggbhdm.
[31] "Datasheet IDH20G120C5," Infineon Technologies AG, 21-Jul-2017.

[32] M. Sekikawa, T. Funaki, and T. Hikihara, "A study on power device loss of DC-DC buck converter with SiC schottky barrier diode," p. 5, 2010, doi: 10/fwmptc.

[33] Brigitte Hauke, "Basic calculation of a buck converter's power stage," Texas Instruments, Application Report SLVA477B, Aug. 2015.

[34] Brigitte Hauke, "Basic calculation of a boost converter's power stage," Texas Instruments, Application Report SLVA372C, Jan. 2014.

[35] J. Schoeck, J. Buettner, M. Rommel, T. Erlbacher, and A. J. Bauer, "4.5 kV SiC junction barrier Schottky diodes with low leakage current and high forward current density," *Mater. Sci. Forum*, vol. 897, pp. 427–430, 2017, doi: 10/ggbxzk.

[36] H. Kitai, Y. Hozumi, H. Shiomi, M. Furumai, K. Omote, and K. Fukuda, "Demonstration of 13-kV class junction barrier Schottky diodes in 4H-SiC with three-zone junction termination extension," *Mater. Sci. Forum*, vol. 897, pp. 451–454, 2017, doi: 10/ggbxzj.

[37] J. B. Casady, V. Pala, D. J. Lichtenwalner, *et al.*, "New generation 10kV SiC power MOSFET and diodes for industrial applications," p. 8, 2015.

[38] Lin Zhu, P. Losee, and T. P. Chow, "Design of high voltage 4H-SiC superjunction Schottky rectifiers," in *Proceedings IEEE Lester Eastman Conference on High Performance Devices*, 2004, pp. 241–247, doi: 10/b9t9k3.

[39] M. M. Alam, D. T. Morisette, and J. A. Cooper, "Practical design of 4H-SiC superjunction devices in the presence of charge imbalance," *Mater. Sci. Forum*, vol. 924, pp. 563–567, 2018, doi: 10/ggbxzn.

[40] G. W. C. Baker, C. W. Chan, T. Dai, *et al.*, "Study of 4H-SiC superjunction Schottky rectifiers with implanted P-pillars," *Mater. Sci. Forum*, vol. 963, pp. 539–543, 2019, doi: 10/ggbxzf.

[41] X. Zhong, B. Wang, J. Wang, and K. Sheng, "Experimental demonstration and analysis of a 1.35-kV 0.92-m$\Omega \cdot$ cm^2 SiC superjunction Schottky diode," *IEEE Trans. Electron Devices*, vol. 65, no. 4, pp. 1458–1465, 2018, doi: 10/ggbx2q.

[42] L. Cao, H. -B. Pu, Z. -M. Chen, and Y. Zang, "Analysis and simulation of a 4H-SiC semi-superjunction Schottky barrier diode for softer reverse-recovery," *Chin. Phys. B*, vol. 21, no. 1, p. 017303, 2012, doi: 10/fxnbcc.

[43] B. Wang, H. Wang, X. Zhong, S. Yang, Q. Guo, and K. Sheng, "Characterization of 1.2 kV SiC super-junction SBD implemented by trench and implantation technique," in *2018 IEEE 30th International Symposium on Power Semiconductor Devices and ICs (ISPSD)*, 2018, pp. 375–378, doi: 10/ggbzbm.

[44] T. Tominaga, N. Kawabata, A. Koyama, *et al.*, "Low resistivity SiC devices with a drift layer optimized by variational approach," *Mater. Sci. Forum*, vol. 858, pp. 765–768, 2016, doi: 10/ggbxzh.

[45] Y. Kobayashi, H. Ishimori, A. Kinoshita, *et al.*, "Evaluation of Schottky barrier height on 4H-SiC m -face $\{1\bar{1}00\}$ for Schottky barrier diode wall

integrated trench MOSFET," *Jpn. J. Appl. Phys.*, vol. 56, no. 4S, p. 04CR08, 2017, doi: 10/gf4jxz.

[46] L. D. Benedetto, G. D. Licciardo, T. Erlbacher, A. J. Bauer, and S. Bellone, "Analytical model and design of 4H-SiC planar and trenched JBS diodes," *IEEE Trans. Electron Devices*, vol. 63, no. 6, p. 8, 2016, doi: 10/gf9vdv.

[47] A. D. Latorre-Rey, M. Mudholkar, M. T. Quddus, and A. Salih, "A charge-based model of junction barrier Schottky rectifiers," *Solid-State Electron.*, vol. 144, pp. 67–72, 2018, doi: 10.1016/j.sse.2018.03.004.

[48] S. Toumi, A. Ferhat-Hamida, L. Boussouar, A. Sellai, Z. Ouennoughi, and H. Ryssel, "Gaussian distribution of inhomogeneous barrier height in tungsten/4H-SiC (000-1) Schottky diodes," *Microelectron. Eng.*, vol. 86, no. 3, pp. 303–309, 2009, doi: 10/bhh3wc.

[49] B. H. Moon, G. H. Han, H. Kim, *et al.*, "Junction-structure-dependent Schottky barrier inhomogeneity and device ideality of monolayer MoS_2 field-effect transistors," *ACS Appl. Mater. Interfaces*, vol. 9, no. 12, pp. 11240–11246, 2017, doi: 10/f9wcsz.

[50] M. Furno, F. Bonani, and G. Ghione, "Transfer matrix method modelling of inhomogeneous Schottky barrier diodes on silicon carbide," 2007.

[51] S. Fichtner, S. Frankeser, J. Lutz, R. Rupp, T. Basler, and R. Gerlach, "Ruggedness of 1200 V SiC MPS diodes," *Microelectron. Reliab.*, vol. 55, no. 9–10, pp. 1677–1681, 2015, doi: 10/gf85mt.

[52] V. Banu, M. Berthou, J. Montserrat, and P. Godignon, "Surge current robustness improvement of SiC junction barrier Schottky diodes by layout design," p. 16.

[53] B. Heinze, J. Lutz, M. Neumeister, R. Rupp, and M. Holz, "Surge current ruggedness of silicon carbide Schottky- and merged-PiN-Schottky diodes," in *2008 20th International Symposium on Power Semiconductor Devices and ICs*, Orlando, FL, USA, 2008, pp. 245–248, doi: 10/d96fcp.

[54] P. Brosselard, N. Camara, V. Banu, *et al.*, "Bipolar conduction impact on electrical characteristics and reliability of 1.2- and 3.5-kV 4H-SiC JBS diodes," *IEEE Trans. Electron Devices*, vol. 55, no. 8, pp. 1847–1856, 2008, doi: 10/bx3tbr.

[55] B. Abdessamad, K. Salah-ddine, and C. E. Mohamed, "Design and modeling of DC/ DC boost converter for mobile device applications," vol. 2, no. 5, p. 8, 2013.

[56] V. P. Galigekere and M. K. Kazimierczuk, "Effect of SiC Schottky and Si junction diode reverse recovery on boost converter," p. 5, doi: 10/cw44wd.

[57] D. I. Zverev, "Switching frequency related trade off's in a hard switching CCM PFC boost convert," p. 6, doi: 10/cnfvxs.

[58] T. Ozpineci and P. Islam, "Testing, characterization, and modeling of SiC diodes for transportation applications," p. 6, doi: 10/btrxzf.

Chapter 8
SiC MOSFETs

Luca Maresca[1], Alessandro Borghese[1], Gianpaolo Romano[1], Asad Fayyaz[2], Michele Riccio[1], Giovanni Breglio[1], Alberto Castellazzi[3] and Andrea Irace[1]

8.1 Introduction

Modern switching components for energy processing widely consist of active semiconductor devices in Si, a mature and well-established technology that is reaching its physical limits. The main limitations of Si concern the blocking voltage capability, the switching frequency, and the operating temperature.

Depending on the voltage range, Si metal-oxide-semiconductor field-effect transistors (MOSFETs) and Si insulated-gate bipolar transistors (IGBTs) are the most common switches adopted for voltage ratings up to 10 kV [1]. More in detail, Si MOSFETs are available for voltage lower than 600 V (super junction (SJ) technology [2]), while Si IGBTs are widely adopted for applications in a voltage range from 600 V up to 10 kV. While the Si MOSFET technology is still suitable for applications up to 600 V, the high switching frequency requirements coming from the market are pushing for the development of faster switching devices for the voltage range from 600 V to 10 kV. The main drawbacks of the Si IGBTs, commonly used in this range, are the very high on-state voltage drop and the switching losses, because of the bipolar nature of these devices.

The aforementioned voltage range includes applications such as rail transportation, industrial medium voltage motor drives [3], AC drives, and grid-connected power converters [4]. The market requirements also push for high-frequency switching devices, for cost and size reduction of power electronic systems.

The introduction of the wide bandgap (WBG) materials for the production of power semiconductor devices led to a revolutionary development. The superior performances of SiC (silicon carbide), compared to Si, quickly became attractive for designing the next generation of power semiconductor devices, as shown in the previous section.

[1]Department of Electrical Engineering and Information Technologies (DIETI), University of Naples "Federico II," Napoli, Italy
[2]Power Electronics, Machines and Control Research Group, University of Nottingham, Nottingham, UK
[3]Nagamori Institute of Actuators, Kyoto University of Advanced Science (KUAS), Kyoto, Japan

Despite the different material properties of SiC, the production of the devices is compatible with most of the technological steps for Si devices. Even if SiC is available in both 6H and 4H poly-types, the properties of the latter are superior compared to the 6H poly-type and all the currently available commercial devices are made of it. Therefore, all the considerations reported in the following refer to the 4H-SiC.

The low losses, the higher temperature capability, and the higher thermal conductivity of SiC, compared to the Si, led to the investment of very large effort in developing high performances active power semiconductor devices. The promising perspective offered by this technology led to the development of the technology suitable for the mass production of wafers for vertical devices. In Figure 8.1, the roadmap of SiC devices applications is reported. The high effort in this technology was confirmed by the introduction on the market of the first Schottky diode in 2001 [1]. Successively, in the following decade, SiC Schottky diodes became available on the market and new generations are constantly proposed. At the same time, the development of SiC switches was carried out. However, the effective adoption of SiC MOSFETs in applications started in 2010 [6], with the introduction on the market of the first trench MOSFET. The long delay between the release of SiC Schottky diodes and SiC switches is due to the technological challenges for the production of a SiC MOSFET. The first SiC unipolar switch proposed on the market was a SiC junction field-effect transistor (SiC JFET) [7], where no SiC/Oxide interfaces are present. This type of device offers the advantage of very low on-resistance and it can operate at high temperatures and high frequencies. The main drawbacks of this technology are the maximum breakdown voltage and the normally-on principle of operation [7]. The second aspect involves the design of new drivers, compared to those adopted for Si MOSFETs and Si IGBTs. These are the main motivations that led the nowadays trend in the development of SiC switches for the 600 V–10 kV application range, focused on the SiC MOSFET, that are normally-off. The maximum achieved breakdown voltage is 15 kV [8] for a SiC MOSFET, with the advantage in terms of switching speed arising from the unipolar

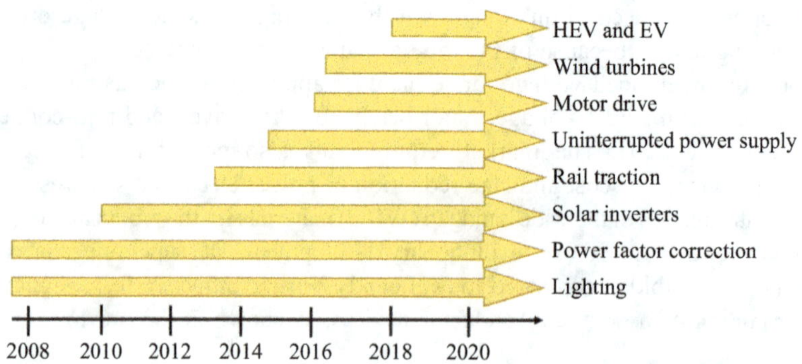

Figure 8.1 SiC power devices application roadmap by Yole. Adapted from [5]

nature of the device. Though a so high blocking voltage has been achieved for a SiC MOSFET, it refers to a prototype device. From the point of view of commercial availability, the existing SiC MOSFETs are in the range of 1.2–3.3 kV, which fit many industrial applications. The main advantage of this technology is the very high switching frequency capability. Recent research results show as the switching frequency of converters can be increased up to 10 MHz [9,10] for 1,200 V SiC MOSFETs. Even if SiC has excellent performances in terms of thermal conductivity, superior thermal performance for a SiC MOSFET, compared to Si devices, has been presented only in 2014 by General Electric [11], with the development of reliable devices up to 200 °C for industrial applications. The upper bound of the modern available commercial SiC MOSFETs lays in the range of 1,700 V, 72 A, and 45 mΩ [12]. In Figure 8.2, the commercially available performances of SiC MOSFET devices are reported [1].

However, in developing a power switching system, the choice between a SiC MOSFET and a Si IGBT solution is still a matter of accurate evaluation. In this range, so far, the IGBT is still the preferred switching device because of its lower cost, compared to the SiC MOSFET technology. Therefore, a high effort has been done from the leading semiconductor companies and research centers in the development of high quality, high speed, and reliable SiC MOSFETs in the 1.2–3.3 kV range.

On the other hand, the SiC MOSFET technology is slowed down by the critical issues mainly related to the defects occurring at the SiC/SiO_2 interface. Basically,

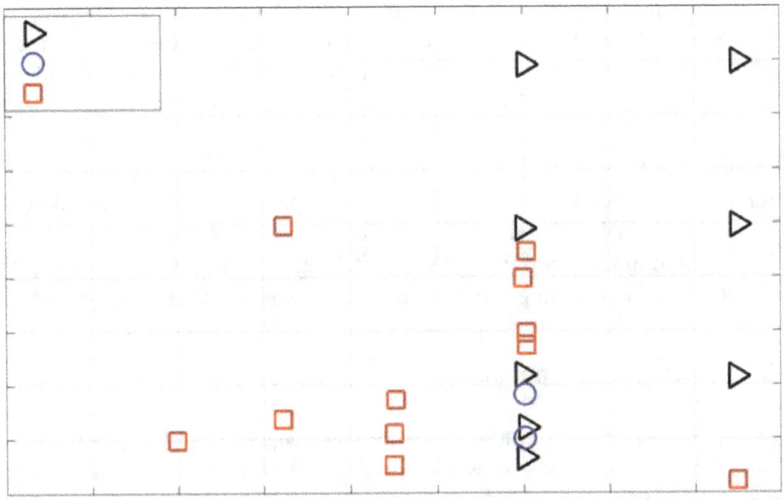

Figure 8.2 Voltage and current rating of commercially available SiC MOSFETs. Adapted from [1]

the presence of the carbon atoms at the interface is the reason for the formation of clusters and dangling bonds that strongly degrade both the electric field-effect mobility of the channel [13] and the stability of the threshold voltage [14]. Moreover, the complexity of the Si/SiO_2 interface is responsible for the yield challenge for wide-area wafers [15].

To overcome the issue related to the Si/SiO_2 interface, sophisticated gate oxide processes and advanced designs have been proposed over the time aimed at mitigating the main limitations of the SiC MOSFETs and planar structures are nowadays adopted for the production of commercial SiC MOSFETs. At the current stage, the main difficulties for the development of the next generation of SiC MOSFETs are as follows:

- The reduction of production costs. The cost of the SiC MOSFETs is still a limitation for large scale adoption of these devices.
- The improvement of the SiC/SiO_2 interface to increase the temperature stability, to increase the channel mobility, and to remove the threshold instability.
- The reduction of the gate driving voltage. Most of the available families of SiC MOSFETs have a driving voltage of 18–20 V, not compatible with a driving voltage of Si MOSFETs and IGBTs (15 V).
- The improvement of the low short-circuit capability.
- The improvement of the avalanche capability. This aspect is relevant since the high switching frequency leads to large voltage overshoot even with low parasitic inductances.
- The increase in the saturation current.

This chapter begins with a brief recall of the design for making modern SiC MOSFETs. Both the technological and physical aspects related to the use of SiC as the bulk semiconductor in the production of a SiC MOSFET will be detailed, with a focus on the open technological challenges. Next, the planar, trench, and super junction MOSFET designs will be examined in terms of performances and data coming from the state of the art. The design of power semiconductor switches is commonly validated when the new device is capable to stand out of safe operating area (SOA). This is achieved by means of specific tests such as the short-circuit (SC) and the unclamped inductive switching (UIS) tests. Therefore, the last research results concerning these aspects will be addressed in Section 8.6.

8.2 Principle of operation

From the technological point of view, the great advantage of SiC is the possibility of growing a high quality stable thermal oxide that allows the formation of an almost ideal MOS interface. Therefore, the design of a modern SiC MOSFET is a derivation of the one for power Si MOSFET. Even if some lateral devices have been presented over time (e.g., in [16]), the most adopted structure is vertical. The principle of operation of a SiC MOSFET is the same as the Si MOSFET. However, the way in which the two devices are implemented is different because of technological and physical differences between the Si and SiC materials. Therefore, in this section, the basic designs nowadays adopted to implement SiC MOSFETs devices are briefly illustrated. The principle of operation will be presented to highlight the critical issues related to the adoption of SiC as the bulk semiconductor. Since some technological issues, such as the SiC/SiO_2 interface quality, are still an open research field, the final part will focus on the main challenges related to the technological challenges of the modern SiC MOSFET devices.

8.2.1 Planar MOSFET

Even if the first commercial SiC MOSFET was a trench gate one, the planar MOSFET design has been adopted to produce most of the commercial families of SiC MOSFETs, recalling the Si MOSFET design. The first reliable generations of SiC MOSFET devices have been made by means of the planar-MOSFET approach [17] and currently, the most widely adopted SiC power MOSFET structure is still the planar one. In Figure 8.3(a), the sketch of the typical cross-sections adopted for the implementation of a planar SiC MOSFET is reported. Its principle of operation is identical to one of the Sis; however, there is a softer transition between the linear and the saturation region due to a lower value of the transconductance. In [8], the authors showed that the high voltage SiC MOSFETs perform better than similarly rated Si IGBTs in applications where low currents and high switching frequencies are required.

The planar design has the advantage of a simple technological implementation in terms of doping profiles and some steps from the Si MOSFET production can be adopted. Basically, P-body, channel, source, JFET, and drift regions are comparable to those of a Si MOSFET device, with a specific reference to the double-diffused

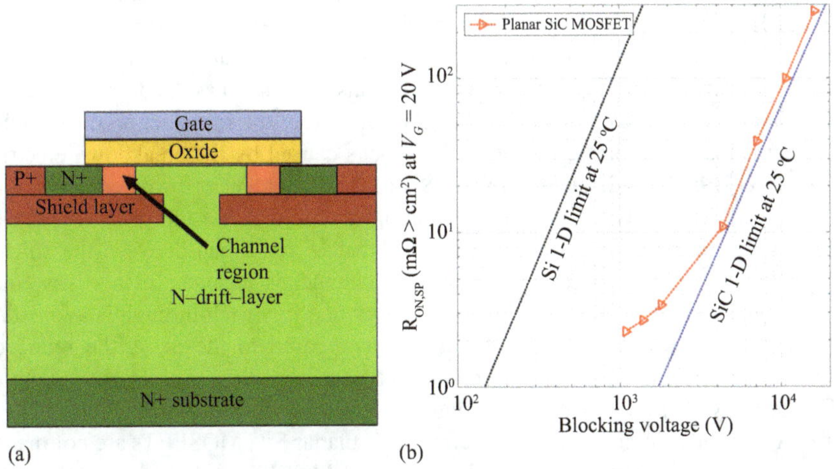

Figure 8.3 (a) A sketch of the planar SiC MOSFET design and (b) the comparison of the real device performances with the ideal one-dimensional limit of Si and SiC. Adapted from [18]

MOS (DMOS) technology. However, since SiC has a different atomic structure and physical properties, compared to the Si, some specific changes in the doping profiles and technological steps have to be made for the production of a functioning and reliable device. From this point of view, three physical aspects of SiC make necessary some crucial variations to the design of the elementary cell. The first aspect concerns the breakdown electric field of 4H-SiC, which is 30 times higher than that of Si. On the one hand, this is one of the major advantages of SiC (the higher breakdown electric field allows to achieve a specific BV with a higher doping concentration of the drift region, leading to a strong reduction of R_{ON}); on the other hand, it has been one of the obstacles in developing a high voltage SiC MOSFET. It impacts on the design of the P-body region of the device. The maximum electric field in 4H-SiC is 9×10^6 V/cm, while the maximum electric field in Si is 3×10^5 V/cm. This affects two aspects: the thickness of the P-body layer and the reliability of the device. Regarding the former aspect, the P-body region should never be completely depleted in any operating condition, even in breakdown conditions. This constraint is easily achievable in Si devices, but it becomes critical for SiC devices. Given that the P-body layer of power MOSFET has a doping concentration of 1×10^{17} cm^{-3} and the oxide thickness varies from 500 to 1,000 angstroms, the threshold voltage varies from 1 to 2 V. To fulfill the aforementioned constraint in a Si MOSFET during breakdown conditions, a thickness of 0.5 μm for the P-body is sufficient. In a SiC device, to prevent the reach-through when the breakdown occurs, a doping concentration of 4×10^{17} cm^{-3} is mandatory to keep the P-body thickness at 0.5 μm. However, a so high doping concentration unacceptably increases the threshold voltage of the device. An alternative could be increasing the depth of the P-body layer while keeping the doping concentration constant at 1×10^{17} cm^{-3}. Again, this change does not solve the problem, since the larger extension of the P-body layer implies an increase of the channel length, which increases the channel resistance up to an unacceptable value. In addition to this, the high breakdown electric field typically occurs close to the oxide layer. Since the maximum reliable electric field in the oxide is 3×10^6 V/cm, if no countermeasure is taken, as soon as the device achieves the breakdown voltage, it inevitably breaks. Therefore, the introduction of a shield layer is mandatory to prevent the two drawbacks caused by the breakdown electric field in SiC [19,20], as detailed in Figure 8.3(a). The shield layer is implemented as a high concentration p-type layer with a doping profile that has the peak below the source diffusion layer and with an extension that leaves a gap between the shield layer and the SiC/SiO_2 interface, where the channel is located. In this way, the electric field peak does not reach the oxide layer and the reach-through is avoided.

The second peculiar aspect of the SiC, compared to the Si, is the reduced diffusion coefficient for dopants. This aspect does not allow the classical double diffusion (DMOS) approach to making the device, with a high criticality in the definition of the channel region. Therefore, SiC planar SiC MOSFETs are not made through the standard processes employed for Si MOSFETs. As an example, differently from the Si DMOS devices, the N+ source and the P-body layers are defined by using photoresist masks instead of the edge of the gate polysilicon. This is one of the technological solutions adopted to avoid the problems related to the very high ion implantation energy needed for SiC. Some other technological details are given in [21].

The third critical aspect of making a SiC MOSFET regards the small mismatch between the conduction band of the SiC and the one of the SiO_2. This property leads to the increase of the hot carrier injection into the silicon dioxide, with the consequent increase of the instability of the threshold voltage. This is still one of the aspects concerning the research field in developing SiC MOSFETs. Despite the better theoretical trade-off between R_{ON} and BV of the trench structures [20], it has been shown that the technological maturity of planar SiC MOSFETs allows building devices with R_{ON}–BV trade-off close to the ideal one-dimensional (1D) limit of SiC (see Figure 8.3(b)) [18]. This is particularly true for devices rated for 3.3 kV and above. Specifically, starting from a voltage rating of 3.3 kV, the channel resistance contributes to the overall resistance becomes negligible compared to the drift layer resistivity. Therefore, the physical limit of the R_{ON}/BV ratio of the material becomes the only factor determining the drain resistance of the device. In terms of the design of planar SiC MOSFETs, the reference elementary cell structure has been almost unchanged over the development of the different families of devices. Most of the effort has been done in improving the technological steps for the reduction of the channel resistance for the different voltage ratings of SiC MOSFETs.

8.2.2 Trench-gate MOSFET

Despite the first SiC MOSFET device used a trench-gate structure, this design has still some drawbacks and their solution is an open research field.

In Figure 8.4, the trench SiC MOSFET (also referred to as U-MOSFET) structure is reported. In principle, as for Si MOSFETs, this design has the great advantage of removing the JFET contributes to the R_{ON} of the device and increasing the channel width, leading to a significant reduction of the R_{ON} of the device. Moreover, since the body region is made by epitaxial growth, the channel region doping concentration can be easily varied. The channel doping profile definition is more complex in planar SiC MOSFETs, since the channel region is

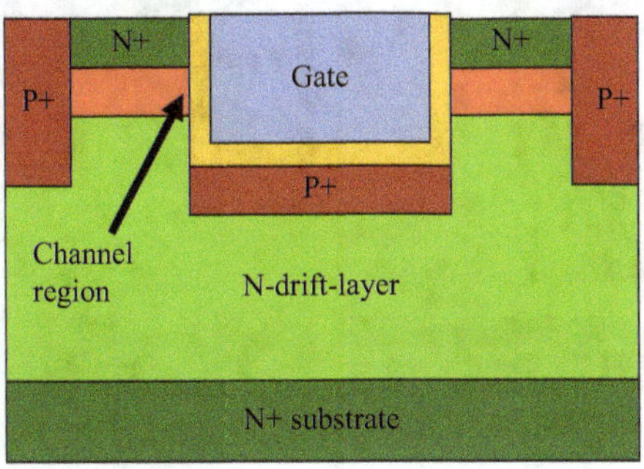

Figure 8.4 A sketch of the trench SiC MOSFET design

made by the implantation of dopants and thermal diffusion, two critical technological steps for SiC. However, more complex technological steps are mandatory to make an operational device. The technological process qualifying this design is the plasma-etched trench that allows the formation of a vertical channel. The same critical aspects presented for the planar SiC MOSFET also affect the trench version and the technological solutions aimed at the mitigation of these issues are mandatory for market validation of this kind of structure. The main issues related to the trench design are the low quality of plasma-etched surfaces and the very high electric field at the corner of the trench when the avalanche breakdown occurs [22].

As previously reported, the SiC material introduces some challenges in the practical manufacturing of reliable trench-gate MOSFETs. The main drawback comes from the critical electric field occurring in the SiC, which is much higher than that of the Si. The main issue associated with this aspect is the very high electric field occurring at the corners of the trench-gate bottom. In blocking conditions, the maximum electric field that the SiO_2 oxide can withstand is achieved, with the consequent severe damage of the gate oxide. Different works focus on the reduction of the electric field at the bottom of the trench-gate [23,24]. Most of the solutions are based on the introduction of a p-doped well in the drift layer below the trench-gate, as reported in Figure 8.4. However, a unique and generally accepted solution to mitigate the electric field at the bottom of the trench-gate is still not well defined. In Figure 8.5, three different solutions proposed in the literature are reported, as an improvement of the basic design in Figure 8.4. In [23], the p-doped layer below the gate of a SiC U-shaped trench-gate MOSFET is surrounded by an n-doped layer. Any narrowing of the electron current path increases the R_{ON} of the device. Therefore, the presence of the p-layer below the trench gate introduces a depletion region toward the gate because of the p-layer/drift junction. The design proposed in Figure 8.5(a) reduces the extension of the depletion region surrounding the p-doped layer below the trench gate, since the higher doping concentration of the n-layer around the p-layer strongly reduces the depletion region. Numerical results coming from [23] show that the R_{ON} decreases by 33.1%

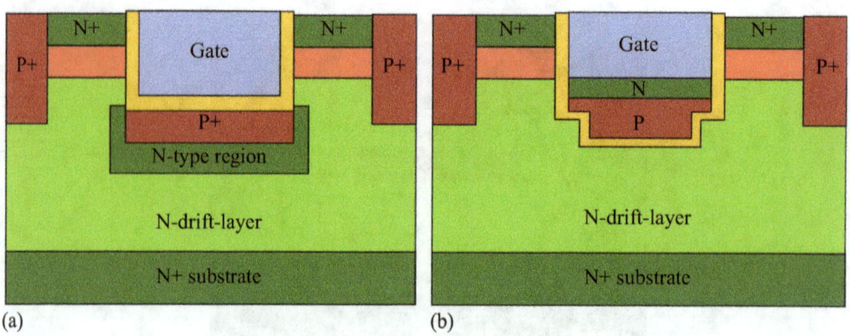

Figure 8.5 Advanced designs for a trench gate SiC MOSFET: (a) adapted from [23] and (b) adapted from [24]

compared to standard devices. In Figure 8.5(b), the structure reported in [24] is sketched. The principle of operation is based on the introduction of a further trench gate bottom region with a reduced lateral extension. This allows the reduction of the JFET effect. On the other hand, this solution introduces an increase in the charge in the gate of the device due to the higher area of the trench gate. Therefore, the introduction of the polysilicon p/n junction in the trench gate reduces the stored charge. Authors demonstrated that the on-resistance reduces by 26.4%, while the gate charge reduces by 34.7%.

8.2.3 Super junction MOSFET

The main limit of Super-junction MOSFET from the point of view of the maximum blocking voltage is related to the unacceptable increase of the R_{ON}. A higher breakdown voltage requires a low doped and thick drift layer, which in turn implies an increasing R_{ON}. Typically, referring to Si, MOSFET devices are adopted for voltage ratings lower than 600 V, since higher voltages devices such as IGBTs have a lower V_{ON}. When the theory of SJ concept has been presented [25], the R_{ON} limitation of Si MOSFETs was overcome. The SJ theory is based on the RESURF (reduced surface field) concept [26] and it consists of making a drift layer where p-doped regions are alternated to n-doped regions. A sketch of a SiC MOSFET is reported in Figure 8.6. A p-doped vertical pillar is connected to the body layer of the SiC MOSFET and it is crucial when the device operates in blocking condition. When the device operates in on-state, the electron current from the source contact flows through the channel and passes through the n-doped layer of the drift region. An optimized n-doped layer allows the achievement of a lower V_{ON} compared to a non-SJ MOSFET, since the doping concentration is higher. Typically, the R_{ON} of the device can be lowered from 5 to 100 times thanks to the higher doping concentration of the n-pillar [2]. Therefore, the presence of the p-doped pillar becomes crucial in blocking conditions. In off-state, when a positive voltage is applied to the drain contact, the n-pillar/p-pillar junction operates in reverse conduction and the depletion region expands. Thanks to the reduced surface (RESURF) effect, when

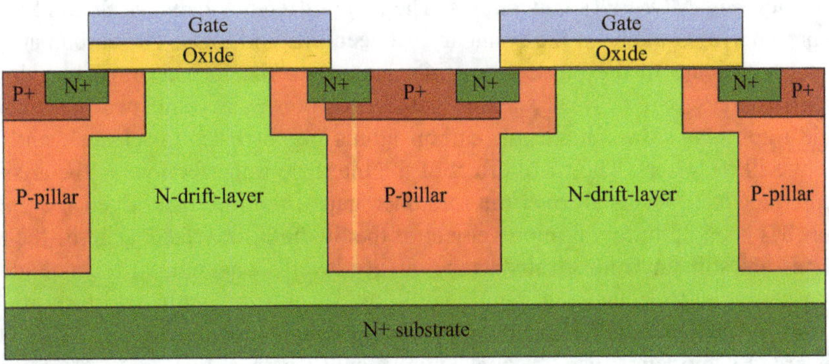

Figure 8.6 A sketch of a super junction MOSFET

the depletion regions reach each other in the y-direction, the 2D electric field distribution becomes flat within all of the drift layer and body region, with a consequent increase of the breakdown voltage (BV) well above the BV of a SiC MOSFET without the SJ design. This approach allows extending the BV up to the maximum one allowed by the material. However, in real-life, the exceptional improvement in terms of R_{ON} and BV can be achieved only with a very careful definition of the n-pillar and p-pillar, since the two regions must have exactly the same doping concentration (charge balance condition). The charge unbalance condition leads to a strong reduction of the BV and a careful definition of the technological steps is mandatory to avoid it. Since SiC MOSFETs show a high R_{ON} for high drain voltage, the SJ concept is becoming attractive to overcome this issue. Recent works, [27] and [28], have shown that the SJ concept can be applied to SiC MOSFETs. These preliminary results show how the SJ concept is applicable to SiC MOSFETs, since the R_{ON} is reduced. However, the definition of the technological steps to achieve a good charge balance requires further investigation.

8.3 SiC/SiO$_2$ interface challenge

The main reasons for the reduced pace in developing new commercially validated SiC MOSFET devices are the high cost related to SiC and the complex technological issues coming from the properties of such material. As for the Si MOSFET development, the most critical part of the design is the quality of the SiC/SiO$_2$ interface. The practical implication of this aspect is the instability of the *I–V* curves of these devices over time and over the operating conditions. Generally speaking, the interface between two materials is the region where the mismatch between the two atomic structures arises, with the consequent variation of the local electrical properties. As an example, in Si/SiO$_2$ interfaces some dangling bonds occur, since the atomic distribution of the Si lattice and the SiO$_2$ are different. This leads to the formation of traps and fixed charges that affect the threshold voltage of the device. However, while for the Si/SiO$_2$ most of the issues related to the trap density have been solved, the SiC/SiO$_2$ interface quality is still a limitation in the development of the new SiC MOSFETs generations. The main difference between Si and SiC is the presence of C atoms at the boundary between SiC and SiO$_2$. The carbon atoms do not contribute to the oxide formation, leading to the formation of two dominating sources of electrically active defects: carbon atoms and near-interfacial oxide traps [29]. More in detail, carbon atoms organize as sp^2-clusters and as graphite-like clusters. The main effect of the interface imperfections is the arising of interface state levels all over the forbidden gap at the interface. Even if a great effort has been put into the improvement of the SiC/SiO$_2$ interface, state-of-the-art devices are still far from a satisfactory quality level. Conventional technological processes able to achieve a high-quality oxide/semiconductor interface in Si MOSFETs are not suitable for improving the quality of SiC MOSFETs [30]. The standard dry oxidation used to make high quality gate oxides in Si MOSFETs yields a high density of interfaces states and the low channel mobility, in the order of 10 cm^2 V^{-1} s^{-1}. Even annealing performed in hydrogen ambient does not improve the interface quality. Therefore, a large number of papers focus on the improvement of the interface states through a careful adjustment of the oxidation process of the SiC. In Figure 8.7, the interface state densities near the conduction and valence band are reported for both n-type and p-type SiC (0001) MOS capacitors, respectively [30]. The reference oxidation process is the dry one and the effect of the nitridation annealing process is reported. More in detail, the state profiles of nitridation in NO or N_2O are reported. Measurements are carried out by the conventional high (1 MHz)-low method. It is clear as the adoption of a nitridation annealing process strongly reduces the interface states density. On the other hand, the nitridation annealing process leads to the presence of a high concentration near the conduction band of very fast interface states. These fast states can respond to a probe frequency of 100 MHz up to frequencies higher than 1 GHz. This introduces a relevant inaccuracy in the characterization of the interface states density when low frequency probes are adopted [31]. The adoption of the annealing in NO or N_2O leads to the accumulation of nitrogen at the SiC/SiO_2 interface. According to the experimental characterizations reported in Figure 8.7, the higher concentration of nitrogen at the interface reduces the states. However, it is not clear how the interface defects are passivated by the presence of nitrogen. Even if one of the most accredited theory comes from the work [29], where the donor-like interface states in the lower half of the bandgap may be originated from carbon clusters. However, a direct link between the carbon clusters and the interface state density is not established and the very high density of interface states close to the conduction band is still not understood. The interface state density (D_{it}) for SiC MOSFETs is two orders of magnitude higher than that of Si MOSFETs. Usually, the interface levels extend all over the forbidden gap of SiC, leading to the variation of the C–V characteristic [32]. From the practical point of view, the presence of a high density of traps in the forbidden gap leads to an unstable behavior of the threshold voltage.

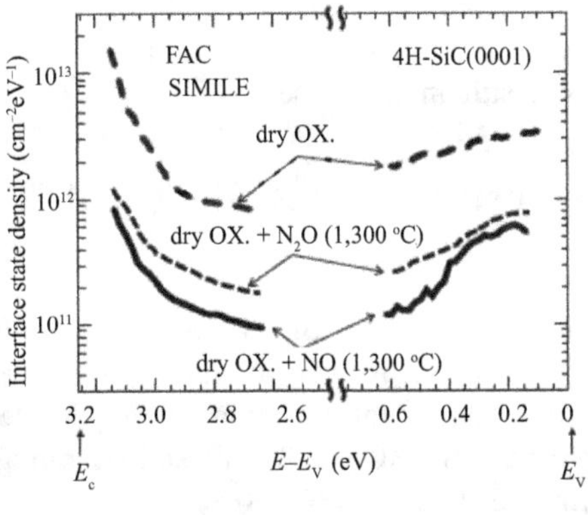

Figure 8.7 Distribution of the interface state density at the edge of the forbidden band for different conditions for a SiC MOS structure. Reprinted, with permission, from [30]

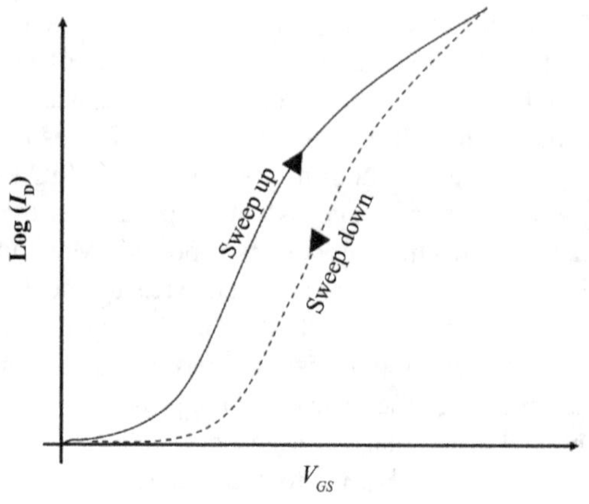

Figure 8.8 A sketch of the threshold voltage instability effect in SiC MOSFETs in the sub-threshold region

In addition, it also determines a very complex behavior of the threshold voltage with temperature. This becomes relevant in highly stressful thermal conditions (such as SC events) or for the reliability during the operation at high temperatures. The latter aspect has been one of the main limitations in the exploitation of the superior thermal properties of SiC [11]. The other effect of the high D_{it} at the SiC/SiO$_2$ interface is the hysteresis effect of the I_D–V_{GS} curve. In Figure 8.8, it is reported a sketch of the hysteresis effect commonly observed in SiC MOSFET devices [33]. Basically, the I_D–V_{GS} curve changes if V_{GS} increases or decreases, leading to a threshold voltage that depends on the derivative of the V_{GS} over the time and the time duration for which the bias is applied. Compared to the first devices introduced in the market, the threshold instability issue has been attenuated by means of complex technological improvements in the fabrication of the oxide layer. Different technological strategies have been proposed to mitigate the instability of the threshold voltage. The main approach is the adoption of nitride gate oxides [34,35]. However, the improvement of the Oxide/SiC interface is still an open research field.

8.4 A comparison between Si MOSFET and SiC MOSFET

As mentioned in Section 8.2, the power SiC MOSFET design has many features in common with the Si MOSFET. However, the high trap concentration at the SiC/SiO$_2$ interface leads to a significant variation of the device behavior over the temperature. Compared to a Si MOSFET with similar ratings, a SiC MOSFET shows the following main differences [36–38]:

- A higher threshold voltage and a higher temperature coefficient of the threshold voltage. This aspect is related to the significant acceptor like trap density at the SiC/SiO_2 interface. While the higher threshold voltage at room temperature is due to the acceptor like traps, the higher temperature coefficient originates from the combination of two phenomena, which are also affected by the traps behavior at high temperature. The first phenomenon is related to the temperature dependence of the contact potential difference between gate and substrate (ϕ_{ms}) [39], that is the main contribution of Si MOSFET. The second phenomenon is due to the high interface traps density occurring in SiC MOSFET devices, since the trapped electrons concur to the increase of the threshold voltage at room temperature. However, at the higher temperature, the inversion electrons are emitted by the traps, with the consequent faster reduction of the threshold voltage. One of the main consequences of the latter phenomenon is a severe increase in the positive temperature coefficient for the I_D, leading to a higher electrothermal instability of the device [36].

- Channel mobility has a positive temperature coefficient for low temperatures, while at high temperatures it has a negative temperature coefficient. The physics related to this effect involves the properties of the SiC/SiO_2 interface, again strongly affected by the high trap density [36]. Two main phenomena determine the mobility dependence on the temperature: Coulomb scattering with the filled traps and acoustic-phonon scattering. While the latter phenomenon is the only one occurring in Si MOSFETs, leading to a negative temperature coefficient of the mobility, the former one is dominant in SiC MOSFETs and leads to a positive temperature coefficient. Again, this is due to the inversion electron emission by the traps at a higher temperature. The lower is the number of filled traps, the lower is the Coulomb scattering of the carriers flowing in the channel. The effect of this peculiar behavior is well visible during the SC test of a SiC MOSFET. In Figure 8.9(a), the SC waveform of a commercial 1.2 kV-rated device [40] is reported for V_{DS} = 100 V and V_{DS} = 600 V, with V_{GS} = 18 V and V_{GS} = 16 V, respectively. Differently from

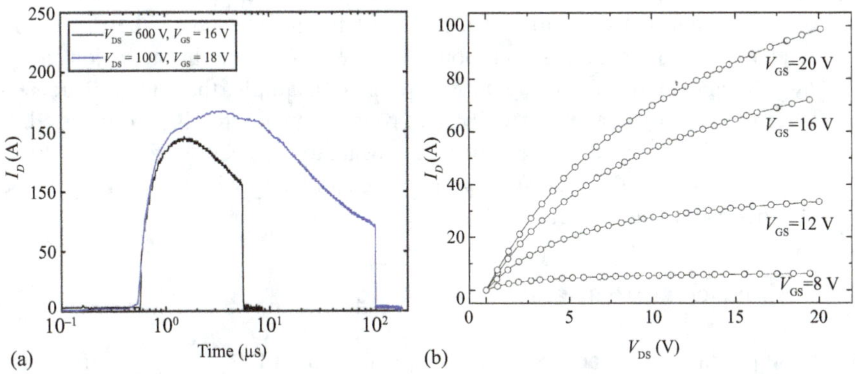

Figure 8.9 (a) Experimental SC waveforms for a SiC MOSFET for V_{DS} = 100 V and V_{DS} = 600 V and (b) typical I_D–V_{DS} curve for 1,200 V SiC MOSFET

a Si MOSFET, first, the saturation current for medium–high gate voltages increases because of both the higher temperature coefficient of the threshold voltage and the positive temperature coefficient of mobility due to the Coulomb scattering phenomenon. When the inner temperature of the device is high enough, the saturation current decreases because the acoustic-phonon scattering becomes dominant.

- A higher drain resistance. One of the important features of SiC material is the critical electric field that is one order of magnitude higher than the one of Si. This allows achieving the same blocking voltage with a much thinner drift layer, compared to a Si MOSFET with the same voltage rating. Moreover, this leads to a strong reduction of the drift layer resistivity. If no other effects are kept into account, then $R_{\text{drift}}(\text{SiC}) = 1/500\, R_{\text{drift}}(\text{Si})$. However, the drain resistance of a SiC MOSFET is still large and it becomes very relevant at high drain voltage. At low drain voltages, as expected, the resistance is low. On the other hand, when the drain voltage increases, its resistance strongly increases. This is mainly due to two effects: (1) the path of the electron narrowing at high drain voltages. When the drain voltage increases, the depletion region between the body and the drift regions expands through the drift layer and shrinks the section orthogonal to the electron current. This occurs in the JFET region of the structure and the narrowing effect is enhanced when the drain region below the gate oxide region depletes. This further reduces the electron current path. (2) Even if SiC has a higher critic electric field, the mobility degrades when the electric field becomes high, further increasing the drain resistance. Since the high drain resistance is mainly due to the resistance of the drift layer, only a small portion of the drain voltage drops across the channel. Therefore, the current characteristics of a SiC MOSFET, compared to a Si MOSFET with a comparable rating, exhibit two main features: (1) since most of the external applied drain voltage drops across the drift layer, when the voltage drop across the channel region achieves the saturation voltage (V_{dsat}), the external applied drain voltage is much higher than V_{dsat}. This effect increases with the increase of the current, that is, with the increase of the gate voltage. The main consequence of this feature is a pronounced extension of the linear region in the I_D–V_{DS} characteristics. In Figure 8.9(b), as an example, the output characteristics of a commercial SiC MOSFET at room temperature [40] are reported. It is well visible how the Triode region extends up to $V_{DS} = 20$ V, for $V_{GS} = 20$ V. (2) Since most of the drain voltage drop occurs along with the drift layer, the channel modulation effect is reduced as well.

8.5 Short-circuit capability

Since the production of the first commercial SiC MOSFET [7], a great effort has been made to develop the last generations of SiC MOSFETs. The last SiC MOSFET families achieved high performances in terms of forward and switching conditions. On the other hand, many literature contributions about the investigation and the characterization of SiC devices highlight that some issues still are not fully addressed. The in-depth investigation of these issues is the key point to push up the boundaries of devices' performances. Therefore, in order to define the device limit in demanding operating conditions, both the SC and the unclamped inductive switching (UIS) tests are adopted to force the device under test (DUT) in highly stressful conditions [41,42]. In this section, the performances of SiC MOSFETs are evaluated at the device level when the device operates out of the safe operating area (SOA), with a specific focus on both the SC and the avalanche conditions.

8.5.1 Short-circuit test

The SiC MOSFETs are becoming a concrete choice in making medium–high power circuits for electrical energy conversion. One of the main application fields is the electrical motor driving, where the load is an inductance related to, for example, the windings in a motor. In these circuits, the SiC MOSFET is adopted to deliver a DC voltage with a timing coming from the control circuit. However, the SC condition can occur during the operation of the system, bringing the device into a very harsh working condition of high voltage and high current at the same time. In Figure 8.10, a sketch of the SC event when the DUT drives an inductive load is reported. Given the very high stressful condition, a power semiconductor switch is clearly capable to stand an SC only for a limited time. The SC test is adopted for the validation of the new generation of SiC MOSFETs and the SC safe operating area (SCSOA) is the maximum time that the device under test can withstand the SC condition. SC events can occur in applications such as motor driving systems, where mechanical events can lead to unwanted conditions [43].

As reported in the International Standard IEC 60747-9, two types of load SC are considered to evaluate the SCSOA [44]. In Figure 8.11, the two types of fault conditions are reported. The first type occurs when the device is subjected to the SC while in normal conduction. In Figure 8.11(a), the first type of SC is simulated by turning on the upper-side switch with a gate voltage higher than the one of the DUT (this leads to a higher saturation current compared to the one of the DUT) while the nominal current flows in the load inductor. In Figure 8.11(b), the second type of SC is simulated by turning on the DUT directly on an SC load. This modality is also

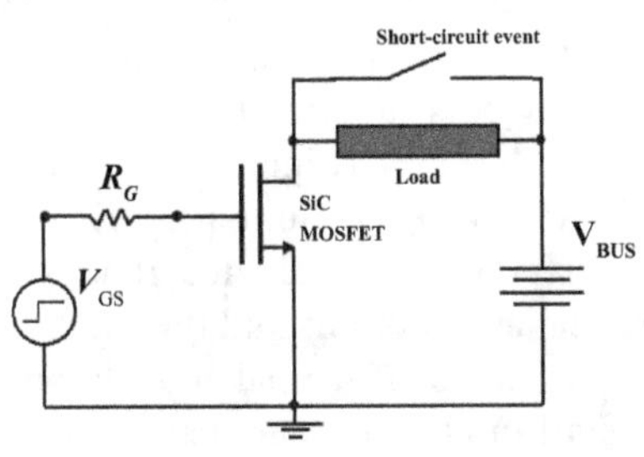

Figure 8.10 A sketch of an inductive load application where an SC event can occur

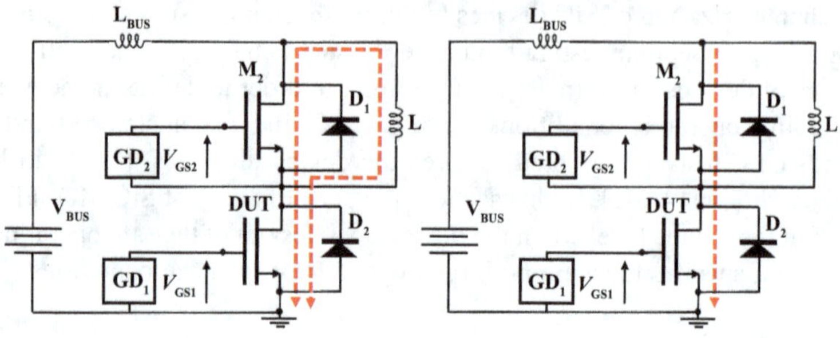

Figure 8.11 (a) A schematic of the load current in SC type 1 and (b) a schematic of the load current in SC type 2

called hard switch fault. The main difference concerns the saturation current achieved during the SC event. Since the Miller capacitance is higher when the device is in on-state compared to the condition of off-state, the saturation current is higher in the first type of SC fault. More in detail, the higher the Miller capacitance, the higher will be the gate voltage increase because of the sudden surge of V_{DS}. Since the Miller capacitance is higher when the device operates in on-state conditions, the SC current is higher. The second type of SC fault is commonly adopted to define the SCSOA in datasheets.

Typically, complex protection circuits are used to avoid the catastrophic failure of the devices during an SC event. The most common solutions include the *resistor sensing,* the *current transformer* and the V_{DS} *sensing* techniques [44]. However, these protection schemes take time in the order of some microseconds to properly protect the device by switching it off. Therefore, the DUT must withstand a minimum time (the typical range is 5–10 μs) in the SC condition to avoid its destruction during an SC event. Therefore, the SC test is usually performed by the simulation of the SC type 1, with a DC voltage that is 66% of the nominal voltage of the DUT. Compared to Si MOSFETs, SiC ones have a lower SC capability [45]. This is due to the higher current density and lower chip size of these devices, compared to those in Si.

8.5.2 Short-circuit failure mechanisms in SiC MOSFETs

The investigation of the SC failure mechanism for a SiC MOSFET must keep into account the internal device structure. Thanks to device symmetry, a half elementary cell of a planar MOSFET (Figure 8.12) is considered for this study and analyzed with the TCAD Synopsys Suite [46]. Even though the structure is calibrated to match the behavior of a real device, it does not represent the actual device structure. Theoretical assumptions and literature data (see [20]) are used to define doping and dimensions. Principal models and corresponding parameters are reported in detail in [13]. For simulation purposes, body and source terminals are physically separated but connected at the same electrical node. Mixed-mode simulations are performed, in which a physically based device is

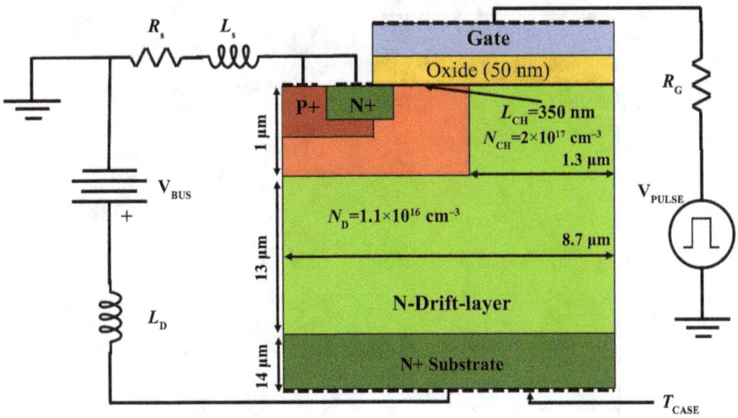

Figure 8.12 A sketch of the TCAD SiC MOSFET elementary cell and the circuit adopted for the mixed-mode simulations

placed alongside a circuit description (in a SPICE netlist format) as depicted in Figure 8.12. Additional components are included to consider the parasitic elements introduced in a real circuit by wires and connections. Specifically, stray inductance and parasitic resistance on the source loop (L_S, R_S) affect the di/dt during the turn-ON phase; stray inductance on the drain (L_D) is responsible for voltage spikes during switching transients. Figure 8.12 shows its estimated values. It is well known that temperature strongly affects the behavior of power devices, and therefore self-heating effects could not be neglected. Accordingly, temperature-dependent parameters are included, and heat generation and transport equations are solved in conjunction with semiconductor equations. The thermal problem is solved by applying the isothermal condition on the back of the device (T_{CASE}) and adiabatic conditions on the remaining edges. To reflect the operation of an actual device, the structure is calibrated obtaining a suitable match with isothermal I_D–V_{GS} characteristics of a 1.2 kV 36 A 80 mΩ commercial device [40], selected as a case study. The experimental curves are measured at $V_{DS} = 20$ V are illustrated in Figure 8.13 for backside temperatures of 300 and 410 K. The calibration procedure required the choice of suitable physical models (e.g., mobility doping dependence, carrier recombination, etc.) and the proper tuning of their parameters. Device behavior is largely dependent on the quality of the oxide–semiconductor interface and could not be correctly reproduced without including fixed charges and trap levels usually present therein. d'Alessandro *et al.* [36] have reported the impact of interface defects and dislocations on MOSFET devices. These trap levels are commonly considered to be acceptor-like above mid-gap energy E_i, that is, negatively charged when occupied. One of the effects is a positive shift of the

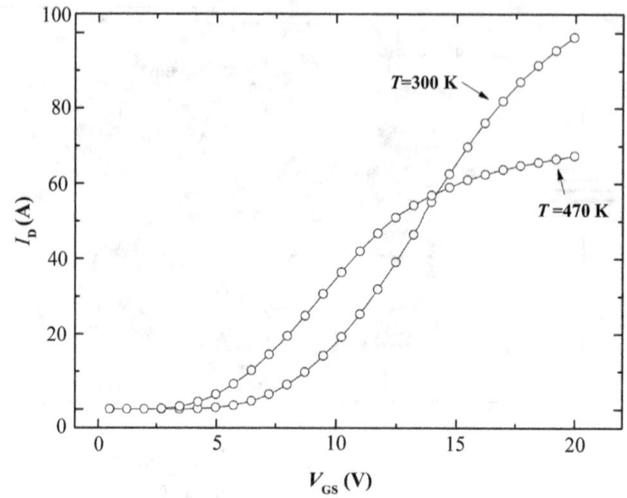

Figure 8.13 Experimental transfer characteristics of the investigate structure

threshold voltage, which can then be analytically expressed as [47]

$$V_{TH} = V_{FB} + 2\phi_B + \frac{1}{C_{OX}}\left(\sqrt{2\varepsilon_S N_D(2\phi_B)} + q\int_{E_i}^{E_i+q\phi_B} D_{it}(E)dE\right) \tag{8.1}$$

where D_{it} is the interface trap density, and other symbols have the common meaning. Furthermore, the filled traps give rise to Coulomb scattering that turns into a mobility decrease of channel electrons flowing close to the surface. The number of filled traps decreases as the temperature increases since trapped electrons tend to be emitted. This leads to a lowering of the threshold voltage. In addition, both a reduction in Coulomb scattering and a higher number of free carriers improve channel mobility. Therefore, there is a temperature range in which mobility actually increases with temperature, until all electrons are released.

SC tests are performed on the device adopted for the calibration procedure for different operating conditions (i.e., V_{DS}, T_{CASE}, V_{GS}, and t_{PULSE}). The SC test setup is made according to the general guidelines in [45]. The device under test (DUT) is placed on a hot plate through which it is possible to set the case temperature. A custom advanced infrared (IR) thermography system, fully described in [48], is used to acquire the surface temperature of the device during the SC test. The aforementioned characterization method, featuring an equivalent time sampling technique, is able to acquire fast transient dynamics, with 1 MHz equivalent frame rate. Thus, it is possible to track the temperature evolution, and therefore the current distribution, during the applied SC pulse. In addition, the system allows a single-shot capture of the temperature map at any desired time instant along with the test. If the device fails, this corresponds to spotting the current distribution right before the failure event, which could lead to useful information about the failure mechanism itself.

The SC failure occurs depending on the electrical conditions. Therefore, two relevant conditions are here investigated in detail to address general behavior:

1. High voltage (≥400 V), short pulses (≤ 20 μs)
2. Low voltage (≤200 V), long pulses (100 μs)

Both the failure mechanisms are related to temperature increase inside the structure, but with different dynamics.

8.5.2.1 High voltage, short pulse tests

From single pulse SC waveforms (Figure 8.14(a) and (b)), the appearance of two phenomena becomes immediately evident. More in detail, the current tends to change the slope at the end of the pulse and current tails, usually present in bipolar devices, which originates after the turn-off.

Generally, it is an uncommon behavior for a power MOSFET, since as a unipolar device, it should not have any current tails, and it should have a negative current slope when biased above the temperature compensation point. These two effects could be considered temperature-related since as the pulse length increases and/or applied voltage and back temperature are higher, they become more and more relevant, up to device catastrophic failure. Physical electrothermal simulations are mandatory to deeper investigate the physics failure during the short-circuit. In Figure 8.15(a), simulated drain current waveforms are depicted, along with average surface temperature, where the same behavior experimentally observed has been reproduced.

A first interesting result can be pointed out (Figure 8.15(b)): the formation of hole current flowing out of the body terminal. It becomes visible when the current starts to change its slope. The second aspect that can be highlighted is that the heat is mainly generated in the JFET region and an extremely high temperature peak value is reached therein. Figure 8.16 depicts the hole current density in different time instants

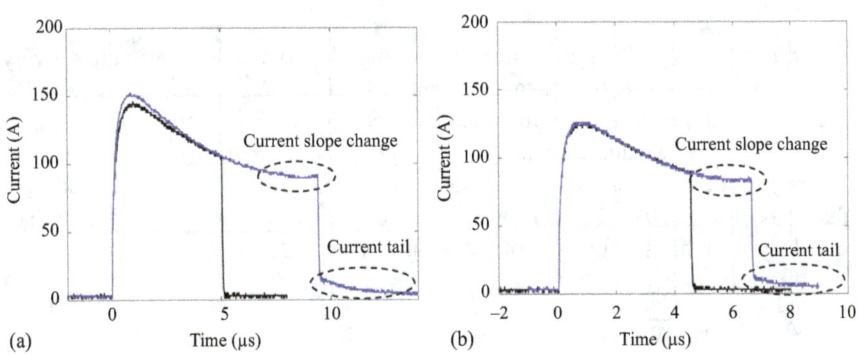

Figure 8.14 Experimental I_D SC waveforms at (a) $T = 75\ °C$ and (b) $T = 150\ °C$

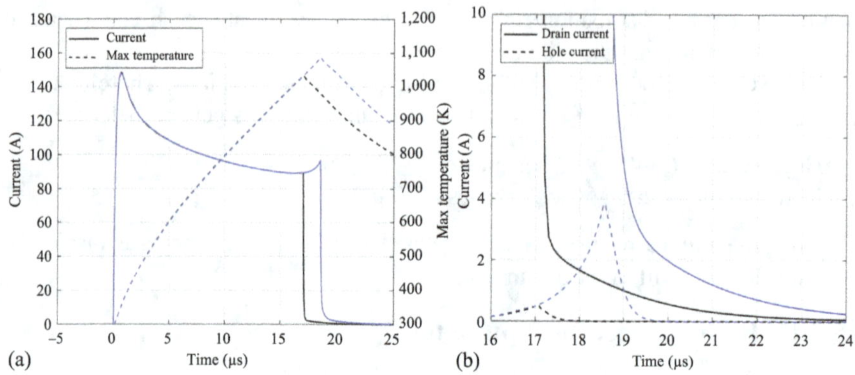

Figure 8.15 Numerical simulations of the SC event: (a) I_D and maximum temperature waveforms and (b) a detail of the hole current density flowing in the body region

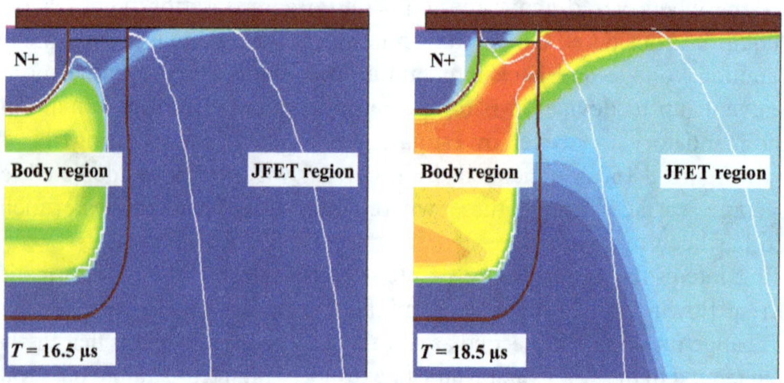

Figure 8.16 Hole current density from numerical simulation of the SC test ($V_{DS} = 400$ V; $V_{GS} = 18$ V; $T_{CASE} = 27$ °C)

along with an 18.5 μs SC pulse. In the beginning, the hole concentration has a very low value and therefore, the leakage current of the body/drift p–n junction is negligible. As the temperature rises, thermal generation increments to the hole level and, consequently, the leakage current keeps on growing gradually. This phenomenon gives rise to the hole current coming out of the body terminal. Using formulas and values reported in [20,49,50], it is possible to carry out an approximate estimation of the leakage current as a function of the temperature:

$$J_S = qn_i^2\left(\frac{D_P}{L_P N_D}\right) \tag{8.2}$$

where D_P and L_P (1–2 μm, [20]) are the diffusivities and the diffusion length respectively, and n_i is given by

$$n_i = 1.7 \cdot 10^{16} T^{\frac{3}{2}} e^{-2.08 \cdot 10^4 / T} \tag{8.3}$$

D_P can be calculated from the mobility μ_p:

$$\mu_p = 125 \left(\frac{T}{300} \right)^{-2.7} \tag{8.4}$$

Combining (8.2)–(8.4) with the assumption of a device area of approximately 3 mm × 3 mm and SiC physical parameters from literature [35], the leakage current can roughly be estimated to be ~30 A at $T = 2{,}000$ K. Far from being an accurate calculation, this result indicates at which temperature range the leakage current is expected to have a value comparable to the ON-state current during SC, that is when the device is experiencing thermal runaway. On the other hand, to get a current tail like the one experimentally observed, the temperature peak value could not be much far from the one obtained in simulation. Thus, holes are thermally generated due to locally elevated temperature increase. The electric field in the drift region drags the generated carriers toward the top of the device. Hole density goes on rising until at a certain point along the body/drift edge, the p–n junction does not exist anymore. This is due to the excessive carrier concentration that "punches through" the junction. Obviously, electrons are thermally generated at the same time and are free to flow from source to drain even when the applied gate voltage is zero. The current tail is indeed built up by the merging of the aforementioned leakage currents. The tail then slowly decreases to zero within a time linked to the one needed to remove all the generated carriers. Nevertheless, the leakage current could approach a level for that thermal runaway takes place leading to device failure. This is a positive feedback phenomenon inducing an uncontrollable increase of the drain current up to MOSFET destruction, even after the turn-off of the device. It is furthermore inferable that these devices do not comply with the usual required SC capability of silicon power devices, which is a minimum withstanding SC pulse of 10 μs with two-third of the rated voltage applied. So that to have a better comprehension of the inner device dynamic preceding the failure event (i.e., during the current tail), the temperature distribution is acquired at the turn-off of an 8 μs SC pulse. To easily accomplish this purpose, the temperature evolution is slowed down choosing $V_{DS} = 600$ V. The thermal map of Figure 8.17 corresponds to the current distribution at the turn-off, thus just before the failure event (as indicated in the figure). It clearly reveals that the failure arises from an excessively high power density shrunk in an extremely confined area (encircled red dot in figure) corresponding to the formation of a hot spot. When there is local growth of leakage carriers, a cluster of adjacent cells might tend to drain more current triggering the thermal runaway event. The current crowds in a limited portion of the total area, activating a self-sustained process that promptly entails the creation of the hot spot. The increase of the

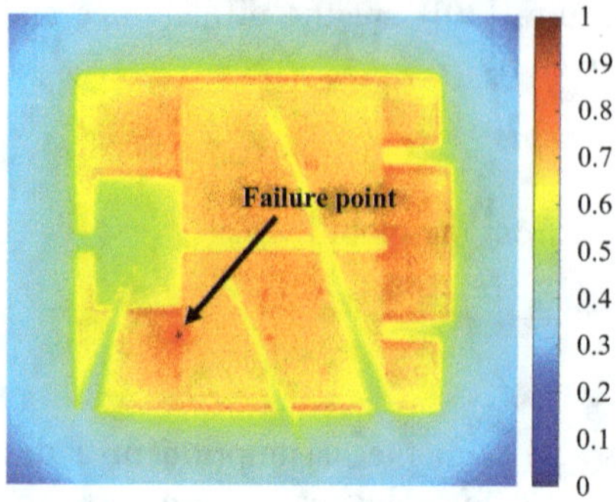

Figure 8.17 Normalized temperature increase at t = 8 μs, before the failure event

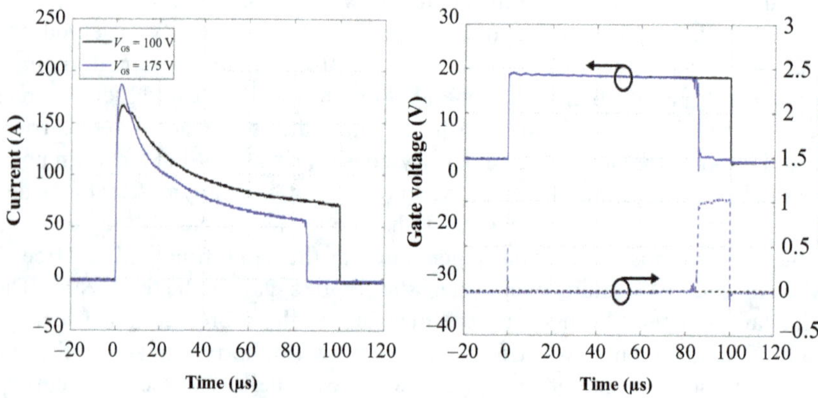

Figure 8.18 Experimental waveforms for SC events with a duration of 100 μs at low applied voltage: (a) drain current and (b) gate voltage and current during the SC test

current at the end of the SC pulse (before thermal runaway takes place) is much more pronounced in simulation than in experimental waveforms. It is a consequence of the used simulation approach, in which just a single cell is investigated. Thus, the electro-thermal interaction with surrounding cells, leading to stronger positive feedback, is not taken into account.

8.5.2.2 Low voltage, long pulse tests

To evaluate the dynamics of the failure at low applied voltage, devices are subjected to SC for long pulse width (100 μs), but with low applied voltage (<250 V). In this way, it is possible to slow down the temperature dynamic, and hence to analyze the device response to long thermal stress. The pulse length is kept constant and the voltage is increased in each step. Resulting current waveforms at different V_{DS} are reported in Figure 8.18, where the effect of the temperature rise on the current profile is clearly visible. When a certain voltage value is reached (175 V in this case), the device is not able to withstand the entire pulse duration and fails after approximately 85 μs, corresponding to the time instant in which the drain current drops to zero. In addition, it is interesting to notice that, at the same time, the gate-source voltage falls down to zero as well, and the gate current suddenly increases. It is then straightforward to assume that the device turned off because a in SC had happened between gate and source, indeed confirmed by the subsequent measure of R_{GS} (<1 Ω). It could be supposed that metallization and/or passivation layers on the top of MOSFET might be melted or somehow corrupted. Hence, in this case, the device does not undergo a catastrophic failure as previously explained, but simply it is not operative anymore because of damaging the gate/source structure.

In the following, an explanation of the different failure mechanisms during a low voltage SC for SiC power MOSFET is given. As the reported results make clear, two separate phenomena might happen when a device fails. It is convenient to indicate them as failure mode I and failure mode II. In the first type of failure, the device experiences a destructive mechanism due to the exponential rise of drain current subsequent to thermal runaway triggering. The second type of failure, on the other hand, involves the degradation of the gate structure, with the subsequent inability of turning the device on, which is why it could be considered a "soft failure." Both are regulated by the temperature increase inside the device, and more precisely by its growth rate. It is useful to define two temperature values:

- T_{DEG}, when surface degradation occurs
- T_{TH_RNW}, when thermal runaway takes place

The value for which top materials get corrupted is related to the temperature at which melting or changing in properties happens to passivation and metallization layers, and it is obviously lower than the triggering point of thermal runaway. Temperature rise is, of course, related to the amount of power that the device is subjected to, and therefore to the applied voltage. When the power applied is low, the temperature has a slow dynamic and might reach T_{DEG}, but could not arrive up to T_{TH_RNW}. If the surface is exposed to T_{DEG} for enough time, permanent damage occurs (failure mode II). The gate/source structure is compromised and therefore the device loses partially or totally its ability to conduct current. On the other hand, higher power turns into a prompt temperature increase. It suddenly reaches T_{TH_RNW}, a huge amount of carriers is then generated and the leakage current goes up to a value for which thermal runaway is activated. Drain current rises uncontrollably and the device blows up (failure mode I). In this case, the time for which the device remains at T_{DEG} is not enough for the surface to be fully damaged. In all other conditions, for moderate applied power, the failure is regulated by the time needed to degrade the device and the one needed to reach the thermal runaway point. When the former is higher, even if the temperature has a value able to produce detrimental degradation, thermal runaway is the predominant mechanism.

8.5.3 Short-circuit aging effect

The SCSOA is a fundamental parameter of a SiC MOSFET since it defines whether a device is suitable for a specific application. On the other hand, in real applications, the SC events occur many times with the consequent intervention of the SC protection circuits. Therefore, the reliability of a SiC MOSFET in a real application can be better estimated by repetitive SC testing. The intervention of the SC protection circuit limits the energy dissipated during the SC event. However, after many SC events, the high stress leads to the degradation of the SiC MOSFET performances, with the reduction of the reliability of the device [51,52]. Therefore, an aging test capable of estimating the degradation trend of the device is mandatory to evaluate the multiple SC events' effects on both the static and dynamic parameters of the SiC MOSFET. While the test conditions for the SCSOA of a power switch are well defined (see Section 8.4.1), the definition of both the test conditions and the degradation trend for multiple SC events is still an open research field. In Figure 8.19, the experimental data of the degradation of the threshold voltage for two families of devices are reported for two different conditions. More in detail, in [51] the aging test conditions are: $V_{DS} = 400$ V, $V_{GS} = -5/10$ V, $T_{SC} = 5$ μs and the device under test is a commercial 1,200 V/19 A. The main result is the decrease of V_{TH} with the number of cycles. On the other hand, in [52] a more extended investigation was carried out by analyzing the threshold voltage variation due to the gate overvoltage at high and low temperatures. For $V_{DS} = 400$ V and $V_{GS} = 25$ V, the V_{TH} increases, differently from the behavior shown in [51]. These results

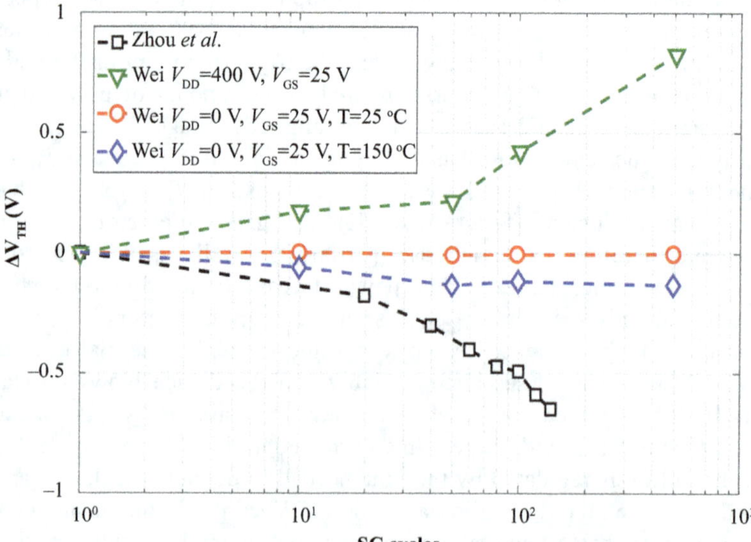

Figure 8.19 Threshold voltage dependence on the number of cycles. Adapted from [49] and [50]

highlight a great variation of the threshold voltage shift depending on the testing conditions. The variation of the threshold voltage leads to the variation of other parameters such as the turn-on and turn-off time. Moreover, degradation affects both turn-on and turn-off energy. In Figure 8.20, the dependence of the switching energy is reported within a wide range of a number of cycles. While the turn-off decreases with the increase of the number of cycles, the turn-on energy increases. Again, these effects come from the variation of the threshold voltage due to the degradation after several SC events. In Figure 8.20, the effect of the aging is reported for the R_{ON} in two different conditions. The R_{ON} depends on the V_{GS} value; the more the V_{GS} increases, the more R_{ON} decreases. However, the aging affects this parameter as well and the effect is more pronounced at high gate voltages. Figure 8.20 shows how the R_{ON} increases for $V_{GS} = 20$ V. As a résumé, it can be addressed that repetitive SC tests lead to the variation of the threshold voltage. Since the threshold voltage shift is related to the SiC/SiO_2 interface properties and the trap levels in the oxide, a framework including experimental, numerical, and theoretical analysis is mandatory to correctly address the phenomena involved in the aging of the device. Therefore, the measurement of the static parameters, TCAD simulations, three-terminal charge pumping (CP) tests, and gate-capacitance versus gate-voltage (CG-VG) have been adopted in [51,52] to address the aging in SiC MOSFETs under repetitive SC tests. In [52], the analysis of the mechanisms leading to the aging of the performances is carried out for high gate voltages ($V_{GS} = 25$ V). The most stressed area is the channel region, where very high electric field and high impact ionization rate due to avalanche occur at the same time. For high gate voltage SC tests, the main degradation mechanisms

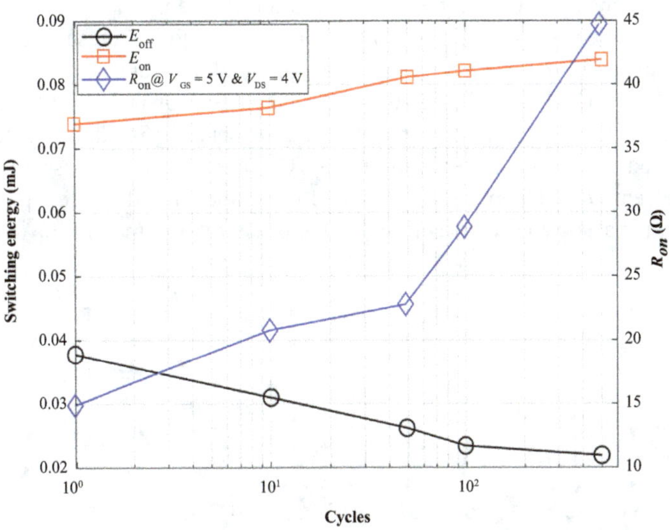

Figure 8.20 R_{ON}, E_{off}, and E_{on} dependency on the number of cycles. Adapted from [50]

are both the injection of hot electrons in the gate oxide and the formation of acceptor traps at the gate-oxide interface. More in detail, when a high voltage is applied to the gate of the device, a strong vertical electric field at the SiC/SiO_2 interface occurs and the direction of the electric field leads to the injection of electrons in the oxide of the device. In addition to the electric field, the high energy of the electrons involved in the impact ionization phenomenon leads to the injection of hot electrons in the oxide. On the other hand, for low gate voltage levels ($V_{GS} = -5/10$ V), the aging mechanism is mainly due to the hot holes injected in the gate oxide [51]. Again, this is attributed to the high electric field and impact ionization rate occurring in the proximity of the channel region. In these conditions, the number of holes injected in the oxide becomes dominant and the threshold voltage decreases with the increase of the cycles, as reported in Figure 8.20.

8.5.4 Short-circuit gate leakage current

In Section 8.4.3, it has been illustrated that depending on the temperature rise time, two different failure types can be triggered in SiC MOSFETs during SC operation. In broad terms, a high rate of increase of the device temperature causes the thermal runaway phenomenon, whereas a slower temperature increase induces degradation of the gate structure. Moreover, the damages produced by these two failure mechanisms are of different nature: while the first is a destructive failure that leads to a SC between the drain and source terminals, the latter is a soft failure which establishes a conductive path between the gate and the source electrodes. Therefore, while the fault produced by the thermal runaway can be classified as fail-to-short (FTS) (i.e., the device loses its voltage blocking capability), the gate degradation can be classified as a fail-to-open (FTO) damage (i.e., the device can no longer be controlled by the gate terminal and it is in a permanent off state). The FTO has never been observed in Si power MOSFETs and it represents a safer kind of failure since the device retains its voltage blocking capability. A consistent number of the SiC MOSFETs tested in literature has shown the FTO failure type (Figure 8.21) [53]. It has been experimentally observed [54] that during the SC events preceding, the occurrence of the FTO, there is an increase of the gate leakage current (I_G). The onset of the surge in I_G appears after a few μs of short-circuit and it is associated with a critical value of energy dissipation (E_C) [54]. Such

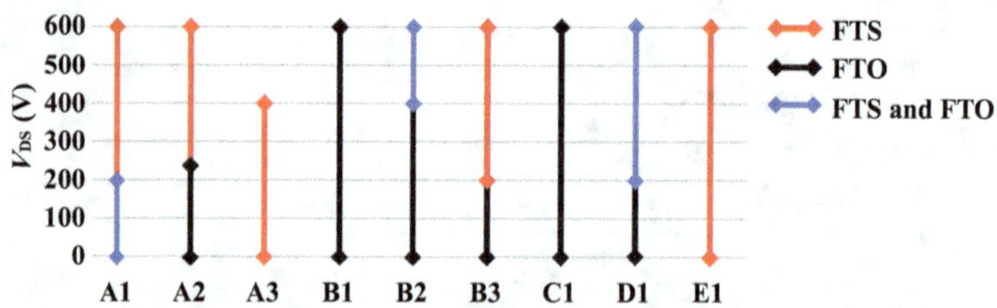

Figure 8.21 Different failure modes for the devices tested in literature. Adapted from [51]

a current does not damage the gate structure if the length of the SC pulse is properly limited [55]. Consequently, the event is easily repeatable with no visible sign of degradation. The dynamic of this current depends on several parameters, such as the gate driving voltage (V_{GG}), the gate resistance (R_G), and the drain-source voltage (V_{DS}) [56]. The physical explanation that best models this current has been studied in and the theory developed in it is recalled in the following paragraph.

Given a dielectric film where the trap-assisted conduction mechanisms can be neglected (i.e., with a low trap density), the two main current sources through the insulator layer are the Schottky emission (SE, also referred to as thermionic emission) and the Fowler–Nordheim tunneling (FN, also referred to as field emission) [57]. In Figure 8.22(a), the sketch of the band diagrams is reported in the presence of both the described mechanisms, while in Figure 8.22(b), a sketch of the current path is reported. The first produces a current density that can be modeled according to (8.5), while the second can be expressed as in (8.6):

$$J_{SE} = A^* T^2 \exp\left[\frac{-q\phi_B + \sqrt{q^3 E/4\pi\varepsilon_r\varepsilon_0)}}{kT}\right] \tag{8.5}$$

$$J_{FN} = \frac{q^3 E^2}{8\pi h \phi_B} \exp\left[\frac{-8\pi\left(2qm^*_{ox}\right)^{1/2}}{3hE}\phi_B^{3/2}\right] \tag{8.6}$$

where T is the absolute temperature, E is the electric field, and all the remaining parameters are summarized in Table 8.1.

Although the FN current density as expressed in (8.6) is not an explicit function of the temperature, this can be introduced by considering ϕ_B as temperature-dependent. Specifically, the barrier height can be expressed as a linearly decreasing function of the temperature [56,59]: $\phi_B(T) = \phi_B + (d\phi_B/dT)T$.

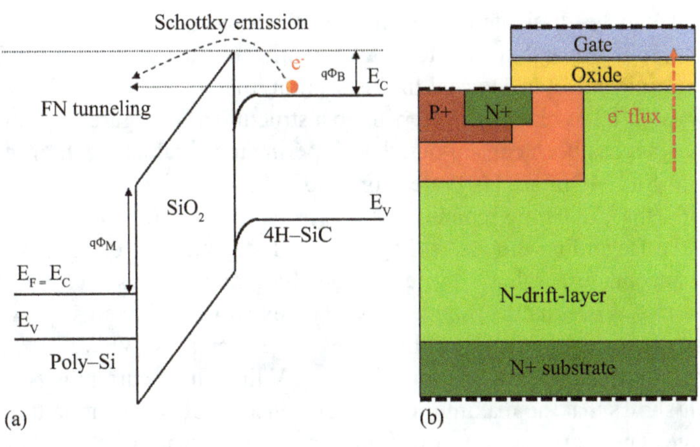

Figure 8.22 (a) Schematic band diagram for $V_{GS} > 0$ V of the MOS structure and (b) the FN tunneling and the Schottky emission are illustrated

Table 8.1 Parameters of the FN and SE current density expressions

Electron charge	q (C)	1.602×10^{-19}
Planck's constant	h ($eV \cdot s$)	4.136×10^{-15}
Semiconductor-oxide barrier height at 300 K	ϕ_B (eV)	2.6/2.7/2.8 min/avg/max [56]
Effective electron mass in the oxide	m^*_{ox} (kg)	3.826×10^{-31} [58]
Vacuum permittivity	ε_0 ($F \cdot m^{-1}$)	8.854×10^{-12}
SiO_2 relative permittivity	ε_r	2.1025
Boltzmann's constant	k ($eV \cdot K^{-1}$)	8.617×10^{-5}
Richardson's constant	A^* ($A \cdot m^2 \cdot K^{-2}$)	$4\pi m^*_{ox} k^2 q / h^3$

Considering a fixed current density, (8.5) and (8.6) can be manipulated to relate the temperature to the electric field. As estimated in [56] for an SC event of 36 μs at $V_{DS} = 250$ V, $V_{GS} = 18.33$ V, and $J_{DS} = 1$ A/cm^2, the temperature reaches approximately 1,500 K, which is better described by SE than by FN.

8.6 Avalanche capability

In applications where the load is an inductor or the parasitic inductances are not negligible, SiC MOSFETs are exposed to avalanche breakdown events when a sudden interruption of the current occurs. Typical applications are snubber-less converters and automotive applications such as antilock systems and engine control units (ECUs). In general, the driving of electrical engines exposes the SiC MOSFETs to avalanche events [60]. All the previous applications have in common an inductive load and if the energy dissipated during the avalanche breakdown event achieves the maximum withstanding capability of the SiC MOSFET, the switching device fails. The UIS test is widely adopted to evaluate the maximum energy that can be dissipated when the DUT operates in avalanche breakdown [61,62] and, together with the SC test, it is a routine test for the design validation of the power semiconductor devices. In this section, the physics of the failure in UIS for a SiC MOSFET is reported. Referring to a structure as in Figure 8.12, in [60] the failure analysis has been carried out. The experimental characterization refers to a commercial SiC MOSFET [40] rated for 1,200 V and 36 A (80 mΩ). The DUT is tested with the following conditions: $V_{DD} = 400$ V, $T_{CASE} = 300$, K and $L = 500$ μH. Regarding the V_{GS} voltage, two off-state voltage driving (V_{GSoff}) are kept into account to evaluate the effect on the maximum test current. More in detail, the V_{GSoff} is equal to 0 and −5 V. The experimental results show that the maximum test current increases as the gate voltage decreases. While for $V_{GS} = 0$ V, the failure current is $I_{AV} = 47$ A, for $V_{GS} = -5$ V the failure current is $I_{AV} = 50$ A. This result shows that the maximum avalanche energy is dependent on the off-state gate voltage. Therefore, TCAD simulations are mandatory to understand the failure physics and the influence of the gate voltage in off-state. Starting from a structure

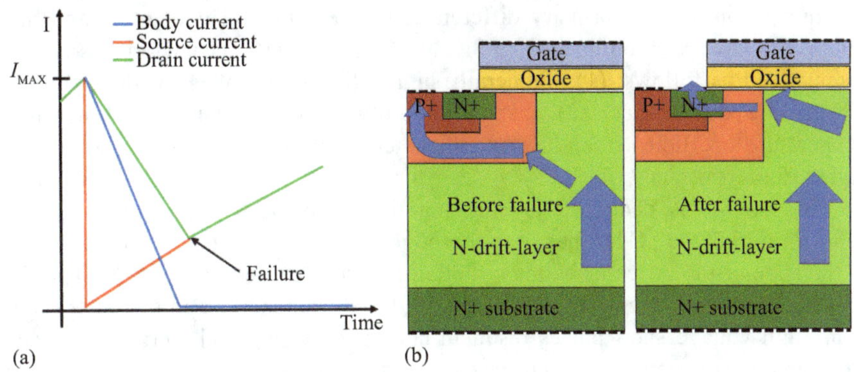

Figure 8.23 (a) A sketch of the failure waveforms of a SiC MOSFET during the UIS test and (b) a sketch of the total current path before and after the failure

like the one in Figure 8.12(a), the UIS test is simulated and the failure mechanism is addressed. In Figure 8.23(a), a sketch of the waveforms of body, source, and drain contacts are reported for a UIS test when the failure occurs. In principle, the current through the source contact should be negligible, since the current should flow only through the body region. However, the high current and high voltage occurring during the UIS test lead to a fast temperature increase in the JFET region, close to the channel. This reduces the threshold voltage and when the temperature achieves a limit, the threshold voltage becomes negative and the device starts to conduct, with consequent failure of the device. In Figure 8.23(b), a sketch of the total current flow before and after the failure is reported. Before the failure event, the current flows mainly at the edge of the p-body region. When the device fails, the current flows mainly through the channel region.

8.7 Summary

The recent performance progress as well as the open issues related to the state-of-the-art 4H-SiC MOSFETs have been reviewed in this chapter. After an introductory historical excursus on the development of SiC devices, the key difficulties for their large-scale adoption, such as the high production cost and the relatively poor quality of the SiC/SiO_2 interface, have been summarized. Afterward, peculiarities of three different vertical MOSFET structures (planar, trench-gate, and super-junction) have been analyzed in terms of technological steps, performance, and design choices.

Subsequently, in Section 8.2, the problems regarding the SiC/SiO_2 interface have been covered. The main effect of the interface imperfections is the arising of interface states in the forbidden gap, causing the exterior issue of the *I–V* curves instability.

In Section 8.3, the primary differences between the behavior of Si and SiC MOSFETs have been outlined. The SiC MOSFETs specific characteristics can be summarized as follows: (1) a generally higher threshold voltage with a stronger temperature dependence, (2) channel mobility showing positive and negative temperature coefficients for low and high temperatures, respectively, and (3) a higher drain resistance.

Section 8.4 is dedicated to studying the out-of-SOA operation. First, the behavior during the SC test has been thoroughly examined. In particular, according to the state-of-the-art contributions, the failure types caused by SC events can be classified into two categories depending on the SC voltage and on the pulse duration. High voltage short pulses result in current crowding and thermal runaway, whereas low voltage and long pulses cause a soft failure due to the degradation of the gate structure (never observed in Si power MOSFETs). Interestingly, an increased gate leakage current due to Schottky emission has been found to be a precursor of the second type of failure. Moreover, it has been observed that the aging effect produced by repetitive SC tests consists of a shift of the threshold voltage. Specifically, the threshold shifts upwards for SC tests with high gate bias, while an opposite trend has been observed for tests at a low level of gate bias. Second, the robustness in the avalanche conditions has been investigated. Particularly, literature data have reported the dependence of the maximum avalanche energy on the off-state gate bias during the unclamped inductive switching test. from TCAD simulations, the reason for the failure during avalanche operation has been found to be related to a strong reduction of the threshold voltage with the increasing temperature.

Note to the reader

This chapter is part of an edited book. To cite or use this chapter as a reference, please use the following format:

Maresca L., *et al.* "SiC MOSFETs." In: Iannuzzo F (ed.), *Modern Power Electronic Devices: Physics, Applications, and Reliability*. Stevenage, UK: IET; 2020. pp. 261–294.

References

[1] She X., Huang A. Q., Lucía Ó., and Ozpineci B "Review of silicon carbide power devices and their applications." *IEEE Transactions on Industrial Electronics*. 2017; 64(10): 8193–8205.

[2] Shenoy P. M., Bhalla A., and Dolny G. M. "Analysis of the effect of charge imbalance on the static and dynamic characteristics of the super junction MOSFET." *Proceedings of the 11th International Symposium on Power Semiconductor Devices and ICs, 1999*. ISPSD'99. IEEE, Toronto, ON, Canada, 1999.

[3] Van Brunt E., Lichtenwalner J. D., Leonard R., *et al.* "Reliability assessment of a large population of 3.3 kV, 45 A 4H-SIC MOSFETs." *29th International Symposium on Power Semiconductor Devices and IC's (ISPSD), 2017.* IEEE, Sapporo, Japan, 2017.

[4] Kumar A., Parashar S., Sabri S., Van Brunt E., Bhattacharya S., and Veliadis V "Ruggedness of 6.5 kV, 30 a SiC MOSFETs in extreme transient conditions." *2018 IEEE 30th International Symposium on Power Semiconductor Devices and ICs (ISPSD)*. IEEE, Chicago, IL, USA, 2018.

[5] Madjour K. "Silicon carbide market update: From discrete devices to modules." Presented at the PCIM Europe, Nuremberg, Germany, 2014.

[6] 1200 V/180 A full SiC power module with integrated SiC trench MOSFET, 2016. [Online]. Available: https://www.rohm.com/documents/11421/3508791/58F6889E_SiCTrenchMOSFET_ss(EN).pdf [accessed on Jan 29, 2019].

[7] Millan J., Godignon P., Perpiñà X., Pérez-Tomás A., and Rebollo J "A survey of wide bandgap power semiconductor devices." *IEEE Transactions on Power Electronics* 2014; 29(5): 2155–2163.

[8] Pala V., Brunt E. V., Cheng L., *et al.* "10 kV and 15 kV silicon carbide power MOSFETs for next-generation energy conversion and transmission systems." *Proceedings of the IEEE ECCE.* IEEE, Pittsburgh, PA, USA, 2014, pp. 449–454.

[9] Guo S., Zhang L., Lei Y., *et al.* "3.38 Mhz operation of 1.2 kV SiC MOSFET with integrated ultra-fast gate drive." *IEEE 3rd Workshop on Wide Bandgap Power Devices and Applications (WiPDA), 2015.* Blacksburg, VA, USA, IEEE, 2015.

[10] Hashimoto K., Okuda T., and Hikihara T. "A flyback converter with SiC power MOSFET operating at 10 MHz: Reducing leakage inductance for improvement of switching behaviors." *2018 International Power Electronics Conference (IPEC-Niigata 2018-ECCE Asia)*. IEEE, Niigata, Japan, 2018.

[11] Losee P., Bolotnikov A., Yu L., *et al.* "1.2 kV class SiC MOSFETs with improved performance over wide operating temperature." *Proceedings of the International Symposium on Power Semiconductor Devices ICs*, 2014. IEEE, Waikoloa, HI, USA, 2014, pp. 297–300.

[12] See https://www.wolfspeed.com/power/products/sic-mosfets/c2m0045170d [accessed Jan 29 2019].

[13] Romano G., Fayyaz A., Riccio M., *et al.* "A comprehensive study of short-circuit ruggedness of silicon carbide power MOSFETs." *IEEE Journal of Emerging and Selected Topics in Power Electronics.* 2016; 4(3): 978–987.

[14] Peters D., Aichinger T., Basler T., Rescher G., Puschkarsky K., and Reisinger H. "Investigation of threshold voltage stability of SiC MOSFETs." *2018 IEEE 30th International Symposium on Power Semiconductor Devices and ICs (ISPSD)*. IEEE, Chicago, IL, USA, 2018.

[15] Banerjee S., Matocha K., Chatty K., *et al.* "Manufacturable and rugged 1.2 kV SiC MOSFETs fabricated in high-volume 150mm CMOS fab." *28th*

International Symposium on Power Semiconductor Devices and ICs (ISPSD), 2016. IEEE, Prague, Czech Republic, 2016.

[16] Spitz J., Melloch M. R., Cooper J. A., and Capano M. A. "2.6 kV 4H-SiC lateral DMOSFETs." *IEEE Electron Device Letters*. 1998; 19(4): 100–102.

[17] Peters D., Friedrichs P., Schörner R., Mitlehner H., Weis B., and Stephani D. "Electrical performance of triple implanted vertical silicon carbide MOSFETs with low on-resistance." *Proceedings of the ISPSD*, Toronto 1999. IEEE, 1999, pp. 103–106.

[18] Allen S., Pala V., Van Brunt E., *et al.* "Next-generation planar SiC MOSFETs from 900 V to 15 kV." *Materials Science Forum.* Vol. 821. Trans Tech Publications, 2015.

[19] Shenoy P. M., and Baliga, B. J. "The planar 6H-SiC ACCUFET: A new high-voltage power MOSFET structure." *IEEE Electron Device Letters.* 1997; 18(12): 589–591.

[20] Jayant B. B. *Silicon Carbide Power Devices.* Hackensack, NJ: World Scientific, 2006.

[21] Shenoy J. N., Cooper J. A., and Melloch M. R. "High-voltage double-implanted power MOSFET's in 6H-SiC." *IEEE Electron Device Letters.* 1997; 18(3): 93–95.

[22] Cooper J. A., Melloch M. R., Singh R., Agarwal A., and Palmour J. W. "Status and prospects for SiC power MOSFETs." *IEEE Transactions on Electron Devices.* 2002; 49(4): 658–664.

[23] Wang Y., Tian K., Hao Y., Yu C., and Liu Y. "An optimized structure of 4H-SiC U-shaped trench gate MOSFET." *IEEE Transactions on Electron Devices*. 2015; 62(9): 2774–2778.

[24] Wang Y., Tian K., Hao Y., Yu C., and Liu Y. "4H–SiC step trench gate power metal–oxide–semiconductor field-effect transistor." *IEEE Electron Device Letters* 2016; 37(5): 633–635.

[25] Fujihira T. "Theory of semiconductor super-junction devices." *Japanese Journal of Applied Physics.* 1997; 36(10R): 6254.

[26] Ludikhuize A. W. "A review of RESURF technology." *Proceedings of the 12th International Symposium on Power Semiconductor Devices & ICs.* (Cat. No. 00CH37094). IEEE, Toulouse, France, 2000.

[27] Kobayashi Y., Kyogoku S., Morimoto T., *et al.* "High-temperature performance of 1.2 kV-class SiC super junction MOSFET." *2019 31st International Symposium on Power Semiconductor Devices and ICs (ISPSD)*. IEEE, Shanghai, China, 2019.

[28] Kosugi R., Sakuma Y., Kojima K., *et al.* "First experimental demonstration of SiC super-junction (SJ) structure by multi-epitaxial growth method." *2014 IEEE 26th International Symposium on Power Semiconductor Devices & IC's (ISPSD)*. IEEE, Waikoloa, HI, USA, 2014.

[29] Afanasev V. V., Bassler M., Pensl G., and Schulz M. "Intrinsic SiC/SiO_2 interface states." *Physica Status Solidi (a).* 1997; 162(1): 321–337.

[30] Kimoto T. "Material science and device physics in SiC technology for high-voltage power devices." *Japanese Journal of Applied Physics*. 2015; 54(4): 040103.

[31] Yoshioka H., Nakamura T., and Kimoto T. "Accurate evaluation of interface state density in SiC metal-oxide-semiconductor structures using surface potential based on depletion capacitance." *Journal of Applied Physics*. 2012; 111(1): 014502.

[32] Schroder D. K. *Semiconductor Material and Device Characterization*. Hoboken, NJ: John Wiley & Sons, 2006.

[33] Lelis A. J., Green R., Habersat D. B., and El M. "Basic mechanisms of threshold-voltage instability and implications for reliability testing of SiC MOSFETs." *IEEE Transactions on Electron Devices.* 2015; 62(2): 316–323.

[34] Yano H., Kanafuji N., Osawa A., Hatayama T., and Fuyuki T. "Threshold voltage instability in 4H-SiC MOSFETs with phosphorus-doped and nitrided gate oxides." *IEEE Transactions on Electron Devices.* 2015; 62(2): 324–332.

[35] Dimitrijev S., Li H., Harrison H. B., and Sweatman D. "Nitridation of silicon-dioxide films grown on 6H silicon carbide." *IEEE Electron Device Letters.* 1997; 18(5): 175–177.

[36] d'Alessandro V., Magnani A., Riccio M., *et al.* "SPICE modeling and dynamic electrothermal simulation of SiC power MOSFETs." *2014 IEEE 26th International Symposium on Power Semiconductor Devices & IC's (ISPSD)*. IEEE, Waikoloa, HI, USA, 2014.

[37] Riccio M., d'Alessandro V., Romano G., Maresca L., Breglio G., and Irace A. "A temperature-dependent SPICE model of SiC Power MOSFETs for within and out-of-SOA simulations." *IEEE Transactions on Power Electronics.* 2017; 33(9): 8020–8029.

[38] Kraus R., and Castellazzi A. "A physics-based compact model of SiC power MOSFETs." *IEEE Transactions on Power Electronics.* 2015; 31(8): 5863–5870.

[39] Klaassen F. M., and W. Hes. "On the temperature coefficient of the MOSFET threshold voltage." *Solid-State Electronics.* 1986; 29(8): 787–789.

[40] *C2M0080120D Datasheet*, Wolfspeed, Durham, NC, USA, 2015.

[41] Trivedi M., and Shenai, K. "Failure mechanisms of IGBTs under short-circuit and clamped inductive switching stress." *IEEE Transactions on Power Electronics.* 1999; 14(1): 108–116.

[42] Chinnaswamy K., Khandelwal P., Trivedi M., and Shenai K. "Unclamped inductive switching dynamics in lateral and vertical power DMOSFETs." *Conference Record of the 1999 IEEE Industry Applications Conference. Thirty-Forth IAS Annual Meeting (Cat. No. 99CH36370)*. Vol. 2. IEEE, Phoenix, AZ, USA, 1999.

[43] Wang Z., Shi X., Xue Y., Tolbert L. M., Blalock B. J., and Wang F. "Design and performance evaluation of overcurrent protection schemes for silicon carbide (SiC) power MOSFETs." *IEEE Transactions on Industrial Electronics.* 2014; 61(10): 5570–5581.

[44] Chokhawala R. S., Catt J., and Kiraly L. "A discussion on IGBT short-circuit behavior and fault protection schemes." *IEEE Transactions on Industry Applications.* 1995; 31(2): 256–263.
[45] ROHM Semiconductor. "Sic power devices and modules application note." June 2013.
[46] Sentaurus^{TM} Device User Guide, Version K-2015.06. Synopsis Inc., Mountain View, CA, USA, June 2015.
[47] Matocha K. "Challenges in SiC power MOSFET design." *Solid-State Electronics.* 2008; 52(10): 1631–1635.
[48] Romano G., Riccio M., De Falco G., Maresca L., Irace A., and Breglio G. "An ultrafast IR thermography system for transient temperature detection on electronic devices." *2014 Semiconductor Thermal Measurement and Management Symposium (SEMI-THERM).* IEEE, San Jose, CA, USA, 2014.
[49] Lutz, J., Schlangenotto, H., Scheuermann, U., and De Doncker, R. *Semiconductor Power Devices: Physics, Characteristics, Reliability.* Cham: Springer, 2018.
[50] Mnatsakanov T. T., Pomortseva L. I., and Yurkov S. N. "Semiempirical model of carrier mobility in silicon carbide for analyzing its dependence on temperature and doping level." *Semiconductors.* 2001; 35(4): 394–397.
[51] Zhou X., Su H., Wang Y., Yue R., Dai G., and Li J. "Investigations on the degradation of 1.2-kV 4H-SiC MOSFETs under repetitive short-circuit tests." *IEEE Transactions on Electron Devices.* 2016; 63(11): 4346–4351.
[52] Wei J., Liu S., Yang L., *et al.* "Comprehensive analysis of electrical parameters degradations for SiC power MOSFETs under repetitive short-circuit stress." *IEEE Transactions on Electron Devices.* 2018; 65(12): 5440–5447.
[53] Boige F., Richardeau F., Lefebvre S., Blaquière J.-M., Guibaud G., and Bourennane A. "Ensure an original and safe 'fail-to-open' mode in planar and trench power SiC MOSFET devices in extreme short-circuit operation." *Microelectronics Reliability.* 2018; 88: 598–603.
[54] Boige F., Richardeau F., Lefebvre S., and Cousineau M. "SiC power MOSFET in short-circuit operation: Electro-thermal macro-modelling combining physical and numerical approaches with circuit-type implementation." *Mathematics and Computers in Simulation.* 2019; 158: 375–386.
[55] Boige F., and Richardeau F. "Gate leakage-current analysis and modelling of planar and trench power SiC MOSFET devices in extreme short-circuit operation." *Microelectronics Reliability.* 2017; 76: 532–538.
[56] Boige F., Trémouilles D., and Richardeau F. "Physical origin of the gate current surge during short-circuit operation of SiC MOSFET." *IEEE Electron Device Letters.* 2019; 40(5): 666–669.
[57] Chiu F.-C. "A review on conduction mechanisms in dielectric films." *Advances in Materials Science and Engineering.* 2014; 2014.
[58] Lenzlinger M., and Snow E. H. "Fowler–Nordheim tunneling into thermally grown SiO2." *Journal of Applied Physics.* 1969; 40(1): 278–283.

[59] Schroeder D., and Avellán A. "Physical explanation of the barrier height temperature dependence in metal-oxide-semiconductor leakage current models." *Applied Physics Letters.* 2003; 82(25): 4510–4512.

[60] Fayyaz A., Romano G., Urresti J., *et al.* "A comprehensive study on the avalanche breakdown robustness of silicon carbide power MOSFETs." *Energies.* 2017; 10(4): 452.

[61] DiMarino C., and Hull B. "Characterization and prediction of the avalanche performance of 1.2 kV SiC MOSFETs." *2015 IEEE 3rd Workshop on Wide Bandgap Power Devices and Applications (WiPDA).* IEEE, Blacksburg, VA, USA, 2015.

[62] Liu S., Gu C., Wei J., Qian Q., Sun W., and Huang A. Q. "Repetitive unclamped-inductive-switching-induced electrical parameters degradations and simulation optimizations for 4H-SiC MOSFETs." *IEEE Transactions on Electron Devices.* 2016; 63(11): 4331–4338.

Chapter 9

GaN metal-insulator-semiconductor field-effect transistors

Shu Yang[1] *and Shaowen Han*[1]

9.1 Introduction: recent progress in GaN power devices and applications

Wide bandgap GaN power devices, by virtue of the low power losses and high switching frequency, are capable of delivering enhanced energy efficiency and increased power density in power electronic systems (Figure 9.1) [1–6] including laptop/smartphone chargers, power supplies for server and data center, on-board chargers for electric vehicles (EVs) and hybrids, solar inverters, etc. Currently, the lateral GaN-based heterojunction devices (or high-electron-mobility transistors, HEMTs) are dominating the scene of the commercialization of GaN power electronics, which have the following striking features: (1) a highly conductive channel based on the two-dimensional electron gas (2DEG) with high density ($\sim 10^{13}$ cm^{-2}) and high electron mobility ($\sim$2,000 cm^2/V·s) at the hetero-interface arising from the spontaneous and piezoelectric polarization effects [7]; (2) nearly one order of magnitude lower charges stored in the output capacitance than that in the latest Si superjunction (SJ) devices (Table 9.1) [8–14]; and (3) zero reverse recovery charges owing to the absence of pn junction. Consequently, low conduction and switching losses and high switching frequency can be realized.

It is compelling and cost-competitive to grow III-nitride epitaxial structure on the Si substrate to implement GaN-based lateral power devices. Not only the low-cost and large-diameter Si substrates can reduce the material cost, but the capability of fabricating the GaN power chips using the fully-depreciated 6- and 8-inch Si fabrication lines can reduce the manufacturing cost and facilitate the monolithic integration of GaN power IC.

The last decade has witnessed tremendous progress in the commercialization of lateral GaN devices as well as the demonstration of GaN technology in a variety of applications. Key GaN power device suppliers GaN Systems, Panasonic, EPC, Transphorm, among others, have released a variety of lateral GaN-on-Si power

[1]College of Electrical Engineering, Zhejiang University, Hangzhou, Zhejiang, China

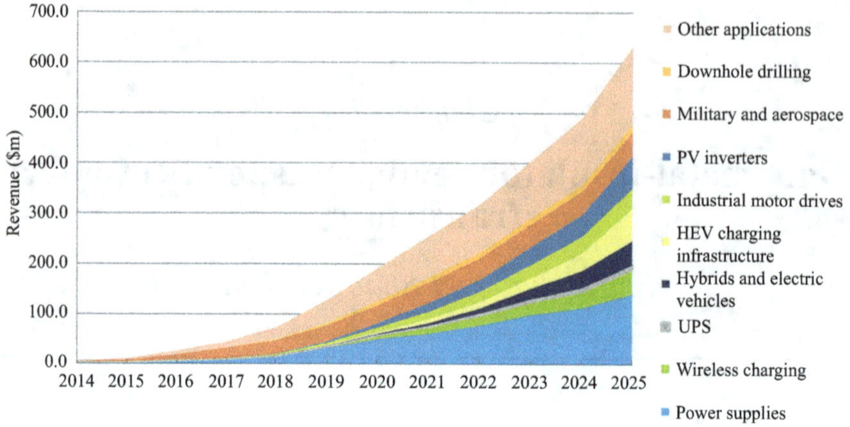

Figure 9.1 Applications and market of GaN power devices. Courtesy of IHS Markit Technology. Adapted from [1]

transistors with a voltage rating up to 900 V and current rating up to 120 A (Table 9.2) [11,12,15,16].

Thanks to the commercialization of GaN power devices, energy-efficient power modules and systems adopting GaN technology have also been demonstrated in recent years (Figure 9.2). In 2015, Yaskawa Electric Corporation [17] launched the world's first 4.5-kW photovoltaic power conditioner using the GaN power semiconductor module, enabling an overall peak efficiency above 98%, 40% reduction in size/weight and three times increase in switching frequency in comparison with that of Si-based modules. Fujitsu [18] developed a 12-W and 15.6-cc energy-efficient AC adapter implemented with GaN transistors for the superfast charging of smartphones and other mobile electronics. In 2017, Yaskawa Electric [19] launched an AC servo motor with a built-in amplifier with unprecedented performance and power density by using 650-V GaN power devices. In 2018, Anker [20] developed a 27-W GaN-based fast charger, the size of which is 40% smaller and frequency is 2.5 times higher than that of the conventional model. In 2018, Toyoda Gosei [21] first developed a full vertical GaN-based DC–DC converter by implementing the state-of-the-art vertical GaN SBD and vertical GaN FET with high current (over 50 A) and high-frequency operation up to several MHz. Delta [22] has demonstrated 3.3 kW/6.6 kW on-board chargers (OBCs) using 600-V GaN power devices, which can realize high switching frequency (0.3~1 MHz), reduced size/weight and 30%~50% higher power density. Diamond Electric [23] has developed a business card-sized, thin isolated bidirectional DC–DC converter, in which the adoption of GaN technology enabled the achievement of high efficiency up to 95% and high switching frequency up to 2 MHz with 75% smaller size than the conventional models.

Table 9.1 Comparison of key figure of merits (FOMs) for Si, GaN, and SiC devices. Adapted from [8]

Device	Rating (V)	R_{DSON} (mΩ)	$R_{DSON}{\cdot}Q_{OSS}$ (mΩ·μC)	$R_{DSON}{\cdot}Q_{RR}$[a] (mΩ·μC)	$R_{DSON}{\cdot}E_{OSS}$ (mΩ·μJ)	$R_{DSON}{\cdot}Q_G$ (mΩ·nC)	V_F@15A (V)	V_{GS}_max (V)
Si SJ [9]	600	56	23.5	336.0	450	3,800	0.9	20
GaN E-mode GIT [10]	600	55	2.2	2.2	350	300	2.7	N/A[d]
GaN E-mode [11]	650	50	2.8	2.8	350	290	3.2	7
GaN cascode [12]	650	52	5	7.0	730	1,460	1.4	18
SiC DMOS [13]	900	65	4.5	8.5	570	1,950	2.7	18
SiC TMOS [14]	650	60	3.8[b]	3.3[c]	540	3,480	3.3	22

[a] Q_{OSS} is included in Q_{RR}; Q_{OSS}, Q_{RR} and E_{OSS} are extracted at $V_{DS} = 400$ V unless otherwise noted.
[b] Q_{OSS} is extracted at $V_{DS} = 400$ V.
[c] Q_{RR} is extracted at $V_{DS} = 300$ V.
[d] GIT is a current driven device typically used with a simple RC network and a standard gate driver. Typical driver output used is 12 V with a maximum value of 20 V.

Table 9.2 Representative commercial E-mode GaN devices

Manufacturer	Voltage rating (V)	Current rating (A)	R_{DSON} (mΩ)	Q_G (nC)	Package (mm)
GaN systems [11]	100	45	15	6.2	GaNPX™
		90	7	12	
		120	5	18	
	650	7.5	200	1.5	
		15	100	3	
		22.5	67	4.4	
		30	50	5.8	
		60	25	12	
		120	12	25	
Panasonic [15]	600	13	140	2	DFN 8 × 8
		26	56	5	
Transphorm [12]	650	16	150	6	TO-220
		20	110	10	
		26	72	14	
		34	49	16	TO-247
		46.5	35	24	
	900	15	170	10	TO-220
EPC [16]	15	3.4	30	0.745	BGA 0.85 × 1.25
	30	90	1.45	19	LGA 6.05 × 2.3
	60	90	2.2	16	
	100	90	3.2	13.2	
	150	48	7	12	BGA 4.6 × 2.6
	200	48	10	8.8	
	350	6.3	65	3.4	BGA 1.95 × 1.95

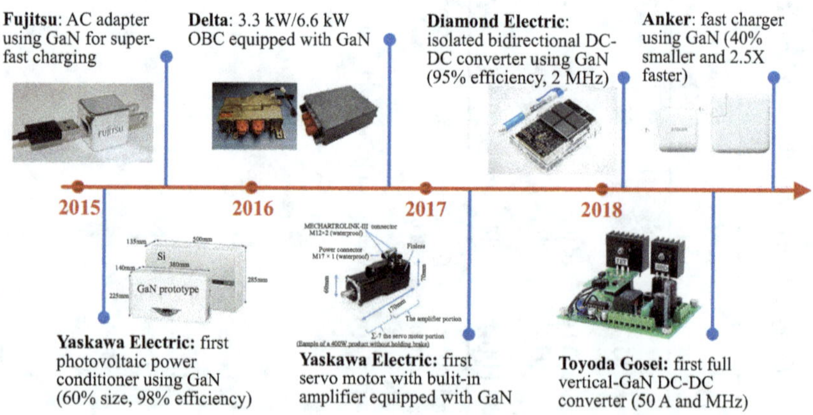

Figure 9.2 Applications/Prototypes employing GaN power devices

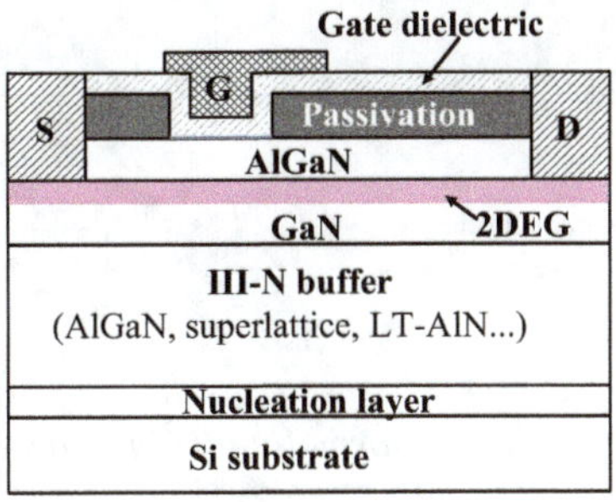

Figure 9.3 Schematic cross-section of the GaN-on-Si power transistors using MIS-HEMT as an example

9.2 Principle of operation

9.2.1 GaN-on-Si power transistor structures

As introduced in Section 9.1, the GaN-on-Si platform is cost-effective and compelling to implement GaN-based power devices. Tremendous efforts have been devoted to epitaxial growth aiming at managing the considerably large lattice- and thermal-mismatch between (0001) GaN to (111) Si [24,25]. The GaN-on-Si epitaxial structure typically consists of an AlN nucleation layer on Si substrate, semi-insulating III-nitride buffer (e.g., graded AlGaN, AlN/GaN superlattice, or multiple AlN insertion layers) [26], GaN channel layer, and AlGaN barrier layer (Figure 9.3).

Because of the unique spontaneous and piezoelectric polarization effects in group III-nitride materials [7], 2DEG with a high sheet carrier density of $\sim 10^{13}$ cm^{-2} and high electron mobility of $\sim$2,000 cm^2/V·s can be induced at the AlGaN/GaN hetero-interface without any intentional doping, forming a highly conductive channel between the source and drain terminals (Figure 9.3). Nevertheless, the 2DEG is inherently a normally-on (or depletion-mode, D-mode) channel that requires a negative gate bias to be pinched off. In power electronic applications, a normally-off power transistor with a positive threshold voltage (V_{TH}) is more desirable, to ensure fail-safe operation during startup or temporarily losing control power and to allow simpler gate driving circuitry. To achieve the normally-off operation, two categories of device technologies have therefore been developed: (1) two-chip hybrid cascode configuration with low-voltage Si MOSFET and high-voltage D-mode GaN metal-insulator-semiconductor-HEMT (MIS-HEMT), and (2) single-chip true normally-off GaN device technologies by implementing p-type (Al)GaN cap, fluorine ion (F^-) incorporation, and fully/partially recessed barrier in the gate region.

9.2.2 Normally-off GaN device technologies

9.2.2.1 Two-chip hybrid cascode configuration

Figure 9.4 shows a cascode configuration [27,28] by co-packaging a high-voltage D-mode GaN MIS-HEMT with a low-voltage E-mode Si MOSFET. The two dies

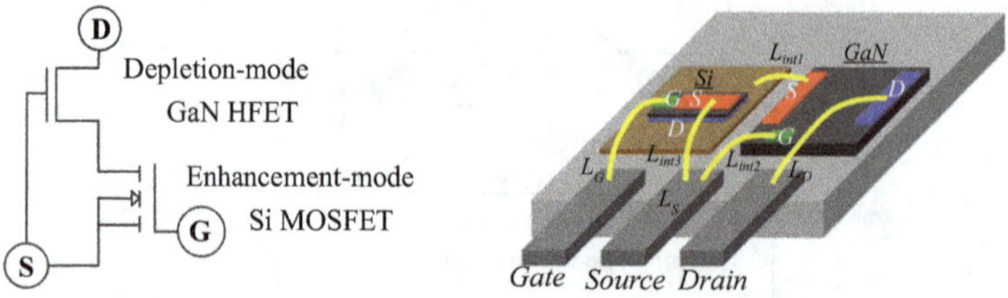

Figure 9.4 Device structure for the normally-off GaN cascode. Adapted from [27] and [28]

are connected in such a way that the output (drain-source) voltage of the MOSFET determines the input (gate-source) voltage of the GaN MIS-HEMT. When turning on the gate of the Si MOSFET, the same ON-state current flows through both Si MOSFET and GaN MIS-HEMT. At OFF-state, the Si MOSFET can sustain the increasing external drain voltage until the GaN MIS-HEMT is turned off and can subsequently block the remaining voltage. In this way, the hybrid cascode device can realize normally-off operation with a V_{TH} of the Si MOSFET, which is particularly favorable from the perspective of gate driving. In the meantime, the switching performance of the cascode device can be influenced by the parasitics inside the package and the junction capacitance matching between the two devices.

9.2.2.2 Single-chip E-mode GaN transistors

A true E-mode GaN power transistor allows a direct gate control and is highly desirable for improved slew rate. A high-performance E-mode GaN power transistor should achieve a normally-off operation while simultaneously maintaining a low R_{ON} to minimize the conduction loss. R_{ON} of an E-mode GaN transistor consists of (1) channel resistance (R_{ch}) underneath the gate electrode that can be modulated by gate bias, (2) gate-to-source/drain access region resistance (R_{ac}), and (3) source/drain contact resistance. In addition to creating a normally-off channel in the gate region, it's essential to make full use of the high conductivity of the normally-on 2DEG in the access region so as to minimize R_{ac}. Therefore, a desirable E-mode GaN power transistor should feature an E-mode channel and D-mode access regions [29].

E-mode GaN transistor technologies can be classified into two categories, depending on whether the starting epitaxial wafer is D-mode or E-mode. The commonly used non-patterned epitaxial growth indiscriminately creates a uniform V_{TH} throughout the entire wafer. Therefore, post-epitaxy V_{TH} modulation techniques are required to locally convert the D-mode gate region to E-mode one if the as-grown wafer is D-mode, or alternatively, convert the E-mode access region to D-mode one if the as-grown wafer is E-mode.

As shown in Figure 9.5, several E-mode GaN transistor technologies have been developed, including p-type GaN [30] or AlGaN cap layer [3,31], fluorine ion incorporation [32,33], and partially [34] or fully [35,36] recessed barrier in the gate region.

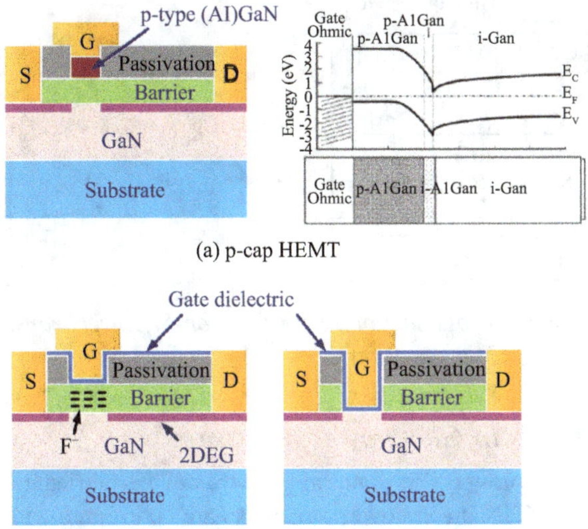

(a) p-cap HEMT

(b) MIS-HEMT w/F⁻ (c) Recessed-gate MIS-FET

Figure 9.5 Typical E-mode GaN transistors employing: (a) p-type (Al)GaN cap layer and the energy band diagram [30], (b) fluorine ion implantation, and (c) barrier recessing techniques in the gate region

By inserting a p-type (Al)GaN layer with a properly designed thickness and doping between the gate electrode and AlGaN barrier, the negative charges in the p-cap can lift up the energy band and consequently deplete the 2DEG only in the gate region, leading to a normally-off operation [30]. In the meantime, the highly conductive 2DEG is preserved in the access region such that a relatively low R_{ON} can be maintained.

Incorporating fluorine ions into the AlGaN barrier in the gate region by plasma treatment or ion implantation is another approach to realizing E-mode in GaN power transistors [32]. Owing to the strong electronegativity, fluorine ions can become negative fixed charges in the III-nitride materials, which can raise the energy band in the gate region and achieve normally-off operation.

V_{TH} modulation toward the positive direction in GaN transistors can also be realized by removing the barrier layer and the positive polarization charges under the gate. To obtain a suppressed gate leakage, enlarged gate swing, and more scalable V_{TH}, the barrier layer is usually replaced by an insulating dielectric layer, forming a recessed-gate GaN metal-insulator-semiconductor field-effect transistor (MIS-FET) (Figure 9.5(c)). The barrier layer can be fully recessed or partially recessed with a remaining barrier layer thickness <1.5 nm. A thin barrier layer kept under the gate can maintain the high electron mobility in the gate-controlled channel, leading to low R_{ON} in the partially recessed MIS-FET. A low-damage, precisely controlled and highly uniform gate recessing technique is critically important to ensure a uniform V_{TH} distribution, high channel electron mobility, and low channel resistance.

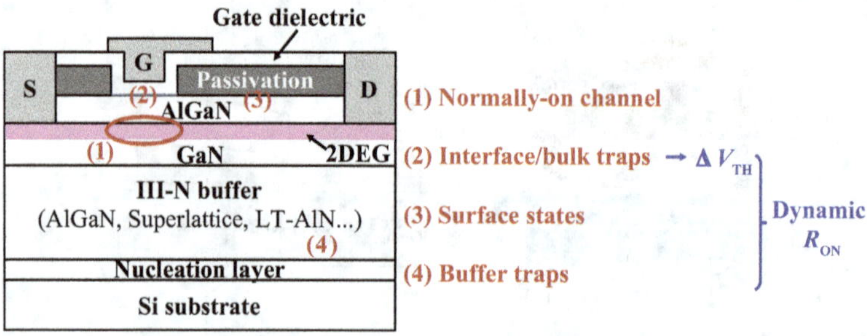

Figure 9.6 Schematic cross-section of the GaN-on-Si power transistor using the MIS-HEMT as an example

9.2.3 Challenges in GaN power transistors

Despite tremendous progress in material growth, device fabrication, and application demonstration, GaN-on-Si power transistors are confronted with several major challenges as illustrated in Figure 9.6, which can be primarily classified into (1) gate instability and reliability and (2) dynamic performance degradation associated with various traps (e.g., interface/bulk traps and hole deficiency in the gate region, surface states in the gate-to-drain access region, and bulk traps in the III-nitride buffer stack). These challenges need to be addressed by developing new processing technologies and structure designs, whereas advanced characterization techniques that are well suited for GaN power devices are of utmost significance toward the in-depth understanding of the underlying physical mechanisms and their impacts on devices' static/dynamic performance.

9.2.3.1 Gate instability and reliability

E-mode GaN transistor technologies except for the p-cap approach, as well as D-mode ones for cascode configuration, prefer MIS gate over Schottky gate, as the former group can provide suppressed gate leakage, enlarged gate swing, and more scalable V_{TH}. However, in MIS-gate GaN FETs, interface traps at the gate-dielectric/III-nitride interface and bulk traps inside the gate dielectric (including the near-interface bulk traps, or border traps) present a grand challenge to devices' gate instability and reliability [37–40]. In particular, at ON-state with a gate overdrive, electrons can spill over and get captured by the interface/border traps in the gate region, leading to V_{TH} shift in MIS-gate GaN power transistors. A positive V_{TH} shift could result in R_{ON} increase, due to the reduced gate overdrive with a preset ON-state gate voltage (V_{G_ON}) (Figure 9.7) [41,42].

On the other hand, it has been found that both positive gate bias at ON-state and negative gate-to-drain bias at OFF-state could cause V_{TH} instability of the p-GaN HEMTs with a Schottky gate contact [43–50]. It is widely observed the positive V_{TH} shift induced by a positive gate bias at ON-state in p-GaN HEMTs, which is attributed to electron injection from the 2DEG channel and getting trapped in the AlGaN barrier layer [45,46], metal/p-GaN space charge region [47], and/or

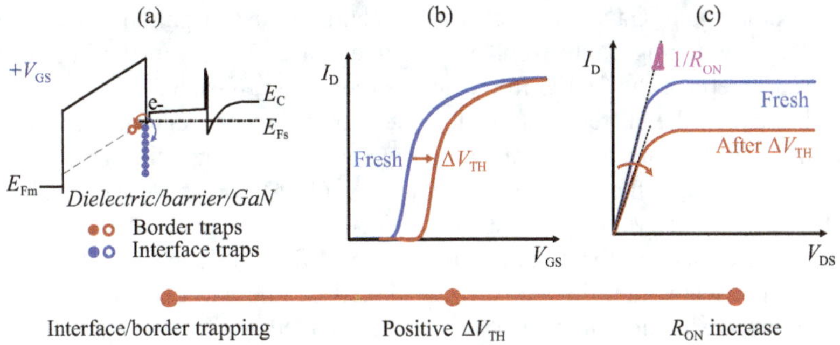

Figure 9.7 Illustration showing the impacts of interface/border trapping on V_{TH} stability and R_{ON} in insulated-gate GaN transistors: (a) schematic energy band diagram in MIS-HEMT at positive gate bias ($+V_{GS}$); (b) and (c) show positive threshold voltage shift (ΔV_{TH}) and the consequent R_{ON} increase due to the interface/border trapping as shown in (a), respectively. Adapted from [42]

p-GaN/AlGaN interface [48]. In addition, the V_{TH} instability of p-GaN HEMTs with a Schottky contact [43,49,50] could also be strongly influenced by the negative gate-to-drain bias stress at OFF-state.

The mechanisms of gate instability, advanced characterization techniques for V_{TH} instability under static/dynamic stress, and long-term gate reliability in GaN power switches will be summarized and discussed in Section 9.3.

9.2.3.2 Dynamic ON-resistance (R_{ON})

In addition to the V_{TH} instability, trapping effects in the access region and III-nitride buffer stack (Figure 9.6) could also adversely influence the dynamic R_{ON} of GaN-on-Si power devices. The high-voltage lateral GaN-on-Si power switching transistors require a high-resistivity III-nitride buffer stack between the 2DEG channel and the conductive Si substrate. A large number of bulk traps exist in GaN buffer layer grown on Si substrate, originating from (1) high-density (10^9–10^{10} cm^{-2}) of crystalline defects and dislocations due to lattice and thermal expansion mismatch between GaN and Si [51,52], (2) residual impurities such as silicon or oxygen introduced during epitaxial growth [53], and (3) intentionally incorporated compensation dopants such as carbon [54] or iron [55]. The reduction of crystalline defects [56] requires further optimization of growth techniques. The intentional dopants (carbon or iron), acting as acceptor traps in wurtzite GaN, can compensate the background donor doping, in order to obtain high resistivity and low leakage current in the buffer layer [57]. Carbon doping, which can provide higher breakdown strength [58,59], has become the most attractive approach to obtaining semi-insulating III-nitride buffer stack for high-voltage power electronics applications. The carbon-doped buffer stack can successfully deliver 650-V voltage rating GaN-on-Si power devices with low OFF-state leakage currents [58,60]. However, it is well acknowledged that carbon occupying the nitrogen-vacancy can act as a deep-level acceptor trap in GaN [61], the dynamic charging/discharging of which during high-voltage switching could result in negative space charges in the buffer stack and partially deplete the 2DEG in the access region at ON-state.

Bulk traps in the buffer stack, surface traps in the access region, and interface/border traps in the gate region can cause dynamic R_{ON} degradation in GaN-on-Si power devices due to the following mechanisms. (1) At OFF-state with high drain bias, the large negative V_{GD} could cause electrons to be injected from the gate electrode and trapped at the polarized III-nitride surface in the gate-to-drain access region [62]. Meanwhile, in the vertical direction, the large positive drain-to-substrate voltage could cause electrons to be injected from the Si substrate and trapped in the GaN buffer layer [63,64]. (2) At high power state during the hard-switching transition, hot electrons can be generated in the channel, then injected to and trapped at the III-nitride surface or in the GaN buffer layer [65,66]. When devices are switched from OFF-state to ON-state, some of the surface states and buffer traps cannot release electrons immediately due to relatively long emission time constants, while the 2DEG in the gate-to-drain access region could remain partially depleted, resulting in increased dynamic R_{ON} (or current collapse) and conduction loss. (3) At ON-state with a gate overdrive, interface/border trapping can occur in the gate region and cause V_{TH} shift in MIS-gate GaN power transistors. A positive V_{TH} shift could also result in R_{ON} increase, due to reduced gate overdrive voltage.

In Section 9.4, with ultrafast dynamic characterization techniques well suited for high-speed GaN power devices, the impacts of V_{TH} shift, buffer trapping, back-gate bias polarity, and different switching conditions on dynamic R_{ON} will be analyzed and discussed, which can provide guidelines for gate driving, buffer stack design, and drain structure engineering. Furthermore, the dynamic performance of lateral GaN-on-Si and vertical GaN-on-GaN power devices will be compared and analyzed.

9.3 Gate instability and reliability

9.3.1 Mechanisms of gate instability

9.3.1.1 V_{TH} shift in MIS-gate GaN transistors

The interface traps at the critical dielectric/III-nitride interface and bulk traps inside the gate dielectric, especially those near-interface bulk traps (e.g., border traps), could capture electrons at positive gate bias (Figure 9.7(a)). These negatively charged states will cause a positive V_{TH} shift (Figure 9.7(b)).

Normally-off GaN MIS-HEMTs/FETs are naturally subjected to positive gate bias stress when their current driving capability is maximized. On the other hand, normally-on GaN MIS-HEMTs in the cascode configuration, though the gate terminal is grounded, could also undergo positive gate stress ($V_{GS} > 0$), as the source potential could become negative during the turn-on transition [67,68].

In fact, similar mechanisms of interface/border trapping also pose a challenge to the stability of Si MOSFETs, which is usually referred to as bias temperature instability (BTI) and is well known as the key reliability issue in Si MOSFETs [69,70].

The origins and mechanisms of BTI are similar in Si MOSFETs and GaN power transistors. However, at the circuit level, BTI has different impacts on Si CMOS digital circuits and GaN power converters [41]. The adverse effects of BTI on Si CMOS logic circuits have been mainly on the degradation of frequency or propagation delay of logic gates, as well as static noise margin [71,72]. It should be noted that R_{ON} is not a major concern in Si CMOS digital circuits, as the static power consumption in a complementary configuration is negligible. By contrast, in GaN power converters, a positive V_{TH} shift could induce R_{ON} increase as a result of reduced gate overdrive, which could raise the conduction loss and reduce the conversion efficiency.

9.3.1.2 V_{TH} shift in p-GaN gate GaN transistors

The p-GaN gate FET with a Schottky gate contact could also exhibit V_{TH} instability. As shown in Figure 9.8 [49], due to the p-GaN layer inserted between the gate electrode and the AlGaN barrier, a p-GaN/gate-metal Schottky junction (J_1) and a p-GaN/AlGaN heterojunction (J_2) would be formed in the gate region. The corresponding junction capacitances are denoted as C_1 and C_2. Figure 9.8(c) illustrates the situation when high drain-to-gate stress is applied, where J_1 is forwardly biased while J_2 is reversely biased. As a result, holes would be injected from the p-GaN layer into the gate electrode through the forwardly biased J_1. Once the drain voltage stress is removed, because of J_1's low reverse leakage current, the

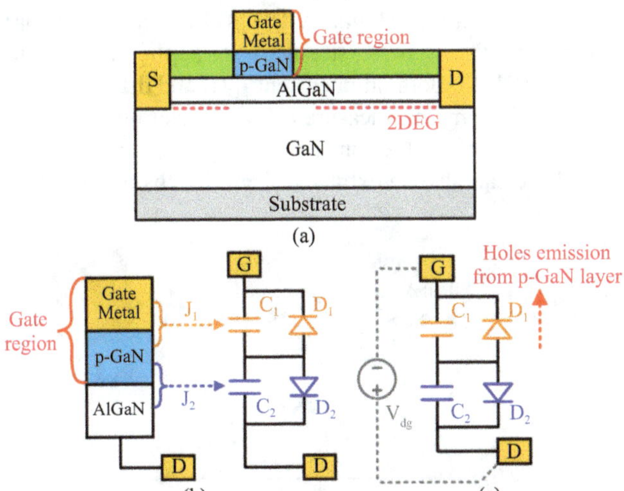

Figure 9.8 (a) Cross-sectional schematic of normally-off GaN transistor with p-GaN gate; (b) the formations of p-GaN/gate-metal Schottky junction (J_1) and p-GaN/AlGaN heterojunction (J_2); and (c) the situation when a high drain-to-gate stress is applied. Adapted from [49]

hole deficiency in p-GaN could not be recovered instantaneously [43], the net negative charges in which could lead to a positive V_{TH} shift.

9.3.2 Characterization techniques

9.3.2.1 Mapping of interface/border trap distribution

For the fully-recessed E-mode MIS-FETs with the dielectric/III-nitride interface located at the channel, the conventional conductance method that has been widely in MOS structures can be directly applied for interface/border trap extraction [73]. By comparison, in the "buried-channel" MIS-HEMTs, the critical dielectric/III-nitride interface is spatially separated from the 2DEG channel by the polarized barrier layer, resulting in underestimation/uncertainty in interface trap extraction if the conventional conductance method is used [39,74]. Instead, multi-temperature/frequency AC-capacitance techniques [39], focusing on the temperature/frequency-dispersion of the second slope of the C–V characteristics, have been developed for interface trap mapping in the MIS-heterojunction GaN devices. The interface trap analysis in GaN MIS-HEMTs was also conducted using the capture emission time (CET) map [38].

9.3.2.2 Evaluation of V_{TH} instability under static/dynamic stress

It should be noted that V_{TH} shift or hysteresis can be severely underestimated in the conventional "slow" quasi-static measurements, due to the recovery effect during the relatively long measurement delay [40,75], as shown in Figure 9.9 [41].

To address such issues, fast dynamic measurements with ultrashort measurement delay have been developed (Figure 9.10) [41] to accurately characterize the V_{TH} shift, in which both DC- and AC-stresses were applied to the gate terminal of the device under test (DUT). Both drain current I_D(t) and drain-to-source resistance R_{DS}(t) can be derived from the measured V_{DS}(t) according to (9.1) and (9.2). Because of the parallel shift in the transient I_D–V_{GS} characteristics, the V_{TH} shift after each stress phase can also be extracted from the change in I_D(t) at the preset

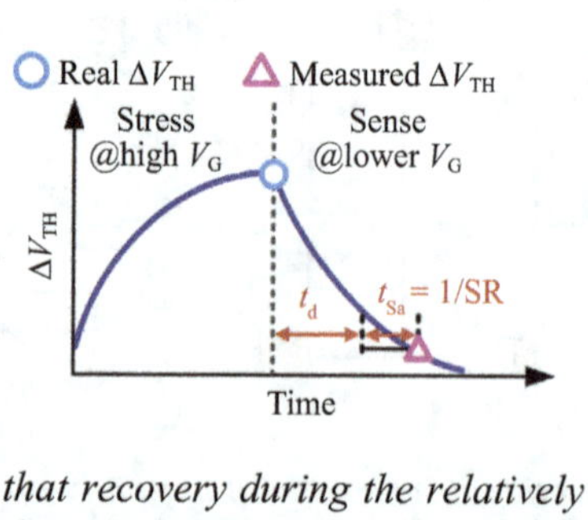

Figure 9.9 Illustration that recovery during the relatively long measurement delay (including the stress-to-sense delay t_d and sampling interval t_{Sa}) in conventional quasi-static measurement could cause underestimation in V_{TH} shift. Adapted from [41]

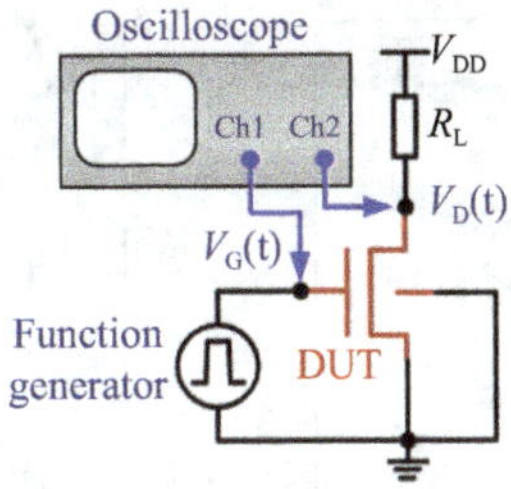

Figure 9.10 Schematic experimental setup for the fast dynamic characterization of V_{TH} shift. Adapted from [41]

gate sensing bias:

$$I_D(t) = \frac{V_{DD} - V_{DS}(t)}{R_L} \tag{9.1}$$

$$R_{DS}(t) = \frac{V_{DS}(t) \cdot R_L}{V_{DD} - V_{DS}(t)} \tag{9.2}$$

The short measurement delay of ~120 ns in [41] can minimize the measurement-induced recovery effect and enable an accurate evaluation of V_{TH} shift under both DC- and AC-gate stresses.

With the fast dynamic characterization techniques, V_{TH} shifts in D-/E-mode MIS-HEMTs/FETs under static (DC) and dynamic (AC) stress have been revealed throughout the entire stress time from 10^{-7} to 10^3 s, as shown in Figure 9.11 [42]. At each stress time, dynamic stress results in smaller V_{TH} shift compared with the static stress, suggesting the conventional static (constant) stress could underestimate the projected lifetime of GaN power switches. As discussed in Section 9.3.1, the frequency dependence of the V_{TH} shift is similar to the dynamic (AC) BTI in Si MOSFETs [76,77], the root cause for which is the widely distributed capture/emission time constants and frequency-dependent occupancy level of traps. The capture/emission time map in GaN MIS-HEMTs was first extracted by Lagger *et al.* [38]. The dispersion between DC- and AC-stress indicates the involvement of border traps [78], which could interact with electrons via tunneling and feature a relatively long capture time constant.

On the other hand, E-mode MIS-FET shows a smaller V_{TH} shift than D-mode MIS-HEMT with the same gate dielectric stack [79]. The underlying mechanisms can be understood by considering the band diagrams in Figure 9.12 [42].

D-mode MIS-HEMT is a buried-channel device with the critical dielectric/III-nitride interface spatially separated from the 2DEG channel by the barrier layer. The strong polarization field inside the barrier layer can raise the energy band at the critical interface and allow interface traps with a broad energy range emerging above the Fermi level at pinch-off [79]. At a positive gate bias, a wide range of interface traps below the Fermi level will be filled with electrons, resulting in a considerable V_{TH} shift in fast measurement.

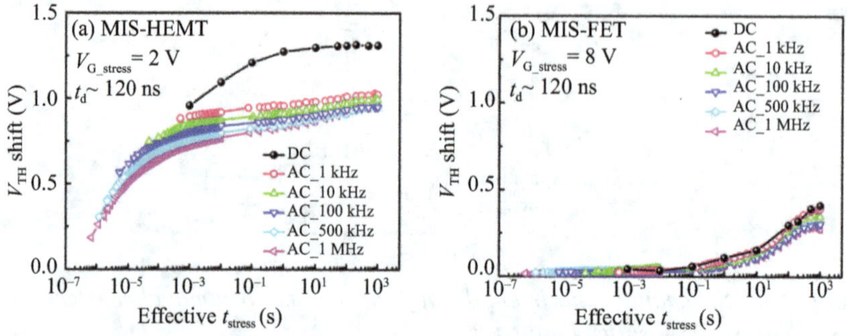

Figure 9.11 Time-resolved V_{TH} shift in (a) D-mode MIS-HEMT and (b) E-mode MIS-FET under static (DC) and dynamic (AC) positive gate stress. Effective stress time (t_{stress}) rather than nominal time is used by considering the duty cycle of 50%. Different starting points with different frequencies in (a) is because V_{TH} shift is extracted after at least one stress half-cycle. Adapted from [42]

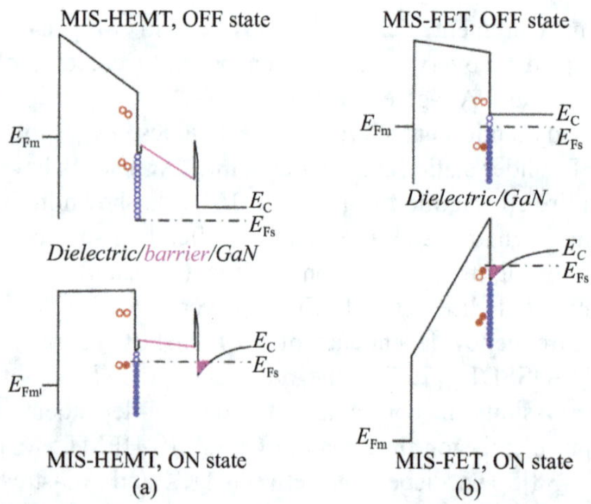

Figure 9.12 Schematic energy band diagrams of (a) D-mode MIS-HEMT and (b) E-mode MIS-FET at OFF-state (upper) and ON-state (lower). 2DEG channel is fully turned on while R_{ON} is saturated in both cases. Adapted from [42]

By comparison, the deeper interface trap levels in E-mode MIS-FET are always buried below the Fermi level and maintain fixed charge states during ON/OFF switching, making no contribution to the V_{TH} shift. Therefore, the V_{TH} shift in E-mode MIS-FET is smaller than that in D-mode MIS-HEMT with the same gate dielectric process.

Similarly, V_{TH} instability in p-GaN HEMTs has been demonstrated under static (DC) and dynamic (AC) stress, as shown in Figure 9.13 [47]. V_{TH} shifts positively

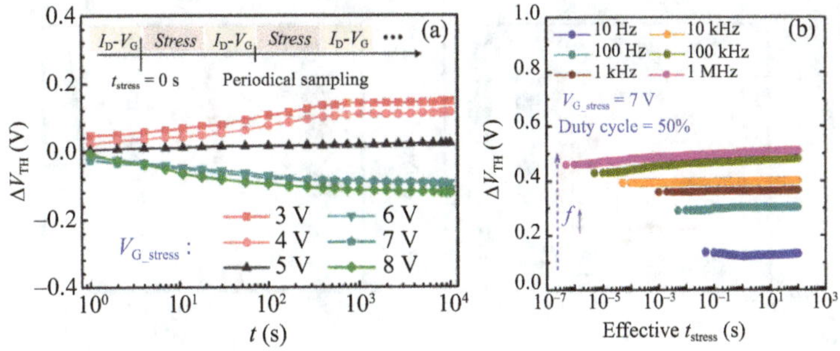

Figure 9.13 Time-resolved V_{TH} shift in p-GaN HEMTs under (a) static (DC) and (b) dynamic (AC) positive gate stress. Effective stress time (t_{stress}) rather than nominal time is used by considering the duty cycle of 50%. Adapted from [47]

with increasing stress time under relatively low static (DC) gate bias stress (i.e., $V_{G\_stress} < 5$ V). By contrast, V_{TH} shifts negatively with increasing stress time under high static (DC) gate bias stress (i.e., $V_{G\_stress} > 5$ V). Under dynamic positive gate stress in Figure 9.13(b), V_{TH} exhibits a positive shift at V_{G_stress} of 7 V in contrast to the negative shift in static stress. Moreover, ΔV_{TH} under dynamic positive gate stress exhibits a monotonous frequency dependence from 10 Hz to 1 MHz.

9.3.3 Time-dependent dielectric breakdown

The time-dependent dielectric breakdown (TDDB) test, which is an evaluation of catastrophic conditions arising after prolonged high-voltage gate stress, has been widely used to evaluate the long-term reliability of the gate dielectric of a MIS-gate device.

The time to breakdown (t_{BD}) was defined at the critical point when the gate leakage increased suddenly. The t_{BD} of a constant stress voltage shows a Weibull distribution [80].

The electric-field- and temperature-accelerated TDDB tests were performed on low-pressure chemical vapor deposition (LPCVD)-SiNx MIS-FETs with an interfacial protection layer in [81], as shown in Figure 9.14. The t_{BD} shows a Weibull distribution with a slope β of 4, indicating a tight breakdown time distribution and small variability in breakdown behavior. The lifetime of the gate dielectric at different V_{GS} with a certain failure rate can be predicted.

9.4 Dynamic performance

9.4.1 Dynamic ON-resistance (R_{ON})

9.4.1.1 V_{TH}-instability-induced dynamic R_{ON}: role of gate overdrive

As discussed in Section 9.3.1, at ON-state, electron trapping at the interface or border traps could occur in MIS-gate GaN transistors and cause V_{TH} shift.

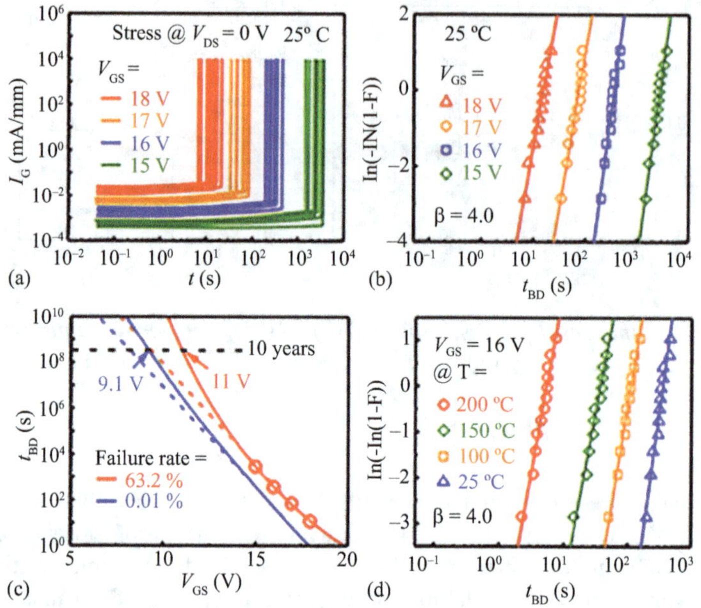

Figure 9.14 (a) t_{BD} of the LPCVD-SiN$_x$ MIS-FETs with interfacial protection layer at forward gate stress of 18, 17, 16, and 15 V at 25 °C. (b) Weibull plot of the electric field-dependent t_{BD} distribution. (c) Lifetime prediction with failure rate of 63.2% and 0.01%, respectively. (d) Weibull plot of the temperature-dependent t_{BD} distribution. Adapted from [81]

Compared with the E-mode MIS-FET, the D-mode MIS-HEMT shows a larger V_{TH} shift (Figure 9.11) but a slightly lower R_{ON} increase (Figure 9.15) [42], suggesting that the D-mode GaN MIS-HEMT can tolerate a considerable V_{TH} instability. There are two major reasons for such distinction in R_{ON}'s susceptibility to V_{TH} shift in D-/E-mode MIS-HEMTs/FETs.

In D-mode MIS-HEMT, the hysteresis (ΔV_{TH}) has a greater influence on R_{DS} in the semi-ON region than in the fully-ON region. Owing to the sufficient gate overdrive (Figure 9.16) [67], R_{ON} in the D-mode MIS-HEMT is fully saturated when $V_{GS} \geq 0$ V, and thus, shows weaker sensitivity to V_{TH} shift.

D-mode MIS-HEMT and E-mode MIS-FET feature different R_{ON} components which also affect R_{ON}'s susceptibility to V_{TH} shift. The overall R_{ON} constitutes (1) channel resistance in the gate region (R_{ch}), (2) gate-to-source/drain access region resistance (R_{ac}), and (3) source/drain contact resistance (R_c). When interface/border trapping occurs in the gate region and induces a positive V_{TH} shift, it is R_{ch} that increases due to the reduced ($V_{G\_ON}-V_{TH}$). In MIS-FET, due to the lower channel mobility in the fully recessed gate region, R_{ch} constitutes a higher percentage of the

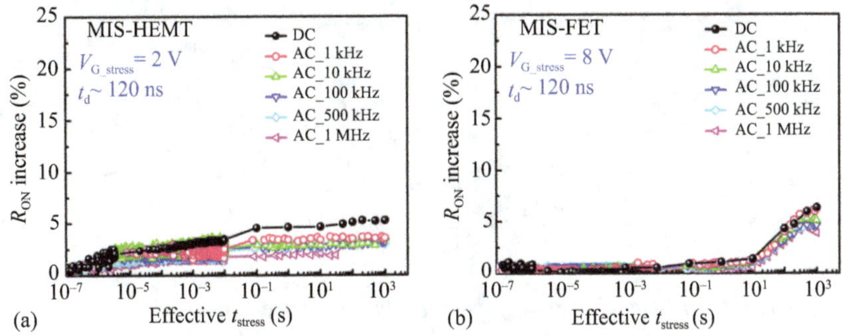

Figure 9.15 Time-resolved R_{ON} increase [$R_{ON}(t)/R_{ON\_fresh}$ − 1] in (a) D-mode MIS-HEMT and (b) E-mode MIS-FET under static and dynamic positive gate stress. Adapted from [42]

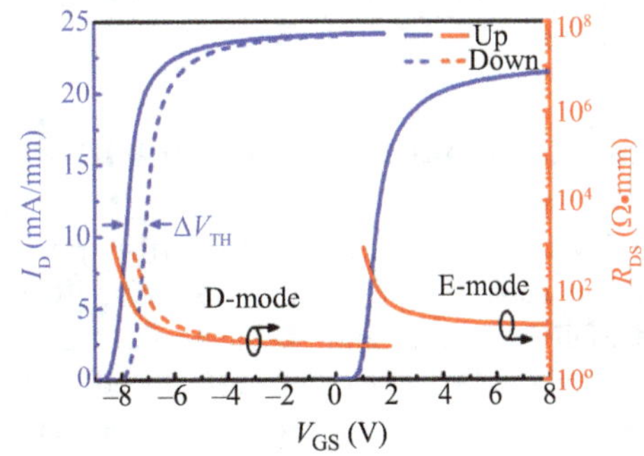

Figure 9.16 I_D–V_{GS} and R_{DS}–V_{GS} characteristics of the D-mode and E-mode GaN MIS-gate transistors. Adapted from [67]

overall R_{ON} [67]. Therefore, R_{ON} in MIS-FET is more susceptible to R_{ch} increase that originates from interface/border trapping in the gate region.

As discussed in Section 9.3.1, the p-GaN HEMT with a Schottky gate contact could also exhibit V_{TH} instability. In order to investigate the V_{TH} stability of the p-GaN device, dynamic transfer characteristics are measured using pulsed I–V (Figure 9.17(a)) [43]. As shown in Figure 9.17(b), V_{TH} shifts positively with higher quiescent bias V_{DSQ}. This positive V_{TH} shift results in a remarkable reduction in the drain current at lower gate voltages ($V_{GS,ON}$) of 3~4 V, and consequently an increased dynamic R_{ON}. Nevertheless, the impact of this positive V_{TH} shift on the drain current diminishes rapidly as $V_{GS,ON}$ increases. Under a $V_{GS,ON}$ of 5~6 V, the positive V_{TH} shift has an insignificant influence on dynamic R_{ON} (Figure 9.17(c)). The investigation of dynamic R_{ON} degradation and V_{TH} instability in p-GaN gate HEMT suggests that a higher V_{GS} of 5~6 V would provide adequate gate over-drive to overcome the V_{TH}-shift-induced dynamic R_{ON} degradation in power switching applications [43].

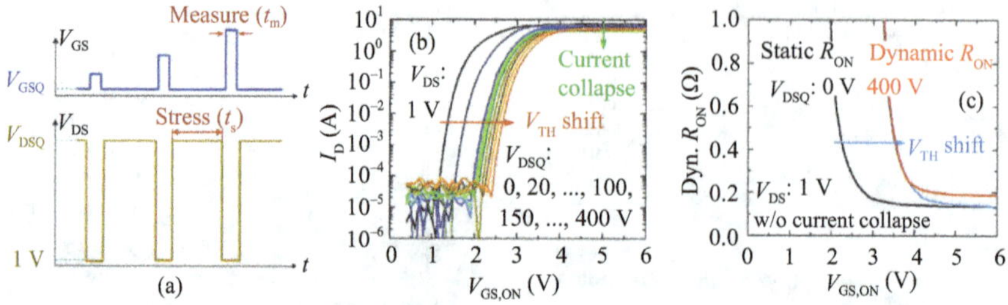

Figure 9.17 (a) Pulsed I–V waveforms for the measurement of transfer characteristics. Quiescent biases V_{GSQ} and V_{DSQ}, as well as pulse width and pulse period are the same as in the dynamic R_{ON} measurement. (b) Transfer characteristics under dynamic switching operations with V_{DSQ} up to 400 V. (c) Dynamic R_{ON} versus $V_{GS,ON}$. Black line: measured static R_{ON} with $V_{DSQ} = 0$ V. Red line: measured dynamic R_{ON} with $V_{DSQ} = 400$ V. Blue line: calculated dynamic R_{ON} with $V_{DSQ} = 400$ V induced only by V_{TH} shift. Adapted from [43]

9.4.1.2 Buffer-induced dynamic R_{ON}: role of buffer stack design

For lateral GaN-on-Si devices, it is necessary for the III-nitride buffer stack to have the high-voltage blocking capability. The semi-insulating III-nitride buffer stack, locating between the 2DEG channel and Si substrate, usually consists of unintentionally doped (UID) GaN layer, carbon-doped (C-doped) GaN layer, super-lattice or graded AlGaN stress relief layer (SRL), and AlN nucleation layer (NL) [26].

As discussed in Section 9.2.3, there are a large number of bulk traps in the III-nitride buffer stack grown on Si substrate. Charge trapping at these bulk traps in the buffer stack [51,63,64], as well as at the surface traps in the access region [62], presents a great challenge for dynamic R_{ON} in the lateral GaN-on-Si power devices. In recent years, advanced dielectric passivation [82,83] and field plate [84,85] techniques have been developed to suppress the surface trapping. However, buffer trapping could still occur and the generated space charges at high-voltage OFF-state or during hard switching cannot recover immediately, leading to dynamic R_{ON} increase (or current collapse) when devices are switched to ON-state.

The 2DEG conductivity can be modulated by applying back-gate bias (V_{Sub}) to the conductive Si substrate, which can distinguish buffer trapping from surface trapping [86,87] and facilitates studies of buffer transport mechanisms [52,88,89].

High-voltage back-gating measurements have been performed to reveal the influence of charging/discharging of buffer traps on the drain current of an AlGaN/GaN-on-Si HEMT. Figure 9.18 shows the measurement setup, in which V_{GS} and V_{DS} were kept at low voltages while V_{sub} was swept to a high negative value [90]. Consequently, the high electric field can drop across the buffer stack and only induce charging/discharging of the buffer traps, while the low V_{GS} and V_{DS} can

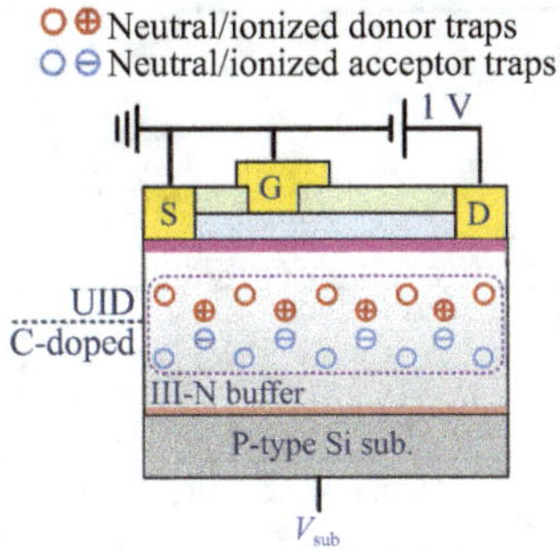

Figure 9.18 Schematic setup of back-gating measurement for a normally-on GaN-on-Si power HEMT. Adapted from [90]

minimize the surface-trapping-induced dynamic R_{ON} degradation and self-heating which could also affect I_D [63,87].

Both acceptor and donor traps usually exist in the buffer stack of GaN-on-Si devices. The UID GaN usually exhibits n-type conductivity with incomplete ionization of donors [91], whereas the C-doped GaN contains deep-level acceptor traps. Accordingly, the buffer stack containing both donor and acceptor traps has been modeled with detailed parameters described in [90], and the dynamics of buffer charging/discharging have been studied and revealed using technology computer-aided design (TCAD) simulations in conjunction with the high-voltage back-gating measurements.

As shown in Figure 9.19(a), compared with the ideal case without buffer trap, the more significant I_D reduction from nearly 0 V to V_1 suggests that the dominant process is the negative buffer charging (e.g., ionization of acceptor traps), which is verified by the simulation as shown in Figure 9.19(b) [92] and can accelerate the 2DEG depletion [87]. Similarly, positive charge generation (e.g., ionization of donor traps) is responsible for the slower I_D reduction (or lower transconductance) from V_1 to V_2, as the positive charges can compensate the increasingly negative V_{Sub} (Figure 9.19(b)).

The opposite doping polarities in the UID and C-doped GaN layers could result in a pn junction at the UID/C-doped GaN interface in the buffer (Figure 9.20) [90], which becomes reversely biased under a negative V_{Sub} with depletion region extending into the two layers and excessive ionization of both donor and acceptor traps. Distinct charging/discharging behaviors within different V_{Sub} ranges are possibly correlated with the nonuniform spatial distribution of traps [53] and the charge redistribution in the buffer stack [93]. The analysis of buffer-related current collapse suggests that dynamic R_{ON} degradation can be suppressed by carefully introducing positive charges (e.g., donor traps) in the UID GaN layer [90], which requires precise control of defects/dislocations and impurities to simultaneously maintain a low OFF-state leakage.

The asymmetric buffer-related current collapse with opposite top-to-substrate bias polarities suggests that increasing the effective barrier at the NL/Si junction or

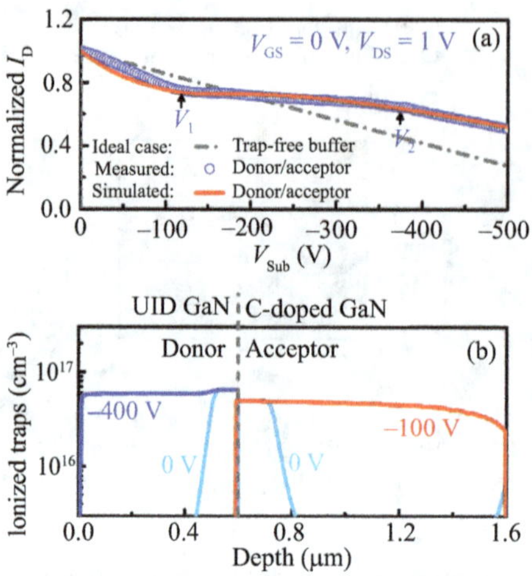

Figure 9.19 *(a) Measured and simulated I_D–V_{Sub} characteristics with V_{Sub} swept from 0 V to −500 V at a rate of 1 V/s. I_D is normalized to the value at $V_{GS}/V_{DS}/V_{Sub}$ = 0/1/0 V. The ideal I_D–V_{Sub} curve with trap-free insulating buffer stack is also shown for reference. (b) Simulated distribution of ionized donor and acceptor traps in the UID GaN and C-doped GaN with V_{Sub} of 0 V, −100 V (∼V_1), and −400 V (∼V_2) during the V_{Sub} sweep. Adapted from [92]*

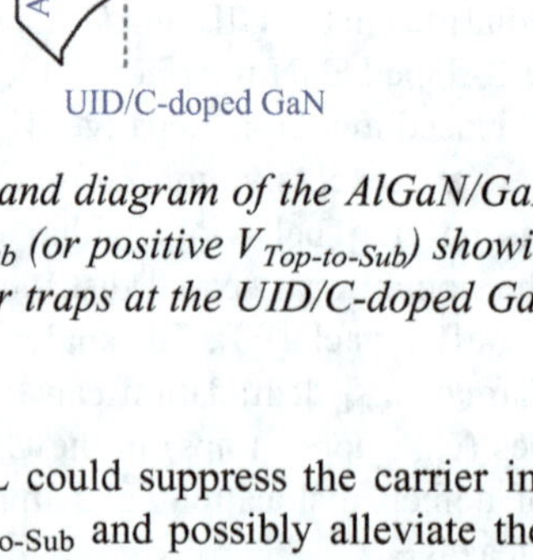

Figure 9.20 *Schematic band diagram of the AlGaN/GaN-on-Si device under negative V_{Sub} (or positive $V_{Top\text{-}to\text{-}Sub}$) showing the ionization of donor and acceptor traps at the UID/C-doped GaN junction. Adapted from [90]*

forming less defective NL could suppress the carrier injection from Si substrate under high positive $V_{Top\text{-}to\text{-}Sub}$ and possibly alleviate the buffer trapping-induced dynamic R_{ON} degradation [90].

Alternative physical models with regard to the charge transport and redistribution within the buffer stack have also been proposed to analyze the buffer-related

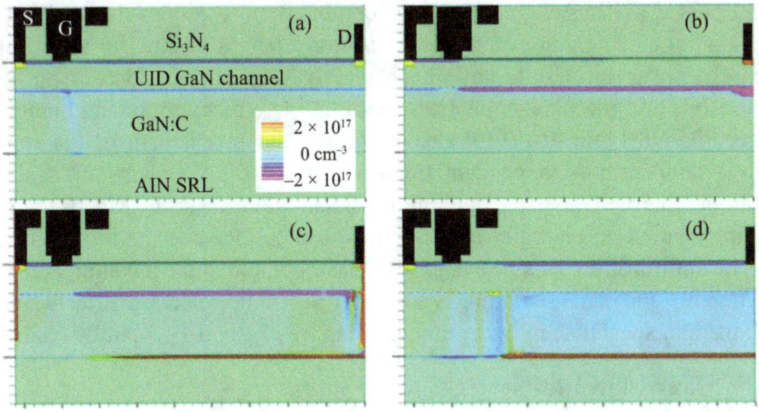

Figure 9.21 Net ionized charge distribution for the simulations with four types of buffer stacks. (a) Insulating buffer stack where there is no charge storage (no trapping). (b) A floating p-type C-doped GaN isolated from the 2DEG by a pn junction (no leakage paths through UID GaN channel). (c) Leakage under the source and drain contacts to the C-doped GaN layer. (d) Leakage through the UID GaN between the 2DEG and the C-doped layer along the entire length of the device. Adapted from [94]

dynamic R_{ON} degradation. Uren *et al.* [94] have found that a trap-assisted band-to-band process in the UID GaN layer can lead to flow of electrons into the 2DEG, whereas the released holes flowing in the C-doped GaN layer can neutralize the ionized acceptor traps and exposing donor traps or act as free holes. As shown in Figure 9.21 [94], the lower resistivity of the UID GaN layer can suppress the formation of negative charges at the top of the C-doped GaN layer and form a positively charged region at the bottom of the C-doped GaN layers, resulting in superior dynamic R_{ON}. It is noteworthy that the exact location of the positive charges varies in different buffer stacks, which depends on the crystalline defects, C doping concentration level, local band bending, and the leakage across each layer.

In summary, buffer trapping-induced dynamic R_{ON} degradation can be possibly suppressed by (1) introducing positive buffer charging through carefully designed donor traps or modified leakage in the UID GaN, and (2) increasing the barrier at the NL/Si junction or forming less defective NL.

9.4.2 Characterization techniques

9.4.2.1 Wafer-level tests

There are mainly two kinds of methods for dynamic R_{ON} measurement currently: wafer-level tests and board-level tests. Wafer-level tests typically feature the pulsed *I–V* measurement using on-wafer test equipment. Such equipment includes Keysight B1505A power device analyzer, AMCAD pulse *I–V* system, Auriga

AU4750 pulsed *I–V* system, etc., or their combination. It allows the study of R_{ON} transients after a switching event over an arbitrary length of time for a wafer-level device. The wafer-level tests have been widely used mainly focuses on device characteristics and mechanism analysis, with the R_{ON} recovery process monitored (e.g., see [95–98]). Jin *et al.* [95] have studied the mechanism of the trapping through a temperature-independent tunneling process on a short timescale and conventional thermally assisted detrapping on a longer time scale. Moens *et al.* [98] have implemented an on-wafer dynamic R_{ON} test to evaluate the buffer structure and do the optimization. Rossetto *et al.* [97] have compared the dynamic R_{ON} under soft-switching and hard-switching with the role of hot-electron effects demonstrated, using a special setup where the overlap of V_{GS} and V_{DS} can be controlled.

9.4.2.2 Board-level tests

Alternatively, custom-designed test boards, which can capture the fast transients within sub-μs and facilitate flexible test setup, have been developed to evaluate the switching and dynamic performance of lateral GaN devices [43,99–105]. It is noteworthy that an accurate assessment of dynamic R_{ON} within hundreds of nanoseconds is of particular interest and importance to GaN devices, which have the capability for high-frequency operation up to Megahertz.

A widely used circuit for dynamic R_{ON} extraction is a double pulse tester (DPT) where the DUT switches under hard switching conditions [43,99,100,106]. This circuit can be used to investigate the impact of key operation parameters on dynamic R_{ON}, including OFF-state time, OFF-state voltage, drain current, temperature, frequency, duty cycle, and gate voltage.

Considering the GaN devices with high switching-speed capability, the challenges for switching characterization evaluation primarily lie in the layout design of DPT board, measurement precision of switching waveforms, and data processing. The high-speed measurement issues in a phase-leg configuration were discussed comprehensively in [102], including guidelines for physical layout, switching voltage and current measurement requirements and the latest probes, the alignment of voltage and current (*V–I*) measurements, the impact of grounding effects induced by probes on the measurement, and the mechanisms causing additional switching losses by cross talk in a phase-leg.

Because the GaN device's turn-on switching loss is much higher than turn-off loss [107], the zero voltage turn-on circuit is preferred for GaN-based converter at high-frequency applications. Therefore, from the application perspective, current DPTs or other hard switching test circuits are not sufficient for a comprehensive investigation of dynamic R_{ON}. Thus, a custom-designed DPT which can simultaneously evaluate the hard and soft switching conditions is highly desirable.

As shown in Figure 9.22 [105], DPT with a half-bridge configuration for hard switching conditions and triangular current mode (TCM) test circuit for soft switching conditions are integrated into one test board to ensure that a GaN device can be tested by different circuits in the same test setup.

Figure 9.23(a) and (b) show the measured dynamic R_{ON} of commercial GaN HEMT under soft switching and hard switching. The measured R_{ON} values are

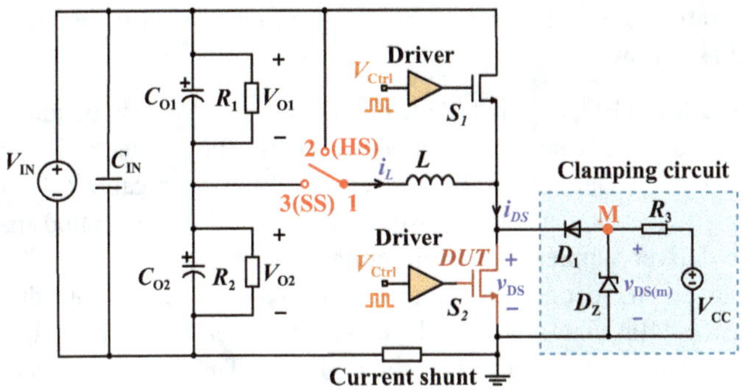

Figure 9.22 Test circuit in half-bridge configuration with clamping circuit. The circuit can realize hard switching (HS) test by connecting 1 to 2 and soft switching(SS) test by connecting 1 to 3. Adapted from [105]

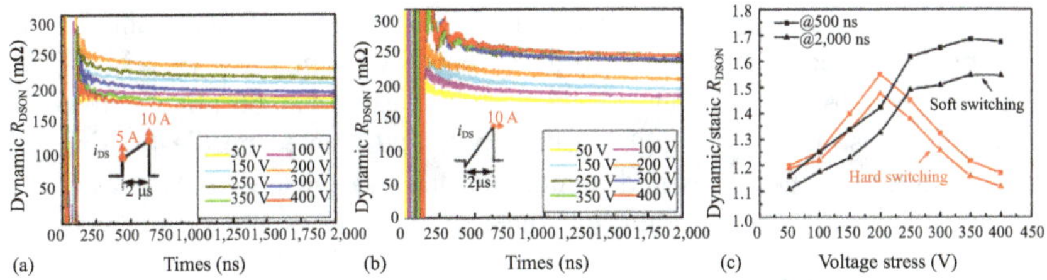

Figure 9.23 Time-resolved dynamic R_{ON} of GaN HEMT under (a) hard and (b) soft-switching conditions with different voltage stresses from 50 V to 400 V with the same drain current. Inset: Drain current variation during the measurement period. (c) Dynamic/static R_{DS} values at 500 ns and 2,000 ns after turn-on from different voltage stresses under hard and soft switching conditions. Adapted from [105]

normalized by the static R_{ON} extracted from output curves measured by curve tracer at room temperature. The extracted dynamic/static R_{ON} values at 500 ns and 2,000 ns after turn-on from different voltage stresses under hard and soft switching conditions are plotted in Figure 9.23(c). The device exhibits different dynamic R_{ON} under hard and soft switching conditions [105]. The device shows a higher dynamic R_{ON} with increasing voltage stress under soft-switching conditions, whereas the dynamic R_{ON} can be suppressed at high voltage stress under hard switching conditions.

9.4.3 Prospects and solutions

9.4.3.1 Dynamic R_{ON} suppression in lateral GaN-on-Si devices

In order to achieve low dynamic R_{ON} while maintaining high breakdown voltage in lateral GaN-on-Si devices, current collapse suppression techniques including the hole compensation and buffer stack optimization have been developed which are summarized as follows.

1. Panasonic Inc. [108,109] has developed a hybrid drain-embedded GIT (HD-GIT) with a p-GaN structure connected to the drain electrode. Injection of holes from the p-GaN region in the vicinity of the drain can effectively release the trapped electrons at the OFF-state. It has been demonstrated that the HD-GIT can deliver suppressed current collapse up to 850 V.
2. As discussed in Section 9.4.1, positive buffer charges by introducing donor traps in the UID GaN layer [90] or optimized leakage across the UID GaN layer [94] can suppress the current collapse. The impact of substrate bias polarity on current collapse also suggests that the buffer trapping-induced dynamic R_{ON} degradation can be possibly alleviated by suppressing the carrier injection from Si substrate with an increased effective barrier at the NL/Si junction or less defective NL [90].
3. FBH [58,110] has demonstrated that the trade-off between high voltage and dynamic performance can be optimized by placing an AlGaN back-barrier between the channel and C-doped buffer, such that the 2DEG channel is less prone to trapped charges during high-voltage switching.
4. It has been found that dynamic R_{ON} degradation can be alleviated with a floating substrate, as the high positive potential created on the floating substrate during OFF-state would reduce the drain-to-substrate bias stress [111]. The cross-talk issue merits further study.

9.4.3.2 Superior dynamic R_{ON} performance in vertical GaN-on-GaN devices

Despite the effectiveness of the current collapse suppression techniques, it is challenging to completely eliminate the trapping effects in the lateral GaN-on-Si devices. The recent emergence of free-standing GaN substrate has enabled the development of vertical GaN-on-GaN power devices. With zero lattice- and thermal expansion coefficient-mismatch between the epitaxial layer and substrate, a larger thickness and lower dislocation density can be realized in the GaN-on-GaN structure (Table 9.3) [112]. Furthermore, the GaN-on-GaN structure exhibits a lower thermal resistance [113], owing to the absence of thermal boundary resistance between the epitaxial layer and the substrate. In addition, the vertical GaN-on-GaN power devices can deliver high current capacity and high breakdown voltage [114,115]. With the vertical current flow, GaN-on-GaN devices are less prone to surface-induced current collapse (Figure 9.24) [116]. In recent years, the fast switching capability has been demonstrated in vertical GaN power devices [117–119].

The dynamic R_{ON} performance of the vertical GaN-on-GaN SBD has been quantitatively evaluated using a double pulse test circuit in [116]. Moreover, it is favorable to compare the vertical GaN-on-GaN diodes with the lateral GaN-on-Si diodes. As the lateral GaN-on-Si diodes are not commercially available, two types of the state-of-the-art E-mode GaN-on-Si transistors at 600 V/650 V ratings were

Table 9.3 Comparison between GaN-on-Si and GaN-on-GaN technologies

Properties	**GaN-on-Si**	**GaN-on-GaN**
Lattice mismatch (%)	17	0
CTE[a] mismatch (%)	54	0
Dislocation density (cm^{-2})	10^8–10^9	10^3–10^6
Max. epi-layer thickness (μm)	~5	≥40
Thermal resistance (°C·mm/W)	~30	~4

[a]CTE: Coefficient of thermal expansion.

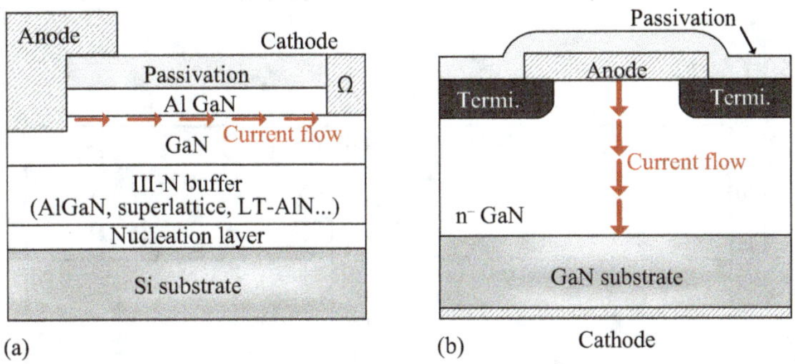

Figure 9.24 Schematic cross-sections of (a) lateral GaN-on-Si and (b) vertical GaN-on-GaN power rectifiers. Adapted from [116]

characterized using the same technique with the gate and source terminals electrically shorted. When operating in the reverse conduction mode, they can work as power diodes. Any possible surface- and buffer-trapping would adversely impact the 2DEG conductivity and dynamic R_{ON}.

For each DUT, dynamic R_{ON} has been characterized under different switching conditions with varying OFF-state bias (V_{OFF}), OFF-state time (t_{OFF}), load current, and measurement temperature (T_m). Dynamic R_{ON} is normalized with respect to the static R_{ON} which is extracted from the forward *I–V* curve of the fresh device.

1. Impact of OFF-state stress (V_{OFF} and t_{OFF})

   Figure 9.25 shows the time-resolved dynamic R_{ON}/static R_{ON} of vertical GaN-on-GaN SBD under different switching conditions. Owing to the ultra-short delay of the fast characterization approach, dynamic R_{ON} can be extracted at an ON-state time (t_{ON}) of only ~200 ns after switching from varying V_{OFF} of 50~500 V. To reveal the time-dependent transients, dynamic R_{ON} values at t_{ON} of 200 and 500 ns are extracted and summarized in Figure 9.25 (b). The vertical GaN-on-GaN SBD is free from dynamic R_{ON} degradation within the entire V_{OFF} range, even at only ~200 ns after switching from a high reverse stress bias up to 500 V.

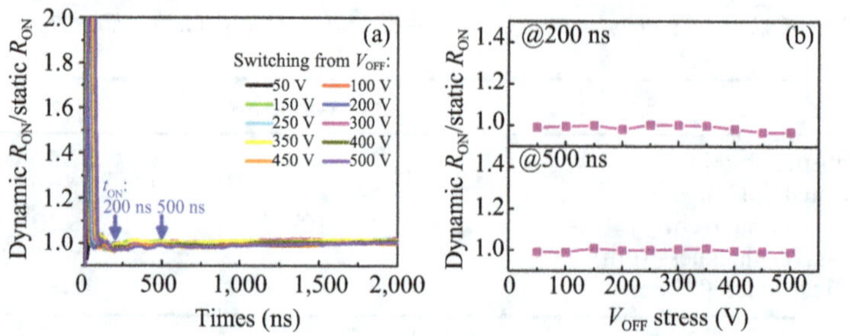

Figure 9.25 (a) Time-resolved dynamic R_{ON}/static R_{ON} of vertical GaN-on-GaN SBD and (b) extracted dynamic R_{ON}/static R_{ON} as a function of V_{OFF} at 200 and 500 ns after turn-on. Adapted from [116]

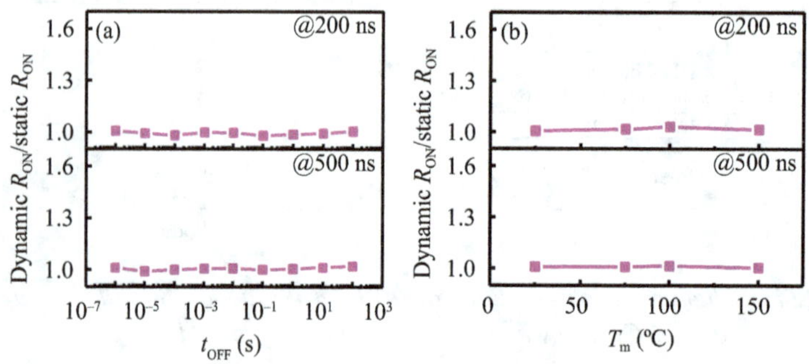

Figure 9.26 Extracted dynamic R_{ON}/static R_{ON} as a function of (a) t_{OFF} and (b) measurement temperature (T_m) at 200 and 500 ns after turn-on. Adapted from [116]

In addition, dynamic R_{ON} has been evaluated with t_{OFF} increased from 1 μs to 100 s (10^{-6}~10^2 s), as shown in Figure 9.26(a). The vertical GaN-on-GaN SBD exhibits negligible dynamic R_{ON} degradation throughout the entire t_{OFF} range (10^{-6}~10^2 s).

2. Impact of temperature

Temperature-dependent dynamic R_{ON} characterizations were carried out. Figure 9.26(b) shows dynamic R_{ON}/static R_{ON} the vertical GaN-on-GaN SBD at an elevated temperature up to 150 °C. The vertical GaN-on-GaN SBD is free from dynamic R_{ON} degradation within the entire temperature range (25 °C~150 °C).

Such superior dynamic performance in the vertical GaN-on-GaN device is attributed to the following factors. (1) In comparison with the lateral GaN devices in which the conductivity of the surface 2DEG channel can be adversely influenced by surface charges, GaN devices with a current flow along the vertical direction are intrinsically less prone to surface trapping effect. (2) The well-designed homo-epitaxial GaN drift layer with high crystalline quality and precisely controlled background/compensation impurities grown on free-standing GaN substrate can lead to minimal bulk trapping.

9.5 Summary

The recent progress and primary challenges in GaN power devices are reviewed and discussed in this chapter. The interface/border trapping at/near the dielectric/GaN interface in MIS-gate GaN transistor, as well as the hole deficiency in p-GaN HEMT with a Schottky gate, could result in the gate instability issue. The negatively charged states in the gate region would cause a positive V_{TH} shift, whereas the reduced gate overdrive voltage ($V_{G\_ON}-V_{TH}$) at a preset V_{G_ON} value could lead to the dynamic R_{ON} increase. The gate instability-induced dynamic R_{ON} degradation can be alleviated with a sufficient gate overdrive. For normally-off MIS-gate GaN transistors, the gate instability-induced dynamic R_{ON} can be further reduced by increasing the channel mobility in the recessed gate region and reducing the channel resistance. A monolithically integrated gate driver is desirable from the perspective of gate reliability for p-GaN HEMT with limited gate drive headroom. Meanwhile, bulk trapping in the III-nitride buffer stack could also cause dynamic R_{ON} degradation in GaN-on-Si devices. The buffer-related dynamic R_{ON} degradation can be suppressed by compensating the negative buffer traps with hole injection, or through buffer stack optimization with carefully introduced positive charges. On the other hand, the emerging vertical GaN-on-GaN power devices are capable of delivering superior dynamic performance, owing to the low susceptibility of the vertical current path to surface trapping and the minimized bulk trapping in the high-quality homo-epitaxial GaN drift layer with precisely controlled background/compensation impurities.

Note to the reader

This chapter is part of an edited book. To cite or use this chapter as a reference, please use the following format:

Yang S. and Han S. GaN metal-insulator-semiconductor field-effect transistors. In: Iannuzzo F (ed.), *Modern Power Electronic Devices: Physics, Applications, and Reliability*. Stevenage, UK: IET; 2020. pp. 295–330.

References

[1] [Online]. Available: https://docplayer.net/33950023-Market-forecasts-for-silicon-carbide-gallium-nitride-power-semiconductors-ihs-technology.html [Accessed Mar 10, 2019]

[2] Baliga B. J. "Gallium nitride devices for power electronic applications." *Semicond. Sci. Technol.* 2013, vol. 28, no. 7, p. 074011.
[3] Ishida M., Ueda T., Tanaka T., and Ueda D. "GaN on Si technologies for power switching devices." *IEEE Trans. Electron Devices*. 2013, vol. 60, no. 10, pp. 3053–3059.
[4] Shenai K. "Future prospects of widebandgap (WBG) semiconductor power switching devices." *IEEE Trans. Electron Devices*. 2015, vol. 62, no. 2, pp. 248–257.
[5] Millan J., Godignon P., Perpina X., Pérez-Tomás A., and Rebollo J. "A survey of wide bandgap power semiconductor devices." *IEEE Trans. Power Electron.* 2014, vol. 29, no. 5, pp. 2155–2163.
[6] Lidow A. "GaN transistors—Giving new life to Moore's law." in *Proc. IEEE Int. Symp. Power Semiconductor Devices IC's (ISPSD)*, Hong Kong, China, May 2015, pp. 1–6.
[7] Ambacher O., Smart J., Shealy J., *et al.* "Two-dimensional electron gases induced by spontaneous and piezoelectric polarization charges in N-and Ga-face AlGaN/GaN heterostructures." *J. Appl. Phys.* 1999, vol. 85, no. 6, pp. 3222–3233.
[8] Chen K. J., Häberlen O., Lidow A., *et al.* "GaN-on-Si power technology: Devices and applications." *IEEE Trans. Electron Devices*. 2017, vol. 64, no. 3, pp. 779–795.
[9] Infineon. *Infineon Power Products*. [Online]. Available: https://www.infineon.com/cms/en/product/ [Accessed Mar 10, 2019].
[10] Infineon CoolGaN Preliminary Values, Infineon, Neubiberg, Germany, 2017.
[11] GaN Systems. *GaN Systems Power Products*. [Online]. Available: https://gansystems.com/gan-transistors/ [Accessed Mar 10, 2019].
[12] Transphorm. *Transphorm Power Products*. [Online]. Available: https://www.transphormchina.com/en/products/ [Accessed Mar 10, 2019].
[13] Cree. *Cree Power Products*. [Online]. Available: https://www.wolfspeed.com/power/products [Accessed Mar 10, 2019].
[14] Rohm. *Rohm Power Products*. [Online]. Available: https://www.rohm.com [Accessed Mar 10, 2019].
[15] Panasonic. *Panasonic Power Products*. [Online]. Available: https://industrial.panasonic.com [Accessed Mar 10, 2019].
[16] EPC. *EPC Power Products*. [Online]. Available: https://epc-co.com/epc/Products/eGaNFETsandICs.aspx [Accessed Mar 10, 2019].
[17] [Online]. Available: https://www.yaskawa-global.com/wp-content/uploads/2015/07/150721_haifu_en.pdf [Accessed Mar 10, 2019].
[18] [Online]. Available: http://www.fujitsu.com/global/about/resources/news/press-releases/2015/1209-01.html [Accessed Mar 10, 2019].
[19] [Online]. Available: https://www.yaskawa-global.com/newsrelease/product/9021 [Accessed Mar 10, 2019].
[20] [Online]. Available: https://www.anker.com/deals/powerport_atom [Accessed Mar 10, 2019].

[21] [Online]. Available: http://www.toyoda-gosei.com/news/detail/?id=183 [Accessed Mar 10, 2019].
[22] [Online]. Available: https://www.energy.gov/sites/prod/files/2016/06/f32/edt067_zhu_2016_o_web.pdf [Accessed Mar 10, 2019].
[23] [Online]. Available: https://www.diaelec.co.jp/cp-bin/wordpress/wp-content/uploads/2018/01/Develops-Business-Card-sized-Thin-1-kW-Isolated-Bidirectional-DCDC-Converter.pdf [Accessed Mar 10, 2019].
[24] Liu L. and Edgar J. H. "Substrates for gallium nitride epitaxy." *Mater. Sci. Eng. R*. 2002, vol. 37, no. 3, pp. 61–127.
[25] Krost A. and Dadgar A. "GaN-based optoelectronics on silicon substrates." *Mater. Sci. Eng. B*. 2002, vol. 93, no. 1–3, pp. 77–84.
[26] Marcon D., Saripalli Y., and Decoutere S. "200mm GaN-on-Si epitaxy and e-mode device technology." in *IEEE IEDM Tech. Dig.*, Washington, DC, USA, Dec. 2015, pp. 16.2.1–16.2.4.
[27] Jones E. A., Wang F. F., and Costinett D. "Review of commercial GaN power devices and GaN-based converter design challenges." *IEEE J. Emerg. Sel. Topics Power Electron.* 2016, vol. 4, no. 3, pp. 707–719.
[28] Liu Z., Huang X., Lee F. C., and Li Q. "Simulation model development and verification for high voltage GaN HEMT in cascode structure." in *Proc. IEEE Energy Convers. Congr. Expo. (ECCE)*, Denver, CO, USA, Sept. 2013, pp. 3579–3586.
[29] Chen K. J. and Zhou C. "Enhancement-mode AlGaN/GaN HEMT and MIS-HEMT technology." *Phys. Status Solidi A*. 2011, vol. 208, no. 2, pp. 434–438.
[30] Uemoto Y., Hikita M., Ueno H., *et al.* "Gate injection transistor (GIT)—A normally-off AlGaN/GaN power transistor using conductivity modulation." *IEEE Trans. Electron Devices*. 2007, vol. 54, no. 12, pp. 3393–3399.
[31] Morita T., Yanagihara M., Ishida H., *et al.* "650 V 3.1 mΩcm^2 GaN-based monolithic bidirectional switch using normally-off gate injection transistor." in *IEEE IEDM Tech. Dig.,* Washington, DC, USA, Dec. 2007, pp. 865–868.
[32] Cai Y., Zhou Y., Chen K. J., and Lau K. M. "High-performance enhancement-mode AlGaN/GaN HEMTs using fluoride-based plasma treatment." *IEEE Electron Device Lett.* 2005, vol. 26, no. 7, pp. 435–437.
[33] Wang R., Cai Y., Tang C.-W., Lau K. M., and Chen K. J. "Enhancement-mode Si_3N_4/AlGaN/GaN MISHFETs." *IEEE Electron Device Lett.* 2006, vol. 27, no. 10, pp. 793–795.
[34] Saito W., Takada Y., Kuraguchi M., Tsuda K., and Omura I. "Recessed-gate structure approach toward normally off high-voltage AlGaN/GaN HEMT for power electronics applications." *IEEE Trans. Electron Devices*. 2006, vol. 53, no. 2, pp. 356–362.
[35] Oka T. and Nozawa T. "AlGaN/GaN recessed MIS-gate HFET with high-threshold-voltage normally-off operation for power electronics applications." *IEEE Electron Device Lett.* 2008, vol. 29, no. 7, pp. 668–670.
[36] Huang W., Li Z., Chow T., Niiyama Y., Nomura T., and Yoshida S. "Enhancement-mode GaN hybrid MOS-HEMTs with Ron,sp of 20 mΩ-cm^2."

in *Proc. IEEE Int. Symp. Power Semiconductor Devices IC's (ISPSD)*, Orlando, FL, USA, Jun. 2008, pp. 295–298.

[37] Huang S., Yang S., Roberts J., and Chen K. J. "Threshold voltage instability in Al_2O_3/GaN/AlGaN/GaN metal–insulator–semiconductor high-electron mobility transistors." *Jpn. J. Appl. Phys.* 2011, vol. 50, no. 11R, p. 110202.

[38] Lagger P., Ostermaier C., Pobegen G., and Pogany D. "Towards understanding the origin of threshold voltage instability of AlGaN/GaN MIS-HEMTs." in *IEEE IEDM Tech. Dig.*, San Francisco, CA, USA, Dec. 2012, pp. 13.1.1–13.1.4.

[39] Yang S., Tang Z., Wong K.-Y., *et al.* "Mapping of interface traps in high-performance Al_2O_3/AlGaN/GaN MIS-heterostructures using frequency-and temperature-dependent CV techniques." in *IEEE IEDM Tech. Dig.*, Washington, DC, USA, Dec. 2013, pp. 6.3.1–6.3.4.

[40] Lagger P., Reiner M., Pogany D., and Ostermaier C. "Comprehensive study of the complex dynamics of forward bias-induced threshold voltage drifts in GaN based MIS-HEMTs by stress/recovery experiments." *IEEE Trans. Electron Devices*. 2014, vol. 61, no. 4, pp. 1022–1030.

[41] Yang S., Lu Y., Wang H., Liu S., Liu C., and Chen K. J. "Dynamic gate stress-induced V_{TH} shift and its impact on dynamic R_{ON} in GaN MIS-HEMTs." *IEEE Electron Device Lett.* 2016, vol. 37, no. 2, pp. 157–160.

[42] Yang S., Liu S., Lu Y., and Chen K. J. "Trapping mechanisms in insulated-gate GaN power devices: Understanding and characterization techniques." *Phys. Status Solidi A*. 2017, vol. 214, no. 3, p. 1600607.

[43] Wang H., Wei J., Xie R., Liu C., Tang G., and Chen K. J. "Maximizing the performance of 650-V p-GaN gate HEMTs: Dynamic R_{ON} characterization and circuit design considerations." *IEEE Trans. Power Electron.* 2017, vol. 32, no. 7, pp. 5539–5549.

[44] Meneghini M., Hilt O., Wuerfl J., and Meneghesso G. "Technology and reliability of normally-off GaN HEMTs with p-type gate." *Energies*. 2017, vol. 10, no. 2, p. 153.

[45] Shi Y., Zhou Q., Cheng Q., *et al.* "Bidirectional threshold voltage shift and gate leakage in 650 V p-GaN AlGaN/GaN HEMTs: The role of electron-trapping and hole-injection." in *Proc. IEEE Int. Symp. Power Semiconductor Devices IC's (ISPSD)*, Chicago, IL, USA, May 2018, pp. 96–99.

[46] Shi Y., Zhou Q., Cheng Q., *et al.* "Carrier transport mechanisms underlying the bidirectional V_{TH} shift in p-GaN gate HEMTs under forward gate stress." *IEEE Trans. Electron Devices*. 2019, vol. 66, no. 2, pp. 876–882.

[47] He J., Tang G., and Chen K. J. "V_{TH} instability of p-GaN gate HEMTs under static and dynamic gate stress." *IEEE Electron Device Lett.* 2018, vol. 39, no. 10, pp. 1576–1579.

[48] Tang X., Li B., Moghadam H. A., Tanner P., Han J., and Dimitrijev S. "Mechanism of threshold voltage shift in p-GaN gate AlGaN/GaN transistors." *IEEE Electron Device Lett.* 2018, vol. 39, no. 8, pp. 1145–1148.

[49] Xie R., Yang X., Xu G., *et al.* "Switching transient analysis for normally-off GaN transistor with p-GaN gate in a phase-leg circuit." *IEEE Trans. Power Electron.* 2019, vol. 34, no. 4, pp. 3711–3728.

[50] Wei J., Xie R., Xu H., *et al.* "Charge storage mechanism of drain induced dynamic threshold voltage shift in p-GaN gate HEMTs." *IEEE Electron Device Lett.* 2019, vol. 40, no. 4, pp. 526–529.

[51] Binari S. C., Klein P., and Kazior T. E. "Trapping effects in GaN and SiC microwave FETs." *Proc. IEEE*. 2002, vol. 90, no. 6, pp. 1048–1058.

[52] Uren M. J., Cäsar M., Gajda M. A., and Kuball M. "Buffer transport mechanisms in intentionally carbon doped GaN heterojunction field effect transistors." *Appl. Phys. Lett.* 2014, vol. 104, no. 26, p. 263505.

[53] Chevtchenko S., Cho E., Brunner F., Bahat-Treidel E., and Würfl J. "Off-state breakdown and dispersion optimization in AlGaN/GaN heterojunction field-effect transistors utilizing carbon doped buffer." *Appl. Phys. Lett.* 2012, vol. 100, no. 22, p. 223502.

[54] Klein P., Binari S., Ikossi K., Wickenden A., Koleske D., and Henry R. "Current collapse and the role of carbon in AlGaN/GaN high electron mobility transistors grown by metalorganic vapor-phase epitaxy." *Appl. Phys. Lett.* 2001, vol. 79, no. 21, pp. 3527–3529.

[55] Choi Y. C., Pophristic M., Cha H.-Y., Peres B., Spencer M. G., and Eastman L. F. "The effect of an Fe-doped GaN buffer on off-state breakdown characteristics in AlGaN/GaN HEMTs on Si substrate." *IEEE Trans. Electron Devices*. 2006, vol. 53, no. 12, pp. 2926–2931.

[56] Hierro A., Kwon D., Ringel S., *et al.* "Optically and thermally detected deep levels in n-type Schottky and p+-n GaN diodes." *Appl. Phys. Lett.* 2000, vol. 76, no. 21, pp. 3064–3066.

[57] Lyons J. L., Janotti A., and Van de Walle C. G. "Carbon impurities and the yellow luminescence in GaN." *Appl. Phys. Lett.* 2010, vol. 97, no. 15, p. 152108.

[58] Wurfl J., Hilt O., Bahat-Treidel E., *et al.* "Techniques towards GaN power transistors with improved high voltage dynamic switching properties." in *IEEE IEDM Tech. Dig.*, Washington, DC, USA, Dec. 2013, pp. 6.1.1–6.1.4.

[59] Uren M. J., Moreke J., and Kuball M. "Buffer design to minimize current collapse in GaN/AlGaN HFETs." *IEEE Trans. Electron Devices*. 2012, vol. 59, no. 12, pp. 3327–3333.

[60] Moens P., Vanmeerbeek P., Banerjee A., *et al.* "On the impact of carbon-doping on the dynamic Ron and off-state leakage current of 650V GaN power devices." in *Proc. IEEE Int. Symp. Power Semiconductor Devices IC's (ISPSD)*, Hong Kong, China, Jun. 2015, pp. 37–40.

[61] Van de Walle C. G. and Neugebauer J. "First-principles calculations for defects and impurities: Applications to III-nitrides." *J. Appl. Phys.* 2004, vol. 95, no. 8, pp. 3851–3879.

[62] Vetury R., Zhang N. Q., Keller S., and Mishra U. K. "The impact of surface states on the DC and RF characteristics of AlGaN/GaN HFETs." *IEEE Trans. Electron Devices*. 2001, vol. 48, no. 3, pp. 560–566.

[63] Zhou C., Jiang Q., Huang S., and Chen K. J. "Vertical leakage/breakdown mechanisms in AlGaN/GaN-on-Si devices." *IEEE Electron Device Lett.* 2012, vol. 33, no. 8, pp. 1132–1134.

[64] Meneghini M., Vanmeerbeek P., Silvestri R., *et al.* "Temperature-dependent dynamic R_{ON} in GaN-based MIS-HEMTs: Role of surface traps and buffer leakage." *IEEE Trans. Electron Devices.* 2015, vol. 62, no. 3, pp. 782–787.

[65] Joh J. and Del Alamo J. A. "A current-transient methodology for trap analysis for GaN high electron mobility transistors." *IEEE Trans. Electron Devices*. 2011, vol. 58, no. 1, pp. 132–140.

[66] Jin D. and del Alamo J. A. "Mechanisms responsible for dynamic ON-resistance in GaN high-voltage HEMTs." in *Proc. IEEE Int. Symp. Power Semiconductor Devices IC's (ISPSD)*, Bruges, Belgium, Jun. 2012, pp. 333–336.

[67] Yang S., Lu Y., Liu S., Wang H., Liu C., and Chen K. J. "Impact of Vth shift on Ron in E/D-mode GaN-on-Si power transistors: Role of dynamic stress and gate overdrive." in *Proc. IEEE Int. Symp. Power Semiconductor Devices IC's (ISPSD)*, Prague, Czech Republic, Jun. 2016, pp. 263–266.

[68] Huang X., Du W., Lee F. C., Li Q., and Liu Z. "Avoiding Si MOSFET avalanche and achieving zero-voltage switching for cascode GaN devices." *IEEE Trans. Power Electron.* 2016, vol. 31, no. 1, pp. 593–600.

[69] Miura Y. and Matukura Y. "Investigation of silicon-silicon dioxide interface using MOS structure." *Jpn. J. Appl. Phys.* 1966, vol. 5, no. 2, pp. 180–180.

[70] Stathis J. H. and Zafar S. "The negative bias temperature instability in MOS devices: A review." *Microelectron. Rel.* 2006, vol. 46, no. 2, pp. 270–286.

[71] Reddy V., Krishnan A. T., Marshall A., *et al.* "Impact of negative bias temperature instability on digital circuit reliability." *Microelectron. Rel.*, 2006, vol. 45, no. 1, pp. 31–38.

[72] Hofmann K., Reisinger H., Ermisch K., *et al.* "Highly accurate product-level aging monitoring in 40nm CMOS." in *IEEE VLSI Technol. Tech. Dig.*, Honolulu, Hawaii, Jun. 2010, pp. 27–28.

[73] Schroder D. K. *Semiconductor Material and Device Characterization*, 3rd ed. Hoboken, NJ: Wiley; 2006.

[74] Yang S., Liu S., Lu Y., Liu C., and Chen K. J. "AC-capacitance techniques for interface trap analysis in GaN-based buried-channel MIS-HEMTs." *IEEE Trans. Electron Devices*. 2015, vol. 62, no. 6, pp. 1870–1878.

[75] Lansbergen G. P., Wong K. Y., Lin Y. S., *et al.* "Threshold voltage drift (PBTI) in GaN D-MODE MISHEMTs: Characterization of fast trapping components." in *IEEE Proc. Int. Rel. Phys. Symp.*, Waikoloa, HI, USA, Jun. 2014, pp. 6C.4.1–6C.4.6.

[76] Zhao K., Stathis J. H., Linder B. P., Cartier E., and Kerber A. "PBTI under dynamic stress: From a single defect point of view." in *IEEE Proc. Int. Rel. Phys. Symp.*, Monterey, CA, USA, Apr. 2011, pp. 4A.3.1–4A.3.9.

[77] Reisinger H., Grasser T., Ermisch K., Nielen H., Gustin W., and Schlünder C. "Understanding and modeling AC BTI." in *IEEE Proc. Int. Rel. Phys. Symp.*, Monterey, CA, USA, Apr. 2011, pp. 6A.1.1–6A.1.8.

[78] Fleetwood D. M. "'Border traps' in MOS devices." *IEEE Trans. Nucl. Sci.* 1992, vol. 39, no. 2, pp. 269–271.

[79] Yang S., Liu S., Liu C., Tang Z., Lu Y., and Chen K. J. "Thermally induced threshold voltage instability of III-nitride MIS-HEMTs and MOSC-HEMTs: Underlying mechanisms and optimization schemes." in *IEEE IEDM Tech. Dig.*, San Francisco, CA, USA, Dec. 2014, pp. 17.2.1–17.2.4.
[80] Hua M., Liu C., Yang S., *et al.* "Characterization of leakage and reliability of SiN_x gate dielectric by low-pressure chemical vapor deposition for GaN-based MIS-HEMTs." *IEEE Trans. Electron Devices*. 2015, vol. 62, no. 10, pp. 3215–3222.
[81] Hua M., Zhang Z., Wei J., *et al.* "Integration of LPCVD-SiN_x gate dielectric with recessed-gate E-mode GaN MIS-FETs: Toward high performance, high stability and long TDDB lifetime." in *IEEE IEDM Tech. Dig.*, San Francisco, CA, USA, Dec. 2016, pp. 10.4.1–10.4.4.
[82] Koley G., Tilak V., Eastman L. F., and Spencer M. G. "Slow transients observed in AlGaN/GaN HFETs: Effects of SiN_x passivation and UV illumination." *IEEE Trans. Electron Devices*. 2003, vol. 50, no. 4, pp. 886–893.
[83] Huang S., Jiang Q., Yang S., Zhou C., and Chen K. J. "Effective passivation of AlGaN/GaN HEMTs by ALD-grown AlN thin film." *IEEE Electron Device Lett.* 2012, vol. 33, no. 4, pp. 516–518.
[84] Xing H. G., Dora Y., Chini A., Heikman S., Keller S., and Mishra U. K. "High breakdown voltage AlGaN-GaN HEMTs achieved by multiple field plates." *IEEE Electron Device Lett.* 2004, vol. 25, no. 4, pp. 161–163.
[85] Chu R., Corrion A., Chen M., *et al.* "1200-V normally off GaN-on-Si field-effect transistors with low dynamic on-resistance." *IEEE Electron Device Lett.* 2011, vol. 32, no. 5, pp. 632–634.
[86] Marso M., Wolter M., Javorka P., Kordoš P., and Lüth H. "Investigation of buffer traps in an AlGaN/GaN/Si high electron mobility transistor by backgating current deep level transient spectroscopy." *Appl. Phys. Lett.* 2003, vol. 82, no. 4, pp. 633–635.
[87] Yang S., Zhou C., Jiang Q., Lu J., Huang B., and Chen K. J. "Investigation of buffer traps in AlGaN/GaN-on-Si devices by thermally stimulated current spectroscopy and back-gating measurement." *Appl. Phys. Lett.* 2014, vol. 104, no. 1, p. 013504.
[88] Moens P., Banerjee A., Uren M., *et al.* "Impact of buffer leakage on intrinsic reliability of 650V AlGaN/GaN HEMTs." in *IEEE IEDM Tech. Dig.*, Washington, DC, USA, Dec. 2015, pp. 35.2.1–35.2.4.
[89] Yacoub H., Fahle D., Eickelkamp M., *et al.* "Effect of stress voltage on the dynamic buffer response of GaN-on-silicon transistors." *J. Appl. Phys.* 2016, vol. 119, no. 13, p. 135704.
[90] Yang S., Zhou C., Han S., Wei J., Sheng K., and Chen K. J. "Impact of substrate bias polarity on buffer-related current collapse in AlGaN/GaN-on-Si power devices." *IEEE Trans. Electron Devices*. 2017, vol. 64, no. 12, pp. 5048–5056.
[91] Poblenz C., Waltereit P., Rajan S., Heikman S., Mishra U. K., and Speck J. S. "Effect of carbon doping on buffer leakage in AlGaN/GaN high electron mobility transistors." *J. Vac. Sci. Technol. B.* 2004, vol. 22, no. 3, pp. 1145–1149.

[92] Yang S., Han S., Sheng K., and Chen K. J. "Dynamic on-resistance in GaN power devices: Mechanisms, characterizations and modeling." *IEEE J. Emerg. Sel. Topics Power Electron.* 2019, vol. 7, no. 3, pp. 1425–1439.

[93] Pooth A., Uren M. J., Cäsar M., Martin T., and Kuball M. "Charge movement in a GaN-based hetero-structure field effect transistor structure with carbon doped buffer under applied substrate bias." *J. Appl. Phys.* 2015, vol. 118, no. 21, p. 215701.

[94] Uren M. J., Karboyan S., Chatterjee I., *et al.* "Leaky dielectric" model for the suppression of dynamic R_{ON} in carbon-doped AlGaN/GaN HEMTs." *IEEE Trans. Electron Devices*. 2017, vol. 64, no. 7, pp. 2826–2834.

[95] Jin D. and del Alamo J. A. "Methodology for the study of dynamic ON-resistance in high-voltage GaN field-effect transistors." *IEEE Trans. Electron Devices*. 2013, vol. 60, no. 10, pp. 3190–3196.

[96] Wang M., Yan D., Zhang C., *et al.* "Investigation of Surface- and Buffer-Induced Current Collapse in GaN High-Electron Mobility Transistors Using a Soft Switched Pulsed *I-V* Measurement." *IEEE Electron Device Lett.* 2014, vol. 35, no. 11, pp. 1094–1096.

[97] Rossetto I., Meneghini M., Tajalli A., *et al.* "Evidence of hot-electron effects during hard switching of AlGaN/GaN HEMTs." *IEEE Trans. Electron Devices*. 2017, vol. 64, no. 9, pp. 3734–3739.

[98] Moens P., Uren M. J., Banerjee A., *et al.* "Negative dynamic Ron in AlGaN/GaN power devices." in *Proc. IEEE Int. Symp. Power Semiconductor Devices IC's (ISPSD)*, Sapporo, Japan, May/Jun. 2017, pp. 97–100.

[99] Badawi N., Hilt O., Bahat-Treidel E., Böcker J., Würfl J., and Dieckerhoff S. "Investigation of the dynamic on-state resistance of 600 V normally-off and normally-on GaN HEMTs." *IEEE Trans. Ind. Appl.* 2016, vol. 52, no. 6, pp. 4955–4964.

[100] Cai Y., Forsyth A. J., and Todd R. "Impact of GaN HEMT dynamic on-state resistance on converter performance." in *Proc. IEEE Appl. Power Electron. Conf. Expo. (APEC)*, Tampa, FL, USA, Mar. 2017, pp. 1689–1694.

[101] Peng H., Ramabhadran R., Thomas R., and Schutten M. J. "Comprehensive switching behavior characterization of high speed Gallium Nitride E-HEMT with ultra-low loop inductance." in *Proc. IEEE Workshop Wide Bandgap Power Devices Appl. (WiPDA)*, Albuquerque, NM, USA, Oct./Nov. 2017, pp. 116–121.

[102] Zhang Z., Guo B., Wang F. F., Jones E. A., Tolbert L. M., and Blalock B. J. "Methodology for wide band-gap device dynamic characterization." *IEEE Trans. Power Electron.* 2017, vol. 32, no. 12, pp. 9307–9318.

[103] Cappello T., Santarelli A., and Florian C. "Dynamic R_{ON} characterization technique for the evaluation of thermal and off-state voltage stress of GaN switches." *IEEE Trans. Power Electron.* 2018, vol. 33, no. 4, pp. 3386–3398.

[104] Li K., Evans P. L., and Johnson C. M. "Characterisation and modeling of gallium nitride power semiconductor devices dynamic on-state resistance." *IEEE Trans. Power Electron.* 2018, vol. 33, no. 6, pp. 5262–5273.

[105] Li R., Wu X., Yang S., and Sheng K. "Dynamic on-state resistance test and evaluation of GaN power devices under hard and soft switching conditions by double and multiple pulses." *IEEE Trans. Power Electron.* 2019, vol. 34, no. 2, pp. 1044–1053.

[106] Yao T. and Ayyanar R. "A multifunctional double pulse tester for cascode GaN devices." *IEEE Trans. Ind. Electron.* 2017, vol. 64, no. 11, pp. 9023–9031.

[107] Huang X., Liu T., Li B., Lee F. C., and Li Q. "Evaluation and applications of 600V/650V enhancement-mode GaN devices." in *Proc. IEEE Workshop Wide Bandgap Power Devices Appl. (WiPDA)*, Blacksburg, VA, USA, Nov. 2015, pp. 113–118.

[108] Kaneko S., Kuroda M., Yanagihara M., *et al.* "Current-collapse-free operations up to 850 V by GaN-GIT utilizing hole injection from drain." in *Proc. IEEE Int. Symp. Power Semiconductor Devices IC's (ISPSD)*, Hong Kong, China, May 2015, pp. 41–44.

[109] Tanaka K., Morita T., Umeda H., *et al.* "Suppression of current collapse by hole injection from drain in a normally-off GaN-based hybrid-drain-embedded gate injection transistor." *Appl. Phys. Lett.* 2015, vol. 107, no. 16, p. 163502.

[110] Hilt O., Bahat-Treidel E., Cho E., Singwald S., and Würfl J. "Impact of buffer composition on the dynamic on-state resistance of high-voltage AlGaN/GaN HFETs." in *Proc. IEEE Int. Symp. Power Semiconductor Devices IC's (ISPSD)*, Bruges, Belgium, Jun. 2012, pp. 345–348.

[111] Tang G., Wei J., Zhang Z., *et al.* "Dynamic R_{ON} of GaN-on-Si lateral power devices with a floating substrate termination." *IEEE Electron Device Lett.* 2017, vol. 38, no. 7, pp. 937–940.

[112] Kizilyalli I. C., Bui-Quang P., Disney D., Bhatia H., and Aktas O. "Reliability studies of vertical GaN devices based on bulk GaN substrates." *Microelectron. Rel.* 2015, vol. 55, no. 9–10, pp. 1654–1661.

[113] Killat N., Montes M., Pomeroy J. W., *et al.* "Thermal properties of AlGaN/GaN HFETs on bulk GaN substrates." *IEEE Electron Device Lett.* 2012, vol. 33, no. 3, pp. 366–368.

[114] Kizilyalli I. C., Edwards A. P., Nie H., Bui-Quang P., Disney D., and Bour D. "400-A (pulsed) vertical GaN p-n diode with breakdown voltage of 700 V." *IEEE Electron Device Lett.* 2014, vol. 35, no. 6, pp. 654–656.

[115] Ohta H., Hayashi K., Horikiri F., Yoshino M., Nakamura T., and Mishima T. "5.0 kV breakdown-voltage vertical GaN p–n junction diodes." *Jpn. J. Appl. Phys.* 2018, vol. 57, no. 4S, p. 04FG09.

[116] Han S., Yang S., Li R., Wu X., and Sheng K. "Current-collapse-free and fast reverse recovery performance in vertical GaN-on-GaN Schottky barrier diode." *IEEE Trans. Power Electron.* 2019, vol. 34, no. 6, pp. 5012–5018.

[117] Ueno M., Yoshimoto S., Ishihara K., *et al.* "Fast recovery performance of vertical GaN Schottky barrier diodes on low-dislocation-density GaN substrates." in *Proc. IEEE Int. Symp. Power Semiconductor Devices IC's (ISPSD)*, Waikoloa, HI, USA, Jun. 2014, pp. 309–312.

[118] Shibata D., Kajitani R., Ogawa M., *et al.* "1.7 kV/1.0 mΩcm^2 normally-off vertical GaN transistor on GaN substrate with regrown p-GaN/AlGaN/GaN semipolar gate structure." in *IEEE IEDM Tech. Dig.*, San Francisco, CA, USA, 2016, pp. 10.1.1–10.1.4.

[119] Yang S., Han S., and Sheng K. "1 kV/1.3 mΩ·cm^2 vertical GaN-on-GaN Schottky barrier diodes with high switching performance." in *Proc. IEEE Int. Symp. Power Semiconductor Devices IC's (ISPSD)*, Chicago, IL, USA, May 2018, pp. 272–275.

Chapter 10
Gallium nitride transistors: applications and vertical solutions

Giorgia Longobardi[1,2]

Gallium nitride (GaN)-based transistors are among the most promising candidates to substitute Si-based solutions in the 600–1,200 V range of applications. This chapter starts with a brief introduction to the main advantages of GaN-based power transistors over those based on silicon or other wide bandgap materials. The main market segments where GaN power devices could have a significant share in the near future will be discussed. GaN lateral devices will be briefly presented. The chapter will then dwell on the vertical GaN technology discussing all the currently available solutions, their advantages, and their limitations.

10.1 Introduction

It is well recognized that improvements in power system performance in terms of efficiency, volume, and weight are strongly driven by improvements in the semiconductor switch performance. Silicon devices have long been the dominant choice for high-voltage switching devices. However, with the continuous demand for higher power density, higher temperature, higher frequencies, and higher efficiency, the Si-based power systems are fast approaching their theoretical limits. Wide-bandgap semiconductors, particularly silicon carbide (SiC) and gallium nitride (GaN), are attracting much attention in the power electronics field because they offer several potential advantages over silicon devices. They are able to operate at high switching frequency (>100 kHz), higher operating temperature (>150 °C), and potentially higher blocking voltages [1,2].

10.2 Advantages of GaN for power devices

When assessing the performance of GaN-based transistors, one has to consider not only the theoretical advantage related to the material properties but also the

[1]Engineering Department, University of Cambridge, Cambridge, UK
[2]Cambridge GaN Devices Ltd, Cambridge, UK

operation of the power transistor in a real application. The advantages of using GaN as a material for power application derive from three main points:

1. Material properties
2. Heterostructure-type device architecture
3. Simplified and more efficient system configuration and external components

All these three aspects will be discussed to provide a complete picture of the potential of GaN.

10.2.1 Material device and system-level benefit of GaN

Table 10.1 shows the material properties of the main semiconductors used for power applications. One of the first advantages that one could notice when comparing the physical properties of wide-bandgap (WBG) semiconductors with those of silicon is the low carrier intrinsic concentration, n_i. The direct consequence of having a low n_i is a much lower off-state leakage current which in turn reduces the off-state power dissipation allowing the operation at a higher temperature without thermal runaway and with reduced cooling requirements. In power supplies, the off-state leakage also affects the standby power supply, an important system parameter. Nevertheless, in GaN devices, it has been demonstrated that the off-state leakage is dependent on n_i only at low off-state biases, while for higher voltages, the leakage is dominated by trap-related mechanisms [3]. However, with the continuous improvement both in material quality and device design for the mitigation of trap induced leakage current [4], it has been possible to re-establish a direct correlation between low n_i and low off-state leakage in GaN and up to 600 V.

Another important benefit of WBG semiconductor is the higher critical electric field (3.3 MV/cm in GaN compared to ~0.3 MV/cm in Si). The high breakdown strength allows, in fact, the design of shorter drift regions for a given blocking voltage when compared to silicon technologies. For unipolar devices, this results in a drastic reduction in the specific on-resistance, R_{on}, accompanied by lower capacitance per current performance. Indeed, the reduction in R_{on} for a fixed breakdown voltage, V_{BR}, leads to a net reduction in the device area, which in turn translates into lower capacitances. The associated switching losses are then reduced, and a higher switching operation frequency is enabled. This argument is applicable to

Table 10.1 Physical properties of different semiconductors for high-voltage devices

Physical property	Si	4H-SiC	GaN	Diamond
Bandgap (eV)	1.1	3.26	3.4	5.45
Intrinsic carrier concentration (cm^{-3})	1.5×10^{10}	8.2×10^{-9}	1.9×10^{-10}	1.6×10^{-27}
Breakdown field (MV/cm)	0.3	3	3.3	5.6
Electron mobility (cm^2/Vs)	1,350	700	1,700	19,000
Thermal conductivity (W/cmK)	1.5	4.5	1.3	20

both lateral and vertical GaN devices. In addition, for lateral devices, the confinement of a large density of electrons (~9×10^{12} cm^{-2}) in a two-dimensional quantum layer, unobstructed by large ion dopants, results in high carrier mobilities (1,600–2,000 cm^2/Vs), which together with the small dimensions of the transistors result in further reductions in the R_{on}. Given the small surface dimensions of the transistors, the input (Ciss), output (Coss), and transfer (Crss) capacitances of GaN transistors are all extremely small (~pF). As a consequence, the turn-off and turn-on times are significantly reduced and so are the switching losses [5,6]. The overall power circuit adopting GaN transistors can, therefore, benefit from both smaller passive components, such as inductors, transformers or capacitors, and smaller (or nonexistent) cooling sinks which enable a significant reduction in the overall weight and volume of the power system.

10.3 GaN applications and market trends

10.3.1 Applications and market value

The majority of the applications currently adopting GaN transistors are focused on the low-voltage DC–DC converters, typically point-of-loads (POL). A huge potential is, however, in the 600 V market, where GaN devices could be used for power supply and power factor correction (PFC) applications. The PFC segment is expected to grow thanks to more stringent regulations for energy efficiency and to the reduction in the cost of such circuits. Given the impressive predictions of 91% compound annual growth rate (CAGR) in 2022, GaN is expected to surpass silicon in high-end 600 V power supply and PFC applications and gain some territory in all of that market with the exception of the very low margins/low-performance end of it. In 2 years, GaN is envisaged to penetrate other rich sectors such as photovoltaic (PV) modules and motor control inverters. Indeed, one of the most wide-spread applications for 600 V HEMTs is in the inverter drive circuitry for motors. Two of the largest potential volume applications are appliances and electric vehicles (EV) or hybrid electric vehicles (HEV). It was shown that in a 400 W motor drive inverter circuit, using GaN HEMTs versus the classical silicon IGBTs, the on-state losses were reduced by a factor of six, while at the same time the switching losses were halved [7]. Such improvements in power handling capability result in a massive increase in the inverter power density (by more than one order of magnitude) and come with a significant increase in efficiency, reduced system volume, weight, and ultimately reduced system price. Figure 10.1 highlights the market segments and the applications in each market segment where GaN-based transistors can be adopted. Figure 10.2 shows the market split by application in 2016 and 2022 [8] clearly indicating an increase in the share of the automotive market. It is important to note how the automotive market is expected to grow due to the increased interest in the development of electric and hybrid electric vehicles as a consequence of the increasing regulations in terms of increased efficiency and reduced CO_2 emission. The overall market share for GaN transistors is growing from virtually nothing today to a staggering \$500 M as shown in Figure 10.3, which depicts the forecasted GaN market values until 2022.

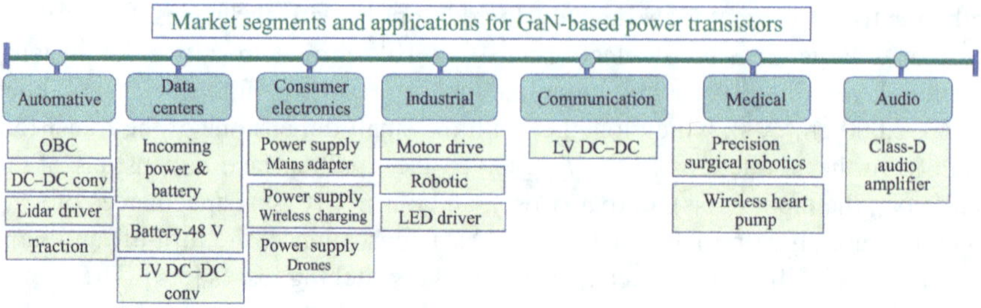

Figure 10.1 Market segments and applications for GaN-based power transistors

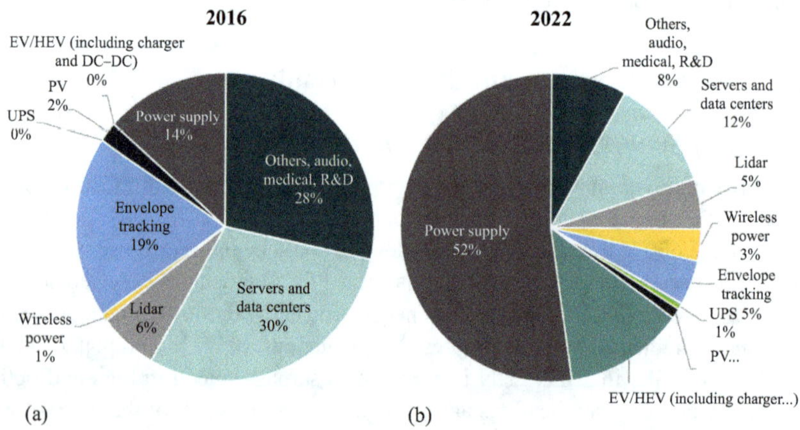

Figure 10.2 Market share for GaN split by applications in 2016 and 2022. Adapted from [8]

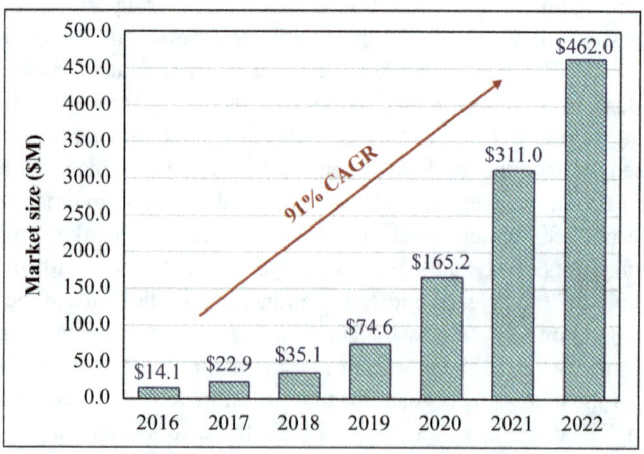

Figure 10.3 Evolution of the market share for GaN devices. Adapted from [8]

10.4 GaN power HEMT

10.4.1 GaN heterostructure-based transistors

The high electron mobility transistor (HEMT), also known as the heterostructure FET (HFET) or the MOdulation-Doped FET (MODFET), is a lateral device based on the physics of the heterostructure. The most basic GaN transistor is a normally-on (or equivalently depletion mode) HEMT, as depicted in Figure 10.4(a). One should notice that the HEMT is a lateral technology where the current flows at the surface of the transistor in the so-called two-dimensional electron gas (2DEG). This 2DEG is formed at the interface between the AlGaN and GaN layer due to the high-density positive piezo-polarization charges (~9×10^{12} cm^{-2}) present at that interface [9,10], and very importantly, it is formed without any intentional doping resulting in extremely high carrier mobility (up to 2,000 cm^2/Vs). The AlGaN-based transition layer (or multiple layers with different mole fractions grown on top of each other, above the nucleation AlN nucleation layer) that separates the AlGaN/GaN top heterostructure from the Si substrate plays a significant role in the operation of the device. The primary role of this layer is to adapt the lattice mismatch and compensate the stress-induced deformation between the top GaN/AlGaN layers and the silicon substrate. However, the composition and thickness of this layer significantly affect the reliability and off-state leakage of GaN transistors [3,11].

Normally-off (or equivalently enhancement mode) devices are preferred to normally-on solutions for the safety of operation, especially for EV/HEV applications. In addition, normally-off devices are significantly easier to control and do not require a negative voltage to keep them in the off-state. A full classification of normally-off designs, their advantages, and electrical performance can be found in

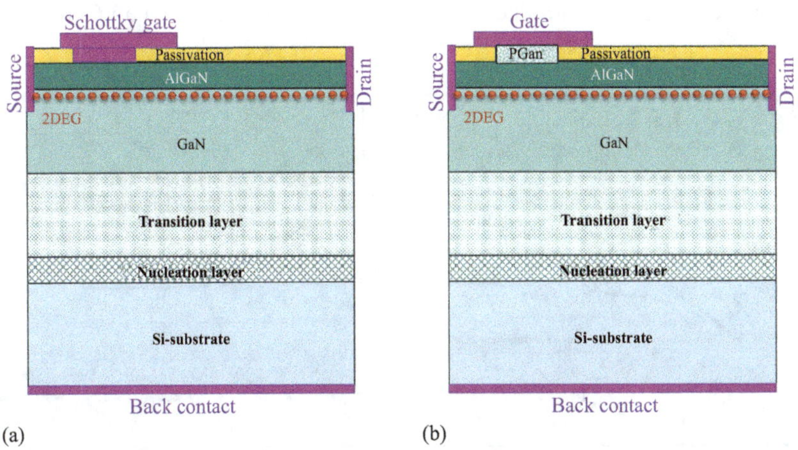

Figure 10.4 (a) Cross-section of normally-on HEMT and (b) cross-section of normally-off p-GaN gate transistors

[10]. The main normally-off devices are the MIS (metal-insulator-semiconductor) FET and the so-called p-GaN gate HEMT. In the MISFET trench gate structure, the AlGaN layer is etched to the AlGaN/GaN interface prior to the deposition of the gate oxide layer and the gate metal. When a positive gate bias is applied, an inversion channel is created underneath the gate oxide joining the existent 2DEG channels along the source-to-gate and gate-to-drain paths. Due to the presence of an insulator layer underneath the gate contact, these devices feature very low gate leakage current, when compared to the other GaN-based solutions. However, the electrical instability of the insulator/GaN interface has hampered the implementation of these devices in real power systems. In the most common of the normally-off technologies, the p-GaN gate, an 80–100 nm p-doped (Mg) layer of GaN is grown selectively on top of the AlGaN layer and in the area beneath the gate contact, as shown in Figure 10.4(b). The p-layer depletes the channel at zero gate bias, enabling the normally-off operation. The doping concentration and activation energy as well as the p-layer thickness are all important parameters that define not only the threshold voltage of the transistor but also the gate leakage and the instability of the gate terminal [12,13]. Recent research has also shown that the nature of the contact of the gate metallization with the p-GaN is equally important. A Schottky contact ensures lower leakage current but is more prone to instabilities of the threshold voltage when the gate swings between high potentials and 0 V [14]. The ohmic contact solves this problem, but threshold instabilities can still appear when the device is exposed to prior high-voltage stress applied to the drain [13]. Figure 10.5(a) and (b) shows the output characteristics and capacitances of a typical p-GaN gate transistor. From the *I–V* characteristics, one can note that in the region of interest (in this particular case around 8 A), the on-state resistance is almost unchanged when the gate voltage is above 5 V. This means that the 2DEG channel under the gate is well-modulated and its contribution to the overall on-state resistance no longer changes when a gate overdrive is applied. In this technology, the gate voltage overdrive is limited to around 7 V, as beyond that the pGaN/AlGaN

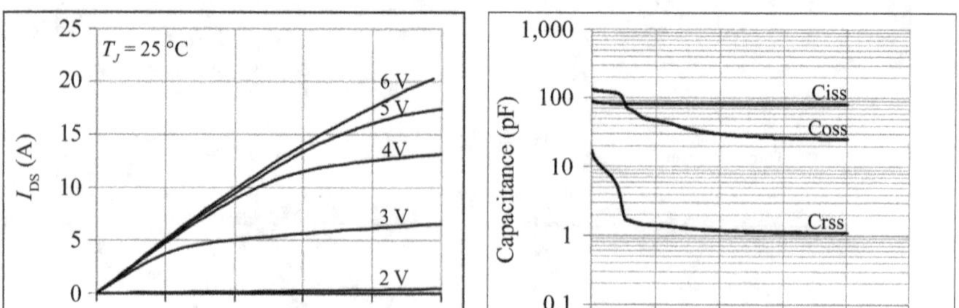

Figure 10.5 (a) Output characteristics and (b) capacitances of a normally-off pGaN gate HEMT

junction opens and very high injection of carriers occur through the gate. This is not the case in MISFETs, where the voltage overdrive is only limited by reliability effects such as time-dependent dielectric breakdown (TDDB). As shown in Figure 10.5(b), the capacitances in HEMTs are particularly lower than in silicon devices for the same voltage class and current ratings. In particular against super junction devices, the Miller and output capacitances in GaN devices present a smoother variation with the drain voltage, which otherwise could create stability problems in the system.

10.5 Vertical GaN transistors

Commercially available GaN transistors are all based on lateral technologies. Their principle of operation is based on the lateral conduction along the *x*-direction of the heterostructure formed between AlGaN and GaN layers, which is located only a few nanometers away from the surface of the transistor (20–50 nm) [10]. In a lateral technology, the scaling of the breakdown voltage (BV) is achieved laterally by increasing the gate-to-drain distance (drift region) and vertically by increasing the thickness of the buffer layer grown on the substrate. In a vertical technology, the current flows vertically from the drain to the source along the *y*-direction. A comparison between lateral and vertical conduction is shown in Figure 10.6.

There are several potential advantages associated with vertical technology:

1. The scaling of the BV goes with the vertical drift layer thickness without sacrificing surface wafer area.

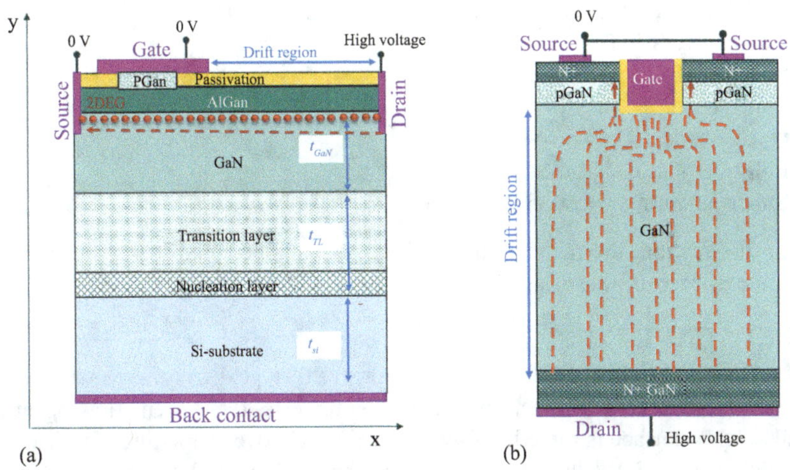

Figure 10.6 Cross-section of (a) lateral p-GAN device and (b) vertical trench-based GaN device

2. The increased distance between the high-voltage terminal and the surface isolation pushes the maximum electric field away from the surface thus reducing the trapping phenomena that affect the electrical performance in switching operations.
3. The heat generation and dissipation are more uniformly distributed avoiding critical hot-spots in the device that could cause premature failure.
4. They have shown to have avalanche breakdown capability that is missing in the lateral devices.

For these reasons, there has been an increased effort in recent years to design and fabricate GaN vertical power devices that could deliver the promised high breakdown voltages in excess of 1 kV and current levels in excess of 150 A. Such specifications would allow GaN to enter the higher voltage applications in automotive and motor control markets where it could compete with silicon carbide and silicon insulated gate bipolar transistors (IGBTs). Today most of the research in GaN vertical devices focuses on demonstrating the potential for high-voltage capability together with high current densities. Nevertheless, high currents cannot be achieved presently because of the poor quality of the large-area GaN wafers (above 2 inches), which impacts on the yield and performance of any vertical GaN devices fabricated in such wafers. To overcome the poor quality of the GaN substrates, some research is conducted toward quasi-vertical devices based on different substrates such as silicon, sapphire, or SiC.

10.5.1 Fabricated solutions for vertical and quasi-vertical GaN FETs

GaN vertical transistors can be classified into the following main categories:

- CAVET (current aperture vertical electron transistor)
- Trench MOSFET
- Slanted HEMT p-GAN gate
- FinFET
- Quasi-vertical GaN-on-Si
- Fully vertical GaN-on-Si
- Super junction vertical GaN device

In this section, the main design features, advantages, and the associated performance of each of these categories of devices will be discussed.

10.5.1.1 CAVET

This structure was first proposed by Mishra in 2004 [15]. As the name CAVET (current aperture vertical electron transistor) suggests, the current flowing at the AlGaN/GaN interface is forced to flow vertically in between apertures fabricated in the GaN layers and it is then collected by the drain. Figure 10.7(a) shows the basic design and the current flow direction for the CAVET transistor. Although extremely innovative, this solution not only required an extra etch-step but was also based on normally-on gate technology. Note also that this solution does not take

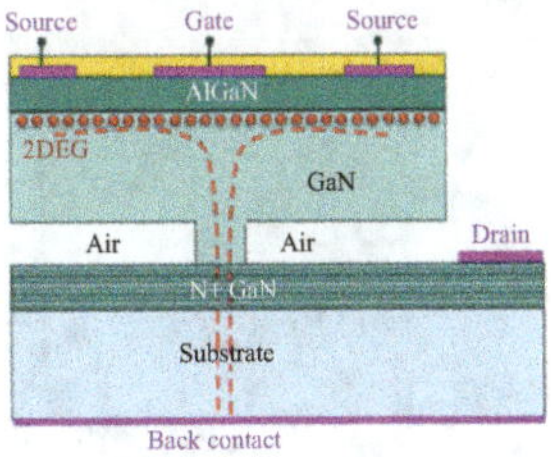

(a) CAVET with air

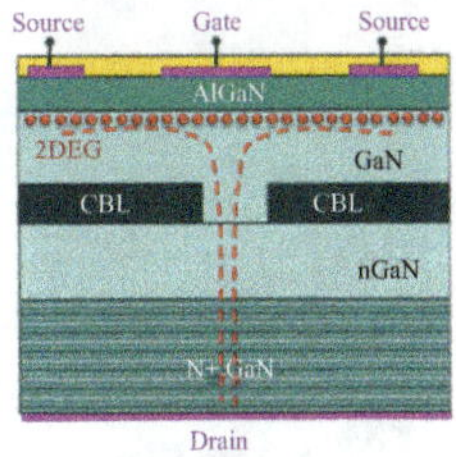

(b) CAVET with CBL (current blocking layer)

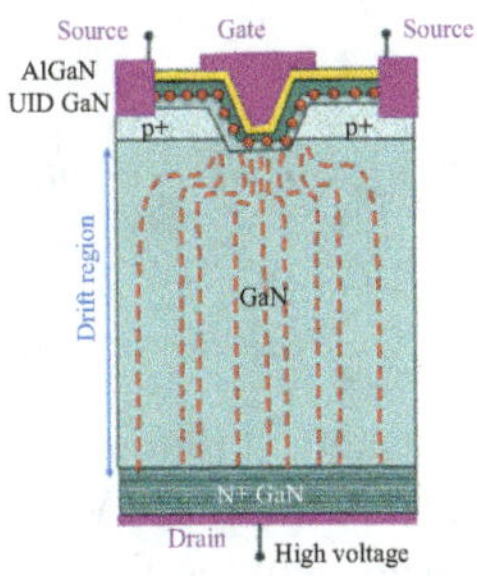

(c) Trench MOSFET 1

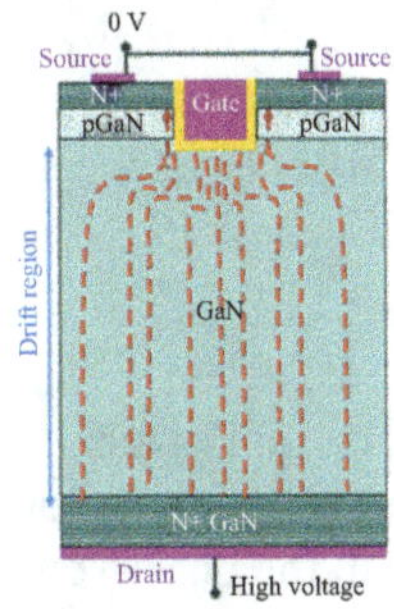

(d) Trench MOSFET 2

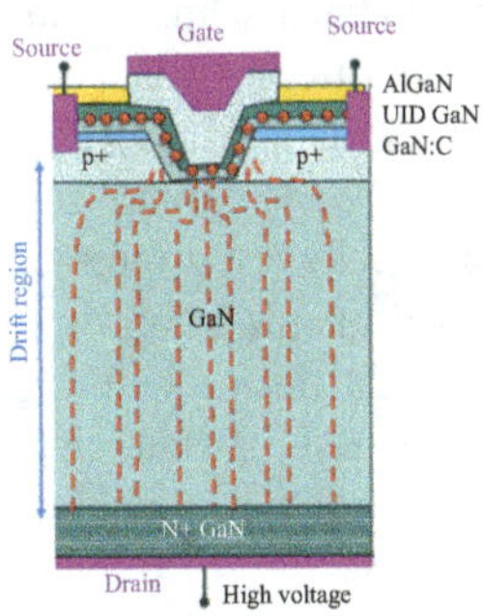

(e) Slanted HEMT P-GaN gate

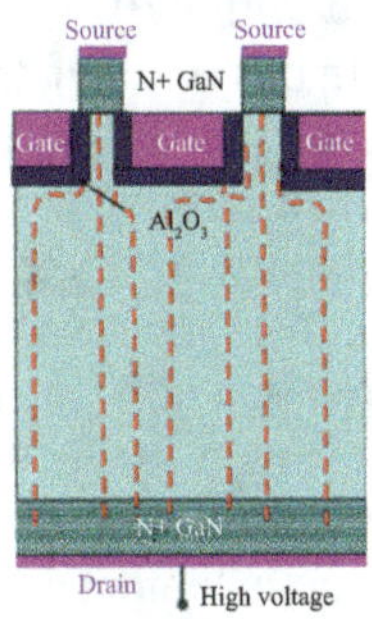

(f) FinFET

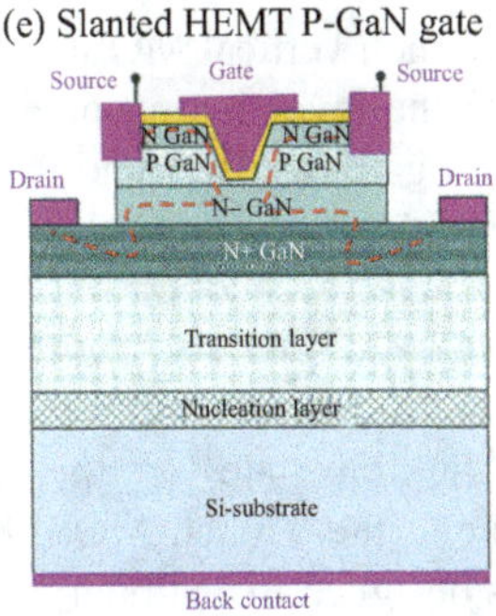

(g) Quasi-vertical GaN

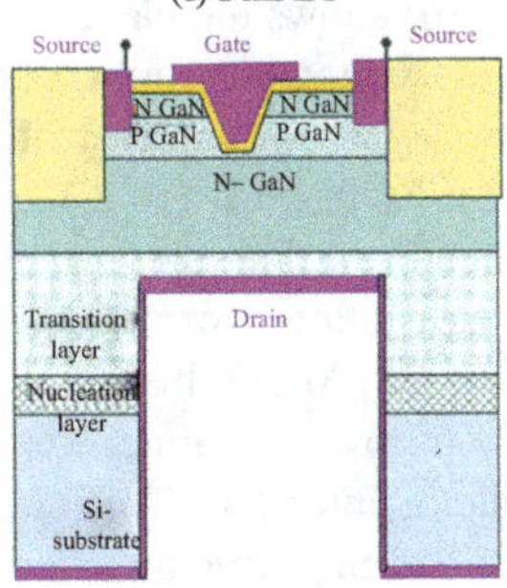

(h) Fully vertical GaN-on-Si membrane

Figure 10.7 Cross-section of the main GaN vertical devices

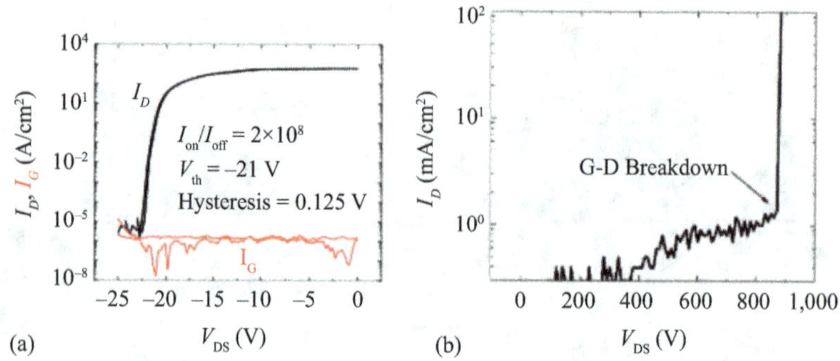

Figure 10.8 (a) On-state and (b) off-state characteristics of the trench CAVET in Figure 10.7 (c). Adapted from [18]

advantage of the full vertical dimension of the device as the current is not collected at the bottom of the transistor. In many ways, this first form of CAVET could be associated with the quasi-vertical designs discussed below. In the following years, more advanced CAVET solutions were proposed. In Figure 10.7(b), the apertures are substituted with CBL (current blocking layers) formed with implantations of Mg dopants. In this solution, the electrons are collected directly at the bottom of the transistor [16,17]. In [17], a BV in excess of 1 kV was reported with R_{on} as low as 2.6 mΩ cm^2 and positive threshold voltage $V_{th} = 0.5$ V. The main drawbacks of these first solutions were the low V_{th} and high gate leakage currents through the gate terminal. A more advanced structure that aimed at mitigating both of these issues was proposed in 2016 and it combines the CAVET design with a gate trench structure [18] (Figure 10.7(c)). In Figure 10.8, on-state and off-state characteristics are shown for the trench CAVET in Figure 10.7(c) are shown. It can be noted that the gate leakage is limited and the BV is in excess of 800 V.

10.5.1.2 Trench MOSFET

Figure 10.7(d) shows the basic cross-section of a trench vertical MOSFET. Here, a positive gate voltage is applied in order to create the inversion channel in the p-doped GaN layer. A thin layer of SiO_2 is used as a gate insulator and its presence guarantees a V_{th} in excess of 2 V. In [19], a BV of 1 kV, $R_{on} < 2$ mΩ cm^2 and $V_{th} >$ 3 V was achieved optimizing the process. The trench MOS solution is also often preferred as it allows to achieve current levels in excess of 10 A with low gate leakage current [20,21]. It should be mentioned that the trench MOSFET structure suffers from lower channel mobility due to the trap-states at the insulator-semiconductor interface. This structure is similar to the Trench MOSFETs used today in Si and SiC structures. The drift region is designed in a similar way to that in SiC with slight punch-through (reach-through) design, to optimize the trade-off between on-state resistance and the BV. By slight punch-through design, one means that the depletion region reaches the n+ GaN layer substrate (just) before breakdown occurs. This structure competes directly with SiC Trench MOSFETs in the same voltage class sharing the same advantages (low on-state resistance, low capacitances) when compared to silicon MOSFETs, but also the same disadvantages (low channel mobility, threshold instability). Nevertheless, the development of the SiC counterpart is much more mature and is now available in the market, competing directly with silicon super junctions and IGBTs.

10.5.1.3 Slanted HEMT PGaN gate

The slanted PGaN gate solution shown in Figure 10.6(e) is a combination of the trench CAVET in Figure 10.6(c) and the P-GaN gate in Figure 10.4(b). The 2DEG channel is formed at the AlGaN/GAN interface both along the polar and semi-polar direction. The p-GaN layer underneath the gate metal guarantees the normally-off operation. The buried p-GaN layers act as semi-insulating layer introducing acceptor-like dopants in the *n*-type dope GAN layer [22]. In [23], the impact of the slanted geometry on the threshold voltage is clearly demonstrated, as also shown in Figure 10.9.

10.5.1.4 FinFET

One of the main drawbacks of the solutions discussed so far is that they either depend on the regrowth of one of the GaN-based layers, which increases the complexity and cost of the epitaxial growth, or on a p-doped layer in the channel region of the transistor, which negatively impacts the channel mobility. The FinFET geometry shown in Figure 10.6(f) requires only *n*-type layers and it does not need a regrown layer. The current in the FinFET is controlled via the narrow fin-shaped n-GaN channels that are characterized by all-around gate metal. At zero bias, the n-GaN fins are depleted due to the field-effect from the gate contact, enabling normally-off operation. A breakdown voltage of 1.2 kV, a specific on-resistance $R_{on,sp} = 1\ m\Omega\ cm^2$ and $V_{th} = 1.3$ V have been reported for a maximum on-state current equal to 5 A on a 681 fins structure [24]. Although the epi growth of this technology is simplified by the absence of regrown layers, the FinFETs still suffer from relatively low V_{th} (maximum reported 1.3 V) and limited on-state current (10 A for 800 V BV [25] and 5 A for 1,200 V BV [24]).

10.5.1.5 Quasi-vertical GaN-on-Si

A quasi-vertical or pseudo-vertical solution is based on a double orientation of the current flow. The electrons follow first the *y*-vertical direction as for a CAVET structure, but they are then collected on the side by the drain terminal, as shown in Figure 10.6(g). The advantage of this solution is the possibility to use a silicon (or sapphire) substrate, therefore reducing the costs associated with the small and expensive GaN bulk wafers. An example of this approach can be found in [26]. The disadvantage of this structure compared to fully vertical GaN devices is the lack of scalability at high voltages (due to the lateral breakdown) and in this respect, it shares the same limitation as that of lateral HEMTs or MISFETs.

All the solutions presented so far have been demonstrated on Sapphire or GaN bulk substrates. Recently, progress has been made in order to achieve a fully

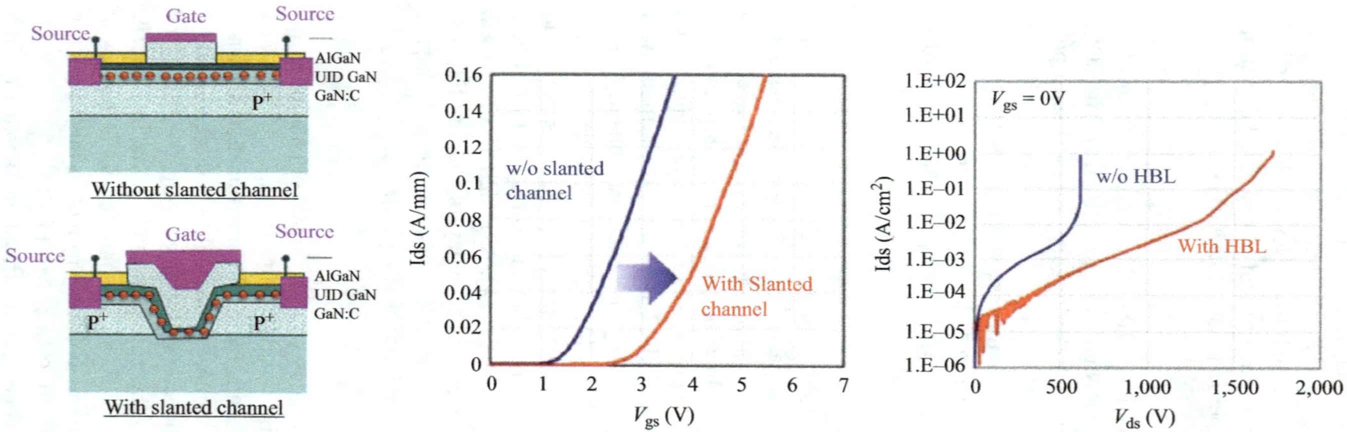

Figure 10.9 *Cross-section, transfer characteristics, and breakdown curve of a vertical GaN transistor with buried p-type layer with and without the slanted geometry*

vertical GaN-on-Si transistor, thus taking advantage of both the vertical technology and the availability of large and cost-efficient silicon substrates.

A recent solution [27] proposes the combination of the trench MOSFET structure with a partly recessed substrate, as shown in Figure 10.6(h). The maximum BV voltage obtained is however only 550 V. A similar approach based on partial Si substrate removal was considered in [28] for the demonstration of a fully vertical GaN-on-Si power diode. This solution, while innovative and already demonstrated in silicon is based on a MEMS final etch step of the silicon substrate. Not only that this MEMS step adds cost but also poses some risks in terms of membrane yield and packaging and it limits the heat dissipation to the substrate.

10.5.1.6 Super junction vertical GaN devices

The on-state resistance of the drift region in vertical devices could be further minimized by employing the super junction concept [29]. Alternate p and n pillars with compensating charge could be formed in the drift region to increase the conductivity of the n-pillar above that predicted by the 1D Poisson theory for a given breakdown voltage. The super junction concept allows a linear scaling of the on-resistance with the breakdown and can achieve ultra-low values of on-state resistance. While the super junction has been tremendously successful in silicon, cutting edge research is now carried out to implement in Silicon Carbide. Simulation studies show that its application in GaN vertical devices could lead to record low values of on-state resistance of 18 $\mu\Omega\ cm^2$ for a BV of 2 kV [30]. Nevertheless, the experimental demonstration and the market penetration of such super junction vertical GaN structures is still some way ahead.

10.6 Summary

This chapter has been concerned with an overview of the GaN transistor applications and an outlook to the research frontiers of the vertical GaN technology. The advantages and disadvantages of the vertical GaN devices against the lateral high-electron-mobility-transistors (HEMTs) are given. State-of-the-art vertical technologies are thoroughly described, with seven possible solutions, each of them presented and discussed together with their advantages and disadvantages.

Despite the numerous innovative solutions proposed for vertical GaN transistors, the only GaN transistors that are commercially available are based on heterostructures with lateral geometries. The main reasons are associated with the poor quality and scare availability of the GaN substrates and to some extent the cost of the final packaged parts. GaN-on-Si lateral FETs and HEMTs are available on 8-inch silicon or silicon-on-insulator substrates and on 6-inch SiC and sapphire substrates, while vertical GaN devices are currently fabricated on lower quality 2–4 inch GaN substrates. Moreover, the lateral technology allows for easier co-packaging of the main FET (HEMT) with the gate driver and smart control circuits due to the presence of all the pads at the surface. It is well known that GaN transistors suffer from high-frequency reliability issues associated to high di/dt and dv/dt signals and therefore the possibility to co-package the power transistor with the control and protection circuit to reduce the parasitic inductances is a significant advantage of the lateral technology when considered in the final application. Furthermore, due to easier isolation, the lateral devices allow for monolithic integration of the drive, sensing and protection circuitry, and possibly some basic form of control circuitry. Half-bridges could also be monolithically integrated or co-packaged using lateral devices and these could be used in applications such as motor control. Nevertheless, the quest for a vertical GaN technology will continue in the years to come. The possibility of using vertical GaN devices for higher voltage (>1 kV) and higher current (>150 A) applications remains an attractive proposition.

Note to the reader

This chapter is part of an edited book. To cite or use this chapter as a reference, please use the following format:

Longobardi G. "Gallium nitride transistors: applications and vertical solutions." In: Iannuzzo F (ed.), *Modern Power Electronic Devices: Physics, Applications, and Reliability*. Stevenage, UK: IET; 2020. pp. 331–345.

References

[1] Flack T. J., Pushpakaran B. N., and Bayne S. B. "GaN technology for power electronic applications: A review." *Journal of Electronic Materials*. 2016;**45**(6): 2673–2682.

[2] Pengelly R. S., Wood S. M., Milligan J. W., Sheppard S. T., and Pribble W. L. "A review of GaN and SiC high electron-mobility transistors and MMICs." *IEEE Transactions on Microwave Theory and Techniques*. 2012;**60**(6): 1764–1783.

[3] Longobardi G., Yang S., Pagnano D., Camuso G., and Udrea F. "On the vertical leakage of GaN-on-Si lateral transistors and the effect of emission and trap-to-trap-tunneling through the AlN/Si barrier." Proceedings of the 29th International Symposium on Power Semiconductor Devices & ICs (ISPSD), Japan. 2017. pp. 227–230.

[4] Longobardi G., Pagnano D., Udrea F., Sun J., Garg R., Imam M., and Charles A. "Suppression technique of vertical leakage current in GaN-on-Si power transistors." *Japanese Journal of Applied Physics*. 2019:**58**(SC): SCCD12.

[5] Lidow A., Strydom J., De Rooij M., and Reusch D. *GaN Transistors for Efficient Power Conversion*. New York: John Wiley & Sons, 2014.

[6] Efthymiou L., Camuso G., Longobardi G., Chien T., Chen M., and Udrea F. "On the source of oscillatory behaviour during switching of power enhancement mode GaN HEMTs." *Energies*. 2017;**10**(3): 407.

[7] Longobardi G., Efthymiou L., and Arnold M. "GaN power devices for Electric Vehicles State-of-the-art and future perspective." 2018 IEEE International Conference on Electrical Systems for Aircraft, Railway, Ship Propulsion and Road Vehicles & International Transportation Electrification Conference (ESARS-ITEC). IEEE, Nottingham, UK, 2018.

[8] Yole Développement Report: Power GaN 2017: epitaxy, Devices, Applications, and Technology Trends. Available from: https://www.electronicspecifier.com/products/power/power-gan-market-expected-to-be-worth-450m-by-2022.

[9] Ambacher O., Smart J., Shealy J., *et al.* "Two dimensional electron gases induced by spontaneous and piezoelectric polarization in undoped and doped AlGaN/GaN heterostructures." *Journal of Applied Physics*. 1999;**85**(6): 3222–3233.

[10] Longobardi G. "GaN for power devices: Benefits, applications, and normally-off technologies." 2017 International Semiconductor Conference (CAS). IEEE, Romania, 2017.

[11] Choi F. S., Griffiths J., Ren C., *et al.* "Vertical leakage mechanism in GaN on Si high electron mobility transistor buffer layers." *Journal of Applied Physics*. 2018;**124**(5): 055702.

[12] Efthymiou L., Longobardi G., Camuso G., Chien T., Chen M., and Udrea F. "On the physical operation and optimization of the p-GaN gate in normally-off GaN HEMT devices." *Applied Physics Letters*. 2017;**110**(12): 123502.

[13] Efthymiou L., Murukesan K., Longobardi G., Udrea F., Shibib A., and Terrill K. "Understanding the threshold voltage instability during off-state stress in p-GaN HEMTs." *IEEE Electron Device Letters*. 2019;**40**(8): 1253–1256.

[14] Wei J., Xie R., Xu H., *et al.* "Charge storage mechanism of drain induced dynamic threshold voltage shift in p-GaN gate HEMTs." *IEEE Electron Device Letters*. 2019;**40**(4): 526–529.

[15] Gao Y., Ben-Yaacov I., Mishra U. K., and Hu E. L. "Optimization of AlGaN/GaN current aperture vertical electron transistor (CAVET) fabricated by photoelectrochemical wet etching." *Journal of Applied Physics*. 2004;**96**: 6925

[16] Kanechika M., Sugimoto M., Soejima N., *et al.* "A vertical insulated gate AlGaN/GaN heterojunction field-effect transistor." *Japanese Journal of Applied Physics*. 2007;**46**(6L): L503.

[17] Chowdhury S., Wong M. H., Swenson B. L., and Mishra U. K. "CAVET on bulk GaN substrates schieved with MBE-regrown AlGaN/GaN layers to suppress dispersion." *IEEE Electron Device Letters*. 2011;**33**(1): 41–43.

[18] Ji D., Agarwal A., Li H., Li W., Keller S., and Chowdhury S. "880 V/2.7 m$\Omega \cdot cm^2$ MIS gate trench CAVET on bulk GaN substrates." *IEEE Electron Device Letters*. 2018;**39**(6): 863–865.

[19] Otake H., Chikamatsu K., Yamaguchi A., Fujishima T., and Ohta H. "Vertical GaN-based trench gate metal oxide semiconductor field-effect

transistors on GaN bulk substrates." *Applied Physics Express*. 2008;**1**(1): 011105.

[20] Oka T., Ina T., Ueno Y., and Nishii J. "1.8 mΩcm^2 vertical GaN-based trench metal-oxide-semiconductor field-effect transistors on a free-standing GaN substrate for 1.2-kV-class operation." *Applied Physics Express*. 2015;**8**(5): 054101.

[21] Li R., Cao Y., Chen M., and Chu R. "600 V/1.7 Ω normally-off GaN vertical trench metal-oxide-semiconductor field-effect transistor." *IEEE Electron Device Letters*. 2016;**37**(11): 1466–1469.

[22] Tanaka K., Umeda H., Ishida H., Ishida M., and Ueda T. "Effects of hole traps on the temperature dependence of current collapse in a normally-OFF gate-injection transistor." *Japanese Journal of Applied Physics*. 2016;**55**(5): 054101.

[23] Shibata D., Kajitani R., Ogawa M., *et al.* "1.7 kV/1.0 mΩcm^2 normally-off vertical GaN transistor on GaN substrate with regrown p-GaN/AlGaN/GaN semipolar gate structure." 2016 IEEE International Electron Devices Meeting (IEDM). IEEE, San Francisco, CA, 2016.

[24] Zhang Y., Sun M., Perozek J., *et al.* "Large-area 1.2-kV GaN vertical power FinFETs with a record switching figure of merit." *IEEE Electron Device Letters*. 2018;**40**(1): 75–78.

[25] Zhang Y., Sun M., Piedra D., *et al.* "1200 V GaN vertical fin power field-effect transistors." 2017 IEEE International Electron Devices Meeting (IEDM). IEEE, San Francisco, CA, 2017.

[26] Liu C., Khadar R. A., and Matioli E. "645 V quasi-vertical GaN power transistors on silicon substrates." 2018 IEEE 30th International Symposium on Power Semiconductor Devices and ICs (ISPSD). IEEE, Chicago, IL, 2018.

[27] Khadar R. M. A., Liu C., Soleimanzadeh R., and Matioli E. "Fully vertical GaN-on-Si power MOSFETs." *IEEE Electron Device Letters*. 2019;**40**(3): 443–446.

[28] Zhang Y., Yuan M., Chowdhury N., Cheng K., and Palacios T. "720-V/0.35-mΩ · cm^2 fully vertical GaN-on-Si power diodes by selective removal of Si substrates and buffer layers." *IEEE Electron Device Letters*. 2018;**39**(5): 715–718.

[29] Udrea F., Deboy G., and Fujihira T. "Superjunction power devices, history, development, and future prospects." *IEEE Transactions on Electron Devices*. 2017;**64**(3): 713–727.

[30] Li Z. and Chow T. P. "Design and simulation of 5–20-kV GaN enhancement-mode vertical superjunction HEMT." *IEEE Transactions on Electron Devices*. 2013;**60**(10): 3230–3237.

Chapter 11

Module design and reliability

Daohui Li[1], *Xiaoping Dai*[2] *and Guoyou Liu*[2]

11.1 Introduction

In the last several decades, power semiconductor devices are becoming dominant in the switching energy applications, that is traction, electrical vehicle (EV), hybrid electrical vehicle (HEV), aerospace, marine driver, industrial drivers, smart energy, solar energy, wind energy, high voltage direct current (HVDC) electrical power transmission and so on. Several kinds of different material semiconductor devices have been widely used. Some of key power semiconductor devices have been manufactured based on silicon material, which includes power diode, thyristors, gate turn-off thyristors (GTO), power metal-oxide-semiconductor field-effect transistor (MOSFET), insulated gate bipolar transistor (IGBT), silicon carbide (SiC) based including Schottky barrier diode (SBD), junction-gate field enhanced transistor (JFET), SiC MOSFET, SiC IGBT, gallium nitride (GaN) based including power transistors, enhancement mode high electron mobility transistors (E-HEMT) and other material based wide bandgap devices, the corresponding icons can be seen in Figure 11.1. There are plenty of different types of power module packaging technologies to provide protection and support for the different semiconductor as mentioned to fit different application requirements [1,2].

Power semiconductor chips are core components for power module products. However, they cannot independently provide the electrical current or voltage as required from the power system. Inter-connections from chips to an external power system are normally utilised. The inter-connection bonding between chips, or to an external system is done via metal material, such as using copper or aluminium wire, ribbon and busbar. During the operation of conduction and switching period, power loss generated from semiconductor devices can heat chip's temperature up and affect the performance of the device. Efficient thermal transfer and thermal cooling are important to guarantee the normal performance of semiconductor devices. To provide good thermal performance, the thermal characteristics of packaging materials, that is thermal conductivity, thermal resistance, thermal capacitance and

[1]Power Semiconductor R&D Center, Dynex Semiconductor Ltd, Lincoln, UK
[2]State Key Laboratory of Advanced Power Semiconductor Devices, CRRC Times Electric Co. Ltd, Zhuzhou, Hunan, P.R. China

Figure 11.1 Different type of power semiconductor devices

coefficient of thermal expansion (CTE), are critical parameters the selection of materials aiming at different applications. CTE value mismatch of the packaging materials can cause the thermal-mechanical performance of the power module. The reliability of the power module relates to all the electrical, thermal and mechanical aspects [1–6].

The power module packaging can provide both internal electrical and thermal connection for single or multiple semiconductor chips and act as external electrical and thermal interface for integrating into power module assembly. At the same time, the package can also provide multi-protection of the chips in respect to electromagnetic (EM), thermal, mechanical and chemical environments. The reliability of power semiconductor can be enhanced with proper packaging technologies via an increase of power density, improvement of heat dissipation capability, increase of thermal/power cycling capability, reduction of dimension and overall costs to customers and so on.

Based on different application scenarios and power levels, there are plenty of different types of module packages, such as transistor outline package as shown in Figure 11.2(a), which includes many types of packaging format for discrete power semiconductor devices with low number of pins; standard module packaging can be seen in Figure 11.2(b) and 11.2(c) with chips wire bonded and soldered onto substrate and baseplate, which will be covered by encapsulating material and plastic frame; and press-pack module, which primarily use pressure contact method to replace wire-bonding and soldering inter-connection (Figure 11.2(d)) [2–6].

This chapter will focus on high power density, high current, high voltage power semiconductor packaging technologies for IGBT and SiC devices. The voltage ranges mainly from 1.2, 1.7, 3.3, 4.5 to 6.5 kV and the current from hundreds of amperes to several thousand amperes. Those types of power modules have been vastly utilised on traction, electric vehicle, energy and industrial applications [2–6].

The manufacture and process steps for power semiconductor devices from silicon wafer to final power module have been demonstrated in Figure 11.3(a–e). One can see that IGBT and FRD chips as indicated in Figure 11.3(b) will be fabricated from silicon wafer as shown in Figure 11.3(a); the chips as indicated in Figure 11.3(b) will be then soldered onto high isolation ceramic substrate as shown in Figure 11.3(c); the substrate and power terminals will be soldered and bonded on to baseplate and substrates, respectively, as indicated in Figure 11.3(d); and the plastic frame will be glued onto baseplate and encapsulate material filled to form the final power module as shown in Figure 11.3(e) [4–6].

Figure 11.2 Different types of power module packages

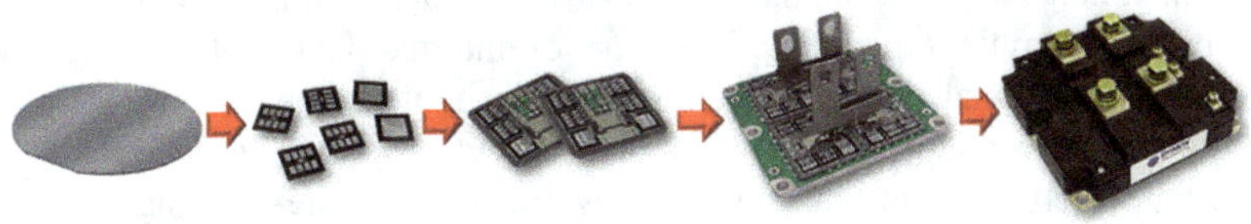

Figure 11.3 From silicon wafer, chips, substrate, assembly till assembled into the whole power module

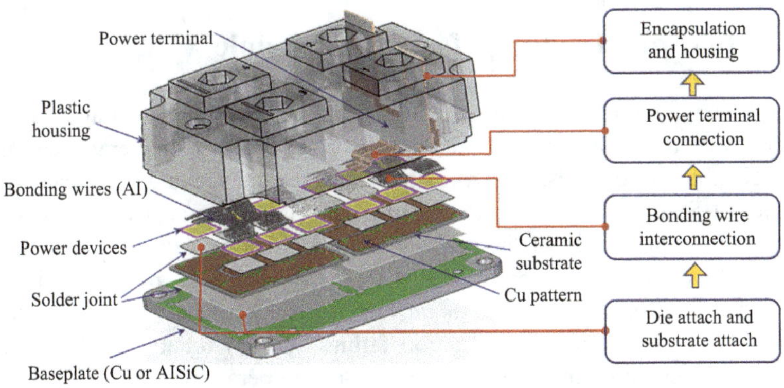

Figure 11.4 Internal structure of standard power module

The typical internal structure of a standard module can be seen in Figure 11.4 for an explosive view. From downside to upside, the key components include (1) baseplate, which is usually copper or AlSiC material; (2) high isolation ceramic substrate, normally alumina, AlN, Si_3N_4 and other material based on different isolation, thermal transfer and mechanical requirements; (3) power semiconductor chips are core devices of the module and silicon IGBT, FRD, SiC MOSFET and SiC SBD are the main devices to supply high power, high current and high voltage as required; (4) copper metal power terminal is critical for electrical connection; (5) plastic housing will be glued onto baseplate to hold silicone gel for high dielectric encapsulation and (6) solder material and bonding wire materials for mechanical and electrical interconnection [1–6].

Also, the four key aspects of standard module packaging have been highlighted in Figure 11.4, such as die attach and substrate attach, bonding wires interconnection, power terminal connection, encapsulation and plastic housing. The power semiconductor devices' performance and reliability will be mainly affected by those four aspects together with the corresponding material selections, processes and mutual influence between the devices and packaging materials during long-term application scenarios.

The main content of this chapter is a module package design and its corresponding reliability. This will include two main sessions: (1) *multi-physics design and optimisation for power module*, which will cover parasitic parameters extraction, circuitry simulation and optimisation of internal layout, thermal network extraction and thermal management, thermal-mechanical simulation and lifetime estimation of power module; and (2) *material and processes to enhance module reliability*, which will introduce some key processes, wire-bonding, soldering, sintering, ultrasonic welding and key packaging materials used such as substrate, baseplate, copper, etc. Also, electrical test and QA-reliability test will include static and dynamic test during production procedures along with reliability test, especially environment-related reliability test.

11.2 Multi-physics design for power module

As power semiconductor devices provide switching mode high voltage, high current energy to drive the applications, there are several key challenges can affect reliability of power devices, such as (1) parasitic inductance and resistance of chip interconnection and control circuit can cause current imbalance, over-shoot voltage at switching on/off stages of the devices; (2) potential damage from Lorentz EM force due to extreme high current at short-circuit stage of power module; (3) thermal deformation due to CTE mismatch of different materials utilised in the package; (4) power terminal's vibration failure and (5) reliability challenges of bonding wire and soldering layer during long-term operation.

To fully understand and solve the challenges of power semiconductor, multi-physics tools are critical to be utilised in the design stage to enhance the reliability of power semiconductor devices. One set of multi-physics simulation technologies for module packaging design, which includes 3D EM, 3D thermal, 3D mechanical and circuitry simulation, will be utilised to verify the key parts design, corresponding structure optimisation, performance anticipation and failure mode analysis. There are several commercially available multi-physics simulation software packages, such as ANSYS, CST, COMSOL and others [7–9].

In this session, some of the simulation examples will be shown during the design stage of power module packaging, such as (1) EM-circuitry simulation verification has been carried out to show advantage of current balance status, low-frequency EM analysis of electro-magnetic field distribution of internal module and parasitic inductance/resistance extraction of whole module packaging for EMC/EMI analysis; (2) electro-thermal simulation and static/transient thermal simulation have been carried out to calculate thermal distribution, power loss of different internal components, thermal resistance, thermal network extraction, etc.; (3) mechanical stress analysis has been carried out for the substrate, soldering layer, busbars and baseplate layer structure; (4) creep-strain analysis has been carried out for different type of solder alloy together with mechanical stress and reliability analysis and (5) mechanical vibration analysis has been carried out for busbars and ultrasonic welding (USW) processes optimisation.

With the assistant of 3D simulation, several challenges regarding module packaging design have been solved and verified, that is, (1) low inductance packaging design and optimisation, where the inductance has been measured via impedance analyser and the design routine can be extended to more advanced packaging design; (2) Lorentz force analysis of high power, where high voltage package has been verified via module explosion test carried out in the test bench [10]; (3) bow change tendency during soldering, where thermal shock test has been simulated, the simulation analysis has indicated the possible reason for one plano-convex baseplate, further experimental test is under-preparing and this simulation routine can help baseplate bow design and (4) power cycling simulation routines where encapsulate materials have been considered along with creep-strain characteristic of solder layer to reflect thermal stress, creep-strain and other more reliability-related challenges.

11.2.1 EM simulation of power module

In this section, some of the key packaging designs will be introduced, such as low inductance design, EM simulation of current and voltage distribution of substrate-busbar assembly and electrical field distribution inside of power.

11.2.1.1 Low inductance design

The IGBT module usually works in switching modes that the parasitic inductance plays as a crucial role to affect the overshooting voltage and switching power loss during the switching-off period. At the switching-off transient period of IGBT, the overshoot voltage due to stray inductance and the current slope di/dt can cause the damage of IGBT chips. In the module design, low inductance busbars are critical to providing a power module with lower power loss and low overshoot voltage.

The standard 190 mm × 140 mm dimension single-switch IGBT module with schematic circuit, as shown in Figure 11.5(a), is one of several key standard modules used in high power applications, that is, locomotive and underground power train. The module is a single-switch circuit, and the assembled full module can be seen in Figure 11.5.

The calculated inductance can be read in Figure 11.6(a), which shows the module inductance for substrate-busbar assembly is about 15.7 nH. The inductance measurement, using an HP4192A LF impedance analyser, has been carried out. The measurement inductance value is about 17 nH for one pair of busbar-substrate assemblies. The difference between simulation and experimental measurement is less than 8%. The module-level inductance is 5.6 nH with three sets connected in parallel.

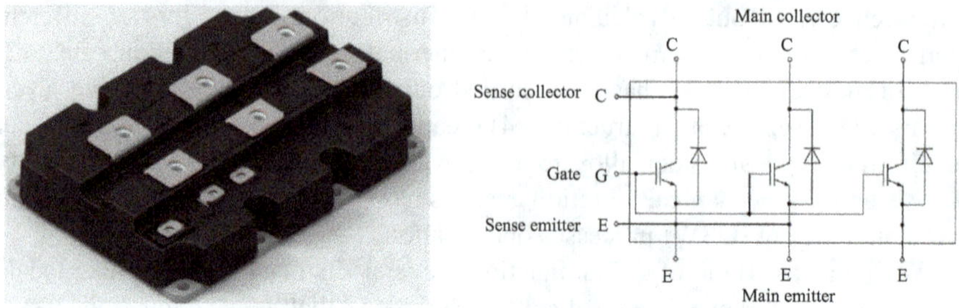

(a) (b)

Figure 11.5 (a) 3D outlook of 140 mm × 190 mm high power module and (b) single-switch schematic circuit diagram

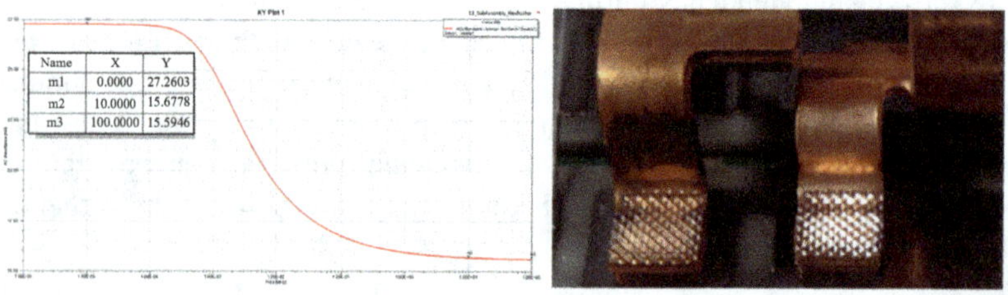

(a) (b)

Figure 11.6 (a) Simulated inductance of busbar-substrate and (b) USW welding of power terminals

11.2.1.2 Low-frequency EM simulation

Some static parameters, such as $V_{ce(on)}$, are very important for the IGBT module; one can use static test equipment to measure the $V_{ce(on)}$ for IGBT at chip level and substrate level, respectively. The minimisation of $V_{ce(on)}$ at substrate level can be tested with the following method. The $V_{ce(sat)}$ at chip level will be decided by the manufacturing process. One can take it as constant during power module packaging design stage.

The $V_{ce(on)}$ at substrate level equals the $V_{ce(sat)}$ at chip level together with the voltage drop due to both the topside bonded wires as shown in Figure 11.7(a) and copper layer on the top side of the ceramic layer. EM simulation from ANSYS can calculate the current/voltage distribution for the wires shown in Figure 11.7(a) and Figure 11.7(b).

It has shown that the voltage of bonded wires which were connected from emitter of chips to the main busbar electrical interface was about 0.07 V with 10 A current via each wire. The simulation results in Figure 11.7(b) have shown that the voltage of copper metal layout without chips was 0.13–0.15 V with 250 A current total from collector to emitter transmitting in per substrate.

Based on the simulation, the voltage added to IGBT's is about 0.2–0.22 V due to bonding wires and topside metal of the substrate. Together with $V_{ce(sat)}$ measured

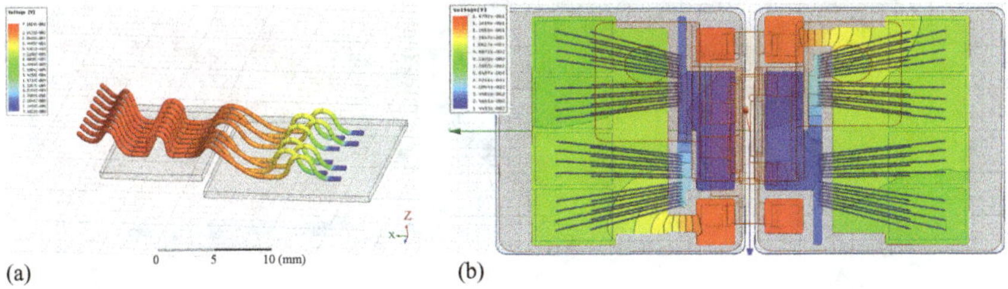

Figure 11.7 (a) Voltage drop due to bonded wires and (b) voltage drop due to copper layer

at the wafer level. the $V_{ce(sat)}$ measured by using proper static test jig is ~2.0–2.1 V for 250 A current going through one substrate. The simulation results fit the measurement results quite well from our internal measurement comparison.

11.2.1.3 Substrate layout optimisation based on EM simulation

Partial discharge is one of the key challenges for the high voltage power module, especially for 3.3 kV and above voltage ranges. The substrate layout will affect the electrical field distribution during the operation of the IGBT module and it is very important to consider the topside metal layout of the substrate to reduce the potential peak E-field area. For the real module, the substrate-busbar assembly will be covered by soft silicone gel with high dielectric strength. Partial discharge is an accumulated effect after long period discharge phenomena at a certain area inside of the module with defects or voids. The repeated high voltage is combined with the operational line voltage together with ripple voltage from the capacitor. There are normally two directions considered to reduce the possibility of breakdown, one way is to reduce the defects by optimising silicone gel filling process, the other way is to optimise the topside copper layout of the substrate to remove the potential high E-field area.

The typical dielectric strength for silicone gel is 22–30 kV/mm. In the EM simulation, adaptive meshes have been utilised. The peak value of E-field increases with the decrease of energy error together with increase in mesh number. The maximum electric field strength is proportional to the minimum mesh.

The E-field contour distributions and their corresponding voltage distributions have been shown in Figure 11.8(a)–(d). Figure 11.8(a) and (b) shows E-field distribution and voltage distribution, respectively, of cross-section for the topside edge moving 0.8 mm from original positions. The peak value of E-field based on simulation is 19 kV/mm. With the same simulation settings, Figure 11.8(c) and (d) has shown the E-field and corresponding voltage distribution, respectively, with 0.5 mm moving for the topside substrate metal edge from its original position. It is found that the peak value of E-field is 20.5 kV/mm. Although the peak E-field relates to the meshing setup, the simulated peak E-field value and peak value area will be taken as a reference to optimise the layout design of the substrate.

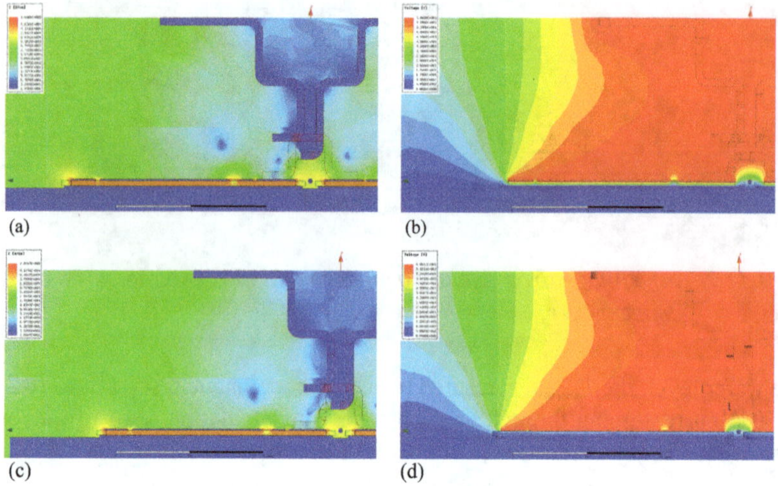

Figure 11.8 (a) Max-19 kV/mm (0.8 mm moving from edge), (b) voltage distribution (0.8 mm moving from edge), (c) max E-field ~20.5 kV/mm (0.5 mm moving) and (d) voltage distribution (0.5 mm moving)

In the other scenario, module explosion or case rupture failure is one of the key challenges faced by high power, high voltage modules in the power train applications for high power modules. The high power module's case rupture can cause direct damages and huge influences on surrounding systems or even cause safety problems depending on its application scenario. Explosion mechanism and how to design explosion proof packaging is one of the biggest challenges in high power module design. New type of 3.3 kV/1,500 A IGBT module with improved explosion performance has been designed and investigated via 3D Maxwell simulation and experiment jointly.

11.2.1.4 EM Lorentz force analysis for short-circuit failure

During the switching operations, IGBT modules can normally afford very high short-circuited current for ~10 μs or longer periods as different application requirements. Control driver circuits are utilised to act once short-circuit happens. Extreme high short-circuited current can cause module failure or even explosion. That is one of the key factors that can cause a power module explosion. The explosion effect normally relates to encapsulating materials, such as silicone gel, plastic frame, or epoxy seal used for the module. The mechanism is rather complicated that once module explosion is unavoidable, customers would like to have minimal effects on the surrounding devices. For explosion experiment, high current pulse >250 kA with 200 μs has been applied to explode 140 mm × 190 mm dimension modules. In the simulation, 70 kA DC current has been applied between DC+ and DC− busbars, then Lorentz forces have been calculated to show some force comparison.

The Lorentz force can cause deformation of the whole module during explosion as high short-circuit current going through busbars. The real measured high current pulse can reach 300 kA during the test. To simplify the simulation, DC current 70 kA is set to going through the busbar-substrate assembly. There are three sets of such busbar-substrate assembly in one module. The current density distribution and magnetic field distribution can be seen in Figure 11.9(a) and (b), respectively.

Lorentz force can be calculated in the EM simulation. The Lorentz force based on the simplified DC current can be seen in Table 11.1. The Lorentz force onto collector and emitter in different directions can be extracted with the assistant of EM simulation. The total equivalent Lorentz force is equal to 365.3 kg under explosion equivalent high current.

The traditional module utilises 2D busbars as indicated in Figure 11.10. It is found that the 2D busbar and substrates have higher current distribution and higher magnetic field distribution than 3D busbars and corresponding substrates. The equivalent force can be estimated as 406.4 kg as indicated in Table 11.2, which is 10% higher than the module with 3D structure busbars. Via the Maxwell simulation with equivalent high current during the explosion test, the module using 3D busbar will experience at least 10% less Lorentz force compared to the module using 3D busbars.

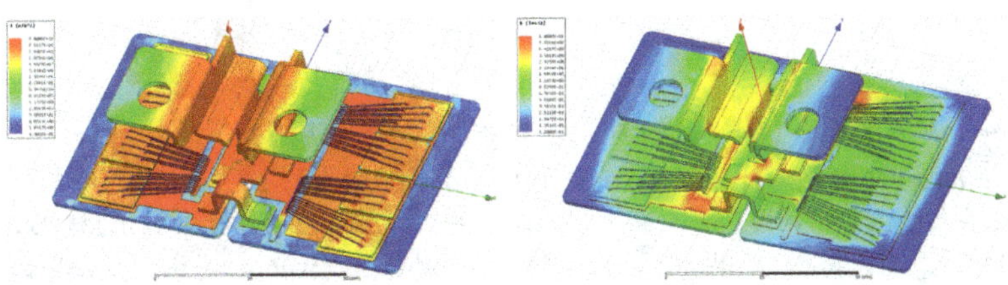

(a) (b)

Figure 11.9 EM simulation results for short-circuit current of new designed structures: (a) E2 – current density distribution (DC 70 kA) and (b) E2 – magnetic field distribution (DC 70 kA)

Table 11.1 Lorentz force calculation for newly designed busbar

ANSYS-Maxwell simulation results	Collector force (*N*)	Emitter force (*N*)
Scalar X (Cross(<Jx,Jy,Jz>, <Bx,By,Bz>))	1,511.7	−1,590.0
Scalar Y (Cross(<Jx,Jy,Jz>, <Bx,By,Bz>))	71.1	−175.7
Scalar Z (Cross(<Jx,Jy,Jz>, <Bx,By,Bz>))	−280.9	−1,268.4
Force in total (*N*)	1,539	2041.5
Equivalent force in total (kg)	**365.3 kg**	

Note: The equivalent force in total = (1539 + 2041.5)/9.8 = 365.3.

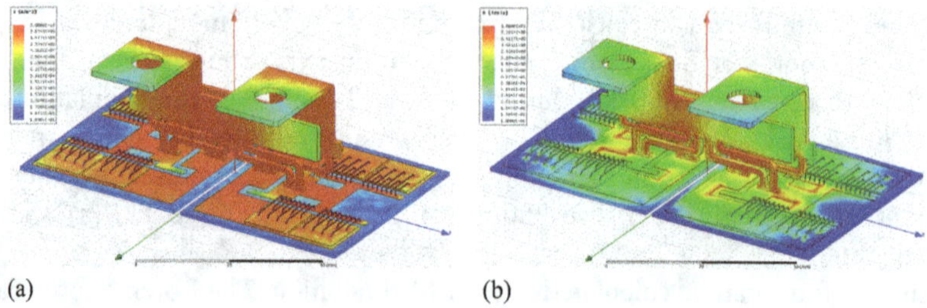

Figure 11.10 EM simulation results of traditional module structure: (a) E – current density distribution (DC 70 kA) and (b) E – magnetic field distribution (DC 70 kA)

Table 11.2 Lorentz force calculation for 2D structural busbars

ANSYS-Maxwell simulation results	Collector force (*N*)	Emitter force (*N*)
Scalar X (Cross(<Jx,Jy,Jz>, <Bx,By,Bz>))	–1372.2	1641.8
Scalar Y (Cross(<Jx,Jy,Jz>, <Bx,By,Bz>))	–638.1	815.8
Scalar Z (Cross(<Jx,Jy,Jz>, <Bx,By,Bz>))	–835.2	–1311.1
Force in total (N)	1728.5	2253.8
Equivalent force in total (kg)	**406.4 kg**	

Note: The equivalent forces in total = (1728.5 + 2253.8)/9.8 = 406.4.

11.2.2 EM-circuitry design in module packaging

To simulate the effect of packaging on the whole module's EM and circuitry performance, ANSYS Q3D has been used to simulate all the self-inductance, coupling inductance, self-resistance and coupling-resistance between all the possible electrical connections throughout the module, such as busbars, auxiliary busbars, current wires and gate wires of the module [11]. The ANSYS Q3D's simulation can calculate magnetic field coupling based on the 3D structure of the module packaging with the connections' real magnetic field coupling and derive their corresponding parasitic inductance and resistance. The 3D structure's connections from parasitic matrix and network are shown in Figure 11.11(a) for inductance and Figure 11.12(a) for resistance, respectively.

One advantage can be seen after first set of simulation results shown in Figure 11.11(b) for inductance and Figure 11.12(b) for resistance. We can easily see that the similar parasitic inductance and parasitic resistance appear in pairs along the horizontal frequency range. Via checking of the curves name, it is found that the pair related to gate1 and gate2 exhibit very symmetrical values. This kind of symmetrical values means the structure of packaging will give very uniform electrical performance as expected.

The parasitic inductance and resistance calculated from Q3D, in Figures 11.11(a) and 11.12(a), can be imported into Simplorer as circuitry block with all the connection

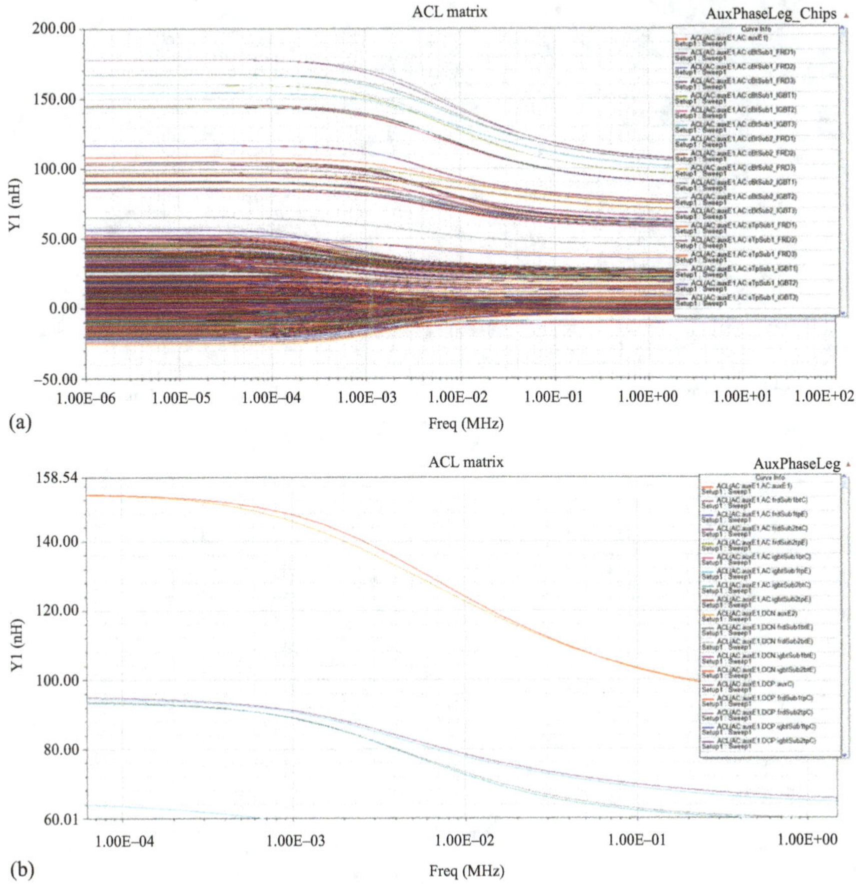

Figure 11.11 Parasitic inductance: (a) overall inductance network and (b) inductance pair for G1 versus G2

pins to one set of chopper circuit. The corresponding IGBT/FRD with 3.3 kV/75 A per chip has been utilised in the circuit.

Figure 11.13(a)–(d) indicates simulation results of double pulse. Line voltage, gate voltage and load current can be seen in Figure 11.13(a) that the current increases up to 450 A till the end of the first pulse, with each IGBT to carry 75 A. Some step changes in voltage and current can be seen during the transient period, respectively. In Figure 11.13(c), during the switch-on period of the second pulse, the transient current going through each chip has shown values ranging from 129.4 to 129.9 A. The 140 mm × 100 mm module has very good balanced current even in the design stage with simulation calculation.

11.2.3 Thermal design and thermal analysis

The typical power module configuration includes the baseplate, substrate, IGBT chips, diode chips and solder layers, which are essential in determining the thermal

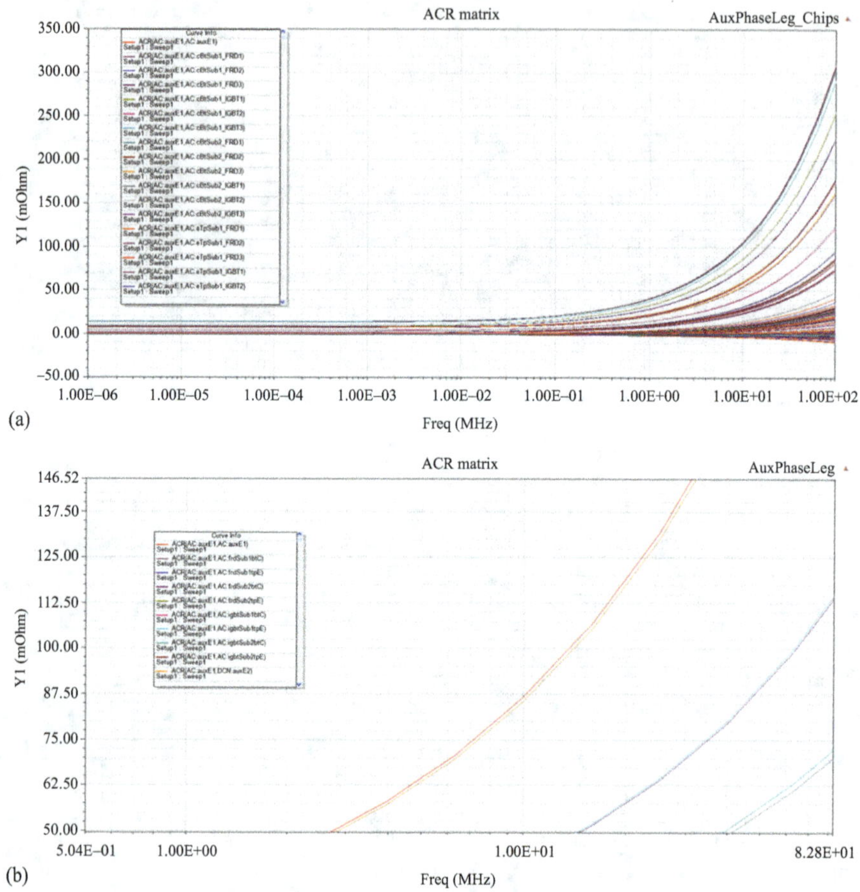

Figure 11.12 Parasitic inductance: (a) overall inductance network and (b) resistance pair for G1 versus G2

dissipation path of a module packaging. The power loss of the IGBTs and diodes are estimated by assuming that the module is used in a three-phase inverter scenario. The switching loss of the IGBTs can be calculated by

$$P_{\mathrm{sw,IGBT}} = \frac{1}{\pi} \times f_{\mathrm{SW}} \times \left(E_{\mathrm{on,IGBT}} + E_{\mathrm{off,IGBT}}\right) \times \frac{I}{I_{\mathrm{nom}}} \times \frac{V_{\mathrm{dc}}}{V_{\mathrm{nom}}}$$

where f_{SW} is the switching frequency of the IGBTs. $E_{\mathrm{on,IGBT}}$ and $E_{\mathrm{off,IGBT}}$ are the loss energy of the IGBTs during turn-on and turn-off, respectively, at the nominal current I_{nom} and nominal voltage V_{nom}. I and V_{dc} are the line current and voltage of the inverter circuit. The conduction loss of the IGBTs is calculated by

$$P_{\mathrm{c,IGBT}} = \frac{1}{2}\left(V_{\mathrm{CE0}} \times \frac{I}{\pi} + r \times \frac{I^2}{4}\right) + M \cos \varphi \times \left(V_{\mathrm{CE0}} \times \frac{I}{8} + r \times \frac{I^2}{3\pi}\right)$$

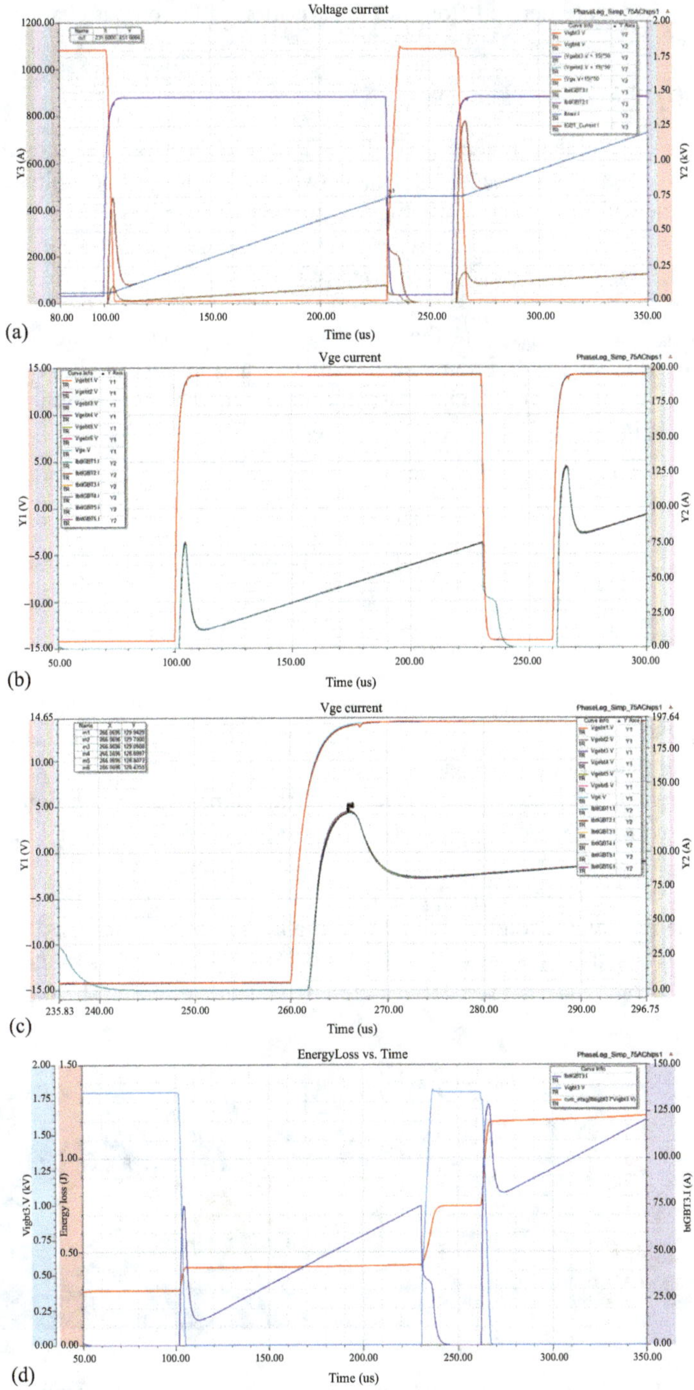

Figure 11.13 *Double pulse simulation results: (a) line voltage and load current for a double pulse, (b) gate voltage and current for IGBT chips, (c) IGBT current for each chip versus time and (d) IGBT 3's energy loss versus time*

where V_{CE0} and r are derived from the linearisation of the output characteristic of the IGBTs.

$$V_{CE}(t) = V_{CE0} + r \cdot I\sin(\omega t)$$

where ω is the angular frequency of the output current wave. M and Φ are the modulation index and phase difference between the current and voltage waves. Similarly, the switching loss of the diodes can be calculated by

$$P_{sw,diode} = \frac{1}{\pi} \cdot f_{SW} \cdot \left(E_{on,diode} + E_{off,diode}\right) \cdot \frac{I}{I_{nom}} \cdot \frac{V_{dc}}{V_{nom}}$$

where $E_{on,diode}$ and $E_{off,diode}$ are the loss energy of the diodes during turn-on and turn-off, respectively, at the nominal current I_{nom} and nominal voltage V_{nom}. The switching on energy of the diode is usually neglected. The conduction loss of the diodes is calculated by

$$P_{c,diode} = \frac{1}{2}\left(V_{CE0} * \frac{I}{\pi} + r * \frac{I^2}{4}\right) + M \cos \varphi * \left(V_{CE0} * \frac{I}{8} + r * \frac{I^2}{3\pi}\right)$$

where V'_{CE0} and r' are derived from the linearisation of the output characteristic of the diodes:

$$V'_{CE}(t) = V'_{CE0} + r' \cdot I \sin(\omega t)$$

Based on the above estimation, the power losses of each IGBT and diode chip are set as 250 and 50 W, respectively. The baseplate temperature is set as 90 °C according to the application environment.

The simulated temperature distribution of the chips is shown in Figure 11.14. The IGBT chip at the centre part of one substrate describes higher temperature those at of the edge, which is due to the thermal coupling effect among chips. By setting monitoring points within different layers, the temperature variations with time for the first two seconds are shown in Figure 11.15.

It is known that the above FEM simulation requires long simulation time, considerable computational resources and massive storage space. The full 3D FEM

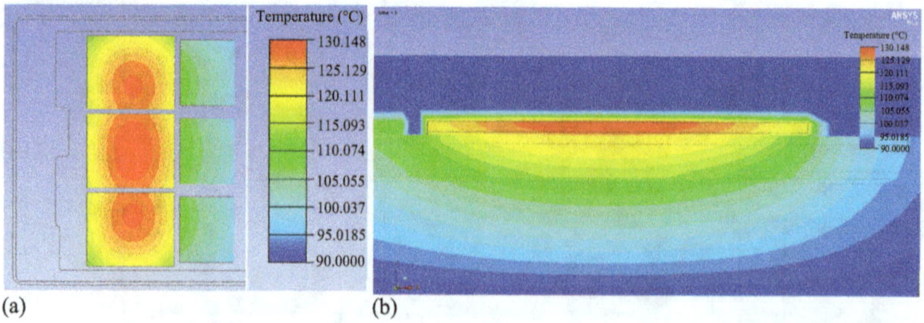

Figure 11.14 (a) Temperature distribution from top view and (b) temperature distribution from a perspective view

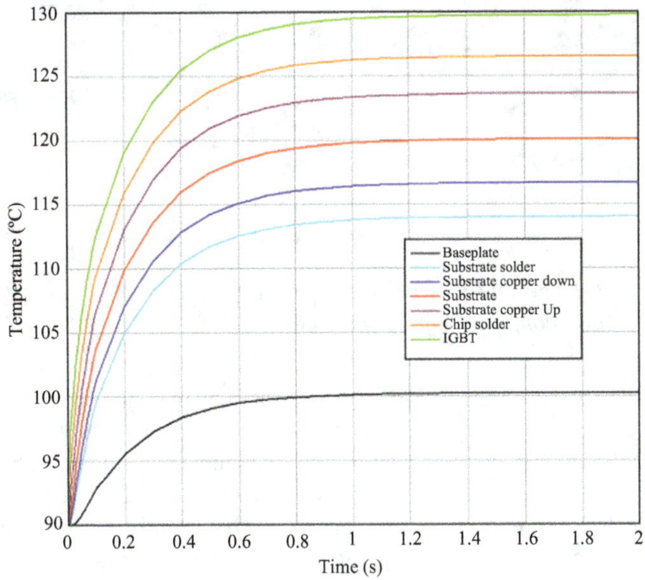

Figure 11.15 Temperature variation of different layers

simulation is unsuitable for parameterised study and optimisation. It is also challenging to interactively coupling the thermal simulation in FEM scenario with other physics such as electrical and mechanical. It is thus a promising practice to extract the 1D thermal network information from the 3D FEM simulation and use a circuit-fashioned way to facilitate the simulation effort.

A sample thermal network built for the module by using Simplorer is shown in Figure 11.16. The red pins of the thermal network represent the input, which is the power loss of chips. This thermal network enables to monitor the temperature variation of components with time-varying loading condition. The pins will be used to connect the thermal network into an electric circuit simulator.

The self-heating and coupling thermal impedance from one IGBT chip to the heat sink is defined by the following equations:

$$Z_{\text{th}}^{\text{self}}(t) = \frac{T_{\text{self}} - T_h}{P_{\text{self}}};Z_{\text{th}}^{\text{coupling}}(t) = \frac{T'_{\text{self}} - T_h}{P_n}$$

where T_{self} is the temperature of the target chip when the loss power of P_{self} is applied to it. T'_{self} is the temperature of the target chip when the loss power P_n is applied to its neighbour. It characterises the thermal coupling effect from its neighbour to the target chip. T_h is the temperature at the heat sink. The self-heating and coupling thermal impedance for IGBT1 is calculated from the thermal network and shown in Figure 11.16(b). The names and locations of the chips are shown in

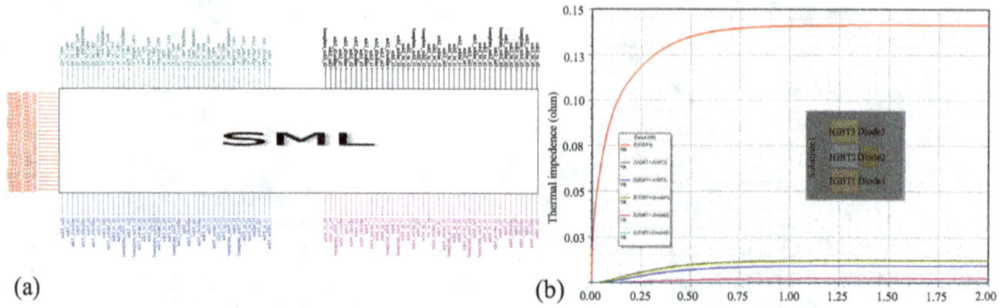

Figure 11.16 (a) Thermal network built for the module and (b) the self-heating and coupling thermal impedance

the inset. Z(IGBT1) denotes the self-heating thermal impedance while Z(IGBT1-IGBT2) denotes the coupling thermal impedance from IGBT2 to IGBT1.

It can be seen that the self-heating thermal impedance of one chip is much higher than the coupling thermal impedances from other chips. The anti-parallel diode of one IGBT contributes the highest thermal coupling rate to that IGBT, compared to other chips. In the substrate layout design, not only the neighbouring IGBTs should be properly distributed, also the thermal coupling from the neighbour diodes need to be considered.

11.2.4 Thermal-mechanical design

One of the key factors in power module packaging is the solder fatigue due to low-cycle thermal-mechanical loading. Thermal stresses and strains in power module packages with nonlinear materials are very difficult to achieve via experimental test. FEM is one of the best methods to investigate solder layer fatigue, thermal loading, jointed with solder material nonlinear mechanical property model. FEM method has been utilised to analyse and obtain numerical results for thermal stresses distribution, deformation distribution and creep-strain distribution in power module packages.

As shown in Figure 11.17(a), ANSYS-workbench listed three main simulation blocks, that is, *Engineering Data* is for material models used in Static Structural simulation, *Geometry* block is 3D CAD model for the simulation and *Static Structural* block is the core of whole mechanical simulation. The list of materials can be seen in Figure 11.17(b). The materials used in structural simulation include copper, aluminium, silicone, AlSiC, AlN ceramic, different types of solder alloy and others. Material properties and proper meshing inside of simulation can determine the accuracy of the FE model. Linear, nonlinear, elastic, plastic, time and temperature-independent and dependent properties have to be included in the FEM model. Key properties of thermal-mechanical material have been included in the workbench, such as CTE values, Young's Modulus, Poisson's ratio, shear modulus, thermal conductivity, etc.

The solder material has been modelled with modified Anand's rate-dependent plasticity model. Anand's model incorporates visco-plasticity, a time-dependent

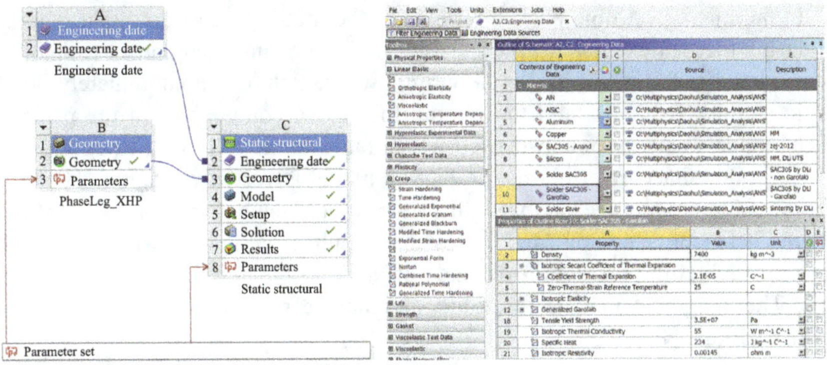

(a) (b)

Figure 11.17 ANSYS-workbench setup: (a) simulation structure in workbench and (b) material list in workbench

plasticity phenomenon, that the development of plastic strains is dependent on the loading ratio. Darveaux provided the solder constitutive relations based on Anand's model for rate-dependent plasticity [12].

The constitutive equation is the flow equation as prescribed in the equation below:

$$\dot{\varepsilon_p} = Ae^{(-Q/RT)}\left[\sinh\left(\xi\frac{\sigma}{s}\right)\right]^{1/m}$$

where $\dot{\varepsilon_p}$ is the inelastic strain rate, A is a pre-exponential factor, Q is the activation energy, R is the universal gas constant, T is the current absolute temperature, ξ is a multiplier of stress, σ is the current tensile stress, m is the strain rate sensitivity and s is the internal state variable (deformation resistance) and the evolution equation can be seen from below:

$$s^* = \hat{s}\left[\frac{\dot{\varepsilon_p}}{A}e^{(-Q/RT)}\right]^n$$

where $\hat{s}$ is a coefficient for deformation resistance saturation value and n is the strain rate sensitivity for s saturation. Anand's model has been shown to provide reasonable results when compared to a combination of plasticity and creep model. To model the material behaviour of solder, nine material constants are needed which are A, Q, ξ, m, h_0, $\hat{s}$, n, a plus the initial value of the deformation resistance, s_0.

The fatigue-life prediction models used to evaluate the thermo-mechanical reliability of the compliant interconnect and solder material. Fatigue life models are used to determine the number of cycles that a package can endure before failing. There are two approaches: high cycle fatigue (HCF) and low cycle fatigue (LCF). In HCF, it is based on stress reversals which are in the elastic regime and have not exceeded the yield point. As for LCF, it is based on strain reversals where plastic deformation dominates.

In the microelectronic packaging, due to the multi-axial loading that is incurred during thermal excursions, the reliability of compliant interconnect technology is considered to fail under low cycle fatigue where failure is predominately due to plastic strains. To understand the fatigue life of the components in the package, fatigue models have been constructed which fall into the following categories.

To build an accurate lifetime prediction model is very challenging. The complexity of 3D module packaging with solder joints includes multi-axial nonlinear material behaviours, complicated joint shapes, multi-axial loadings and others. The accuracy of lifetime prediction depends on understanding and careful considerations. The analytical models for solder fatigue are generally based on Manson-Coffin relations [13]:

$$N_f = \left(\frac{\Delta\varepsilon_{pl}}{\varepsilon_f}\right)^{1/C}$$

The equation uses cyclic inelastic strain amplitude or similar equation for strain energy dissipated in a cycle. It is not possible with easy means to determine multi-axial axial strain stages or the elastic parts of strain. The strain is calculated based on simplifications. In most cases, the multi-axial strain stage is simplified to pure shear strain stage with constant amplitude and all strain is assumed to be a plastic strain. More effort required to understand and utilise the fatigue lifetime prediction model.

An improved experimentally supported model for determining the acceleration factor $N_{\text{test}}/N_{\text{field}}$ was developed by Norris and Landzberg [14]:

$$N_{\text{field}(x)} = N_{\text{test}(x)}\left(\frac{\Delta T_{\text{test}}}{\Delta T_{\text{field}}}\right)^{c1}\left(\frac{f_{\text{field}}}{f_{\text{test}}}\right)^{1/3}\exp\left[1414\left(\frac{1}{T_{\text{field}}}-\frac{1}{T_{\text{test}}}\right)\right]$$

where $N_{\text{field}(x)}$ is the number of field cycles at which a percentage x of connections has failed, for example, $N_{\text{field}(50)}$ mean cycles to failure at which 50% of connections have failed; $N_{\text{test}(x)}$ is the analogous number of test cycles; $c1$ is the empirical constant, applicable is $c1 = 1.9$; f is the cycling frequency expressed in number of cycles per day and T is the maximum cycling temperature in K.

Lifetime estimation based on mission profiles seems to be an appropriate method to validate system designs for lifetime requirements of applications. The operating conditions as start together with dynamic characteristics of load can determine the electrical parameters as the mission profile of the system. There are several bond-wire models listed for future references, that is LESIT model, Norris–Landzberg model, CIPS08 model and Scheuermann's silver-sintered model. The lifetime estimation is not the focus of this session [13–17].

11.3 Enhancement of power module reliability

The design and simulation of power semiconductor packaging in multi-physics aspects have provided technical guidance and acted as efficient analytical tools at different stages during the power module's development. To achieve high reliability for high power density, high voltage power module products, there are more areas to be investigated, such as packaging materials, manufacture processes, electrical test and reliability test [18]. As mentioned in the beginning of this chapter, thermal characteristics of packaging materials, such as thermal conductivity, thermal resistance, thermal capacitance and CTE are critical and affect product reliability. The packaging material selection for different application scenarios needs to be varied even for the same kind of power modules. The CTE values and thermal conductivity values of typical packaging materials used in high voltage/high power modules can be seen in Table 11.3 for a reference. The thermal CTE mismatch is one of the key factors to influence long-term reliability of the power module in the switching energy applications. The CTE differences during different temperature conditions can affect the mechanical bonding from wire bonding to chip soldering layers and substrate soldering layers. The stresses generated from the mechanical deformation can cause strain fatigue of material in long-term operation.

This session will focus on the material, process and test of packaging. Main contents will cover materials and corresponding processes, high insulation materials and the curing processes, static/dynamic electrical tests of the power module and accelerated tests for reliability.

11.3.1 Bonding materials and processes

Power modules are the core devices to provide high current and high power to drive the switching applications, that is railway, renewable energy, HEV/EV, more electric aircraft (MEA) and industry. One of the key elements for power semiconductor devices is to provide highly efficient reliable electrical power. For high power modules with electrical current ranging from tens of Amperes, hundreds of Amperes, up to several thousand Amperes, the bonding wires are normally utilised for current conduction from topside of chips to metallisation layer of insulated substrate and copper busbars are utilised to conduct so high current to external power terminals. The bonding wire, busbar material selection and their

Table 11.3 CTE and thermal conductivity of packaging materials. Adapted from [12] and [19]

Material	CTE (ppm/K)	Thermal conductivity (W/m.K)
Aluminium (Al)	24	235
Copper (Cu)	16	390
Silicon (Si)	2.6	130
Silicon carbide (SiC 4H)	5	120
Aluminium SiC (AlSiC)	7	200
Al_2O_3 (alumina)	6.5	30
Aluminium nitride (AlN)	4.5	180
Silicon nitride (Si_3N_4)	3.3	29

corresponding bonding processes affect the reliability of the power module in the field operation. This kind of different bonding connections can be considered as mutual chips and topside chips connection aspect in power module packaging technologies.

11.3.1.1 Wire bonding

The ultrasonic oscillation frequency ranges normally from 40 to 100 kHz for wedge wirebonding. The oxide coating of the aluminium wire is removed and forms reliable bonding with a chip or topside metal layer of the ceramic substrate. There are two different types of ultrasonic bonding used, thin-wire bonding (99% aluminium and 1% silicon, a gauge of 17 to 100 μm) and heavy-wire bonding (99.99% aluminium, a gauge of 100 to 500 μm). For power semiconductors focused in this, extreme thick aluminium wires are normally selected to meet high current conducting requirements, that is 10 thou (254 μm), 15 thou (375 μm) or even 20 thou (510 μm) as typical aluminium wire diameters. The aluminium wires are bonded onto the topside of silicon chips and copper metallisation layer of the ceramic substrate via ultrasonic wedge bonding process.

For each chip with ~100 mm^2 active regions, there are normally around 10 or even more wires bonded in parallel onto each chip. The CTE values for Si chips, Al wire and copper are quite different as 2.6, 24 and 17, respectively. This CTE mismatch contributes to bonding wire lift-off failure, which is one kind of well-known reliability problems in high current, high power density power modules [20,21]. The CTE mismatch together with a high temperature in operation can result in wires fatigue and lift-off at bonding contact area. To increase power cycling capability, the wedge bonded aluminium bonding feet can be encapsulated within epoxy or polyimide type of materials with the intention to increase bond lifetime under repetitive thermal load as indicated in Figure 11.18 [20]. Further to encapsulate both wire bond and bond foot areas in this extra insulation material may be seen to introduce either detrimental or beneficial thermal reliability effects,

(a) (b)

Figure 11.18 Glob-topping to power electronics die surface: (a) black epoxy material and (b) yellow colour PI are applied to bonding wire foot areas only dependent on the physical properties of materials. The aluminium wires are normally cheap and simple to bond. To increase power cycling capability of topside wire bonding connections, aluminium wires can be replaced by aluminium ribbons, which can provide higher current carrying capabilities. Compared to aluminium wire, aluminium ribbon is one cost-effective topside interconnection solution with less mechanical flexibility [22].

As the electrical conductivity and thermal characteristics are limited for aluminium wires, there are several kinds of different materials or topside bonding process aiming to provide longer life-time and enhance power cycling capabilities of different kinds of power modules. Compared to aluminium material, copper has a higher melting point, higher current conductivity, higher thermal conductivity, higher mechanical resistance to thermal stress under power cycling operations and lower CTE. The copper wires or copper ribbons are good material candidates to provide longer lifetime interconnection in high power switching energy applications [23,24]. For the same current ratio, thinner copper wire can replace the heavy aluminium wire with high reliability. Copper bonded power module packaging can fit the future requirement of higher junction temperature, such as 175 °C and high current ratios with the same package outline dimensions [24].

As copper wire-bonding requires copper metallisation layer for power semiconductor chips, which is rather an expensive solution for wafer fab compared to conventional metallisation layer for power semiconductor chips. There is aluminium coated copper wire (CuCorAl) [25], the CuCorAl material mixes aluminium and copper and try to utilise the advantages for both materials, such as excellent electrical conductivity and thermal conductivity of copper, better bondability of aluminium wire due to the softness of aluminium coating layer outside of the copper wire. The long-term lifetime for both power cycling tests and passive temperature cycling tests has been improved compared to pure aluminium bonding wires [26]. CuCorAl structure can provide reliable bonding windows without changing chip metallisation. The bonding wire can use conventional aluminium wire bonding equipment. For mass production requirements of power semiconductor product, CuCorAl wire bonding can achieve higher reliability with relatively lower cost compared to full copper wirebonding requirements for both process and power semiconductor chips [25].

11.3.1.2 Soldering

Soldering is one of the common backside interconnection processes for power semiconductor chips. The IGBT/FRDand SiC MOSFET/SBD chips are in vertical conducting structure and their backside interconnection will handle high current ratio and transfer most of the powerful heat generated during the chip operation. Their electrical performance of chips is related to the junction temperature. For power cycling and passive temperature cycling tests, power semiconductor soldering layer and substrate to baseplate soldering layer are critical. The soldering materials, soldering process and soldering equipment can influence the quality and future reliability of power modules. For traditional solder material, Sn–Pb-based solder has been selected for production of large-area interconnection, that is chips to ceramic substrate and substrate to the baseplate. In recent years, lead-free solder materials have been used in a wider range corresponding to environmental challenges. During the soldering process, the CTEs of chips, ceramic substrate and baseplate are different and they are joined together by melting solder materials with different deform and stress in the soldered area.

For the soldering processes, solder paste with a huge amount of flux can be easily soldered with good wettability of joint materials. However, flux residual needs to be removed after the soldering process. The flux residual may affect the lifetime of power semiconductor devices, especially for high voltage packaging, which requires high insulation encapsulation material to avoid partial discharge or potential high electric field area. For washing process after paste soldering, water-soluble detergent or ultrasonic waves are usually applied. Some chlorinated materials need to be avoided as washing materials in case of any extra potential damage for plastic frame or power semiconductor chips in long-term high voltage operation [27].

Fluxless preform soldering has been one of the options to solve the flux residue problems. For the fluxless soldering process, there is no solder paste required and just solder alloy preforms. The formic acid vapour soldering process is one of the most ideal reaction gases for fluxless process. The formic acid vapour is active at typical soldering temperature and the operation is relatively simple. The formic acid decomposes at the high end of reflow temperature, which means it will not contaminate the soldered parts. During the formic acid vapour process, formic acid vapour will be activated once the temperature of the bubbler containing HCOOH acid is higher than 150 °C. A channel of pure nitrogen will be utilised and combined with the formic acid vapour to adjust the concentration density of formic acid vapour.

Besides the formic acid vapour soldering process, there are so-called stand-off wire spacers to improve soldering reliability [28] and copper mesh embedded soldering preform to be utilised for high reliability requirement [29]. The general idea of wire spacer or embedded mesh into a soldering layer is to obtain a guaranteed solder thickness for substrate attach by bonding wires or embedded mesh onto the baseplate. The bonding wire can be used mainly for AlSiC baseplates and punched extrusions or coining is utilised for the copper baseplate and the copper mesh embedded preform solder can be used onto both AlSiC and Cu baseplate. The X-ray pictures of stand-off spacer and embedded copper mesh have been shown in Figure 11.19, respectively.

11.3.1.3 Silver sintering

With increased long-term reliability demand for higher junction temperature and higher working scenario temperature in high power density power modules, it is a big challenge to utilise traditional soldering alloy materials and processes. For traditional soldering material, the junction temperature of silicon chips is normally up to 150 or 175 °C. For wideband gap chips, such as SiC and GaN, which aim to have 200 or 300 °C junction temperature, the traditional solder alloy may melt at

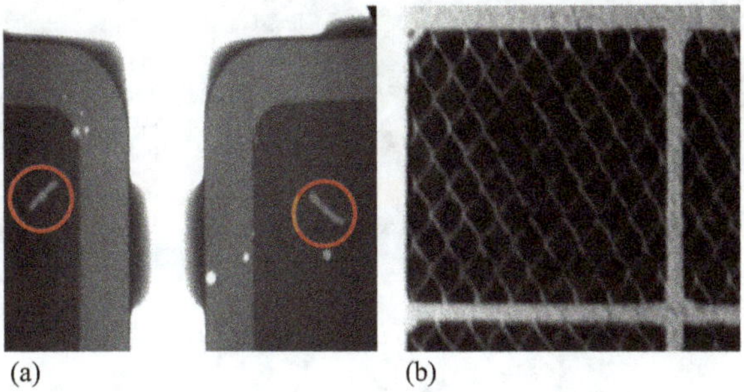

Figure 11.19 X-ray image of soldered substrate with (a) stand-off bonding wires and (b) embedded copper mesh

such high temperature. Silver sintering can be processed below 300 °C, which are fit for a so-called low-temperature process and high-temperature application, with its high melting point (961°C), excellent thermal conductivity and electrical conductivity. Silver sintering can become a good candidate for future higher temperature power module interconnection technology [30,31].

For basic sintering process, most silver sintering materials are in the form of a paste, which needs to be printed onto the substrate and then get the silver paste dried out. There are several companies, which can supply pre-printed and dried silver film. Normally, it may be not in the right side of the die, but a piece of silver film with a large area. Then, a die normally will be placed on a paste containing nanoscale particles. High temperature and pressure press are applied as the following process steps to make sure the good and sufficient connection between die, paste and substrate via silver-ion diffusion. Due to the lower processing temperature than the material's melting temperature, silver sintering layer is a porous structure which is not indicated in Figure 11.20.

The main challenge of the press sintering process is how pressure is utilised. The reason is that multi-chips need to be pressed simultaneously without any differences for the sintering layer thickness, shape and size. The biggest challenge is the uniformity of the applied pressure. Chip placement and alignment to the sliver sintering paste pad is critical as well. Silver sintering material doesn't perform like the solder alloy which will melt and flow around. The solder alloy melting flow can be stopped by solder resistance material. However, silver sintering paste will not flow because of its low processing temperature characteristics. The process temperature for silver sintering is below 300 °C far lower than its melting temperature, about 961 °C. During the sintering process, pressure needs to be applied with extreme care, and displacement may cause stress and die crack and substrate crack. For mass production, automatic sintering machine is required together with pick-place facilities due to the small dimension of SiC or GaN chips. Also, precise jig and fixture are required to keep the dies inaccurate place.

For reliability of silver sintering, bonding mechanism and failure mechanism are a very important aspect to be investigated. Bend test, which has also been used by Semikron [32], can easily indicate whether silver sinter bonding is good or not.

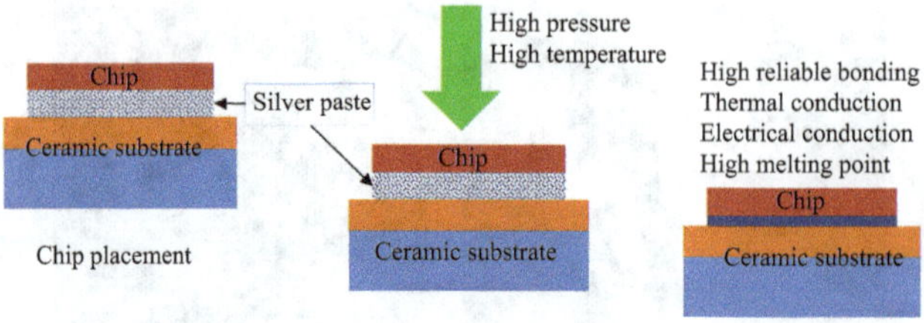

Figure 11.20 Basic silvering sintering processes steps

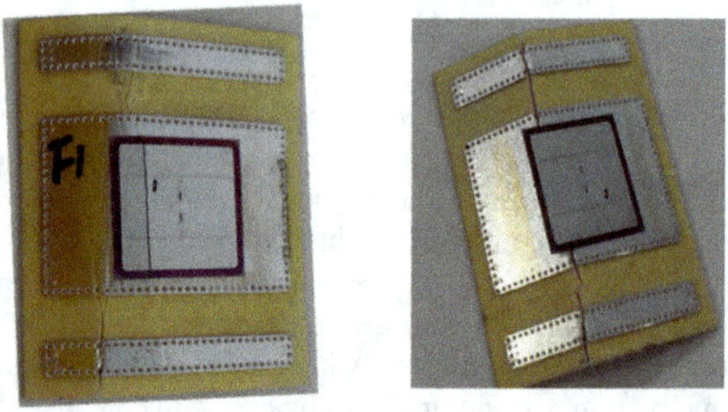

Figure 11.21 Bend test of sintered die to substrate

Figure 11.21 shows examples for obvious comparison between a good and a bad sintering bonding via the bend test. For the good bonding, the chip crack area will not split from the sintered substrate as indicated in the left picture of Figure 11.21. For a bad bonding, as shown on the right-side picture of Figure 11.21, one can see the crack area of the chip has separated from the substrate. However, the bend test cannot tell the bonding strength of the sintering process.

Proper chip shear test can tell the sintering bonding strength. The shear test facility from XYZTEC has die shear tooling as shown in Figure 11.22(a). As die is very thin, the die area is very large relatively, which may result in the stresses created by the shear tool being higher than the yield strength of the die causing it to shatter before the bond is tested. For the bad example, the bonding strength is low, when the shear strength is 3–4 MPa, the die was completely sheared off without crush. For the good sintering bond sample, when the applied shear strength was up to 20 MPa, the die was crushed at first as shown in Figure 11.22(b). The test can indicate that the bonding strength is higher than 20 MPa, which is comparable to soldered die shear strength.

Compared to the soldering process, sintering has its advantages on thermal conduction, electrical conduction can afford high-temperature operation of device even over 300 °C. The silver sinter can be potentially taken as environment-friendly material compared to the traditional lead soldering materials.

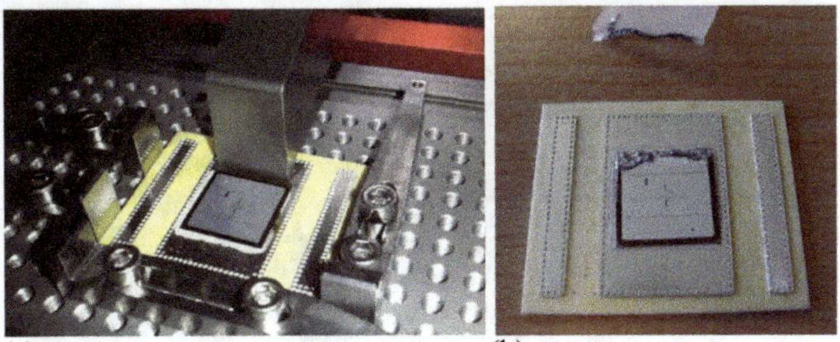

(a) (b)

Figure 11.22 (a) Shear test facility and (b) the chips after the shear test

11.3.1.4 Ultrasonic welding

For high current, high power density power module and copper busbars (power terminals) are widely used to carry high current with its good electrical conductivity. The busbar connection to metallisation layer of substrate is usually one critical interconnection for reliability due to the high current carrying through and mechanical vibration in the real field applications. Traditionally, the soldering process has been utilised to joint copper busbar with a metallisation layer of the ceramic substrate. The package's thermal shock capability, vibration capability and current-carrying capability of soldered busbar are difficult to fit the future higher current operation in long-term reliability requirement.

In recent years, the so-called ultrasonic welding process as shown in Figure 11.23, which has been proven to be one of the most reliable bonding methods in power modules. The baseplate will be fixed mechanically during the welding process, the part which is not included in Figure 11.23. The main emitter/collector busbars and auxiliary busbars are made of copper, which can be directly welded onto the surface of copper metallisation substrate by using a Sonotrode as shown in Figure 11.23. The sonotrode vibrates mechanically at ultrasonic frequency range when exposed to high frequency. It can link the connections among ultrasonic generator, the copper busbars and substrate metallisation layer. The ultrasonic vibration can displace and remove most of the contamination and oxidation layers of the metal surface. Once the contamination and oxidation layers are removed, the welded parts are clean metal layers, which can be bonded together firmly. The USW can provide greater current conduction capability, stronger mechanical connection and reduce thermal load [33].

USW has been widely utilised in power module build and the typical cross-section of USW bonded busbars can be seen in Figure 11.24(a). The picture has shown the cross-section of busbar, ceramic copper metallisation layer, ceramic layer, solder layer and baseplate layer from topside to downside. Normally, there is a nickel plating layer for substrate or oxidation layer for bare copper layer metallisation of the substrate. The plating layer or oxidation layer can be effectively removed via the ultrasonic vibration force to generate reliable copper-copper bond as indicated in

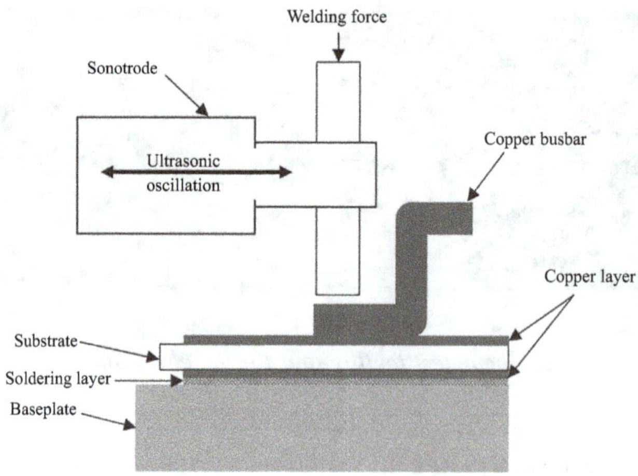

Figure 11.23 Ultrasonic welding for auxiliary and main power terminal connections

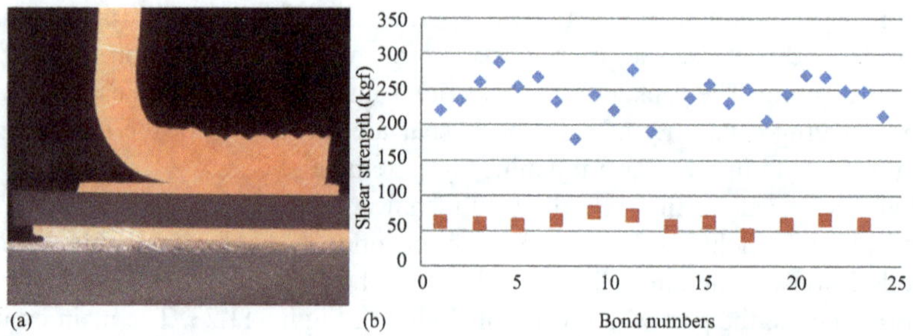

Figure 11.24 (a) Cross-section of good USW bonded busbar and (b) USW bonding shear force compared to soldered busbar

Figure 11.24(a). The metal residue or particle generated during the USW bonding need to be cleaned after the bonding via different cleaning method. The extra attention needs to be paid onto the change for the ceramic substrate layer and soldering layer to see any mechanical effect after USW bonding process. Any crack or damage on the ceramic material and solder layer can affect the substrate's insulation capability and soldering material's inner mechanical stress. USW with proper bonding parameters is critical to providing long-term reliability of the power module.

Shear test of busbar feet has been utilised to compare the strength of USW bonding to traditional soldered busbars. The shear test of USW bonded busbar feet can reach more than 200 kgf sheer force, that is around 60–80 kgf of traditional soldered busbars as indicated in Figure 11.24(b). The average USW bonding force is typically three to four times of soldered busbars. USW bonded copper busbar is one of the key processes to the enhanced reliability of module packages. One can say that the module using USW bonding could provide much stronger and reliable bonding than the traditional soldered busbars module [33].

11.3.2 High insulation material and processes

11.3.2.1 Ceramic substrate

For high voltage and high current power module, high insulation ceramic substrate is essential to carry on the main isolation voltage via ceramic material and high current via metallisation layer. Some key functions have been listed below:

- To provide electrical insulation between the potential at power components and the potential at baseplate or cooling structure, the ceramic material is critical to provide isolation requirement with relatively high thermal conductivity;
- To conduct current via topside metallisation layer tracks, usually, copper or aluminium can be used with thin thickness which fit current conduction requirements;
- To provide good thermal conduction path between semiconductor chips and cooling structure;
- To provide high level reliability in power cycling with good interconnection to topside chips and downside baseplate or cooling structures.

For typical high voltage applications such as 3.3, 4.5 and 6.5 kV, aluminium nitride (AlN) ceramic is selected as the dielectric insulator. AlN has low thermal resistance compared to other insulating materials, which is about 180~200 W/m.K and superior thermal capacity and thermal spreading properties with the copper metallisation. CTE value of AlN is ~4.7 ppm/K, which provides a better match to the silicon semiconductor (~4.0 ppm/K) than other types of substrates based on metal or plastic materials. The metallisation layer of the insulating substrate can conduct high current with a relatively low voltage drop – typically, the ohmic voltage drop across the metallisation should not exceed one-tenth of the IGBT's V_{CESAT}. This ohmic voltage drop can be one of simple verification rule to test the ohmic voltage drop due to the metal layer layout of the substrate. The voltage drop needs to be designed as low as possible via parasitic resistance optimisation introduced in Section 11.2.1.2.

Currently, copper-AlN-copper substrates are one of the main options by considering general soldering processes and ultrasonic welding of copper busbars to fit electrical, thermal, mechanical and chemical requirements in high voltage IGBT module. Copper and AlN material are usually typical materials from ceramic substrate suppliers worldwide. The very thin plating layer on top of the copper layer is very critical to the reliability of the power module assembly. The plating layer can affect soldering processes and USW processes.

Substrates with stronger bond force are needed to obtain a longer running lifetime requirement of the power module. Some substrate suppliers provide active metal brazing (AMB) substrates. AMB process utilises a brazing alloy to form a layer between the copper and ceramic by heating the filler metal above its melting point and it spreads between the two jointed parts by capillary action. The filler metal is heated slightly to liquidus in the vacuum environment. It is quite similar to soldering, except that the temperature used is much higher for the brazing process. AMB substrates are common to be used in the latest high voltage modules in recent years. Compared with the traditional direct bonding copper (DBC) process, AMB has several advantages, such as (1) better adhesion between metal and ceramic, (2) better current-carrying capability and (3) better thermal-fatigue capability.

For high isolation module, that is, typical baseplate, substrates structures are used for high voltage packaging design and manufacturer. For high voltage power module, that is 3.3, 4.5, and 6.5 kV voltage ranges, the isolation is 6 and 10.2 kV AC voltage. The ceramic layer of the substrate is the main insulation material to hold on such high voltages. Any void or crack from the substrate manufacturing process or power module package assembly process will influence the isolation capability of the module. The results can be immediately breakdown during isolation test or long-term breakdown failure during module operation. For low isolation package, that is, 1.2 and 1.7 kV modules, integrated substrates can be considered for future low thermal resistance and longer lifetime reliability. Some new types of integrated substrates structure to replace the traditional substrate-solder layer-baseplate structures have appeared in publications and samples. Some integrated substrates with flat baseplate will be mounted onto the thermal sink with thermal grease and another solution is to integrate ceramic substrate directly onto the thermal sink and thermal grease will not be used in this kind of structures.

11.3.2.2 Baseplate

In a high power module, baseplate acts as a mechanical support to all the substrates, busbars, plastic frames and insulation encapsulant. It can absorb heat from substrates during power transient period and transmit heat from the substrates to the heat sink. The baseplate material needs to be highly thermal conductive so that it can dissipate the generated heat efficiently. The mechanical property of baseplate needs to be strong to experience strong mechanical stress from the thermal load and mounting force during operation. The baseplate material for high power applications should be compatible with the ceramic substrate material. There are critical requirements in thermal, mechanical aspects, that is high thermal conductivity, compatible CTE with the insulating substrate, high tensile strength, high flexural strength, easily shaped and bent, low density and weight to minimise damage due to mechanical shock, low moisture absorption rate, low toxicity material and so on.

There are a number of different baseplate materials that can be utilised in power electronic applications. One of the main choices is half-hardness oxygen-free high conductivity (OFHC) copper. Most other high performance baseplates are fabricated by using composite materials that vary not only with the percentage of component but also with particle sizes of constituent powders and the alloy components of metal. Copper baseplate can be used for low voltage package, that is, 1.2 and 1.7 kV voltages. Aluminium silicon carbide (AlSiC) baseplate has been widely used in high voltage IGBT modules for 3.3, 4.5 and 6.5 kV. SiC is thermally conductive ceramic inside of AlSiC. Both the composite materials and the processing of composite materials can affect the performance of AlSiC products. Inside AlSiC material, aluminium is taken as an exposed element, they usually need a nickel plating layer to strengthen baseplate's solderability and to protect baseplate from corrosions. The AlSiC materials are different from their components percentage with different manufacturing processes and provide different mechanical/thermal properties to fit different application requirements.

The baseplate can affect the soldering process, assembly process, thermal and mechanical stress level of the power module. Baseplate bow level is one of the key parameters to be considered besides material selection and IGBT power module is usually mounted to the heat sink via mounting holes on the baseplate. Different baseplate bows will cause different stress levels to the soldered substrates and semiconductor chips. High stress will damage chips during long-term operation. The initial bow from baseplate supplier will change after soldering process and module assembly processes, and the bow will also be affected by different types of solder alloy, solder processes, etc. Typical bow specifications are generated mainly based on baseplate material, assembly processes used and the solder types, that is lead solder or lead-free solder. Not only during the module build process, the power module's bow will relax to a certain extent after module finish, storage and shipment, but it makes the prediction of the shape and bow value of the baseplate very difficult [34].

Also, the bow value of module will normally vary during power cycling, passive thermal cycling test and thermal shock test as well as in the real field applications. The key for module bow in the consideration is to get as flat as possible at high operating temperature. To achieve this target, the material selection, soldering material/process, packaging materials combination, baseplate mounting force and even operation conditions are key elements to affect the bow. As introduced in Section 11.2.4 in this chapter on simulation and design investigation into mechanical deformation, creep-strain, von-Mises stress, principal stress and shear stress of the assembled components, that is solder layer, ceramic substrate, baseplate, busbars and others. Thermal-mechanical investigation at the design stage can help to guide and achieve better performance power modules.

Although careful selection of packaging materials, there is still CTE mismatch for real module build. The mismatch causes so-called bimetallic effect due to thermal variations during the soldering process and module operation period [35]. The bow changes mainly relating to soldering processes and mounting or possible operating scenario. There are several key elements can be further considered, such as topside plastic frame mounting force and the filled silicone gel or other materials as insulation encapsulant. Especially in the plastic frame, the CTE of plastic material is rather high that it will affect the bow during module operation or different tests that the whole module experienced. To enhance the power module's reliability, the baseplate is always one of the most critical factors.

11.3.2.3 Encapsulant and process

For high power module, high insulation materials have been widely utilised to provide electrical, chemical, moisture, electromagnetic protection of semiconductor chips and interconnection bonds. The encapsulant provides insulation protection between conduction tracks via increasing electrical field breakdown levels. Encapsulant provides resistance to water moisture from the surrounding environment, which can protect and delay the corrosion procedure to the chips. For high power application scenarios in different voltage and current ratios, several different types of encapsulants are candidates to fit different purposes. Typical materials include silicone gel, polyimide (PI), epoxy mounting compound (EMC), etc.

Silicone gel, for example electrically insulating material, is a good choice to coat the power semiconductor chips, bonding wires and substrates. The silicone gel offers several good characteristics, such as wide operating temperature ranging from −60 °C up to 200 °C and low toxicity for environment-friendly module product. For high voltage module, there are several key aspects to be considered, such as (1) high dielectric strength, (2) water resistance ratio, (3) excellent adhesion to packaging materials, that is chips, bonding wires, substrate, baseplate, plastic frame and (4) long-term reliability under severe environments, that is extreme cold, extreme hot or high humidity. Due to the softness of silicone gel, it adds very little stress to the components encapsulated and it can absorb stress generated by thermal expansion of the surrounding materials. Apart from the material itself, the curing process and curing equipment have a big impact on the performance of silicone gel as well.

For high voltage such as 3.3, 4.5 and 6.5 kV power module, one challenge is the packaging's partial discharge (PD) capability. During the test, all the topside electrical interconnections will be short-connected and high voltage will be applied during the test. Based on *IEC Standard 61827 Part 1 – Characteristics and Test Methods*, the PD test procedure is as follows: (1) the isolation voltage can be calculated via $\left(2U_m/\sqrt{2} + 1\text{kV}\right)$, which is 10.2 kV AC for 6.5 kV module and 6.0 kV AC for 3.3 kV module, respectively. The isolation voltage will be applied for 60 s. The discharge level will be measured at the last 10 s. No breakdown means the isolation test is passed if the discharge is less than 10 pC and both the isolation and PD tests finish with PASS; (2) the PD test voltage can be calculated via $1.5U_m/\sqrt{2}$, which is 6.9 kV AC voltage for 6.5 kV module and 3.5 kV AC voltage for 3.3 kV modules, respectively. PD test will last for 60 s. The discharge is measured at the final 10 s period. If the discharge level is less than 10 pC, the test will be finished with a pass. Otherwise, the AC voltage directly moves down to $1.1U_m/\sqrt{2}$, which means 5.1 kV AC voltage for 6.5 kV module and 2.6 kV AC voltage for 3.3 kV module. The PD test voltage lasts for another 30 s. If the discharge level is less or equal to 10 pC, the module passes the PD test. Otherwise, the module fails for the PD test [2].

For 1.2, 1.7 and 3.3 kV voltage level module, silicone gel with high dielectric strength can guarantee the PD test with proper design of substrate layout. For high isolation packaging, especially 4.5 and 6.5 kV voltage level, the ceramic substrate can match the isolation requirements as thick ceramic material, such as AlN substrate. The PD challenge needs an extra type of encapsulant, that is polyimide to coat critical substrate area as indicated in Figure 11.25(a). The side edge of the topside metallisation layer is a critical area that can affect PD value during the test. There are normally small sharp regions at the side edge of the topside layer, which are difficult to be fully coated with only silicone gel. Any air void residue at the side edge of topside metallisation is easy to generate a very high electric field region. The high electrical field region may fail the PD test and it can discharge the surrounded silicone gel during the field application. Once the discharged region grows, the breakdown may happen and the whole power module can fail the operation. The solution is to use polyimide or other high dielectric material to cover the most critical region as shown in Figure 11.25(a). For the real PI coating, Figure 11.25(b) shows the cross-section of the PI coated substrate. The thick PI coating layer has covered the side edge of the substrate. The PI coated substrate with silicone gel-filled module has shown stronger PD capability than the module without PI coating especially for 6.5 kV voltage level module [36].

There are several kinds of high dielectric encapsulants suitable for power module semiconductors packaging, such as epoxy resin, epoxy mounting compound, paralyene and even more. The selection of encapsulant depends on the reliability requirement and applications.

11.3.3 Electrical and reliability test

Reliability enhancement for power package needs combine design/simulation, material selection, assembly process and test verification into a full development cycle. The four parts should be treated as one integrated that cannot be split from each other. Reliability is always the guidance and orientation of whole development in power semiconductor packaging.

With reference to *IEC60747-9* and *IEC60747-15*, electrical test will be carried out at different stage of module build, that is wafer-level semiconductor chips test to screen out the disqualified chips. The wafer process and wafer production target to increase yield level of chip fabrication, provide more consistent performance and reduce the cost of chips. Once semiconductor chips have been bonded onto the ceramic substrate via soldering or sintering, and wirebonding of topside interconnection bonds, static and dynamic test at substrate level will normally be carried

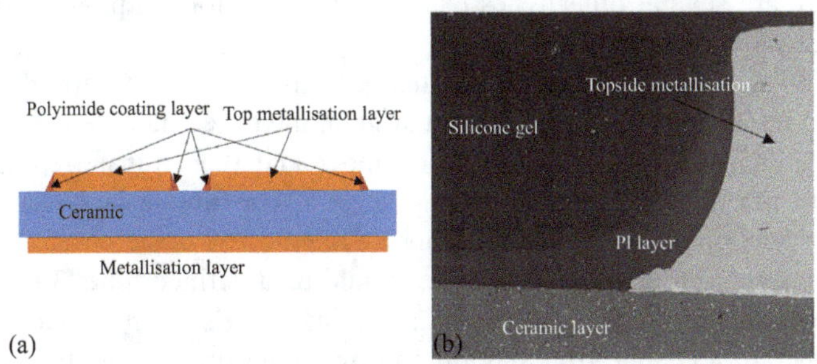

Figure 11.25 Substrate coated with polyimide to enhance PD capability of high isolation module: (a) substrate edge with coated polyimide and (b) cross-section of coated substrate

out to check key electrical performance at both room temperature and high temperature. The substrate level test mainly includes conduction current/voltage of chips and leakage current at different blocking voltage. Dynamic test at substrate level includes (1) single pulse short-circuit test at high temperature to check the maximum current carrying capability during over-current stage, (2) reverse bias safe operating area (RBSOA) test to check the substrate turn-off capability at high current conducting period where the blocking voltage cannot exceed the blocking voltage at turn-off stage and (c) double pulse chopping test at room temperature and high junction temperature.

The substrates will be built into the whole module package once it passes the substrate level test. The static and dynamic electrical test will be carried out with the pass criteria setup by customers or internal standards referring to IEC standards. The product datasheet will be generated from the key electrical parameters. Once the module has passed all the electrical tests, it can be continued for a reliability test. For module production, it will normally have a qualification test before the device can be taken as standard products.

There are three main reliability tests aiming to check power semiconductor chips quality in power modules, which are listed as below:

1. The **HTRB** (high-temperature reverse bias) test is carried out at a rated voltage for a certain period of time and the target is to study the blocking behaviour at 80% to 90% of its normal blocking voltage and at maximum operating junction temperature. The leakage current I_{CES} is measured during this procedure.
2. The HTGB (high-temperature gate bias) test is conducted at a maximum junction temperature with a certain period of time with gate bias applied. The gate is applied to positive and negative voltage (±20 V). No voltage applied between emitter and collector. The gate current is measured to compare with the reference gate current before the test.
3. The H3TRB (high temperature, high humidity, reverse bias) test is conducted at a defined temperature 85 °C and humidity 85%. A blocking voltage of up 80% of blocking voltage is reduced to 80 V to test the leakage current compared with reference leakage current.

   There are several other types of reliability tests for the power module and those are listed as below:
4. The TST (thermal shock test) or temperature cycling is carried out for a defined number of cycles. In the test system, the tested module is taken to low temperature (i.e., −50 °C) in one chamber and then transferred to another chamber at a higher temperature (i.e., 150 °C). The time in each chamber must be long enough for the temperature inside the module to stabilise. The transfer time between hot and cold chambers should be a defined time, 30 s typically. The test can speed up the large area soldering layer's fatigue under stress due to temperature change from day time to overnight time. It can be seen as low cycle fatigue (LCF) for the solder layer reliability. The test can check the weak point or early failure mode of the packaging, such as traditional soldered busbar on high voltage, high power module. For typical industrial application,

the typical temperature ranges are −50 °C to 150 °C or −45 °C to 125 °C and the pass criteria is 100 cycles or several hundred cycles.

5. The **TC** (thermal cycling) or passive cycling is conducted with shorter temperature cycle times in the range of several minutes. The heating source of the module may be passive. The cycling temperature range is smaller than that for the thermal shock test. The large are solder layer's fatigue under thermal stress is normally taken as high cycle fatigue (HCF) for solder layer reliability. Typical ΔT is 60 or 80 K.
6. The PC (power cycling) test is performed to heat up power module via active constant DC current going through chips and to be cooled down for a certain number of cycles. The cycling time, that is 2 s < conduction period T_{on} < 6 s, is set to test the reliability of bonding wires and longer power cycles, that is T_{on} > 10 s to test the reliability of soldering layer under chips. Power cycling test aims to stimulate the stress for different layers of the whole power module under normal operating conditions. Different ΔT can be tested to investigate the reliability of the power module. Higher ΔT test can speed up the cycling test.
7. Vibration test and mechanical shock test are two kinds of accelerated tests to verify weak mechanical properties inside of power module. Referred to *IEC60068-2-6*, vibration test runs at a certain frequency range, from 10 Hz to 500 Hz with 10 g acceleration force as the test condition. The mechanical shock test will be normally carried out referred to *IEC60068-2-27* with half-sine pulse, 25–30 g acceleration rate in *x*-, *y*- and *z*-directions. During the tests, weak points and weak parts of power terminal busbars, auxiliary pins, long bonding wires and their bonding interconnections can fail earlier than other parts. The reliability improvement task will focus on optimising the corresponding weak parts for further vibration or mechanical shock tests.
8. The low-temperature storage (LTS) and high-temperature storage (HTS) are two tests for the packaging material's durability under extreme low-temperature and extreme high-temperature conditions. For some applications, it may require the module to afford low temperature −50 °C for over 1,000 h and 150 °C HTS for over 1,000 h. For those two tests, silicone gel is one of the most important parts to be affected in high power modules.

Most of the above reliability tests are different kinds of accelerated ageing tests. They are general reliability tests for most of the applications. For different applications under different operating environments, even the same power module will normally show different working lifetime.

11.3.4 Environment test

There is a significant portion of high voltage and high power applications of semiconductor modules working under harsh environment scenarios and the long-term operations of power modules are prone to be affected by high humidity, extremely hot or extremely cold weather. The humidity is one of the common harsh environmental challenges during the operation. It is one critical factor in affecting the reliability of power modules, the failure mechanism, test methodology and material and structural improvement regarding humidity which are hot research topics in recent years.

To identify and understand failure mechanism, traditional test methods are limited, that is the H3TRB as mentioned before. H3TRB test has long been utilised to test modules at 85 °C and 85% relative humidity conditions with bias voltage at 80 V. It can accelerate water moisture penetration into the package material and passivation material of chips' edge. However, for high voltage power module, the electrical field distribution inside the package is much higher than the applied 80 V for H3TRB test. The H3TRB test is limited to provide proper reliability information for high voltage power modules. So, 60%–80% of high blocking voltage has been applied by combining with high temperature, that is 85 °C and high humidity, 85% to investigate high voltage HV-H3TRB. HV-H3TRB is one kind of accelerating test to investigate the failure mechanism of the power module. To further speed up the test, different blocking voltage, different high temperature and different relative humidity can be utilised in the test. The leakage current is the key parameters to be monitored during the test. The moisture level change rate and voltage level change rate during the test may influence the reliability of the power module.

The chips' termination edge degradation process has been explained by Zorn and Kaminski [37,38] where 65% and 90% of blocking voltage for 1.2 and 1.7 kV IGBT module tests have been carried out and the corresponding acceleration models have been reviewed. As high voltage and high power module works normally at high-temperature conditions, that is, up to 125 °C, 150 °C or higher, water moisture will be dissipated by heat generated from the inside chips. If the module is working on partial load, the water moisture can be absorbed under long time idle operation [39]. Packaging encapsulants of power module can absorb water moisture till saturation from surrounded humidity as there is no perfect shield cabinet for the inverters. The increase and reduction of moisture level inside the power module will affect the dielectric strength of encapsulants. Once water moistures gather in the silicone gel proximity to semiconductor chips, the passivation layer (PI layer) of chips can absorb moisture as well to form high electrical field region at the guard ring termination [40]. The water moisture level changes under high temperature and high voltage, and the high electrical field may cause metal corrosion and electrochemistry migration (ECM) that can affect the reliability of power modules [37,40].

By considering high humidity conditions, all the material properties will be changed to some extent, especially the silicone gel and passivation materials of semiconductor chips. Silicone gel or encapsulants with better water resistance are crucial to prevent partially or delay water moisture penetration process into the proximity area of chips and metallisation interconnection bonds. Some extra coating layer onto the bonding area, that is wire bonding contact and PI coating layer to the ceramic substrate can improve water resistance capability of the package. For passivation layer structure improvement as described by Papadopoulos *et al.* [40], failure mechanism due to humidity-induced electrical field change has been verified and it also provides one set of environmental robust high power IGBTs design methodology.

11.4 Summary

Reliability is a core requirement for semiconductor devices in power energy switching applications. In this chapter, reliability-oriented design, simulation, packaging material and process have been covered, especially for high voltage, high current and high power density semiconductor devices.

For packaging structure design, parasitic inductance and resistance connection of main power and auxiliary control connection will affect the electrical performance of semiconductor devices. 3D ANSYS EM simulation on parasitic extraction and circuitry simulation has shown how to design and optimise the layout design in this chapter. During electrical switching period, power loss from semiconductor chips will heat up layers of packaging materials, that is substrate, soldering layer and baseplate. Thermal design and simulation are crucial to transfer heat efficiently. Thermal optimisation is another key part of layout and structure design. As highlighted in the chapter, multi-physics coupling simulation of EM, circuitry and thermal are key design, optimisation and failure analysis tools in power packaging development.

CTE mismatch between layers of different packaging materials may cause failure and affect the lifetime of the power module. Material selection and corresponding bonding processes are a key area for reliability improvement as introduced in this chapter. Some new types of interconnection methods, that is silver sinter, copper wirebonding, ultrasonic welding of busbar and PI coating of wirebonding contact area will strengthen the reliability of high power semiconductor devices.

Static/dynamic electrical tests, isolation test and reliability tests as mentioned in this chapter are typical methodologies to study the quality and reliability of high voltage power modules. Only via tests, failure mechanism and improvement methods can be investigated and verified. The modification and improvement will depend on the optimisation of design and material/process modification. For power modules applied under harsh environments, reliability tests and procedures will develop into more environmental-related, such as high humidity or extreme cold conditions. HV-H3TRB has contributed to reliability or robustness in improvement of the latest reliability enhancement of high voltage power module.

Acknowledgements

The authors thank the colleagues from Power Electronics R&D Center, Dynex Semiconductor Ltd, UK, Zhuzhou CRRC Time Semiconductor Co. Ltd and State Key Laboratory of Advanced Power Semiconductor Devices, CRRC Times Electric Co. Ltd, China.

Note to the reader

This chapter is part of an edited book. To cite or use this chapter as a reference, please use the following format:

Li D., Dai X., and Liu G. "Module design and reliability." In: Iannuzzo F (ed.), *Modern Power Electronic Devices: Physics, Applications, and Reliability*. Stevenage, UK: IET; 2020. pp. 347–384.

References

[1] Lutz J., Schlangenotto H., Scheuermann U., and Doncker R. *Semiconductor Power Devices, Physics, Characteristics, Reliability*. Heidelberg: Springer, 2011.

[2] Volke A. and Hornkamp M. *IGBT Modules-Technologies, Driver and Applications*. 2nd edn. Munich: Infineon Technologies AG, 2012.

[3] Wintrich A., Nicolar U., Tursky W., and Reimann T. *Application Manual Power Semiconductors*. 2nd edn. Nuremberg: Semiktron, 2015.

[4] Qi F., Li D., Wang Y., *et al.* 'Status and trend of SiC device power module packaging'. *International Conference for Power Conversion, Intelligent Motion in Asia (PCIM Asia) 2018*, Shanghai, China.

[5] Wang Y., Dai X., Liu G., Li D., and Jones S. 'An overview of advanced power semiconductor packaging for automotive system'. *International Conference on Integrated Power Electronics Systems (CIPS) 2016*, 3–6 Mar 2016, Stuttgart, Germany.

[6] Mumby-Croft P., Li D., Dai X., and Liu G. *Disruptive Wide Bandgap Semiconductors, Related Technologies, and Their Applications*. London: Intech Open Science; 2018.

[7] See https://www.ansys.com/ [Accessed 08 Nov 2019]

[8] See https://uk.comsol.com/comsol-multiphysics [Accessed 08 Nov 2019]

[9] See https://www.cst.com/ [Accessed 08 Nov 2019]

[10] Gekenidis S., Ramezani E., and Zeller H. 'Explosion tests on IGBT high voltage modules'. *International Symposium on Power Semiconductor Devices & IC's (ISPSD) 1999*, 26–28 May 1999, Toronto, Canada.

[11] Li H., Zhou W., Wang X., *et al.* 'Influence of paralleling dies and paralleling half-bridges on transient current distribution in multichip power modules'. *IEEE Transactions on Power Electronics*. 2018; 33: 6483–6487.

[12] Darveaux R. 'Solder joint fatigue life model'. *Proceedings of TMS Annual Meeting*. Orlando FL, February 1997, pp. 213–218.

[13] Manson S. S. *Thermal Stress and Low Cycle Fatigue*. New York, NY: McGraw-Hill; 1966.

[14] Norris K. C. and Landzerg A. H. 'Reliability of controlled collapse interconnections'. *IBM Journal of Research and Development*. 1969; 13(3): 266–271.

[15] Held M., Jacob P., Nicoletti G., Scacco P., and Porch M.H. 'Fast power cycling test for IGBT modules in traction application'. *International*

Conference on Power Electronics and Drive Systems (PEDS) 1997, May 1997, Singapore, pp. 425–430.

[16] Bayerer R., Herrmann T., Lutz T., Lutz J., and Feller M. 'Model of power cycling lifetime of IGBT modules: Various factors influencing lifetime'. *International Conference on Integrated Power Electronics Systems (CIPS) 2008*, March 2008, Nuremberg, Germany.

[17] Scheuermann U. 'Power cycling lifetime of advanced power modules for different temperature swings'. *International Conference for Power Conversion, Intelligent Motion (PCIM) 2002*, Nuremberg, Germany.

[18] Hudgins J., Simin G., Santi E., and Khan M. 'An Assessment of wide bandgap semiconductors for power devices'. *IEEE Transactions on Power Electronics*. 2003; 18(3): 907–914.

[19] See https://www.accuratus.com/ [Accessed 08 Nov 2019]

[20] Packwood M., Li D., Mumby-Croft P., and Dai X. 'Thermal simulation into the effect of varying encapsulant media on wire bond stress under temperature cycling'. *ICEPT2018*. Shanghai, China, 2018.

[21] Loh W., Corfield M., Lu H., Hogg S., Tilford T., and Johnson M. 'Wire bond reliability for power electronic modules – Effect of bonding temperature'. *Eurosime 2007*.

[22] Jacques S., Leroy R., and Lethiecq M. 'Impact of aluminium wire and ribbon bonding technologies on D2PAK package reliability during thermal cycling applications'. *Microelectronics Reliability*. 2015; 55: 1821–1825.

[23] Guth K., Heuck N., Stahlhut Ch., et al. 'End-of-life investigation on the.XT interconnect technology'. *PCIM Europe 2015*. Nuremberg, Germany.

[24] Heuck N., Bayerer R., Krasel S., Otto F., Speckels R., and Guth K. 'Lifetime analysis of power modules with new packaging technologies'. *27th International Symposium on Power Semiconductor Devices & IC's (ISPSD), 2015*, Hong Kong, China.

[25] https://www.heraeus.com/ [Accessed 08 Nov 2019]

[26] Schmid R., Scheuermann U., and Milke E. 'Al-clad Cu wire bonds multiple power cycling lifetime of advanced power modules'. *International Conference for Power Conversion, Intelligent Motion (PCIM) 2012*, Nuremberg, Germany.

[27] See https://www.renesas.com. Application notes on 'Power MOSFET IGBT – Attention of handling semiconductor devices'. R07ZZ0010EJ0600 [Accessed 08 Nov 2019].

[28] Hayashi K., Izuta G., Murakami K., Uegai Y., and Takao H. 'Improvement of fatigue life of solder joints by thickness control of solder with wire bump technique'. *Electronic Components and Technology Conference (ECTC) 2002*, San Diego, USA.

[29] Booth J., Varley M., Slack D., *et al*. 'High-reliability large area substrate solder interconnect by embedded mesh technique'. *International Conference for Power Conversion, Intelligent Motion (PCIM) 2017*, Nuremberg, Germany.

[30] Zhao Y., Wu Y., Evans K., Jones S., and Dai X. 'Evaluation of Ag sintering die-attach for high-temperature power module applications'. *International*

Conference on Electronic Packaging Technology (ICEPT) 2014, Chengdu, China.

[31] Buttay C., Masson A., Li J., *et al. Die Attach of Power Devices Using Silver Sintering - Bonding Process Optimization and Characterization.* Oxford: *IMAPS International Conference and Exhibition on High Temperature Electronics Network (HiTEN)*, 2019, Oxford, UK.

[32] Scheuermann U. 'Power module design without solder interfaces-an ideal solution for hybrid vehicle traction applications'. *IEEE Applied Power Electronics Conference*. 2009, pp. 472–478.

[33] Packwood M., Li D., Dai X., and Jones S. 'Vibration investigation in power module busbar design'. *International Conference on Thermal, Mechanical and Multi-Physics Simulation and Experiments in Microelectronics and Microsystems (EuroSimE) 2017*, 3–5 Apr 2017, Dresden, Germany.

[34] Poecha M. H. and Eisele R. 'A modelling approach to assess the creep behaviour of large-area solder joints'. *Microelectronics Reliability*. 2000; 40: 1653–1658.

[35] Dalleau D., Paul I., and Broermann F. 'First principle optimisation of power module baseplate'. *International Conference for Power Conversion, Intelligent Motion (PCIM) 2011*, Nuremberg, Germany.

[36] Morshed M., Islam A., Roose T., *et al.* 'Control of partial discharge with high-temperature insulating polymer for high voltage IGBT module application'. *PCIM* 2018.

[37] Zorn C. and Kaminski N. 'Temperature humidity bias (THB) testing on IGBT modules at high bias levels'. *International Conference on Integrated Power Electronics Systems (CIPS) 2014*, Nuremberg, Germany.

[38] Zorn C. and Kaminski N. 'Acceleration of temperature humidity bias (THB) testing on IGBT modules by high bias levels'. *International Symposium on Power Semiconductor Devices and ICs (ISPSD)*. Hong Kong, 2015.

[39] Jormanainen J., Mengotti E., Soeiro T., *et al.* 'High humidity, high temperature and high voltage reverse bias: A relevant test for industrial applications'. *International Conference for Power Conversion, Intelligent Motion (PCIM) 2018*, Nuremberg, Germany.

[40] Papadopoulos C., Corvasce C., Kopta A., Schneider D., Pâques G., and Rahimo M. 'The influence of humidity on the high voltage blocking reliability of power IGBT modules and means of protection'. *Microelectronics Reliability*. 2018; 88–90: 470–475.

Chapter 12
Switching cell design

Eckart Hoene[1] *and Kirill Klein*[2]

12.1 Introduction

The continuous improvement in semiconductor technologies encourages the reduction of the size and cost of devices by increasing pulse frequencies and reducing losses. Most applications incorporating passive components can benefit from higher frequencies by smaller magnetic components and capacitors while in drive applications the effort for heat sinks can be reduced by higher efficiency. Nevertheless, handling high switching speed requires special care in designing the direct environment of the semiconductors where voltages and currents change extremely fast and switching behavior is dominated by parasitic electromagnetic effects.

This central element of energy processing is also called a switching cell. In most power electronic circuits it comprises two semiconductors and a DC link capacitor.

The role of the DC link capacitor is to buffer the supply voltage and to provide and absorb energy in the switching moment. The semiconductors are typically two active switches or a switch and a diode. The geometry of the switching cell and therefore the interconnections between the components is very important for switching performance. It influences the first relevant property, the DC link inductance, represented in Figure 12.1 by ESL and L_σ as lumped inductances. In reality tracks, interfaces, bond wires, capacitor geometry, and design form a loop generating a magnetic field, which is represented by a DC link inductance value.

The DC link inductance has a significant influence on switching behavior while switching off and on. At switch off, it impedes the fast change of current and generates overvoltages across the semiconductor (Figure 12.2).

This overvoltage is one reason to use semiconductors with the voltage rating clearly above the operating voltage.

For fast switching components, which are especially unipolar semiconductors like metal-insulator-semiconductor field-effect transistors (MOSFETs) or high-

[1]Fraunhofer IZM, Berlin, Germany
[2]School IV - Electrical Engineering and Computer Science, Research Center for Microperipheric Technologies, Technical University Berlin, Berlin, Germany

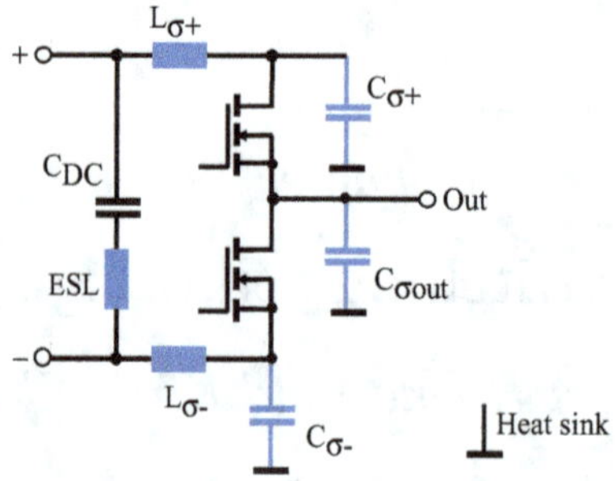

Figure 12.1 Equivalent circuit of a basic switching cell with parasitic properties in blue

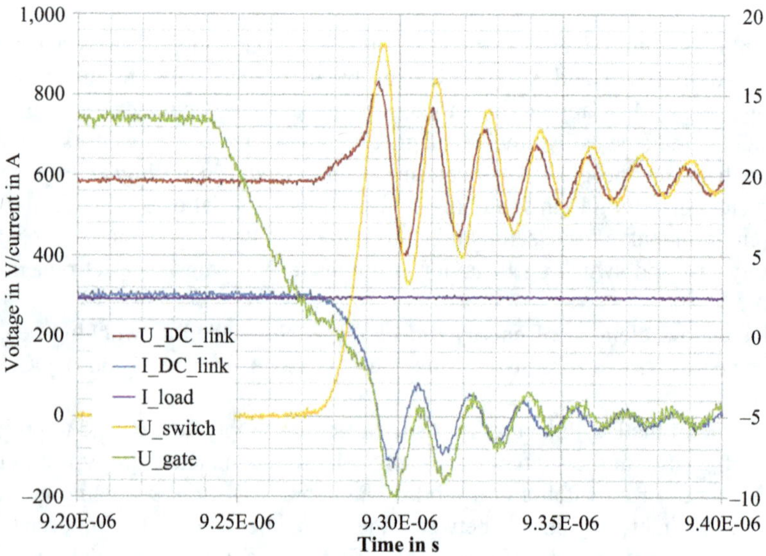

Figure 12.2 Overvoltage of a silicon carbide (SiC) semiconductor at switch off, DC link inductance 9 nH

electron mobility transistors (HEMTs), the switch-off losses are low compared to the switch on losses. Figure 12.3 gives an idea for the reason for this.

Switch on starts in phase t_{f1} with a rise of current in the switching semiconductor and at the same time a decrease of the voltage, which is necessary to get the current flowing in the switching cell. In this case, the switching speed is high and significant part of the voltage drops over the DC link inductance. Therefore the breakdown in DC link voltage is significant. The transition to phase t_{f2} takes place when the current in the switch reaches the value of the load current. As in this example, SiC-MOSFETs are used, the freewheeling diodes show a reverse recovery behavior. The time to extract the reverse recovery charge defines the length of t_{f2}.

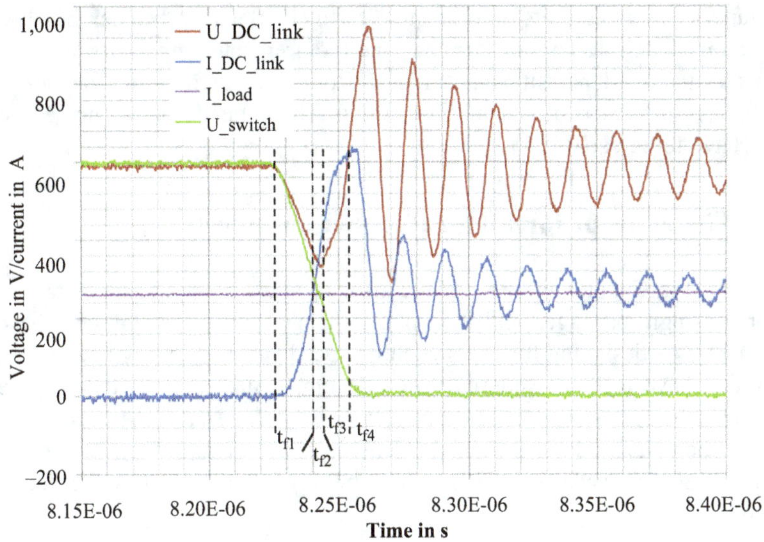

Figure 12.3 Switch on of a SiC module with a DC link inductance of 9 nH

As during this period, the voltage across the freewheeling MOSFET is still zero, the DC link voltage continues declining. After this, phase t_{f3} starts with the charging the semiconductor capacitances. Now the voltage rises across the freewheeling MOSFET and also in the DC link, while the current continues to increase. In this example additionally, there seems to be a parasitic switch-on of the freewheeling MOSFET, which increases the current peak. With DC link voltage reaching its steady-state value t_{f4} begins. The DC link inductance is in this moment charged with much higher than load current and pushes voltage beyond supply voltage and initiates extensive ringing.

This example gives an impression of the influence of DC link inductance. As the whole switching cell loop has to be considered to define its value and semiconductors are mostly packaged without capacitors, the terminal between both adds significant inductance. But also other parasitic properties get very relevant at high switching speed. As mentioned in the example shown the Miller capacitance of the freewheeling MOSFET causes undesired short conduction during switch on. It appears, when the voltage across the semiconductor increases, the charge is injected in the gate and the driver cannot discharge it fast enough due to gate path impedance. Despite the gate resistance as a cause for gate path impedance also inductance is a major issue. To avoid parasitic switch on when fast switching it should stay within a one-digit nH value.

The next topic to care about is electromagnetic compatibility (EMC). In higher power levels semiconductors have to be cooled by a heat sink which is mostly connected to protective earth. The semiconductor is insulated through a thermal pad or a ceramic substrate, nevertheless, this creates parasitic capacitance from the circuit to the heat sink, depicted as C_σ in Figure 12.1. Reloading these parasitic capacitances causes currents on protective earth (PE), which then can be observed as common mode interference current. One requirement to minimize interference currents is to avoid parasitic capacitance from the out node to the heat sink. Nevertheless, a further rule has to be fulfilled. The parasitic inductance and capacitance have to be equal [1].

12.2 The concept for integrated switching cell

As described in Chapter 11, state of the art is to package semiconductors in modules or as single packages and to connect the DC link capacitor and the other components. As designing for fast switching speed gets very tricky due to parasitic effects that have to be considered in the design, it makes sense to integrate the switching cell as one readymade unit. This also enables inexperienced users to adapt the benefits of fast switching. Generally, an integrated switching cell or switching cell in package (SCiP) would be favorable if it has the following properties:

- The electrical interface on the one side, the thermal interface on the other
- Insulation to the heat sink
- Local DC link with damping to enable the lowest inductance (<2 nH)
- Low inductive electromechanical interfaces to bus bar/printed circuit board
- Integrated gate booster for low inductance gate path
- EMI encapsulating design
- Ensuring the creepage distances

A concept of the power module fulfilling these requirements is shown in Figure 12.4.

The module is built up starting with a ceramic substrate to provide safety insulation and good thermal performance. The substrate is patterned with two equally sized areas for DC+ and DC−. Two SiC chips are placed on the substrate, one regular mounted and one flipped (Figures 12.5 and 12.6).

Figure 12.4 Integrated switching cell with 1,200 V SiC MOSFETs by Fraunhofer IZM/ECPE

With these first steps, several of the requirements are fulfilled already: good thermal properties and safety insulation are provided by the ceramic substrate, EMI encapsulating design is achieved by avoiding AC potential on the substrate and symmetric layout for DC+ and DC−.

In the next step, the top side interconnects are manufactured using a printed circuit board (PCB) technology. Therefore, FR4 prepregs are laminated around the chips with cavities, a 100 μm thick prepreg covers the chips and a 5 μm copper foil comes on top. The top side connects of the chips are done by laser drilling onto the copper-plated chip surface and galvanic copper deposition in the next step (Figure 12.8). Photoresist, imaging, developing, and etching finish the first layer manufacturing (Figure 12.7).

This AC track is shielded to the heat sink by DC+ and DC−. Continuing the setup with the next layer is shown in Figure 12.9, manufactured in the same way as before.

This now is the top layer closing the switching cell loop and offering the possibility to mount surface mounted device (SMD) components. These are preferable the local DC link capacitors with damping resistors and boosters for gate

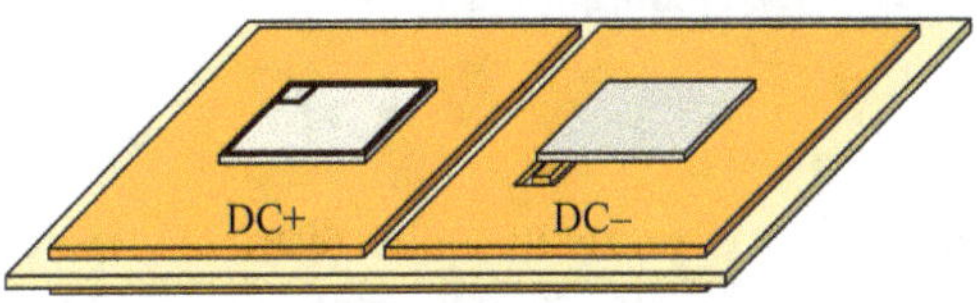

Figure 12.5 Substrate with one chip-mounted regular and one flipped

Figure 12.6 Manufacturing steps: a substrate with silver plating, chips mounted, and underfill for flipped chip

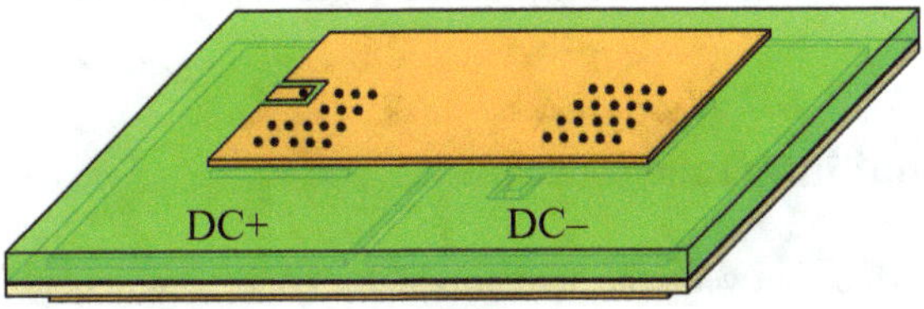

Figure 12.7 AC connection of the half-bridge in PCB technology

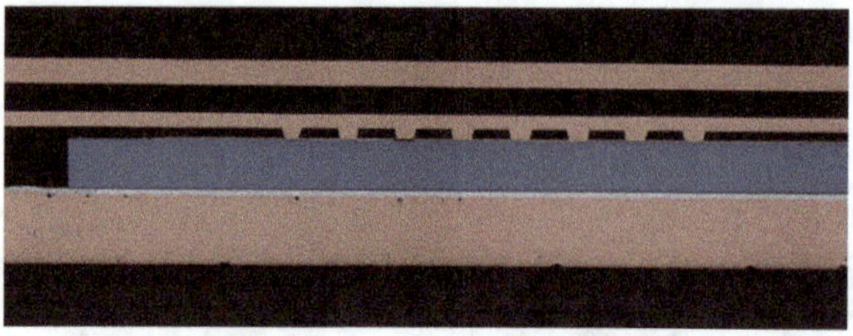

Figure 12.8 Cross-section of a chip embedded in PCB material

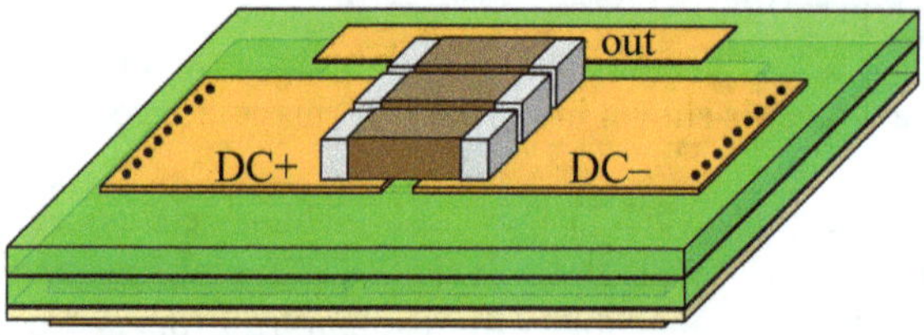

Figure 12.9 Next layer closing the switching cell with local capacitors

driving. As the surface area is normally quite restricted, there is only space for interconnects for power and gate signals remaining. The module shown in Figure 12.4 uses SMD spring contacts for gate signal and a "lying spring" contact for high current connections. This is one suggestion of how to enable a low inductance high current electrical contact. It provides a certain number of contact points to minimize contact resistance and offers many current paths in parallel to minimize inductance and electric resistance.

This example gives an idea for a switching cell concept. To establish such a concept on the market requires a strong performance benefit, as supply chains will have to change. Up to now semiconductor manufacturers seldom integrate their products with passive components by heterogeneous integration technologies. This technological development and the change of mind by customers and suppliers always slow down the adaption. Nevertheless, fast switching can be the enabler in terms of performance benefit and change of mind.

The following sections introduces the important aspects of the realization of integrated switching cells, promising technologies for heterogeneous integration, and some examples.

12.3 Thermal interface

The thermal interface between the semiconductor package and heat sink or cooling media strongly influences the achievable performance of the device. The type of interface can be differentiated by two main properties:

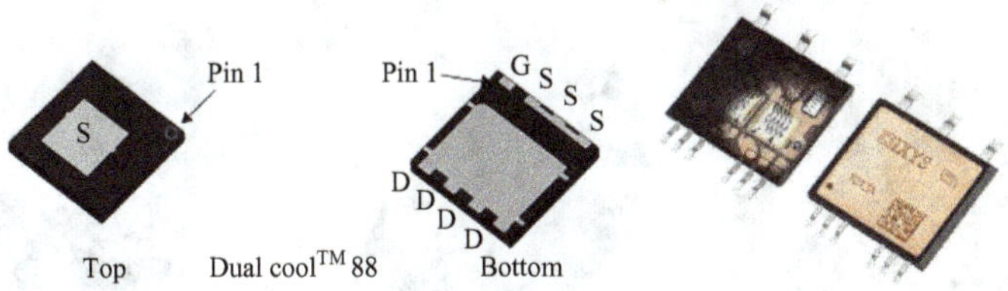

Figure 12.10 Dual cool™ package by Onsemi, SMPD package by IXYS

- Rigid or detachable
- Insulated or not

The highest cooling performance can be achieved by modules soldered or sintered to the heat sink, as the thermal impedance of these metallic interconnections is low. Of course, this requires an insulating substrate in the module, which is in higher power modules mostly a ceramic substrate due to its good thermal performance and high voltage insulation capability (see Chapter 11). Nevertheless, substrates using organic insulators like IMS substrates (insulated metal substrates) or printed circuit boards with layers with thermally better conductivity gain market share due to performance improvements. As organic insulators put less mechanical stress on the setup, they can be combined with thicker copper for heat spreading to compensate the lower thermal performance. Also, in systems without grid connection, the insulation requirements are lower, which works for IMS substrates.

In the case of a noninsulated package, mainly lead frame-based packages, the insulation mostly is provided by a thermal pad. Other opportunities are thermally conducting adhesives with filler particles of defined size to avoid electrical contact or the lamination of the module to the heat sink using a thermally conductive organic laminate.

The decision on the selection of the thermal interface type is strongly influenced by the construction of the device. At lower power, it is very favorable to keep SMD assembly as long as possible, as this keeps production cost low. Nevertheless, there are only very few SMD packages available with cooling on the top side and even less with integrated insulation (Figure 12.10). The challenge with this kind of setup is ensuring good thermal contact, for example, with thermal grease over a lifetime and even mechanical pressure of the PCB to the cooler. Higher power modules in air-cooled applications are mostly assembled with a detachable thermal interface to the heat sink and screwed electrical interface (see Chapter 11). With water cooling the thermal resistance of a detachable thermal interface reduces performance too much and the heat sink is soldered or sintered to the heat sink. This is a common strategy for the traction inverter in electric vehicles and others (Figure 12.11). In the case of water cooling also solutions with direct substrate cooling are very favorable, as the short thermal path is beneficial for low thermal resistance. In this case, the short thermal path compensates for the spreading of the missing heat.

Figure 12.11 Module with the metallic interconnection between the substrate and cooler for automotive traction (Danfoss Silicon Power)

Figure 12.12 Power module with direct substrate cooling: the substrate with four chips and the cooling structure in the mirror image (left) and the transfer molded module metalized and with SMD components on top (right). Adapted from [2]

Figure 12.12 shows a module with direct substrate cooling. In this concept, the surface to be cooled is small, therefore it is mandatory to ensure an effective coolant flow to achieve competitive thermal resistance.

Generally, it can be seen that at low power SMD semiconductor packages are designed with thermal and electrical interfaces on the same side. This has disadvantages in terms of thermal performance and electrical routing. Coming to a higher power the thermal and electrical interfaces of a module are placed on different sides of the package. For most applications, this is a sufficient solution. Packages with double-sided cooling are available as well, but it seems to be a small niche. High-performance single side cooled setups already can achieve a cooling performance that makes the mechanical extra effort for double-sided cooling obsolete.

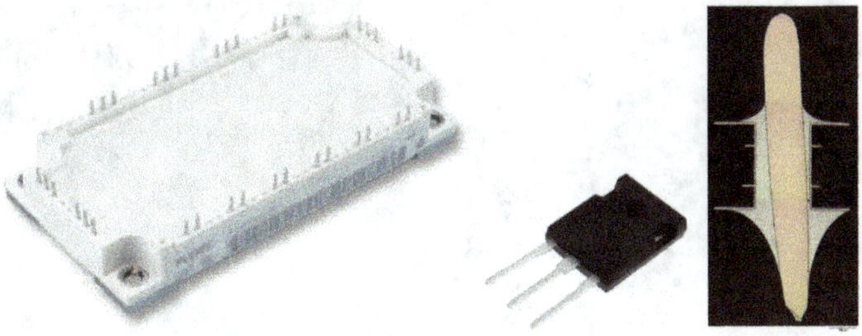

Figure 12.13 Module with solder pins, TO247 as THT component, a cross-section through a THT pin

12.4 Electrical interfaces

Electrical interfaces provide a rigid or detachable electric contact of a module or switching cell to a bus bar or PCB. Generally, electrical interfaces take a significant volume for power modules, especially in the case of screw terminals (see power modules shown in Chapter 11). The relevant properties of electrical interconnects are current carrying capability, their inductance, and their rated voltage capability.

Depending on the current level, there are several common solutions. For small components, soldering in an SMT process is the most effective solution, nevertheless, the thermal interface is the same for most SMD packages, despite only the little number of packages (Figure 12.10).

The through-hole technology (THT) has proven its suitability and reliability in power electronics for a long time. It is applied for power modules as shown in Figure 12.13.

The detachable interface solutions are press-fit contacts, spring contacts, and screws. The press-fit contacts create the interconnection through a mechanical force when pressed into the PCB hole or bus bar (Figure 12.14).

These connections are beneficial in production, as no thermal process is required (instead of a high force pressing step) and is capable of handling high current if the contacts are designed suited.

Spring contacts as power connectors are not so common up to now, one example is shown in Figure 12.4. They provide a low inductance power connection as described in Section 12.2.

For high power, the screw contacts are state of the art. The drawback of these interconnects is their high inductance. A classical screw interface adds about 10 nH inductance. For faster switching, this can develop to a bottleneck and also the screws require a certain module height.

12.4.1 Insulation: clearance/creepage distances

The requirement to fulfill creepage and clearance distances for electrical insulation strongly influence the design of a module. Clearance distance describes the minimum distance between terminals or to the heat sink required to avoid arcing.

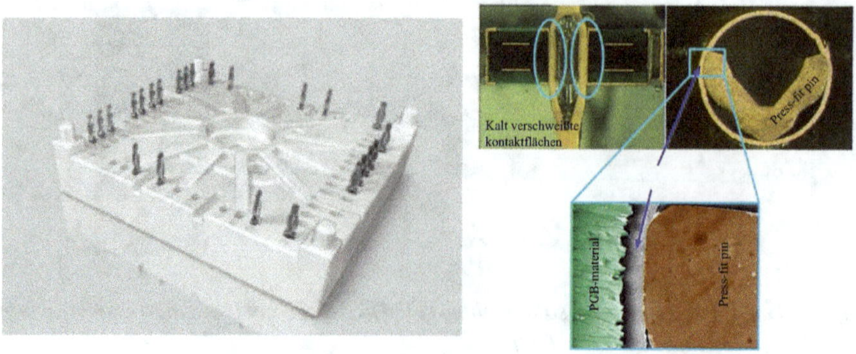

Figure 12.14 Module with press-fit contacts (Semikron) and cross-sections of press-fit contacts

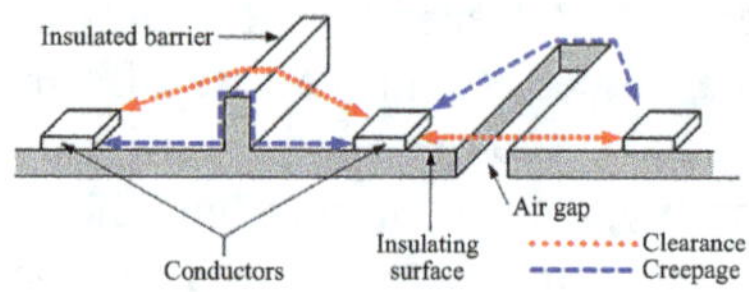

Figure 12.15 Visualization of creepage and clearance distances

Creepage distance is the shortest way between two conductors on an insulating surface (Figure 12.15).

This required clearance distance, of course, is dependent not only on the operating voltage but also on the altitude above sea level. Figure 12.15 gives some ideas about how to enlarge clearance in practical setups. The value of clearance distance is specified by standards, for example, DIN EN 60664 in Europe.

The required creepage distance depends despite the voltage on the pollution level the device is operated in and the property of the insulating material. This property is defined by the CTI-index (comparative tracking index, DIN EN 60112) and strongly influences the required distance (see also DIN EN 60664).

For highly integrated modules these insulation requirements have a significant effect on the size, as all modules have electrical interfaces and often insulation to PE. While the electrical interfaces have to be exposed by definition, the creepage distances for components like integrated drivers or others can be reduced by coating them with a suited resin. In classical power modules, silicone is an established solution for PCBs coating resins which are common. The coating resins have to fulfill insulation requirements according to DIN EN 60664-3. Typically, they are sprayed to the PCB, sometimes dams are applied to avoid the resin to flow into connectors. The resins often are luminescent for UV to enable easier process control.

12.5 Mechanical interfaces

Semiconductors and modules using a detachable thermal interface need mechanical fixing to apply pressure to the thermal interface material (TIM) or pad. To minimize the thermal resistance it is important to ensure a thin TIM layer, especially beneath the semiconductors. Furthermore, it has to be ensured that the TIM is not pumped out over a lifetime.

As shown in Chapter 11, the most power modules and some discreet components use a copper base plate that is screwed to the heat sink as a mechanical interface. Big modules often are manufactured with a convex bow to ensure pressure also in the middle of the module (Figure 12.16).

The module in Figure 12.14 has no base plate; a ceramic substrate is pressed by the plastic housing and a screw in the middle to the heat sink. Compared to the base plate modules this follows a different strategy: a base plate spreads the heat before it has to pass the TIM. This is beneficial to compensate the high thermal impedance of the TIM. Without the base plate, it is very important to ensure a thin TIM layer beneath the chips, as the spreading effect is not so distinct. This is achieved by a mechanical structure inside the module pressing the ceramic to the cooler around the chip. Even better performance can be achieved if the mechanical force is applied directly onto the chip. This, of course, is only possible if the chip is not contacted by wire bonds.

For PCB mounted discrete components and switching cells in package, it is also desirable to press each package to the heat sink. While TO-247 like packages offer a screw hole for mounting in case of SMD components, there is nothing like this. The PCB has to be pressed by screws or spring type fixtures, which creates a loss of PCB space and requires a significant number of screws in order to ensure even the thermal performance of each chip. Mechanically stiff housings can also be used to provide the required pressure.

As an alternative gluing, thermally conductive glue can be considered.

Figure 12.16 Convex base plate of a base plate power module

12.6 DC link design

As was already mentioned in Section 12.1, one of the most critical aspects while the design of the power module is the design of the DC link and gate driver interconnections. The proper design of the switching cell allows reducing switching losses and using the benefit of modern fast switching semiconductors. This section shows the state-of-the-art DC link design and proposes a design suited to evolve the potential of fast switching semiconductors by enabling fast switching with reducing oscillations within the switching cell. For simplicity reason, a half-bridge topology as a current switching unit will be discussed in this section because in most cases it is the main building block of most power modules.

12.6.1 State-of-the-art DC link design

The energy converted by switching power supply comes from a source, for example, voltage supply source as it is shown in the case of Figure 12.17. The inductance of the connection to the voltage supply source is much larger than DC link inductance: $L_{supp} \gg L_{c.dclink}$. The energy is transported to output (node connecting high side source and low side drain potentials) and transferred to further transformation. The energy at least for one switching cycle is stored in the DC link capacitor. Most of the applications accept the voltage drop over the DC link capacitor of 5%.

While switching event, the current i flows from the DC link capacitor to the semiconductor and creates a magnetic field in all (stray) inductances in the current path in-between: ESL of DC link capacitor $L_{cdclink}$, the stray inductance of power module interconnection to the DC link capacitor L_c, and the stray inductance of the interconnections within power module L_{switch} (Figure 12.17). Whereby L_{switch} for low inductive power modules is usually in the range of 1..5 nH and L_c for low inductive setups is in the range of 10..20 nH.

While the turn-off of the current, the magnetic field energy stored in stray inductances ($1/2(L_c + L_{switch})i^2$) releases and transforms to the current flowing

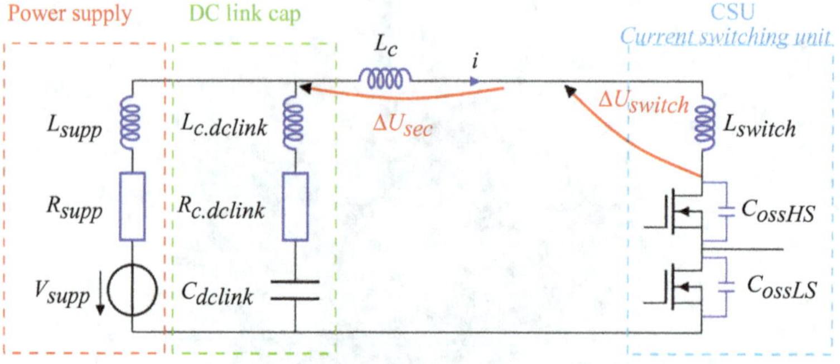

Figure 12.17 Simplified schematic of the classic switching cell with connection to the power supply and parasitic elements

toward semiconductor, loading parasitic layout capacitances, which are usually negligible, and semiconductors output capacitance C_{oss} (Figure 12.17). The C_{oss} of fast semiconductors is quite small and stray inductance can be high enough to create an overvoltage ΔU, which in superposition with DC link voltage could exceed the breakdown voltage of semiconductor and destroy it. Under the assumption of the nearly steady voltage potential at the DC link capacitor, the overvoltage can be calculated, corresponding to (12.1).

$$\Delta U_{\text{max}} = i \times \sqrt{\frac{L_c + L_{\text{switch}}}{C_{\text{oss}}(V_{\text{ds}}, V_{\text{gs}})}} \tag{12.1}$$

12.6.2 DC link design for fast switching power modules

The design of the switching cell with smaller inductance would reduce the overvoltage and solve the problem. For this purpose, an additional capacitor close to semiconductors (primary DC link capacitor, C_{snubb}) or RC-snubber (see Sections 12.6.3 and 12.6.4) should be inserted (Figure 12.18). It splits the switching cell into two parts. The primary DC link capacitor builds a new low inductive commutation loop, where all fast di/dt and dV/dt transients occur. The value of C_{snubb} can be small enough to absorb only the electrical energy released from the stray inductance L_c at switch off. Consequently, C_{snubb} has a small geometrical size and therefore can be placed close to the semiconductors. The high nominal value DC link capacitor becomes in a new topology, a name addition "secondary" and has the same functionality as before.

The overvoltage in the switching moment in the proposed design is the superposition of the $\Delta U_{\text{switch}}(t)$ and $\Delta U_{\text{sec}}(t)$. The value of C_{snubb} compared to C_{oss}-value is much higher, thus $\Delta U_{sec}(t)$ is reduced significantly:

$$\Delta U(t) = \Delta U_{\text{switch}(t)} + \Delta U_{\text{sec}(t)}$$

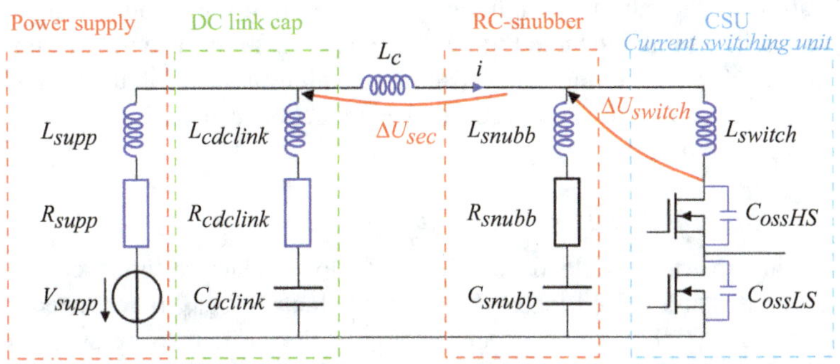

Figure 12.18 Simplified schematic of the proposed DC link design including power supply and parasitic elements

Insertion of a capacitor C_{snubb} would result in nearly undamped oscillation between C_{snubb} and L_c. The behavior can be predicted only if an additional damping resistor R_{snubb} is added in series to the C_{snubb} building an RC-snubber. To be able to design primary DC link correctly, it is necessary to understand the boundary conditions and understand the behavior and properties of used devices.

12.6.3 Design rules for capacitor C_{snubb}

To realize fast and clean switching without ringing two major criteria should be considered.

The first one is preventing an overvoltage during the turn-off of the current caring semiconductor caused by parasitic inductances. The second one is the damping of oscillation while turn-on of semiconductor and reloading of parasitic capacitances in the switching moment.

Considering the second parameter is important to reduce EMI and ringing in the switching moment, such design rules lead to high damping resistor values and increase the overvoltage.

To fulfill the first parameter, the primary DC link capacitor value should be high enough to store the energy from the secondary stray DC link inductance.

The energy stored in stray inductance L_c has to be completely absorbed by primary DC link capacitor C_{snubb}. The minimal value of the capacitor results to

$$\frac{1}{2}L_c \times i^2 = \frac{1}{2}C_{\text{snubb}} \times \Delta U_{\max}^2 \tag{12.2}$$

$$=> C_{\text{snubb}} = \frac{L_c \times i^2}{\Delta U_{\max}^2} \tag{12.3}$$

Oversizing the capacitor value, on the one hand, decreases overvoltage peak; on the other hand, it can increase the module size, integration effort, and costs.

The value of C_{snubb} in (12.3) matches the overvoltage condition for all switching speeds under the assumption of low R_{snubb} values.

Derating of the capacitor value at given operating conditions must be considered as well. As it is shown in Figures 12.19 and 12.20, some types of capacitors have only 25% of their nominal capacitance value at 80% of their maximum operating voltage. Further 20% capacitance reduction can be caused by higher operation temperature.

12.6.4 Damping resistor R_{snubb} design

The damping resistor should be in ideal case low inductive, good cooled, and capable to withstand pulsed loads. The upper value of R_{snubb} is restricted by the overvoltage $\Delta U_{\max}$ that would be generated for the current to be switched off. Lower R_{snubb} values decrease the damping factor and change the phase of the voltage over semiconductor. The optimum value can be achieved by impedance matching to the characteristic impedance of the C_{snubb} and the DC link inductance.

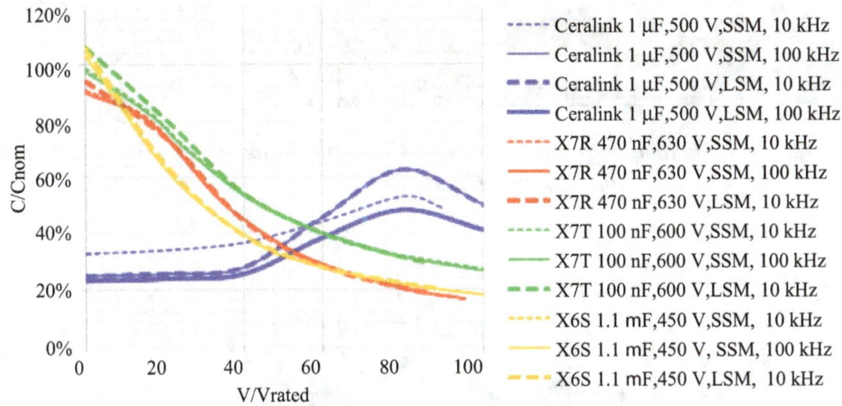

Figure 12.19 Voltage dependency of capacitance for small signal (thin) and large signals (thick) depending on capacitor type and AC frequency (dashed: 10 kHz; solid: 100 kHz)

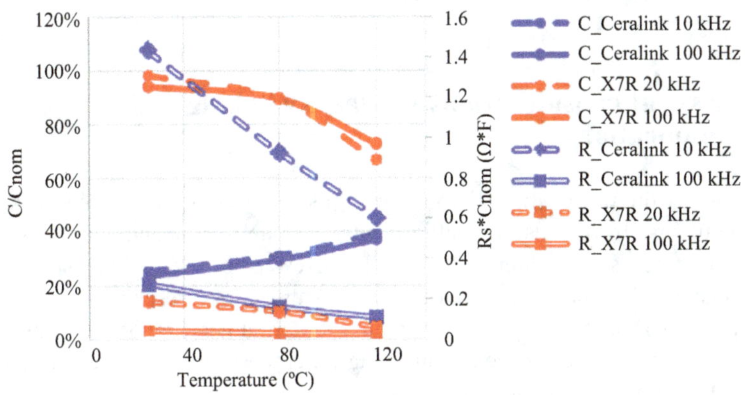

Figure 12.20 C_{snubb} *(single line) and* R_{snubb} *(double line) of EPCOS Ceralink (blue) and X7R (red) for different temperatures and frequencies*

The value with the lowest overvoltage can be found if the resistor value is half of the characteristic impedance as detailed in [3]:

$$R_{\text{snubb opt}} = \frac{1}{2} \times \sqrt{\frac{L_c + L_{\text{cdclink}} + L_{\text{snubb}}}{C_{\text{snubb}}}} \tag{12.4}$$

Snubber resistors absorb the energy stored in the stray inductance between RC-snubber and DC link capacitor. The dissipated power of the resistor is proportional to switching frequency and quadrat of the switched current. The energy absorbed at turn-on and turn-off can be calculated using (12.5):

$$R_{\text{Rsnubber}} = \left(f \times (L_c + \text{Lcdclink}) \times i^2\right) \tag{12.5}$$

Figure 12.21 Principle of bus bar design. Example value: 30 mm width, 100 mm length, and 0.5 mm distance

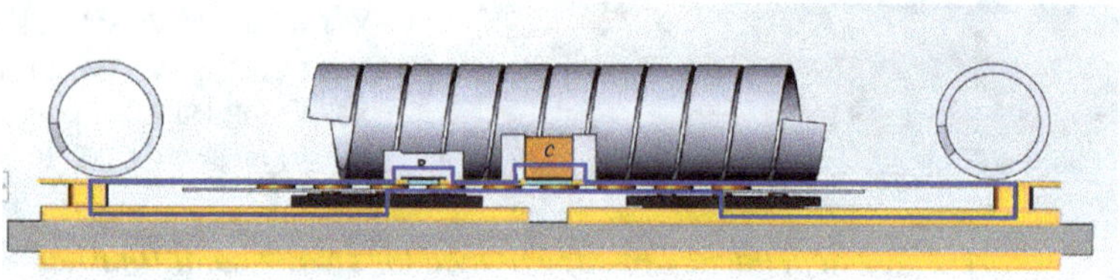

Figure 12.22 Schematic DC link current path (blue line) of the module, which is detailed in Section 12.2

12.7 Layout considerations for fast switching applications

The layout of the power module must be adjusted to the switching speed of used semiconductors. In the case of gallium nitride (GaN) semiconductors, switching speeds can reach several hundred V/ns. Thus few picofarads of parasitic capacitance cause displacement currents of several Ampere, and respectively, can cause EMI issues, especially in the gate drive circuitry. It is very important to be aware, that planes and areas referenced to jumping potential cause displacement currents to any other remaining steady potentials.

The further important aspect is that any current path has parasitic inductance, which not only slowing down signals and reducing switching speeds but also causes parasitic ringing circuits with other surrounding capacitances.

This section summarizes the most important layout considerations.

12.7.1 Low inductive bus bar design

A common strategy for low inductance in DC links is the stripline or bus bar geometry. It relies on two phenomena: on the one hand, a wide conductor creates a long path for the magnetic flux to close around it. The resulting magnetic resistance is high and the inductance is low. On the other hand, the return path of the current is brought closer in order to cancel the magnetic fields (Figure 12.21).

Implementing bus bar design into a power module with vertical power devices requires three layers for high current rooting. Two basic possibilities are depicted in Figures 12.22 and 12.23.

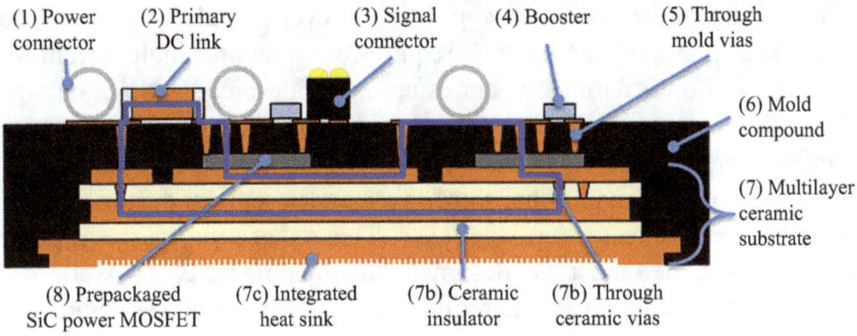

Figure 12.23 DC link current path for the module is shown in Figure 12.12

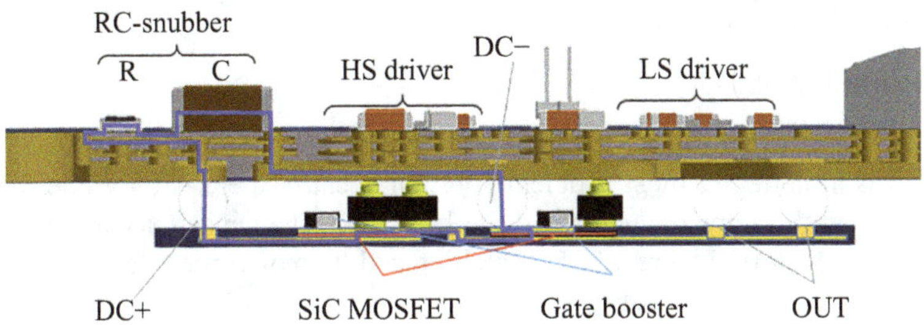

Figure 12.24 DC link current path for SiC power module with low inductive connection to the system PCB

1. Two layers above the semiconductors, one beneath in the thermal path: in this case one layer is provided by the ceramic substrate and two layers above the semiconductors in PCB technology. SMD components on the module close the switching cell.
2. Two layers in the thermal path, one above the semiconductors: with a two layer thermal substrate only one is required above the semiconductors.

Another approach is to shift RC-snubber to system board PCB and to create a low inductive interconnection between PCB and system board, for example, with spring contacts as it is shown in Figure 12.24. The contacts should be (1) low profile but (2) mechanically flexible in order to compensate CTE induced XY-mismatch and Z-distance variations between semiconductors and system board, and (3) capable to care high current.

All three strategies allow low inductive integration of the gate drive path.

Despite a bus bar configuration, conductors besides and close to each other allow low inductance as well. Nevertheless, geometries like this only reduce inductance for high frequency by proximity effect, at a low frequency, the current loops formed still show high-value inductance.

Implementing of the bus bar design into a power module with lateral devices on isolating substrate like GaN-on-SiC require two layers for high current rooting. But lateral devices on isolating substrates are not widespread and not common.

Lateral devices on conductive substrates, like GaN-on-Si or GaN on semi-insulating SiC, require a defined substrate potential. Thus from the packaging point of view, they require an additional third layer under semiconductors with a connection to the semiconductor top-side layer. This connection must not be able to carry high current, but must be designed and manufactured as well. As consequence, three horizontal layers are required for GaN-on-Si power devices, similar to in the case of SiC MOSFETs, discussed above.

At the same time, parasitic capacitance from the GaN HEMT substrate to the heat sink causes displacement currents in switching moment and consequently EMI identical to the situation with vertical devices.

There several approaches for the design of the switching cell current path loop for lateral devices:

1. Vertical layout and
2. Hybrid layouts.

This is a thermally most efficient solution because there is only one thermal layer between the semiconductor and the heat sink. The capacitor can be directly over semiconductors (Figure 12.25, switching cell loop of vertical layout including RC-snubber) or on the side (Figure 12.26).

1. Embedded layout. An additional layer with low thermal conductivity under semiconductors (in the thermal path) offers an implementation possibility for embedded layout (Figure 12.27). Additional layer carries the current, spreads the heat and shields the out potential from the heat sink by DC+ or DC− potential. For EMI critical applications, it is a good compromise. Further, the snubber resistor can be cooled efficiently.

   Adding the fourth routing layer gives a possibility of the design, where RC-snubber and driver components are placed on the top layer (Figure 12.28). DC link inductance becomes minimally larger, but gate loop inductance is significantly lower than in other cases. Such layout is good for manufacturing in

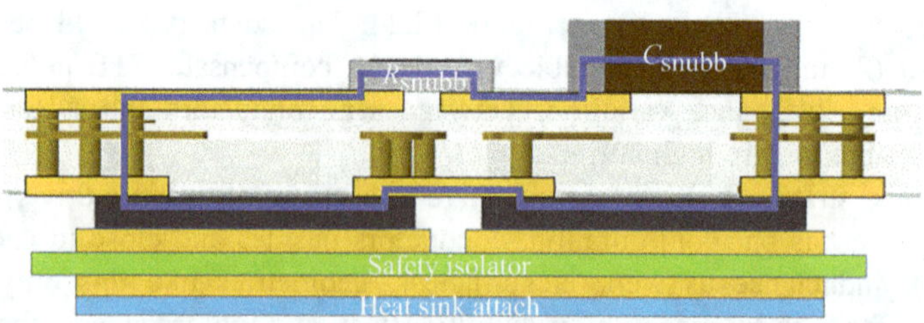

Figure 12.25 Switching cell loop of vertical layout including RC-snubber

Figure 12.26 Switching cell of the hybrid layout including RC-snubber

Figure 12.27 Switching cell of the embedded layout including well cooled RC-snubber

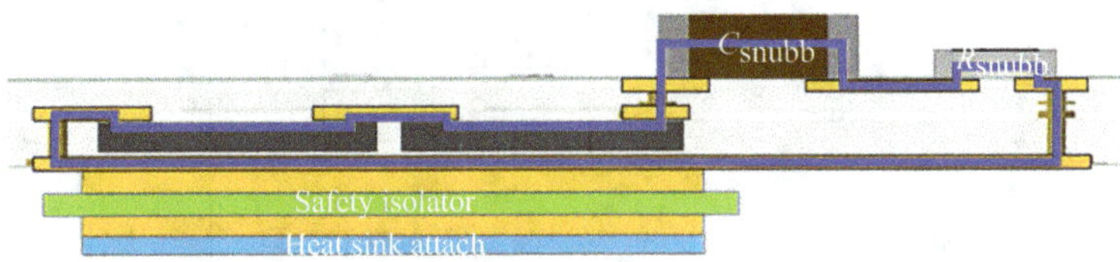

Figure 12.28 Switching cell of the four-layer embedded layout without cooling of RC-snubber

PCB embedding technology, where fast switching and EMI are high of importance.

2. Minimal inductance layout. Adding the fourth routing layer offers further a so-called minimal inductance layout as shown in Figure 12.29. The inductance of such a loop is minimal. It requires cooling from both sides. Both semiconductors are between two heat sinks in a sandwich-like construction. Such layout is optimal in (1) low-power high-current applications, where DC link inductance is extremely important and (2) in cases with a strategy to create as low as possible losses and the heat sink can be left out. Both side high-performance cooling has high integration complexity regarding reliability requires a complex design with optimized CTE matching, electromechanical interfaces, and fulfilling isolation requirements of the full assembly.

12.7.2 Parasitic turn-on

A very critical aspect when using fast switching semiconductors in a half-bridge configuration is the parasitic turn-on (also called as dV/dt-induced turn-on,

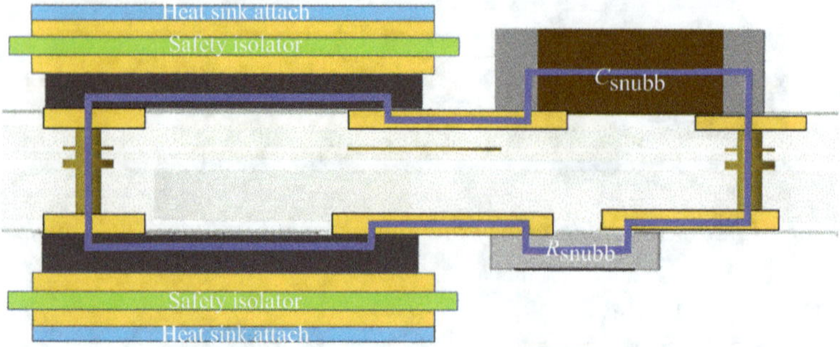

Figure 12.29 Switching cell of the minimal inductance layout including RC-snubber

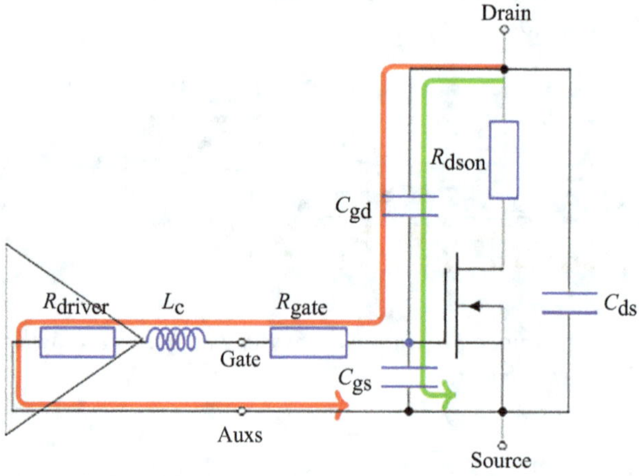

Figure 12.30 Parasitics of the transistor and driver influencing the parasitic turn-on

parasitic switch-on). It creates significant losses, heats the semiconductors, and spoils the performance and energy conversion efficiency of the converter.

The technical background of parasitic turn-on will be explained exemplarily on parasitic turn-on of low side (LS) transistor shown in Figure 12.30.

During switching of the HS transistor to the on-state, the output- and consequently drain-contact potential of the LS transistor rapidly becomes equal V_{DC+}-potential. A high voltage slope over LS transistor causes according to Figure 12.30 displacement current through Miller capacitance C_{gd}:

$$I_{Cgd} = C_{gd} \times dV_{gd}/dt$$

This current can flow as the green current path into $C_{gs.}$ C_{gd} and C_{gs} are a nonlinear function of V_{ds} and V_{gs}. At high $V_{ds,}$ C_{gd} is usually much smaller than C_{gs}.

At low V_{ds}, C_{gd} increases and for some semiconductors achieve a magnitude of C_{gs} and consequently increase the current though C_{gd}. Injected displacement current causes a gate-source voltage drop according to capacitive voltage divider:

$$\Delta V_{gs} = \Delta V_{ds} \times C_{gd}/(C_{gd} + C_{gs})$$

At very high dV/dt, the voltage drop of V_{gs} becomes bigger than the voltage difference between voltage delivered by driver and V_{th}. At this moment, both transistors in half-bridge become conducting, causing through-shoot. The so-called parasitic turn-on is happening.

In the same way, a parasitic turn-on occurs at HS transistor, during LS transistor is turning on.

12.7.3 Gate drive path layout

One of the most challenging tasks while power module design is the design of the gate drive path. During switching transition, every coupling and voltage noise on the gate pin becomes amplified on the drain pin of the transistor. Minimization of the coupling of the gate drive path with high current paths is essential for clean and failure-free switching.

The bus bar design technique, described above, is very effective for the gate drive path as well. Graphical drawing of the closed turn-on and turn-off gate current loops often helps to find an optimal layout with the lowest gate driving path inductance, as shown in Figure 12.31. All signal traces and especially analog signal traces must be as short as possible. Buffer capacitors of the driver should be placed as close as possible to their load.

Rule of thumb, well matching around 1 GHz is: bandwidth $\approx 0.35/t_{rise}$. Driver signals have rise times in the ns range. Bandwidth in the range of GHz turns traces to components, which often oscillate with each other in the switching moment. The usage of curves instead of right-angle routings reduces signal reflections on corners and, respectively, ringing.

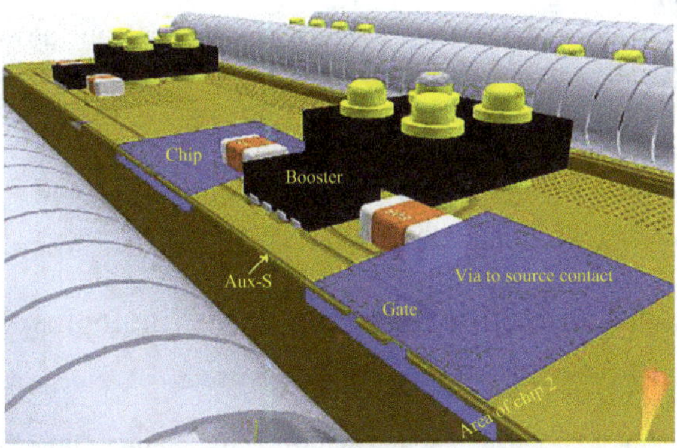

Figure 12.31 Turn-off (blue) and turn-on (yellow) current paths between semiconductor and booster

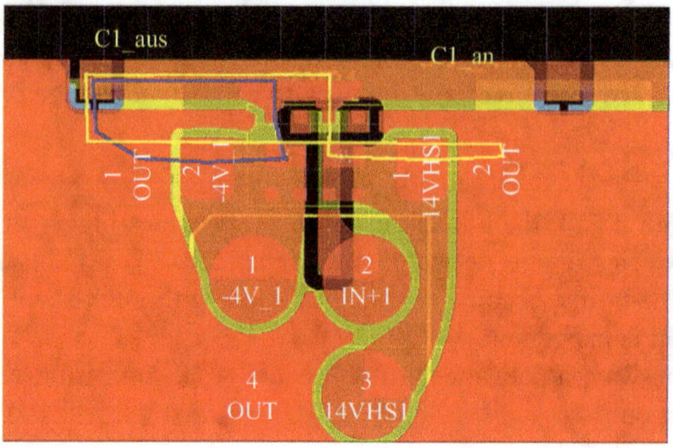

Figure 12.32 3D view of SiC power module. Two SiC MOSFETs (blue) are parallelly driven by a PNP/NPN booster. Gate connected as bus bar

Any parasitic capacitances between drain and gate must be avoided. In less bad cases, they will cause oscillations in bad case parasitic turn-on of the transistor.

In order to reduce displacement currents in the switching moment, the digital ground plane has to not overlap with power planes. In case of overlapping, digital traces have to be decoupled from analog by a reference potential plane under it.

Figure 12.32 shows an example, where a metalization of high-side source contact acts as a reference plane for high-side gate current path.

The self-inductance of vias is unneglectable if the gate inductance in the nH range is required. The approximation of the inductance of the via can be estimated by its length h (in mm) and diameter d (in mm):

$$L = 0.2h\left[\ln\left(\frac{4h}{d}\right) + 1\right]\text{nH}$$

Paralleling of several vias reduces inductance. Thus in Figure 12.32, six vias to gate contact can be seen.

Multichip power modules require additional attention for power distribution and signal propagation times within the module. It was noticed that low inductive interconnections between parallel chips help to reduce the switching load distribution asymmetry.

Usage of the auxiliary source connection is mandatory and is in-between a state of the art for fast switching power modules.

12.8 Alternative top side chip contact technologies

12.8.1 PCB embedding

Although wire bonds are used in the majority of modules, there are promising alternatives that enable more freedom in electromagnetic design and integration. The first one is the PCB embedding technology, which was used for the module

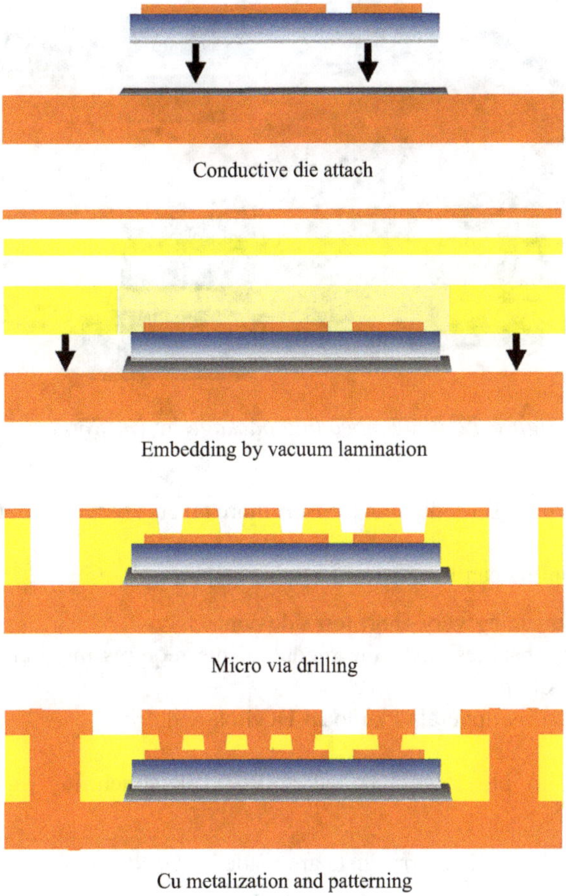

Figure 12.33 Process flow chip embedding

shown in Section 12.2. Using PCB embedding technology chips are laminated into a printed circuit board and contacted by micro vias. An example of a production process is shown in Figure 12.33.

The setup starts by attaching the chip to a copper sheet, for example, by sintering or transient liquid phase bonding. Then a semi-cured layer of PCB material so-called prepreg is prepared by creating cavities for the chip and laid onto the copper sheet. Another prepreg layer and a copper sheet are added and everything is laminated under mechanical pressure, temperature, and vacuum. To contact the chips in the next step, holes are drilled into the top copper and cured prepreg usually using a laser drill. In order not to destroy the chips and to be compliant with the following processes, the top side of the chips has to be metalized with copper. Then the setup is plated in a galvanic copper process creating the micro vias and increasing copper thickness. The next step is applying photoresist, exposing, and

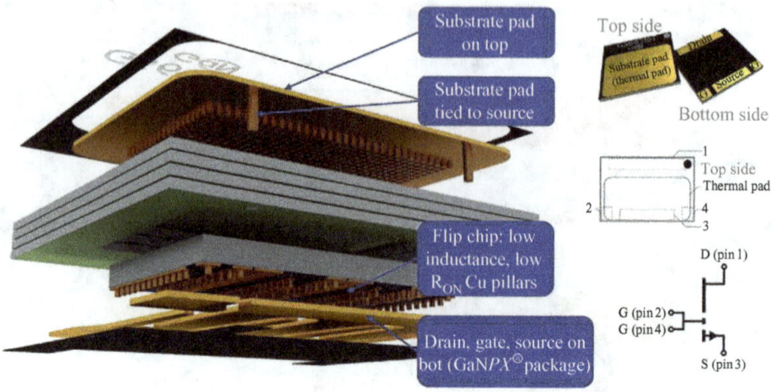

Figure 12.34 Embedding package by GaNSystems

etching as regular PCB production steps. If more layers are required, these steps are repeated.

Several benefits can be achieved by this technology:

- Low inductance by very short top side contacts
- Possible bus bar design in the module. This requires three electrical layers, which is easily possible
- Integration of peripherals like local DC link and driver
- Reliability
- Production in big lots on standard PCB manufacturing equipment (cost reduction)

Commercially available products are made by GaNSystems (Figure 12.34).

12.8.2 Metal clips and metalized transfer mold

The transfer mold technology is standard for encapsulation of small electronic components like the ones shown in or bigger modules like in Figure 12.11. It is based on a highly with SiO_2 filled epoxy resin, which is pressed under temperature and pressure into a cavity (Figure 12.35).

Typically top side chip contacts in this kind of packages are done by wire bonds. An alternative is metal clips as used for the Dual Cool package shown in Figure 12.10. The metal clips are soldered to the top side and show benefits in better possibilities for cooling and thermal impedance (Figure 12.36).

Another opportunity is briefly introduced by the module in Figure 12.12. This is based on a two-layer ceramic substrate with chips assembled, which is shown in Figure 12.12, encapsulated by transfer mold (Figure 12.37).

First, the module is molded (left picture), then holes are drilled to the (copper metalized) chip top side and substrate. By sputtering copper, a seed layer for the following galvanic copper deposition is created. This module covered with the copper surface is then processed like a PCB with photoresist, exposing, etching,

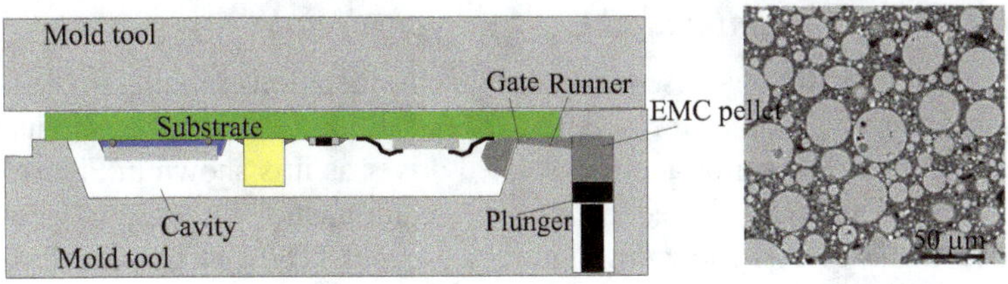

Figure 12.35 *Cross-section of a mold tool with a substrate to be molded and epoxy mold compound (SiO_2 filled)*

Figure 12.36 *The top side metal clips in a transfer mold SMD package (Onsemi)*

Figure 12.37 *Production steps for module shown in Figure 12.12: molding, laser drilling, copper sputtering, and galvanic copper deposition*

and solder resist. The module now is prepared for SMD assembly as shown in Figure 12.12.

As a two-layer ceramic is used for this module, the single layer on top of the module is sufficient to fulfill the requirement of three routing layers for low inductance.

12.9 Examples

To give some impression, which way of thinking is ongoing while design of the power module, this chapter shows two further examples of integrated power modules, with the switching cell design that does not limit switching speeds of the semiconductors. The one and only limiting factor is the semiconductor itself. Both examples show that using modern packaging technologies enables the usage of fast switching properties.

12.9.1 IMS/PCB embedded GaN power module

Example in this section discloses main design considerations running in the head of the engineer on the way to the solution for half-bridge GaN power module with integrated DC link capacitor and nonisolated driver as it is shown in Figure 12.38.

The start conditions based on project requirements for design were PCB embedded GaN devices, shown in Figure 12.34, and TDK EPCOS Ceralink capacitor as DC link capacitor which must be electromechanically connected to system PCB on top and cooled to the heat sink on the bottom.

Concept: Due to the fact that chosen transistors are SMD components with electrical pads on top and thermal pad on the bottom, it is intuitive to use a PCB as an interconnection layer and a thermally conductive, but electrically isolating substrate on the bottom. Isolation between PCB and substrate could be made by encapsulation with silicon or better underfiller.

The first step is always to create a good cooling path. Based on experience from past projects, ceramic substrates combined with PCB on it show very difficult predictable bending behavior. The bending is caused by CTE mismatch between ceramic core (e.g., $CTE_{AlN} \approx 4.5$ ppm/K), CTE of its metalization ($CTE_{Cu} \approx 18$ ppm/K), and PCB on it (e.g., $CTE_{PCB} \approx 14$ ppm/K). The IMS substrate was a promising candidate with better CTE matching ($CTE_{IMS} = 12 \ldots 23$ depending on the type). To get better heat spreading, PCB embedded GaN transistors are soldered to the structured thick copper side of the IMS substrate. IMS structuring allows the isolation of high and low side transistors in the last step as well. The smallest distance between transistors without significant thermal crosstalk was determined by thermal simulation. This allows the DC link loop with the lowest DC link inductance (later measured as ≈ 2.1 nH). To define the soldering process, a solder stop mask must be applied on IMS copper. Additional fiducials on the solder mask allow automated paste application on the pads.

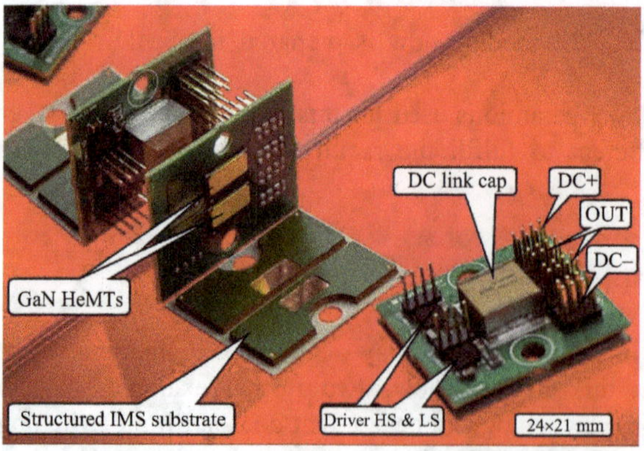

Figure 12.38 Integrated GaN power module: PCB embedded/IMS, integrated DC link, and driver

The second step is the design of the PCB. It starts with the design of the DC link connection. GaN devices have lateral high current flowing inside the device and consequently both high current contacts (Source and Drain) on the top. Closing the DC link loop on the top of the PCB allows the lowest DC link inductance. High current output current path can go to the side because only one chip per switch has to be integrated.

Then a driver gate path is designed as it is described in Section 12.7.3. Both drivers create very low inductive bus bar connection to gates, the bottom layer of the PCB builds a reference plane and auxiliary source connection for drivers minimizing inductive couplings, signal current paths have nearly right angle to power current paths, all routings are rounded, and two separate gate resistors for turn-off and turn-on allow independent switching speed optimization for both switching events. Negative and positive driver voltage requires four pins signal connection to system PCB.

The tiny size of the module demands the usage of additional surface isolation with resin on the PCB to ensure clearance and creepage distances in the application. In laboratory conditions, the creepage distance of 1 mm was sufficient to ensure 650 V functional isolation without additional resin finish.

To ensure safety insulation to the heat sink, the distance between PCB and IMS substrate including corners and 2 mm border is filled with silicone.

Based on the considerations, above mechanical interface to the PCB must fulfill the following requirements:

- Stacking high of 4.2 mm,
- Two times four pin contacts for driver signal and power supply, and
- High current contacts able to carry 8 A continuous current for OUT and, respectively, $8\ \mathrm{A}/\sqrt{2} = 5.7$ A for DC− and DC+ contacts.

The most simple and small solution in this particular case was the usage of 1.27 mm pitch header contacts. Such header could be used for driver contacts and eight pins in parallel for high current contacts. Other solutions would enlarge the power module size.

Mounting to the heat sink is solved with two M3 screws, pressing on PCB. Manufacturing design is worked out as follows:

1. All SMT components are soldered in a reflow oven in a classical way to PCB;
2. Then using a special tool, structured IMS substrate is soldered to GaN HEMTs bottom side; and
3. Silicon filling.

While soldering process, the CTE mismatch between IMS and PCB caused small convex bending of IMS substrate directly in the middle. This must be considered while mounting the module to the heat sink. At the same time, the TIM paste under the module becomes pressed to the sides, thus the resulting TIM layer is thinner than assumed in simulation in the beginning. Exemplarily turn-off curves in Figure 12.39 shows fast and oscillation-free switching.

12.9.2 Full PCB SiC power module

This section shows another example of the power module development with the following objective: a PCB only embedded SiC power module for three-phase 100 kW industrial inverter with ultra-low DC link inductance for 800 V DC link voltage with a size defined as 45 × 75 mm.

PCB technology is cheap and flexible, and allows integration of further functionalities under reasonable price. In this case, RC-snubber, NTC, current sensing resistors, gate booster, and isolated driver were placed directly on the surface of the module (Figure 12.40).

Organic prepreg insulator on the bottom side is chosen as an alternative to the ceramic substrate due to good CTE matching with the whole setup under considerations similar to in the example above. The thermal performance of organic prepreg is much worse than of the ceramic substrate, consequently, only one isolation layer is acceptable between chip and heat sink. A further consequence is, that two routing layers for DC link must be over the chip, and that is easily possible in PCB technology. To achieve sufficient thermal performance, the chips are sintered on a 500 μm thick copper substrate for heat spreading. For the estimation of thermal resistance R_{th}, a thermal simulation uses three simplifications: (1) losses are generated in full chip body and chip is a single block of SiC material sintered on substrate; (2) no material around the chip and consequently no heat spreading over

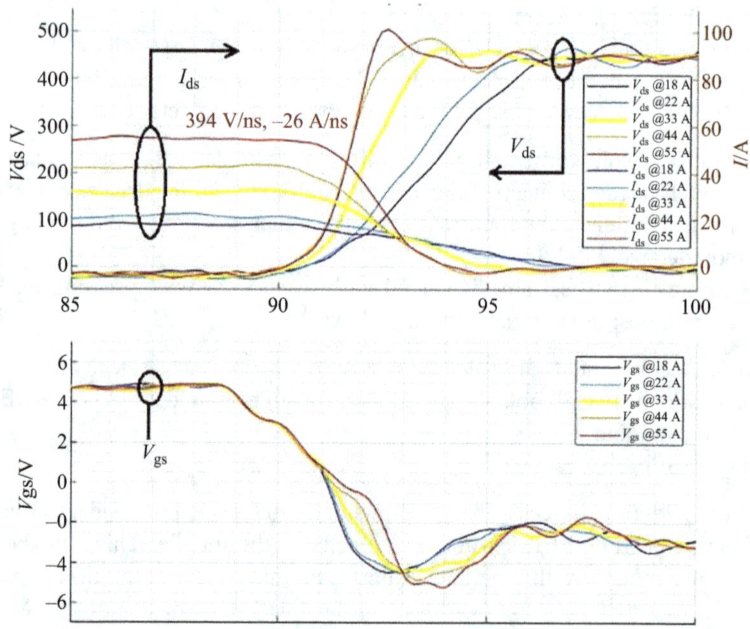

Figure 12.39 Turn-off waveforms of GaN InPM with $R_g = 0\ \Omega$

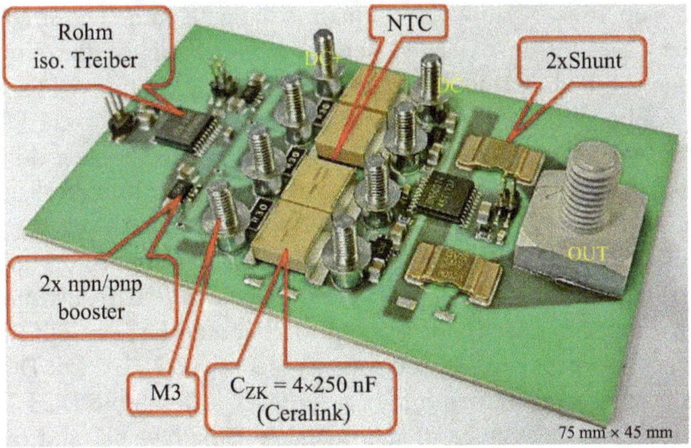

Figure 12.40 SiC InPM: PCB embedded, integrated DC link, and isolated driver

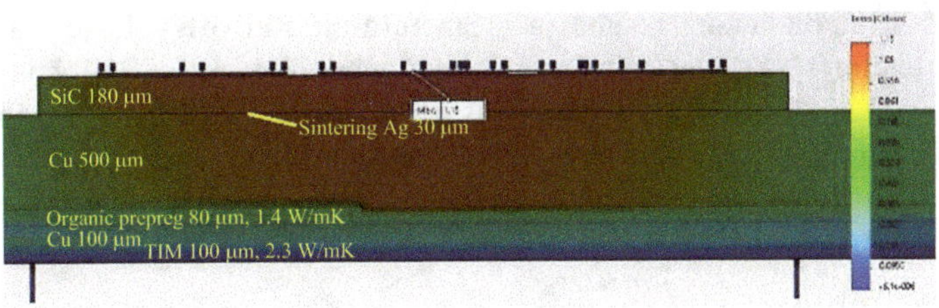

Figure 12.41 Thermal stack and simulation of single-chip R_{th} of SiC InPM including TIM layer

it, makes simulation simpler and faster; (3) the bottom surface is a 100 μm TIM layer with 2.3 W/mK, that is the worst-case, where a standard TIM material is applied by hand. Resulting thermal resistance $R_{\text{th}} = 1.15$ K/W, shown in Figure 12.41, comes in a range of Al_2O_3-ceramic substrate.

For further improvement of the thermal performance, the following trick is designed with eight SMT M3-connectors placed directly over chips. These act at the same time as electrical connections to system PCB and as mechanical connections press TIM under the chips to the range of ≈10 μm and expect significantly improve thermal resistance. Additional plastic bar over system PCB levels the mechanical forces on it.

In the next step, electrical losses are determined using a classical spreadsheet to 19.2 W per chip. With dT = 40 K overtemperature and $T_j = 125$ °C, 8 × 25 mΩ SiC MOSFET dies (four in parallel) in the half-bridge configuration are enough to deliver desired power.

According to DC link design rules, $C_{\text{DClink}} = 780$ nF is required. Thus four capacitors, a 250 nF each with 0.1 Ω damping resistor are placed on the top of the module over the semiconductors. Primary DC link inductance of resulting power module was measured later to be as low as $L_{\text{DClink}} = 0.97$ nH.

The next step after DC link current path is the gate path. The strategy is again similar to the case above: extended source metalization of the chip acts as a reference plane for its driver, multiple vias to chips gate contact reduce inductance, and bus bar connection between gate/auxiliary source and booster. The isolated driver handles two NPN/PNP boosters and each NPN/PNP booster drives two SiC MOSFETs, thus the signal distribution is very symmetrically and thanks to properties of bipolar devices coupled over emitter voltages to each other.

Classical headers in 1.27 mm pitch can be used as connectors to the system PCB for driver signal and power supply. They are available with any mating height.

The high current connection to system PCB should be low inductive as well. Thus the connection between M3 screws and secondary 75 μF foil DC link capacitor was realized as a bus bar with four layer PCB. Alternative DC+/DC− layers allow to achieve the inductance of the connection of few nH and overall interconnection inductance as low as 8.6 nH including M3 pins (Figure 12.42).

Comparing to the switching forms of the power module with DC link inductance of 9 nH shown in Figures 12.2 and 12.3, adding of primary RC-snubber closely to semiconductors results in clean turn-on and off voltage waveforms, shown in Figure 12.43. Occurring overvoltage while turn-off of 220 A is <100 V, thus operation at 800 V DC link voltage is possible using 1,200 V semiconductors. Gate and drain voltages, measured directly on the module surface, show no significant ringing and coupling, and no parasitic turn-on is evident.

12.10 Summary

In this chapter, strategies were shown how to adapt packaging technologies to catch up with improving semiconductor properties. Defining the switching cell as the "heart" of power electronic circuits with the highest changes in voltage and current and integrating it helps to influence and design parasitic electromagnetic properties intentionally. In the examples shown above, an optimization of the geometric setup of switching cells is demonstrated using advanced manufacturing technologies and at the same time fulfilling practical requirements for electric, thermal, and

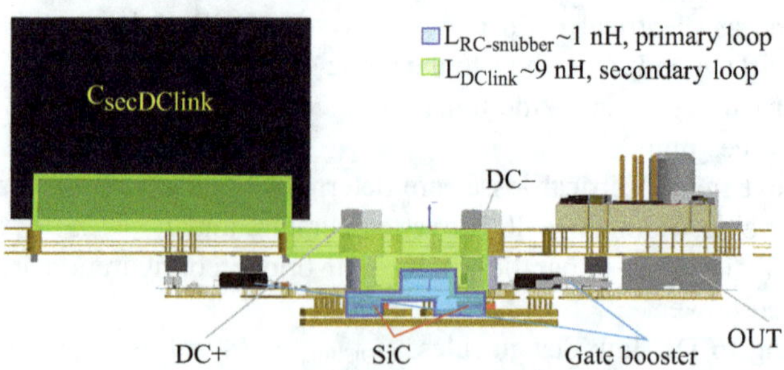

Figure 12.42 Current loops of primary and secondary DC link inductances for SiC InPM

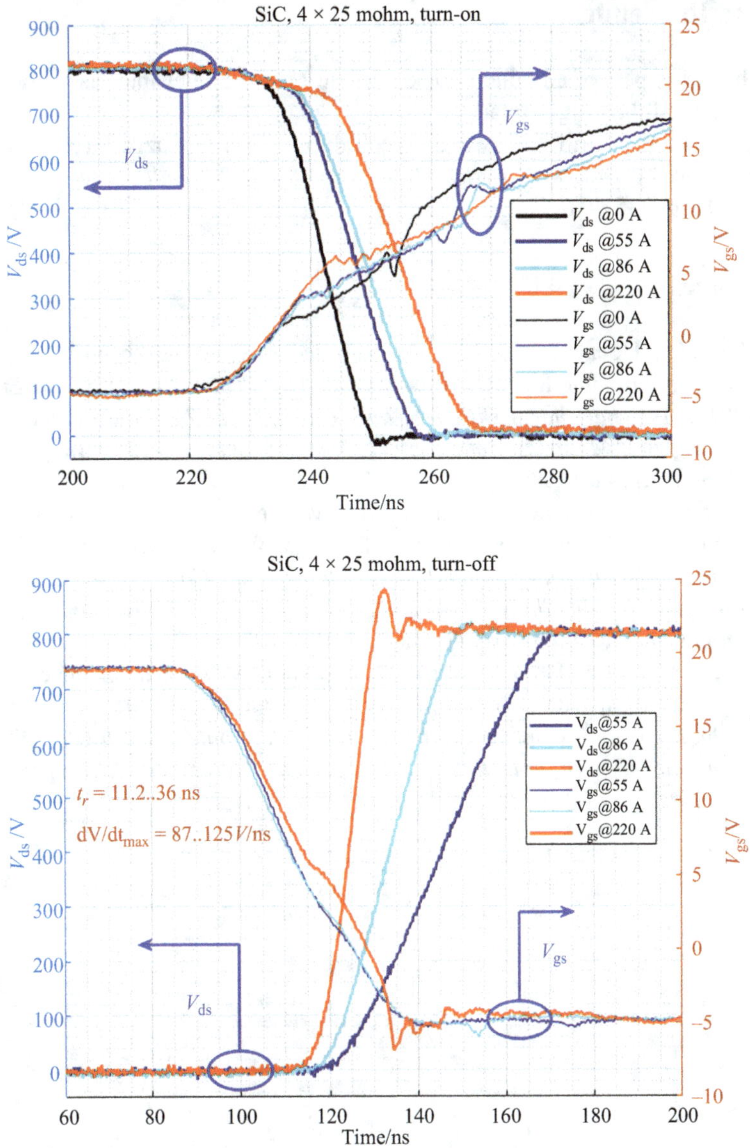

Figure 12.43 Turn-on (top) and turn-off (bottom) waveforms of SiC InPM for different currents at $V_{DClink} = 800$ V

mechanics interfaces. The layout rules given for low inductance design in the gate loop and DC link are proven to allow the fastest switching. The examples shown above give an impression on a possible solution for next-generation switching cells and the adaption of different aspects of these solutions in commercial production will remain interesting to observe.

Note to the reader

This chapter is part of an edited book. To cite or use this chapter as a reference, please use the following format:

Eckart H. and Kirill K. “Switching cell design.” In: Iannuzzo F (ed.), *Modern Power Electronic Devices: Physics, Applications, and Reliability*. Stevenage, UK: IET; 2020. pp. 385–416.

References

[1] Feix G., Hoene E., Zeiter O., and Pedersen K. “Embedded very fast switching module for SiC power MOSFETs.” *Proceedings of PCIM Europe 2015; International Exhibition and Conference for Power Electronics, Intelligent Motion, Renewable Energy and Energy Management*, Nuremberg, Germany, 2015, pp. 1–7.

[2] Marczok C., Hoene E., Thomas T., Meyer A., and Schmidt K. “Low inductive SiC mold module with direct cooling.” *Proceedings of PCIM Europe 2019; International Exhibition and Conference for Power Electronics, Intelligent Motion, Renewable Energy and Energy Management*, Nuremberg, Germany, 2019

[3] Klein K., Hoene E., and Lang K. “Comprehensive AC performance analysis of ceramic capacitors for DC link usage.” *PCIM Europe 2017; International Exhibition and Conference for Power Electronics, Intelligent Motion, Renewable Energy and Energy Management*, Nuremberg, Germany, 2017, pp. 1–7.

Chapter 13

Modern insulated gate bipolar transistor (IGBT) gate driving methods for robustness and reliability

Haoze Luo[1], *Wuhua Li*[1] *and Francesco Iannuzzo*[2]

13.1 Introduction

After almost three decades of development, insulated gate bipolar transistors (IGBTs) are widely used in many high-power industrial applications [1]. The reliability issues have been studied by employing solutions in active and passive components, mechanical structures, packaging designs, and control strategies. Meanwhile, the complex and harsh working conditions are demanding for higher reliability of the power conversion systems. According to a recent industrial survey indication, the semiconductor and solder failures in devices account for 34% of failures in the power conversion systems, which is the largest portion in all failure cases [2]. Therefore, it is still necessary to investigate the failure mechanisms and appropriate alternative methods for improving the reliability of IGBTs.

The high-power IGBT module FZ1500R33HE3 rated at 3.3 kV/1.5 kA from Infineon is taken as an example to demonstrate the semiconductor and package characteristics, as shown in Figure 13.1 [3]. In order to increase module current capacity, the modern high-power IGBT module is usually achieved by multichip in parallel connection. Concerning the wire-bonded IGBT modules, connected by aluminum bond wires, copper layers, power connection terminals, and direct copper bonded, the parasitic parameters are inevitably in the modules. Accordingly, as in Figure 13.1, L_{aC}, L_{ca}, L_g, L_{ek}, and L_{kE} represent the parasitic inductors between the external power terminals (including power terminals C and E; control terminals c, g, and e) and internal semiconductor chips, R_{in} is the internal gate resistor which increases driver loop damping. Considering the semiconductor characteristics, C_{ge} is gate capacitor, C_{gc} is gate collector capacitor, and C_{ce} is the parasitic capacitor across collector and emitter terminals. From the module outside, the power terminals C and E connected to the main circuit are mainly used for the power flow. The auxiliary terminals c, g, and e are usually connected to the gate driver and used for

[1]College of Electrical Engineering, Zhejiang University, Hangzhou, PR China
[2]Department of Energy Technology, Aalborg University, Aalborg, Denmark

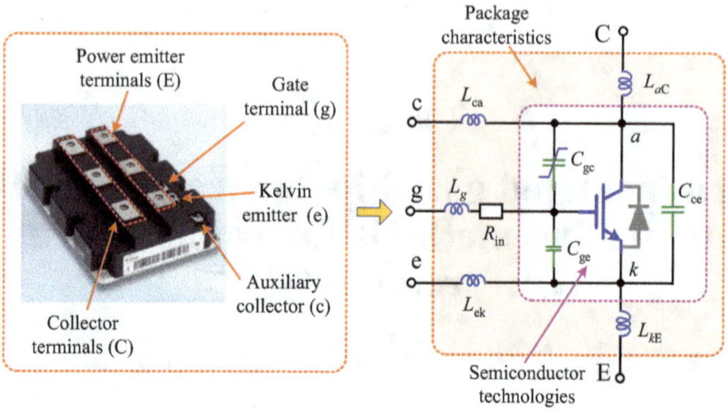

Figure 13.1 High-power IGBT module considering package and semiconductor characteristics

gating and monitoring IGBT modules. During the operations, the collector voltage and current of IGBTs can be reflected by the internal parasitic inductors. Furthermore, by monitoring the signals from the auxiliary terminals, the optimization and protection methods for enhancing IGBTs' reliability can be achievable in gate drivers.

Along with the development of IGBT modules, gate drivers have been improved dramatically over the years, significantly contributing to reliability improvement. In fact, as an important interface between IGBT modules and microcontrollers, modern gate drivers do not only can provide the optimal switching signals, but also monitor the operation status of IGBT modules themselves.

From the reliability engineering point of view, system failures can be roughly divided into three main stages: early failure period, middle random failure period, and wear-out period [4]. In all three stages, modern gate drivers can play a key role. Stage I (early failure period): most of the design and manufacturing defects can be detected before the component is shipped to the end user. In case that a defective component is delivered to the end user, it will incur an early failure in Stage I, and the gate driver can stop the system and assert a failure indication. Stage II (middle random failure period): the failure root causes are mainly the random excessive electrical stresses (e.g., short circuit or avalanche breakdown) [5]. If properly designed, the gate drivers can be used to monitor different electrical parameters and prevent major damages to the IGBT, such as short-circuit event. Stage III (long-term wear-out failure period): approaching the end of life, the package-related degradations begin to deteriorate the module's electrical and thermal characteristics [6,7]. In this period, the gate driving strategy can be implemented, for example, reducing the temperature variation to mitigate the thermo-mechanical stress and keep it in operation until maintenance. Owing to its intrinsic characteristics of slow and regular degradation, IGBT wear-out failure is the most commonly monitored and predicted by modern gate drivers during the long-term services.

13.2 Operation principle of IGBTs

To get the desired output voltage waveform at a specific level, pulse width modulation (PWM) technique is used to encode the control algorithms into a series of pulsing waveforms and then regulate the electrical energy delivered to a load [8]. Under PWM operations, the running IGBT is always switched between ON and OFF states under high frequency. The typical IGBT turn-off switching waveforms under inductive load (e.g., collector-emitter voltage and collector current) together with the gate voltage are plotted in Figure 13.2. As shown in Figure 13.2, the turn-off transitions can be divided into four stages.

Turn-off stage 1: At the beginning of the turn-off operation, the gate voltage v_{ge} decreases exponentially due to the gate capacitor discharging via gate resistor R_g. The depletion layer expansion under the gate region begins when v_{ge} reduces to the Miller plateau voltage v_{gp}. The time from t_1 to t_2 can be defined as the gating delay time. Collector voltage v_{ce} rises slightly and the load current I_L remains constant. Since v_{ge} is constant during the Miller plateau, all the gate current i_g is supplied by i_{gc} during this stage.

Turn-off stage 2: The gate-collector capacitance C_{gc} consists of the variable depletion layer capacitance C_{dep} in series with the fixed oxide capacitance C_{ox}. The gate collector capacitance C_{gc} decreases quickly in connection to the decrease of depletion capacitance C_{dep} as v_{ce} increases. There is negative feedback from the rising collector voltage v_{ce} to the gate voltage v_{ge}. An extra current caused by the rising collector voltage v_{ce} will flow toward the gate through C_{gc}, preventing the reduction of v_{ge}. Both the gate voltage v_{ge} and gate current i_g decrease slightly. Essentially, the gate current i_g is provided by the charging of C_{gc}, and the collector voltage v_{ce} increases near linearly.

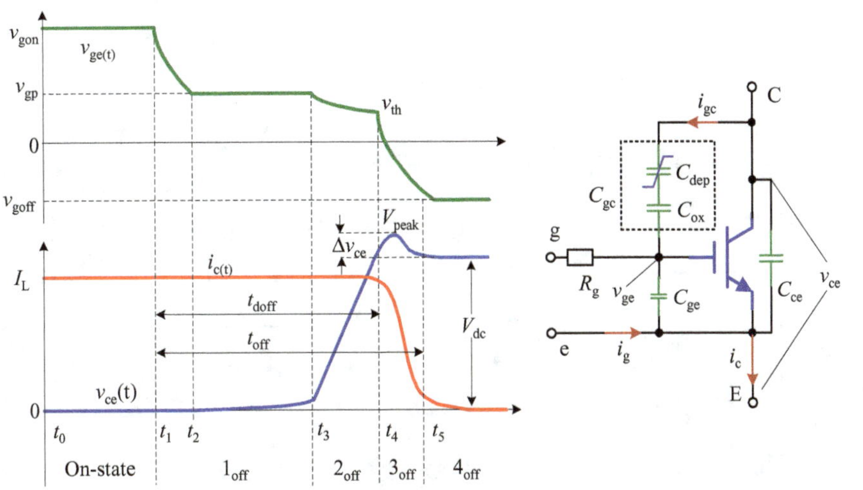

Figure 13.2 Typical IGBT turn-off transient waveforms

Turn-off stage 3: At t_4, when v_{ce} reaches the bus voltage V_{dc}, the load current I_L is diverted through the so-called freewheeling diode (not reported in Figure 13.1). At this stage, the collector current i_c decreases rapidly. The collector current gradient di_c/dt induces the overshoot voltage Δv_{ce} across the loop parasitic inductances in the current commutation loop. This overshoot Δv_{ce} can lead to a high turn-off peak voltage V_{peak} and trigger the overvoltage failure. The duration from t_1 to t_5 is defined as the turn-off time.

Turn-off stage 4: At t_5, the collector voltage v_{ce} reaches the bus voltage, but the collector current i_c takes some time to fall to zero. This phenomenon is well known and is called "tail current," which contributes to the turn-off power losses. It is worth to note that the tail current and its duration are only related to the device characteristics and cannot be optimized by the gate driver.

Similarly to the turn-off transition, the IGBT turn-on process under inductive load can also be divided into four stages, as shown in Figure 13.3. During the off state, the IGBT withstands the high bus voltage V_{dc} and no collector current i_c flows through it.

Turn-on stage 1: At t_1, the gate voltage v_{ge} starts to increase from off-state gate voltage v_{goff}. This stage from t_1 to t_2 can be defined as the turn-on delay time t_{don}. With respect to this period, because the gate voltage v_{ge} is lower than the threshold voltage v_{th}, the collector current is zero.

Turn-on stage 2: As soon as the gate voltage v_{ge} reaches the threshold voltage v_{th}, the collector current i_c begins to flow through the emitter (E). The duration from t_1 to t_2 is defined as turn-on delay time t_{don}. After t_3, the reverse recovery current of the commutation diode starts to flow through the IGBT. The collector current i_c consists of both load current I_L and the reverse recovery current. It is remarkable that the freewheeling diode remains in on state with a negligible

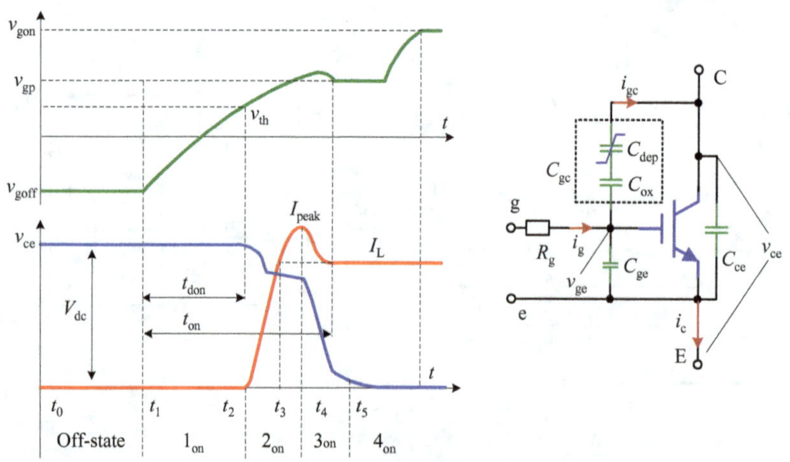

Figure 13.3 Typical IGBT turn-on transient waveforms

forward voltage. Therefore, the collector current rises rapidly while the collector voltage v_{ce} is maintained at a high level.

Turn-on stage 3: At t_4, the turn-on collector current i_c reaches its peak value I_{peak}. Then the diode reverse recovery current begins to decrease so that the turn-on i_c starts to decrease to load current from I_{peak}. Since the commutation diode can withstand the bus voltage, the collector voltage v_{ce} shows a significant decrease in this stage. The time from t_1 to t_5 is defined as the turn-on time t_{on}.

Turn-on stage 4: The freewheeling diode starts to withstand the whole bus voltage in this stage. The turn-on process is completed.

The details of transient stages and potential failure mechanisms are summarized in Table 13.1. Different types of failure cases occurring in specific stages of the turn on/off transitions are also summarized in Table 13.1. Typically, abnormal operations can be detected by the gate driver if they differ from the normal turn on/off transitions. According to the characteristics of different stages, opportune optimization measures can be adopted and implemented in gate drivers. Nowadays, more optimization and protection measures can be adopted with modern gate drivers rather than passive based mitigation solutions.

Table 13.1 Analysis of turn-on/off transitions

Turn-off transitions			
Stages	**Description**	**Features**	**Negative impacts and potential failure mechanisms**
Stage 1	Gate voltage fall to the threshold	$v_{ge}>v_{gp}$	Limiting switching speed
Stage 2	Collector voltage rise	dv_{ce}/dt	Lower dv_{ce}/dt, larger losses. This duration can be shortened by gate current injection
Stage 3	Collector current fall	di_c/dt	Potential failure mechanisms: over-voltage breakdown, turn-off latch-up, and turn-off avalanche breakdown
Stage 4	Tail current	Tail current	Depends on device technology. The longer tail current, the larger switching losses
Turn-on transitions			
Stage 1	Gate voltage rise to the threshold	$v_{ge}<v_{gp}$	Limiting switching speed
Stage 2	Collector current rise	di_c/dt	Lower di_c/dt, larger losses. Higher di_c/dt, higher I_{peak}. Moreover, higher di_c/dt may trigger the avalanche breakdown of freewheeling diodes
Stage 3	Collector voltage fall	dv_{ce}/dt	Lower dv_{ce}/dt, larger losses. The peak current should be suppressed for avoiding thermal runaway breakdown
Stage 4	Final gate voltage rise	Discharge C_{gc}	Faster extraction speed of C_{gc} charge will result in less loss

13.3 Basic IGBT gate driving methods

Normally, IGBT can be regarded as a kind of voltage-controlled device, which only requires a constant voltage on the gate side to maintain device conduction. This means that the gating requirements for gate driver are lower than that of bipolar junction transistors (BJTs). In this section, the basic design considerations for modern IGBT gate driver are discussed.

13.3.1 Voltage-source gate drivers

The typical voltage-source gate driver schemes are usually adopted for IGBT modules, as depicted in Figure 13.4. The modern IGBT gate driver consists of three main parts: isolation units for control signal and the required gating voltages, power booster stage, and the gate resistor part. In the two-level half-bridge leg, the high side IGBT emitter will be switched between the high bus voltage and the low ground voltage. Hence, as an interface between the microcontroller and high-power electronics, the input control signal and the related output gating voltages should be separated in the gate driver. Usually, the control signal is isolated by the opto-coupler or optical fiber. A dual-output DC–DC power supply is applied to the gate driver for the driving power sources.

Nowadays, the standard on-state gate voltage v_{gon} is +15 V for IGBTs. The off-state gate voltage v_{goff} is in the range of −10 to −15 V according to the design requirements. For the switching transitions, the IGBT parasitic capacitances are charged and discharged frequently by the gate driver. To enhance the gate driving ability, the control signal will be amplified by means of a booster stage. The conventional booster circuit is usually achieved by the push–pull circuit with bipolar junction transistors (BJTs) or MOSFETs. Concerning the low-cost BJT-based booster circuit, the voltage drop on the PN junction in BJT is around 0.7 V, which decreases the terminal gate voltage v_{gon}/v_{goff}. To overcome this shortcoming, the booster circuit is achieved by the low-resistance MOSFETs with a higher cost,

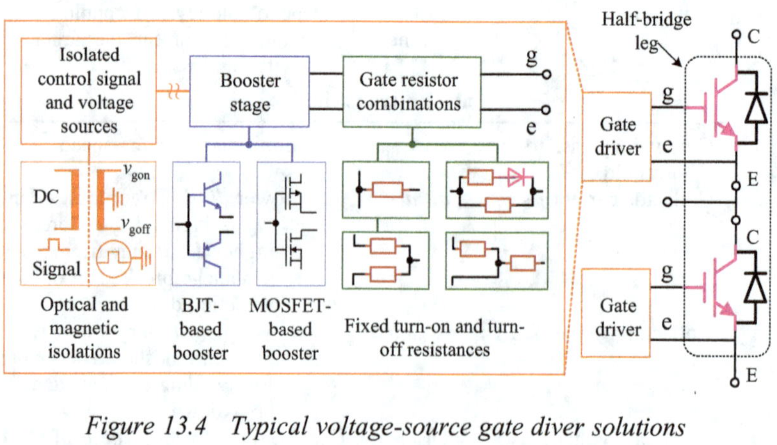

Figure 13.4 Typical voltage-source gate diver solutions

which can fulfill the rail-to-rail low impedance output. On top of that, the gate resistance selection is of great importance to IGBT switching characteristics. In terms of some low-voltage and low-power IGBTs, there is only one gate resistance for the turn-on/off operations. In practice, gate drivers are required to adjust the turn-on/off switching speeds of IGBTs by means of different gate resistor combinations. Generally speaking, the selection of appropriate gate resistance requires careful consideration on the tradeoff of voltage overshoot versus switching power losses.

13.3.2 Current-source gate drivers

With the increase of switching frequency, the switching power losses account for a large proportion of the overall power losses. To solve this problem, the current-source gate drivers are proposed to reduce both the device losses and gate driver's losses and enhance the switching frequency. The typical current-source gate driver and the related gating current are depicted in Figure 13.5 [9]. Referring to Figure 13.5(a), the current generator is implemented by an H-bridge circuit with an inductance L_{gc}. The gate current comparison between the typical voltage-source driver and current-source driver are shown in Figure 13.5(b). The current-source gate driver behaves nearly like a constant current source so that the rise and fall times of the gate-emitter voltage v_{ge} are faster in contrast with the voltage-source gate drivers. According to the summary in Table 13.1, the constant gate current can speed up the charge and discharge processes of IGBT parasitic capacitance, including C_{gc}, C_{gc}, and C_{ce}. Hence, the turn-on and turn-off times significantly decrease significantly so that the switching losses may decrease as well. In general, most of the commercial IGBT gate drivers are voltage-source drivers. Hence, many optimizations and protection methods have been developed on the basis of voltage-source derivers. Due to the advantages of switching losses reduction and accelerating switching speeds, the idea of controllable gate current is also introduced in modern gate drivers.

13.3.3 Optimization and protection principles

In addition to the fundamental driving functions, many optimization and protection functions are gradually integrated into the modern IGBT gate drivers. Though the

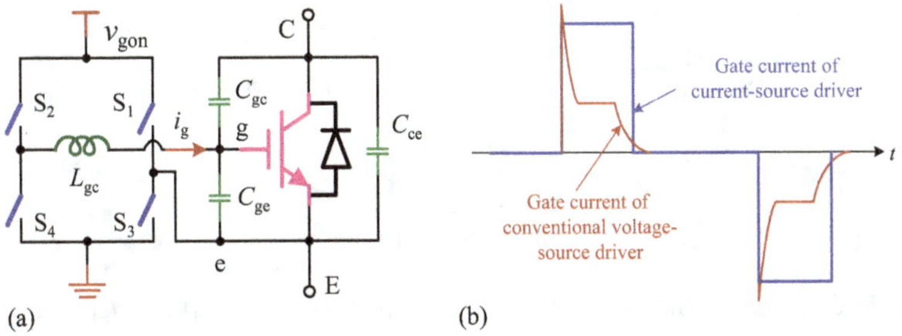

Figure 13.5 Current-source gate driver and related gate current

additional optimization and protection units will create more cost, the overall functions and reliability of the power conversion system can be improved. The detection of abnormal operation is a precondition for the optimization and protection of running IGBTs. Compared with the normal working conditions or the predetermined criterion, the detection unit will provide the gate driver with the current IGBT status. After this, the optimization or protection actions are carried out subsequently.

The typical four types of IGBT optimization and protection approaches are depicted in Figure 13.6 [10]. According to the sampling signal sources from IGBT module, the advanced protection and optimization solutions can be divided into (1) collector voltage v_{ce}-based approach, (2) gate signal-based approach, (3) collector current i_c-based approach, and (4) collector current rate di_c/dt-based approach. For a better readability, the abbreviations of protection and optimization functions are listed as follows: (1) overload protection (OLP), (2) short-circuit protection (SCP), (3) current overshoot suppression (COS), (4) voltage overshoot suppression (VOS),

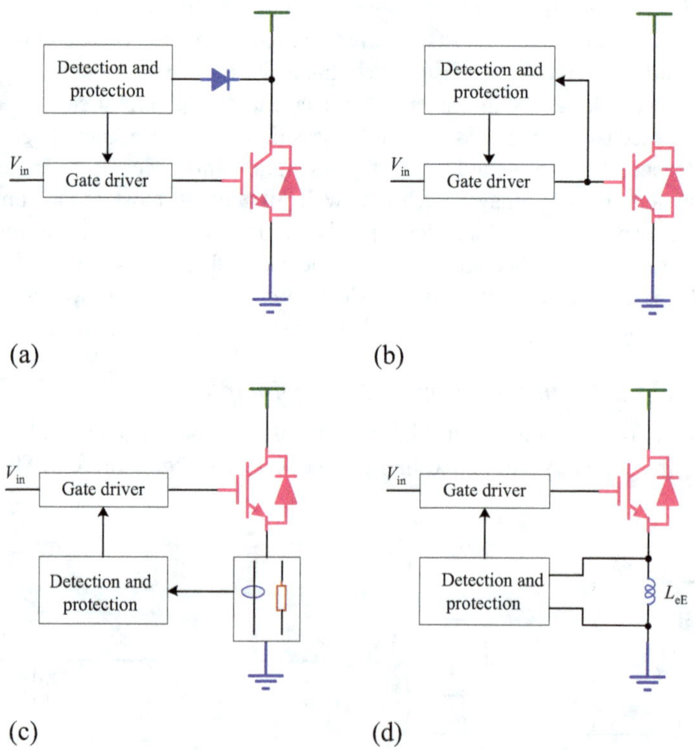

Figure 13.6 Four basic approaches of gate driver-related fault detection and protection circuits: (a) v_{ce}-based approach, (b) gate signal-based approach, (c) i_c-based approach, and (d) di_c/dt-based approach

(5) power losses optimization (PLO), (6) switching trajectory control (STC), and (7) junction temperature measurement (JTM).

For the first type approach depicted in Figure 13.6(a), one high-voltage biased diode and sampling circuit are required. The biased diode is used to protect the detection and protection unit from the high collector voltage. In the case of short-circuit or overload operations, the on-state collector voltage v_{ce} under normal load current can be measured via the diode. The abnormal collector voltage v_{ce} can be measured and compared with the safety settings. Then, further operations in the gate driver will be carried out, such as decreasing gate driving voltage v_{gon}. Besides, since on-state v_{ce} is related to junction temperature under fixed collector current, these relations can be used to estimate the junction temperature for the given collector current and v_{ce}.

Since some abnormal working conditions can be reflected by the gate-side signals, the gate signal-based approach is used for IGBT protection, as shown in Figure 13.6(b). Usually, the gate voltage v_{ge}, gate current i_g, and gate charge Q_{gc} can be used to perform OLP, SCP, PLO, and JTM.

Because the highly focused operations are directly related to collector current, such as short circuit and over current, the collector current i_c-based approaches have been developed, as shown in Figure 13.6(c). The instantaneous collector current i_c can be measured by mirror current or Rogowski coil. Considering the mirror current approach, the scaled-down instantaneous i_c can be obtained by a part of the unit cells of IGBT. However, the mirror current function depends on manufactures. Rogowski coil-based method can be regarded as an isolation approach, which has been used to estimate the collector current and closed-loop control.

Another available electrical parameter for IGBT protection is the change rate of collector current di_c/dt. The common application is shown in Figure 13.6(d). Taking advantage of parasitic emitter inductance L_{eE}, the value of di_c/dt can be obtained using some low voltage logic and sampling components. The di_c/dt-based approach has been used in the protections of SCP, COS, PLO, STC, and JTM.

On top of that, almost all the electrical parameters associated with IGBT operation can be measured from the collector side, gate side, and emitter side. Moreover, by combining the four basic approaches, more optimized and sophisticated protection methods can be achieved according to application requirements and cost considerations. In the following sections, the practical implementations and operation principles will be introduced and discussed in detail.

13.4 Fault detection and protection methods

13.4.1 Voltage and current overshoot

13.4.1.1 Voltage overshoot

Due to the existence of parasitic inductance, IGBTs are required to withstand hazardous voltage overshoot during the normal turn-off transition. Usually, the voltage overshoot is proportional to the parasitic inductance and turn-off collector current. Therefore, in the case of a short-circuit event, the turn-off overvoltage also should be suppressed by means of soft turn-off methods. The typical soft turn-off methods consist of active clamping method, multilevel method, and di_c/dt feedback method.

In practice, active clamping methods with passive components have been used widely in high-power IGBT modules. Two industrial implementation schemes of active clamping methods are plotted in Figure 13.7(a) and (b). A simple single-loop clamping method implemented with a series of transient voltage suppressors D_z is shown in Figure 13.7(a). Once the collector voltage v_{ce} exceeds the breakdown voltage of D_z, a breakdown current I_z flows directly to the gate capacitor C_{ge}. As a result, the gate voltage v_{ge} would be raised again and the current change rate di_c/dt is suppressed to a safe value. There are two main drawbacks in the single-loop active clamping method. First, the breakdown current I_z may raise v_{ge} temporarily to a critical value higher than v_{gon}. Although the gate voltage can be limited, there is a high voltage drop across R_g. Next, the additional energy injected by I_z will be consumed in R_g and v_{goff} transistor after the clamping process. In order to overcome the shortcomings of single loop active clamping, another advanced dual-loop active clamping method is shown in Figure 13.7(b). The breakdown current I_z is separated into gate capacitor injection current I_1 and gate voltage adjustment current I_2. The advantages of the dual-loop active clamping method are the following: the gate capacitor injection current I_1 is limited by R_1, the turn-off gate voltage is adjusted by I_2 after the detection of overvoltage. Consequently, the high voltage drop across R_g and power losses in R_g and D_z can be reduced for enhancing the reliability of passive and active components in active clamping loops.

The active clamping needs to be triggered when peak overvoltage V_{peak} exceeds a specific high value. However, V_{peak} can be suppressed during v_{ce} rising stage by using multilevel control of gate voltage. In the multilevel methods, the turn-off gate voltage will be divided into several levels. Taking the two-level turn-off method as an example, the control principle is depicted in Figure 13.8. During the turn-off process, the gate voltage v_{gon} falls to an intermediate level V_{int0} and holds for a short time T_{int0} then shutdown. In turn-off stage 2 (v_{ce} rising stage shown in Figure 13.2), the extraction of gate current is controlled at a slow rate to

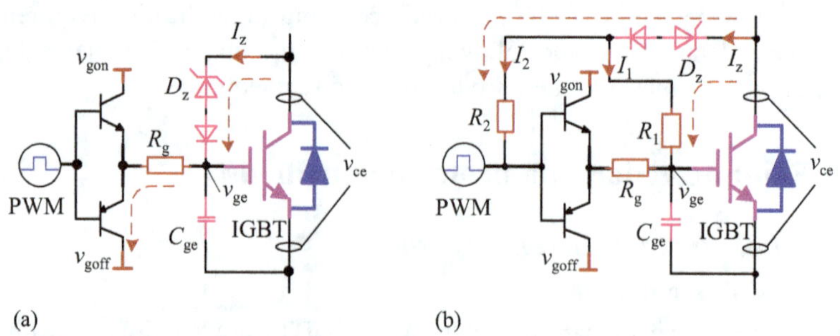

Figure 13.7 Two implementation schemes of active clamping methods: (a) single-loop active clamping and (b) dual-loop active clamping

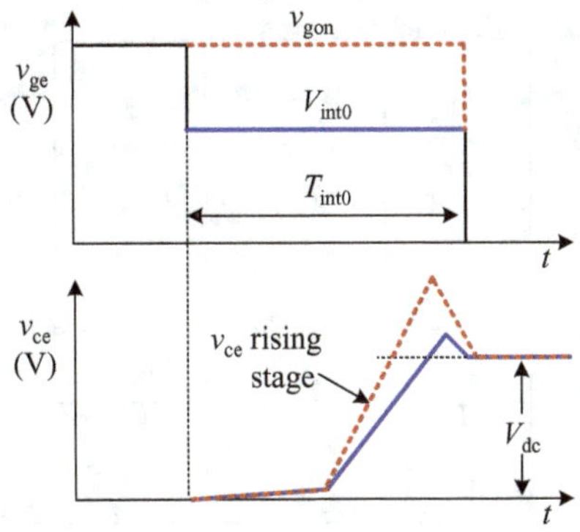

Figure 13.8 Simplified two-level turn-off switching waveforms

adjust dv_{ce}/dt. By this means both dv_{ce}/dt and V_{peak} can be controlled as specified value due to the lower V_{int0}. More details of multilevel control of v_{ge} can be found in [11,12], and the time sequence and charging speed of the gate capacitance have great impacts on the performance of V_{peak} suppression.

Since excessive V_{peak} is induced by the rapid di_c/dt, the voltage overshoot can also be reduced by real-time control of di_c/dt. The di_c/dt feedback method is one of the features of trajectory control. Hence, the principle of di_c/dt feedback method will be introduced in Section 13.4.2.

Concerning the current overshoot, the maximum turn-on collector current is composed of the peak reverse recovery current of the freewheeling diode and the load current. Since the turn-off speed of the freewheeling diode can be controlled by IGBT switching speed, the current overshoot can also be reduced by the gate driver. Similarly, the idea of multilevel control methods can also be used to suppress turn-on current overshoot. Compared with complex multilevel control, a simple two-level turn-on method has been applied for current overshoot suppression [13]. The simplified waveforms of drive voltage and related collector current at turn-on are plotted in Figure 13.9. The value of the intermediate voltage must be higher than the threshold voltage v_{th} to ensure the internal transistor of IGBT is turned on. Meanwhile, V_{int} should be chosen lower than v_{gon} to limit turn-on di_c/dt. The duration of intermediate voltage T_{int1} is required to cover the turn-on transition. When the reverse recovery stage of diode ends, and collector current i_c equals the load current I_{load}, then the gate driver voltage rises up to v_{gon}. The turn-on current overshoot decreases with decreasing V_{int} as shown in Figure 13.9. However, the turn-on current overshoot may not be controlled once V_{int} gets higher than a specific value. For a given device, the uncontrolled V_{int} should be determined by testing procedures. Besides, the turn-on current overshoot can also be reduced by di_c/dt-based method.

Compared with the passivity-based method, the voltage and current overshoot suppression methods can provide more flexibility in changing the dynamic characteristics of IGBT. Most of the existing industry optimization methods can be

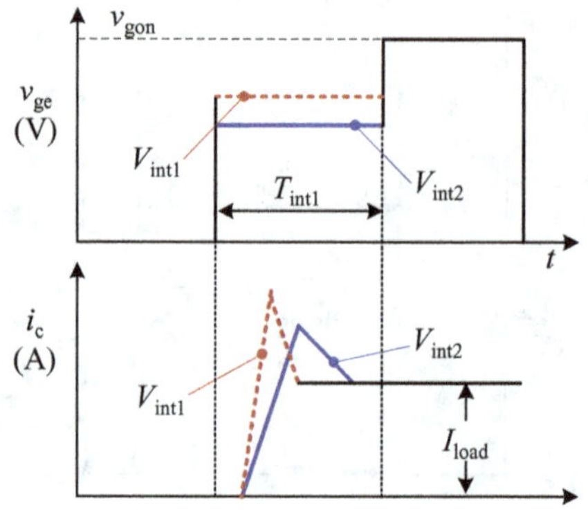

Figure 13.9 Simplified waveforms of two-level turn-on drive voltages and collector currents

summarized as part of the "open-loop methods," such as the Miller-plateau-based method [11] and feed-forward control method [14]. In order to eliminate the negative effects of variable working conditions of the IGBT, the closed-loop control methods for real-time control have been introduced in advanced IGBT gate drivers.

13.4.2 Overload and short-circuit event

Short-circuit operation is often characterized as the most familiar and serious failure in IGBTs. IGBT short-circuit modes can be classified into two scenarios: short-circuit I (hard switching fault, HSF) is a direct turn-on short-circuit condition; and short-circuit II (fault under load, FUL) is the occurrence of a short circuit during the on-state of IGBTs, the rising rate and peak current are lower than that of HSF. So far, most of the studies focus on the detection and distinction of the two short-circuit operations. Besides, the gate voltage oscillations and multichip influence under short-circuit events [15] are research topics when considering robustness. The impact of gate driver on the short-circuit ruggedness of IGBT is discussed in [16]. When the IGBT device withstands the short-circuit event, two failure mechanisms could be triggered by the fast gate turn-off operation. The first one is the high voltage overshoot induced by high di_c/dt in the presence of the parasitic inductors. The peak voltage overshoot V_{peak} during the turn-off transition may exceed the rated voltage of IGBT. In this very dangerous case, the so-called dynamic self-clamping phenomenon occurs. This clamping mechanism can be unstable and result in the forming of current filaments that can cause thermal runaway failure. The other one is the gate breakdown caused by v_{ge} overvoltage induced by the fast collector transients. As the collector voltage v_{ce} is increasing during short circuit, the high dv_{ce}/dt will induce displacement current i_{gc} through the reverse transfer capacitor C_{gc}. Hence, the gate voltage v_{ge} becomes higher than v_{gon} and could cause gate breakdown failure. Besides, another cause of gate overvoltage is the parallel connection of multichip [17]. It has been studied that the gate voltage would oscillate and exceed +20 V in multichip module packaging under short-circuit operation [18]. In Figure 13.10, the evidence of gate oscillations varies with the DC-link voltages are demonstrated, and the maximum value of oscillation voltage reaches a dangerous +25 V. It can be observed that the starting time of the oscillation varies with the DC-link voltages, which must be considered when selecting the protection solutions.

The most common commercial solution for short-circuit protection is the on-state voltage drop v_{ce} measurement. The equivalent circuit of IGBT module with parasitic parameters and the curves of output characteristics including saturation current are plotted in Figure 13.11. The principle of short-circuit protection method is as follows: when the IGBT is turned on, the collector voltage v_{ce} falls to a low voltage drop from the high bus voltage level, and collector current i_c could rise to the load current level. During the short-circuit fault, the short-circuit current I_{sc}

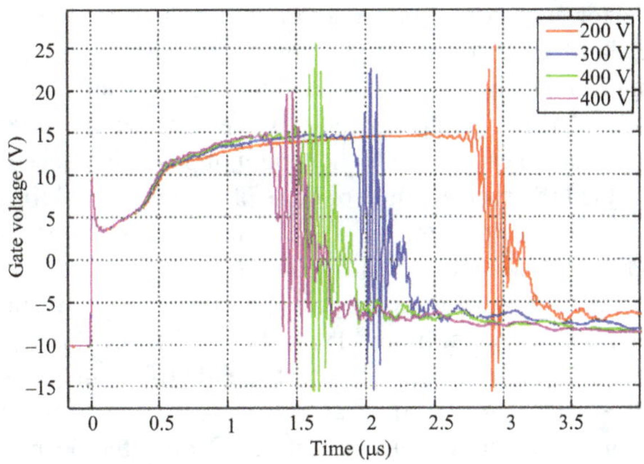

Figure 13.10 Gate voltage oscillation under short circuit with different DC-link voltage

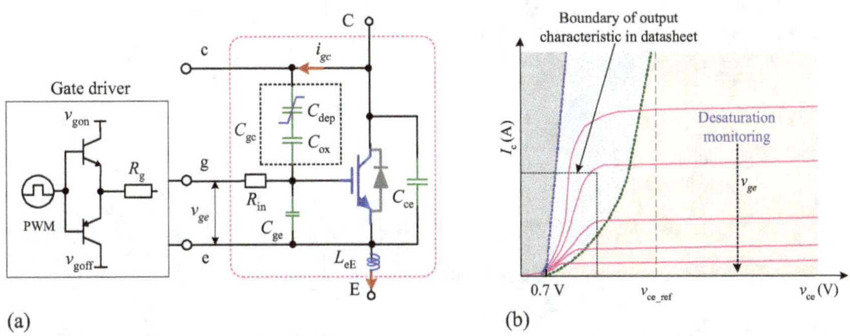

Figure 13.11 (a) Equivalent circuit of IGBT module with parasitic parameters and (b) static curves of output characteristics

increases up rapidly to its saturation level which depends on the applied v_{ge}, and then the collector voltage will also increase up to the bus voltage level. Once the measured v_{ce} is higher than a specific value v_{ce_ref}, the IGBT can be identified as short-circuit statues. Further, gate voltage v_{ge} needs to be reduced for desaturation operation to limit the short-circuit current, as depicted in Figure 13.11(b).

Another typical solution for short-circuit detection is monitoring the change of di_c/dt [19]. Whether HSF or FUL, the short-circuit loop contains only circuit parasitic inductors and excludes the huge load inductor. Hence, the collector current rises very quickly so that di_c/dt can be used as a short-circuit indicator. One commonly used detection principle by means of di_c/dt is shown in Figure 13.12. In terms of high-power IGBT modules, there is a parasitic inductor L_{eE} between the Kelvin emitter (e) and power emitter terminals (E), as shown in Figure 13.12(a). In the case of constant L_{eE} under switching operation, the current rate of change di_c/dt can be extracted by the induced voltage v_{eE} across L_{eE}. Note that during the on-state period, the collector current is constant and therefore veE is zero. As a result, v_{eE} induced by the short-circuit event can be contrasted with the normal v_{eE} for fault detection. In Figure 13.12(b), the solid lines show the conventional switching i_c and related normal v_{eE}. The positive part of v_{eE} can be considered as the rising time of collector current t_r. In the case of HSF, the length of positive induced voltage t_{r_HSF} would be longer than the typical value of t_r, as plotted by dotted lines. In addition, once the unexpected voltage jump v_{eE_FUL} occurs during the on-state operation, short-circuit FUL can also be detected by monitoring v_{eE}.

Compared with the on-state voltage v_{ce} drop detection method, the di_c/dt approach has a faster response time. Typically, the saturation v_{ce} drop method requires about 10 μs blanking time to detect the short-circuit faults for avoiding noise disturbance and thermal runaway. However, by means of di_c/dt-based approaches, the inspected IGBT can be turned off after the occurrence of short-circuit faults within 1 μs [20].

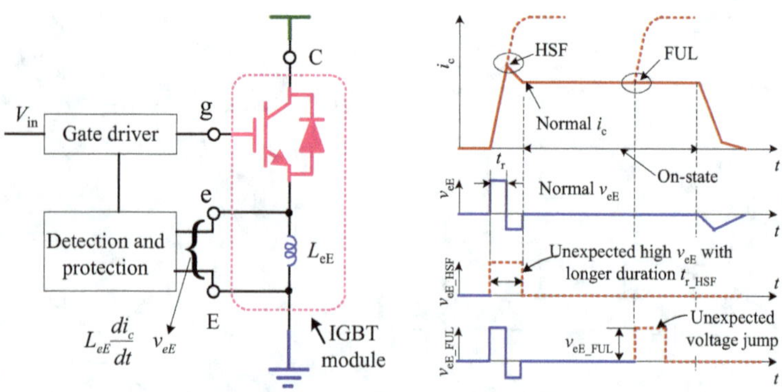

Figure 13.12 Detection principle of di_c/dt and comparisons of short-circuit faults: (a) detection method of di_c/dt and (b) simplified curves of short-circuit current and related v_{eE} voltages

For overload protection, the overload current can reach very high but v_{ce} is lower than that of short-circuit condition. Hence, common overload protection can be monitored by the load side current sensor. Practically, the on-state voltage v_{ce} drop method requires a blanking time (e.g., 10 μs) to detect short-circuit failure. Besides, the performance of di_c/dt-based detection method is also determined by the value of parasitic L_{eE}.

To solve the shortcomings, short-circuit faults can be detected by the gate charge characteristics without blanking time and not related to L_{eE}, as shown in Figure 13.13 [21]. In Figure 13.13(a), v_{gon} and v_{gp} represent the steady-state turn-on gate voltage and Miller plateau voltage, respectively. Unlike normal operation, there is no Miller plateau in the case of HSF, as shown by the solid line. The gate charge Q_g increases as the gate voltage v_{ge} increases, but the amount of gate charge under HSF conditions Q_{ref1} is less than that under normal conditions Q_n when gate voltage reaches v_{gon}. Therefore, both the gate voltage v_{ge} and gate charge should be monitored for HSF detection. For FUL conditions plotted in Figure 13.13(b), gate voltage v_{ge} becomes higher than v_{gon}, while gate charge Q_g is smaller than normal Q_n. As long as $v_{ge} > v_{gon}$ and $Q_g < Q_{ref2}$, the FUL condition of the evaluated IGBT can be determined. The short-circuit faults by applying the gate charge characteristics can be detected within about 3 μs [21].

Moreover, the open-circuit fault can also be detected by gate charge characteristics [22]. When the open-circuit fault occurs, the value of the collector

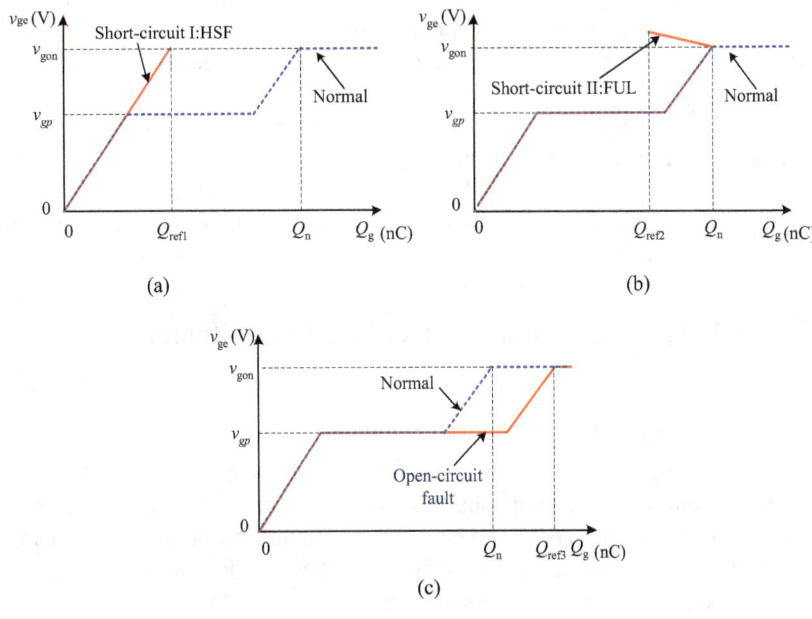

Figure 13.13 Gate charge characteristics under abnormal conditions: (a) HSF condition, (b) FUL condition, and (c) open-circuit condition

voltage v_{ce} is kept to zero. Hence, the capacitance C_{gc} under open-circuit fault is larger than that of short-circuit faults. The turn-on gate voltage comparison between the open-circuit fault and the normal condition is plotted in Figure 13.13 (c). The time duration of the Miller plateau voltage during normal condition is shorter than during the open-circuit condition, and the normal Q_n is also less than the charge Q_{ref3} under open-circuit fault. Based on the idea of gate charge characteristics, the charging time constant of turn-on transition can also be used for short-circuit detection [23].

13.4.3 Gate voltage limitations

The gate voltage v_{ge} could exceed the gate oxide limitation in the case of turn-off displace current i_{gc} and the short-circuit gate voltage oscillation, as shown in Figure 13.10. Normally the gate voltage limitation schemes are usually implemented by the passivity-based methods. The gate voltage limitation methods can be divided into four categories: (i) constant gate voltage clamping, (ii) lower impedance for displace current i_{gc}, (iii) layout improvement, and (iv) ferrite core for increasing resistive impedance. Accordingly, the typical v_{ge} limitation implementations are depicted in Figure 13.14 [24].

In Figure 13.14(a), a clamping diode D_c is used so that the maximum v_{ge} can be clamped to the constant voltage source v_{cc}. Another gate voltage clamp is achieved by a Zener diode D_z depicted in Figure 13.14(b). The excessive gate voltage can be clamped to the breakdown voltage of D_z. Figure 13.14(c) and (d) demonstrates two approaches to provide a lower impedance for displace current from capacitor C_{gc}. The diode D_c in Figure 13.14(c) and the transistor T_c in Figure 13.14(d) provide a lower impedance when the voltage across R_g reaches a specific value (around 0.7 V).

Apart from the applications of a single IGBT module, the excessive gate voltage would occur in the multimodule applications. In Figure 13.14(e), the gate voltage oscillation can be caused by the parasitic inductance L_{ee} between the Kelvin emitters (e). Moreover, two external ferrite cores are inserted into the gating loops to increase the high-frequency impedance, as shown in Figure 13.14(f).

13.5 Active gating methods for enhancing switching characteristics

13.5.1 Closed-loop control methodology

Usually, the dynamic characteristics depend on many factors in terms of bus voltage, load current, junction temperature, and circuit layout. On the basis of semiconductor knowledge and control theory, the switching trajectory of IGBT can be achieved by real-time adjustment of driving strategies. The implementation elements and classification of IGBT switching trajectory control are summarized in Figure 13.15. First, the selected samplings (e.g., v_{ce} and i_c) should be chosen carefully according to the optimization requirements and sensor locations. The

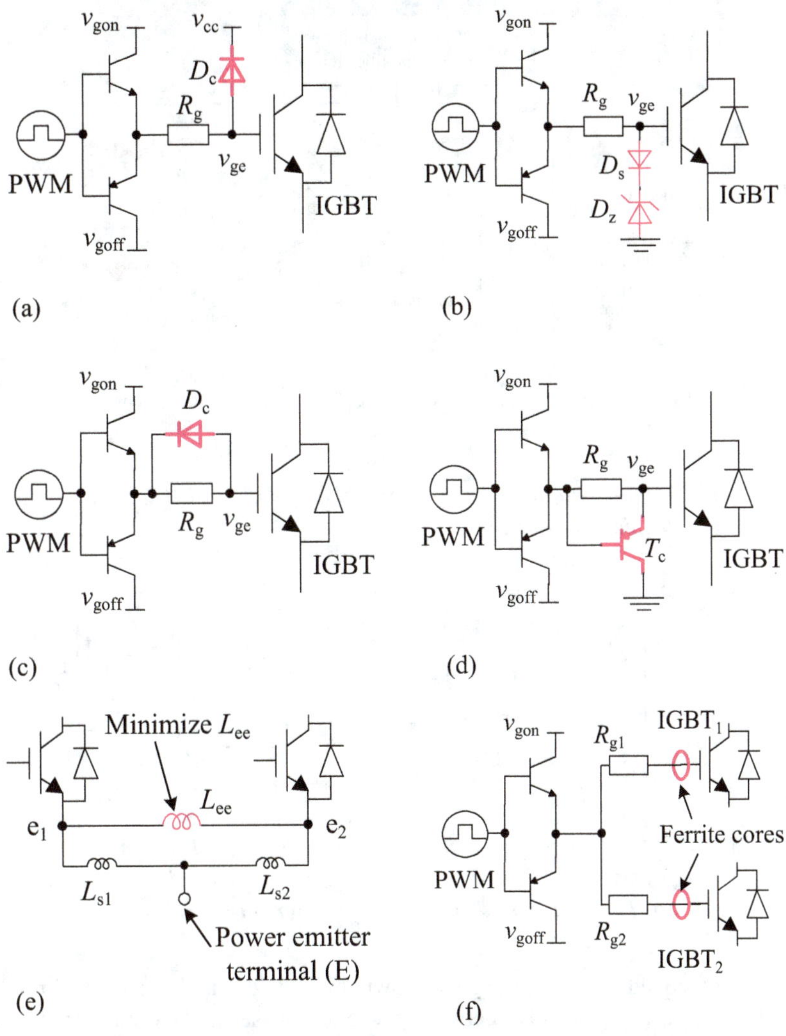

Figure 13.14 Six kinds of gate voltage limitation approach using passive components

main approaches consist of gate resistor change, gate current change, and gate voltage change.

These three approaches are even the fundamentals for more complex closed-loop control strategies. In terms of closed-loop control, the general control objects are the gradients of voltage and current and their maximum value. Accordingly, the switching losses can also be controlled by the adjustment of the overlap of v_{ce} and i_c. The more sophisticated closed-loop control schemes are instantaneous collector voltage/current control. The instantaneous collector voltage or current can be

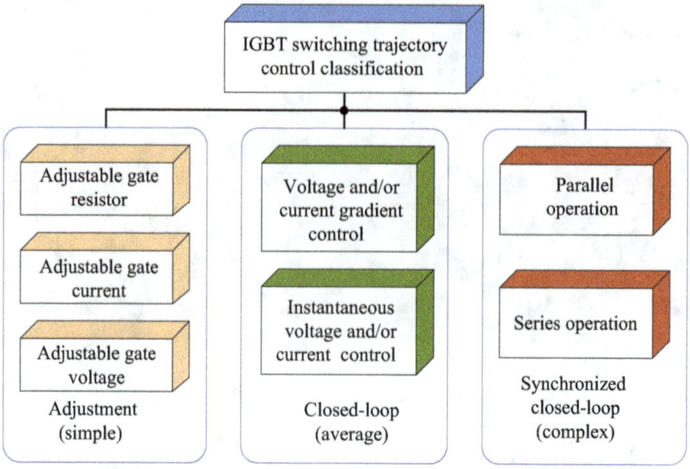

Figure 13.15 Implementation elements and classification of IGBT switching trajectory control

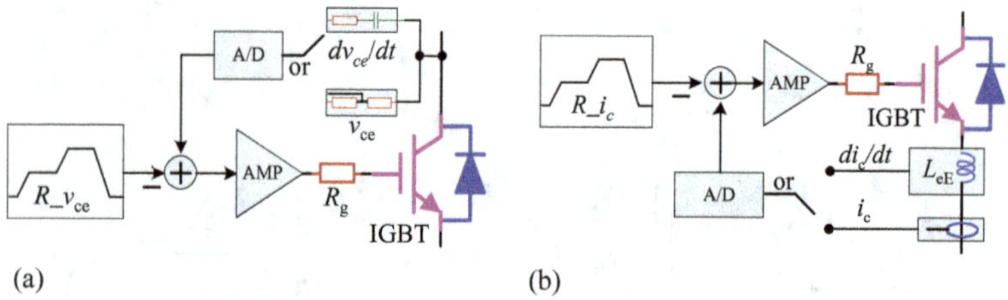

Figure 13.16 Simplified single variable closed-loop control with a reference presetting: (a) v_{ce}-based closed-loop control schematic and (b) i_c-based closed-loop control schematic

controlled independently during specific switching stages, and this control method can reach the best compromise between switching losses and overshoots. Finally, the synchronized closed-loop methods can also be applied to parallel operation and series operation in high-power conversion systems. Both the static and switching behavior differences can be eliminated by the closed-loop control methods in gate drivers.

13.5.2 Closed-loop control implementations

With the aid of closed-loop control, multiple objectives can be optimized within one single switching transition. In Figure 13.16(a), a primary v_{ce} closed-loop diagram using operational amplifier (AMP) is depicted [25,26]. The instantaneous v_{ce} and gradient of v_{ce} can be measured by an A/D converter to be in agreement with the reference R_v_{ce}. As a single variable feedback method, the v_{ce}-based method is mainly used in the series connection control of IGBTs. Similarly, the collector current i_c related optimization and parallel operation can be realized only by the i_c feedback control. The instantaneous i_c acquisition can be achieved by the isolated PCB Rogowski coil. Besides, di_c/dt is usually extracted by the voltage v_{eE} across on L_{eE} as pointed out in Figure 13.16(b), which is a nonisolation solution.

An enhanced optimization can be achieved by a combination of i_c and v_{ce} feedback loop and two independent i_c and v_{ce} references are needed. In Figure 13.17(a), a closed-loop di_c/dt and dv_{ce}/dt control scheme is depicted [27]. The switching speed can be adaptively adjusted according to the preset references. As a result, I_{peak}, V_{peak}, and switching power losses can be effectively optimized with the instantaneous i_c and v_{ce} feedback control. The features of control signals can be illustrated in Figure 13.17(b). The PWM gate signal duty ratio is used to control the device's on-state and off-state time as the same as the conventional gate driver. In this case, the amplitudes of PWM gate signal can be used to control the turn-on and turn-off speeds, respectively. This kind of closed-loop method can be classified as the dual variable feedback method.

Concerning the dual variable feedback methods, the gate-related information has not been sampled and adjusted directly. Since the switching characteristics are affected by the gate current i_g and gate voltage v_{ge}, gate-related information can be implemented into closed-loop control for faster response and more accurate performance. Remarkably, in order to realize the desired optimization objectives, costly high-speed A/D units, D/A units and data processing units (e.g., FPGA) are required. In Figure 13.18, three variable feedback (v_{ce}, i_c, and gate-related i_g or v_{ge}) control methods are depicted. In Figure 13.18(a), the current-source gate driver is applied in a digital gate driver [28]. Accordingly, the instantaneous variables v_{ge}, v_{ce}, and i_c are extracted for closed-loop control. Similarly, for the voltage-source driver, the gate current i_g is extracted and controlled instead of v_{ge}, as shown in Figure 13.18(b) [29].

The features of overall closed-loop control method consist of current/voltage overshoot suppression, gradient control, switching losses optimization, and over-current protection. However, the control bandwidth and closed-loop design related to parasitic parameters in the IGBT module are currently hot challenge in multi-variable feedback methods.

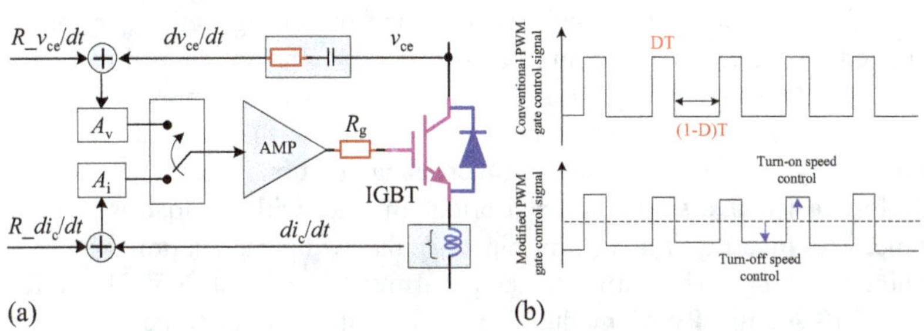

Figure 13.17 (a) Schematic dual closed-loop di_c/dt and dv_{ce}/dt control concept and (b) conventional and advanced PWM gate control signals

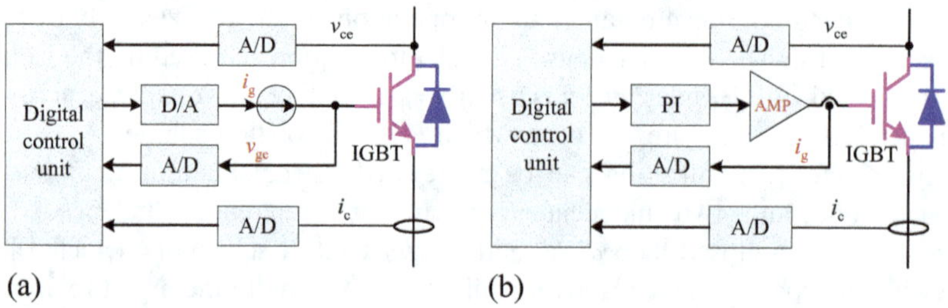

Figure 13.18 Advanced gate driver unit comprising closed-loop strategies and integrated microcontroller: (a) gate current i_g adjustment with feedbacks of v_{ce}, i_c, and v_{ge} and (b) gate voltage v_{ge} adjustment with feedbacks of v_{ce}, i_c, and i_g

13.6 Active thermal control methods using IGBT gate driver

13.6.1 Principles for thermal mitigation method

The widely used wire-bonded IGBT module is made of several layers with different materials. In fact, because of the different thermal expansion coefficient of several materials in IGBT modules, the accumulation of thermo-mechanical stresses could eventually lead to long-term wear-out failures [30]. Recently, some new package technologies for increasing thermal cycling are flourishing, such as ribbon and sintered connections [31]. Anyhow, the IGBT module remains a multilayer structure with different thermal expansion coefficients. The well-known thermal related wear-out mechanisms in IGBT modules are bond wire lift off, solder joint fatigue, bond wire heel cracking, and aluminum reconstruction. All these effects reduce the maximum current capability and number of cycles with operation time.

Indeed, thermal factors are regarded to have a primary impact on long-term reliability issues. The typical experimental time (or cycles) to failure as a function of temperature swing ΔT_j and mean temperature T_{jm} is plotted in Figure 13.19 [32]. The number of cycles N_f is an empirical function that mostly relates to the temperature swing ΔT_j and mean temperature T_{jm}. It can be observed that the smaller ΔT_j and T_{jm}, the longer the useful time. Since the change of T_{jm} is primarily determined by the environment and thermal dissipation, ΔT_j control can be more likely to be implemented by means of the active thermal control methods. An appropriate active thermal control method can increase the expected lifetime and improve the system reliability without increasing additional cost.

In Figure 13.20, a single-phase H-bridge inverter with RL load is taken as an example. The inverter is driven by bipolar pulse-width modulation with 10 kHz switching frequency. The output fundamental frequency is set to 50 Hz. Referring to Figure 13.20, the PWM modulation is from the controller according to the control strategies. In the case of AC output, the power losses and thermal performance of S_1 are plotted in Figure 13.21. It is shown that T_j fluctuation frequency of

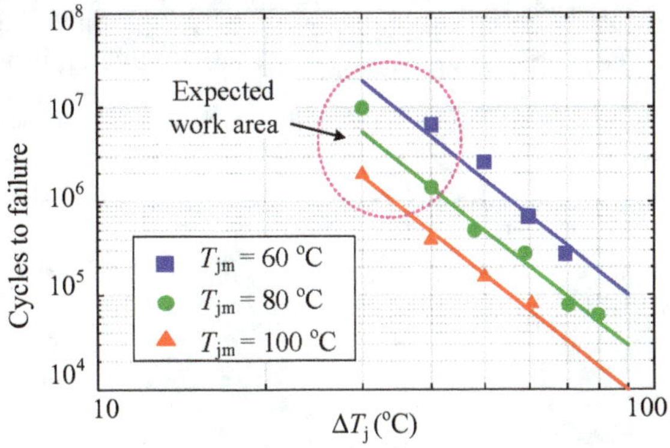

Figure 13.19 Power cycling curves as established for power modules in dependency of different mean temperature in the LESIT study

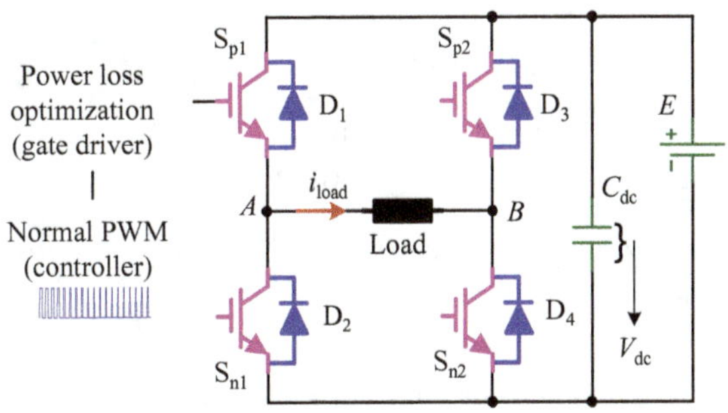

Figure 13.20 Single-phase H-bridge inverter with RL load

S_{p1} is in agreement with the output current i_{load}. Since the thermal network can be modeled as an inertia element, the outline of T_j lags behind the AC load current. For the sake of completeness, it is worth noting that there is a switching-frequency related saw tooth variability of junction temperature. Because of the low swing of such variation, the thermal stresses due to power cycling are mostly attributable to the fundamental frequency [33]. Generally, the huge ΔT_j mainly comes from two aspects: the fundamental AC output mechanism and the variable electric power requirement. Concerning the thermal mitigation principle, how to regulate the device power losses against the power requirement changes is the key.

13.6.2 Thermal mitigation methods

This study focuses on active thermal control methods by means of gate driver. In view of gate driver, the junction temperature can be adjusted individually regardless of system-level factors. In typical AC applications, the temperature fluctuation consists of switching period related small T_j variability and fundamental output

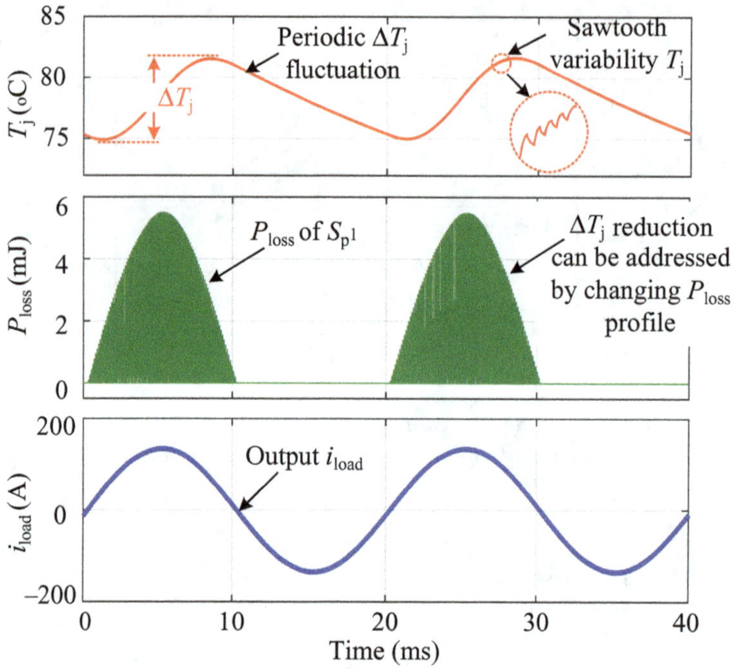

Figure 13.21 Thermal performance of S_{p1} and related power losses at sinusoidal i_{load} output

related large T_j variability. Accordingly, the thermal mitigation methods can be classified into short-term time scale and long-term time scale methods. The short-term time scale methods are characterized by junction temperature changing in one switching period. Theoretically, both the switching power losses and conduction power losses can be changed by the gate driver. However, the conduction power loss changing should be achieved by the gating voltage adjustments, which would increase the driver complexity and reliability risk. Unlike the short-term time scale methods, the long-term time scale methods focus on the junction temperature regulation in one or several fundamental periods.

13.6.2.1 Gate resistance adjustment (short-term time scale method)

In order to avoid interrupting the original PWM strategy, such as the switching frequency and the modulation index, the power loss regulations can be achieved by the gate resistance adjustment. The voltage-source gate driver with a switchable gate resistor network is depicted in Figure 13.22.

In Figure 13.23, the switching strategy of gate driver regarding the load current amplitude is shown. To better illustrate the operating concept, one study case is selected for easier control requirements. The AC output i_{load} is divided into two levels according to the current amplitude I_L: large I_L region and small I_L region.

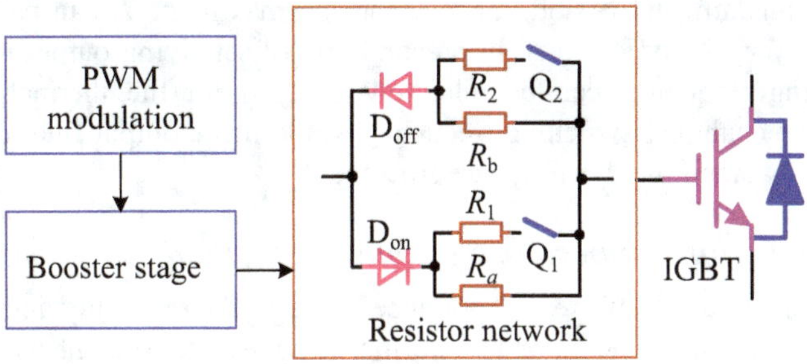

Figure 13.22 Conventional voltage-source gate driver with switchable gate resistor network

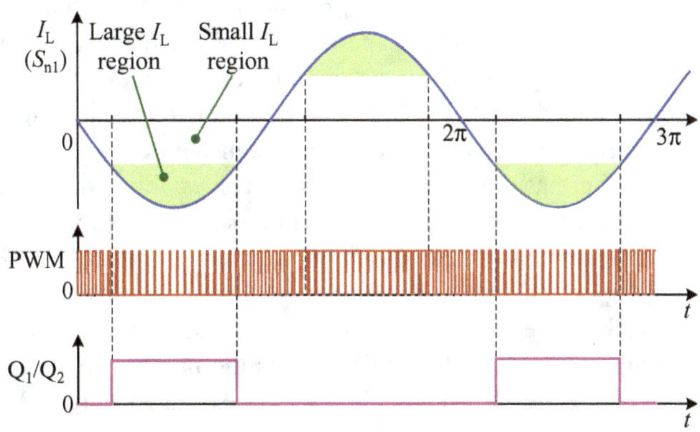

Figure 13.23 Gate resistor control strategy in accordance with output load current under bipolar modulation strategy

Due to the symmetry, the control method of the active gate driver is analyzed on the first positive half cycle. The PWM control signals are used to implement the modulation strategy as usual. The default gate turn-on and turn-off resistors are R_a and R_b, which are the maximum values provided by the gate driver, respectively. Since the switching power losses are proportional to I_L under fixed bus voltage, the minimum gate resistances $R_1//R_a$ and $R_2//R_b$ are selected at large current region. Remarkably, the thermal control method is independent of control strategies and does not need the knowledge of internal T_j [34].

13.6.2.2 Switching frequency adjustment (long-term time scale method)

The most reported and common studies on active thermal control methods are realized by changing the switching frequency (f_s), as depicted in Figure 13.24 [35,36]. The desired modulation wave is modulated under high-frequency carrier f_{s1} from t_0 to t_1. Since the accumulative switching power losses can be reduced within one fundamental period, the ΔT_j and the maximum T_j can be reduced by means of lower f_{s2} from t_1 to t_2. In precondition of satisfying output requirement, the switching frequency can be relatively reduced for the thermal mitigation. However, the changing switching frequency would make output filter design more difficult and worsen the dynamic performances.

13.6.3 Junction temperature estimation methods

Junction temperature is of great importance to the IGBT reliability and robustness. Moreover, junction temperature monitoring provides an efficient way to realize active thermal control for high-power converters, hence it is a potential approach to strengthen system reliability. Nowadays, the existing IGBT junction temperature measurement approaches can be categorized into the optical-based, physical contact-based, and thermo-sensitive electrical parameter (TSEP)-based solutions [37]. Infrared camera measurement is a representative optical-based method, but it is expensive and restricted in practical applications. The thermocouple and build-in thermistor are the typical physical contact-based methods, which are cost-effective and widely applied in industrial applications. But their dynamic response is relatively slow thus cannot detect the dynamic junction temperature fluctuations. By using die itself as a thermal sensor, TSEP methods can establish the correspondence between external electrical parameters and internal junction temperature. It is therefore concluded that TSEP methods afford the most promising and feasible way to obtain fast temperature from high-power IGBT modules. Considering the acquisition approaches, the common TSEPs are divided into static parameters and dynamic parameters. Static TSEPs are defined as the parameters extracted during the on-state or off-state, while dynamic TSEPs are extracted during the turn-on or turn-off transitions.

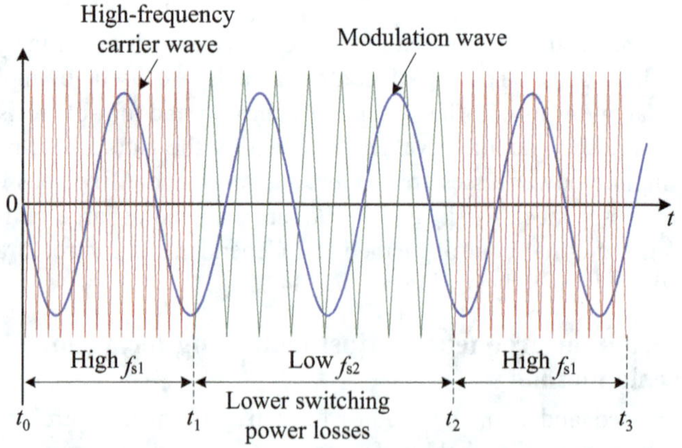

Figure 13.24 Variable switching frequency strategy for thermal mitigation

13.6.3.1 Static TSEPs

Static TSEPs are defined as the parameters extracted during the on-state or off-state operation. In Figure 13.25, the classifications of static TSEPs are listed. For the on-state IGBTs, the measurable TSEPs are mainly the saturation short-circuit current and the on-state voltage drop within the rated collector current.

The measurement circuit for the short-circuit saturation current is depicted in Figure 13.26. Usually, the bond wire parasitic resistance in the IGBT module can be used as the current shunt. According to the IGBT output characteristic curves shown in Figure 13.11(b), the short-circuit saturation current is only related to junction temperature under fixed gate voltage. However, when short-circuit current saturation is reached in a deteriorated IGBT, the power module is susceptible to catastrophic.

In order to avoid the extreme working case, the on-state collector voltage V_{ce} under normal working conditions is developed. The typical bipolar characteristics of IGBT device under different temperatures are plotted in Figure 13.27. Around a particular current, close to the load current, V_{ce} has the property of both negative and positive temperature coefficients (NTC/PTC) which depend on the current rating. For a certain current, V_{ce} declines as T_j rises in the NTC region, but increases in the PTC region. It can be seen that the closer to the intersection point, the lower the TSEP sensitivity ratio. There is a "blind region" for the on-state voltage-based TSEP near intersection A, where it is difficult to determinate the junction temperature when only utilizing on-state V_{ce}.

The problem about the blind area can be overcome by the voltage at the low current injection method. Correspondingly, a practical on-state measurement method for saturation voltage V_{sat} under low current injection is depicted in Figure 13.28. During the turn-on period, a low constant current is injected to the IGBT through two diodes D_{H1} and D_{H2}. When the inspected IGBT is turned off, the diode D_{H1} blocks the high V_{ce} voltage to protect the measurement board. Remarkably, the two diodes should have a similar forward voltage temperature coefficients under the particular low current. Assuming the two diodes D_{H1} and D_{H2} are identical, the measured V_{ce} during the on-state period can be calculated by $V_{ce} = V_b - V_{DH1} = V_b - (V_a - V_b) = 2V_b - V_a$.

An auxiliary current-source injection circuit is required to ensure the current is low enough to avoid self-heating effects in V_{sat}-based TSEP method. The saturation

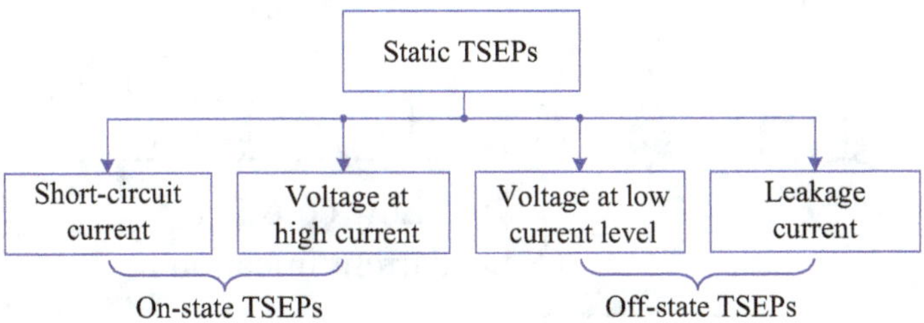

Figure 13.25 Classifications of static TSEPs

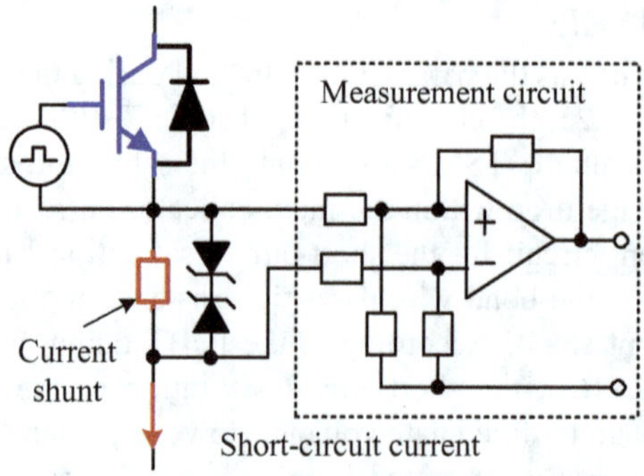

Figure 13.26 Short-circuit current measurement method for T_j estimation

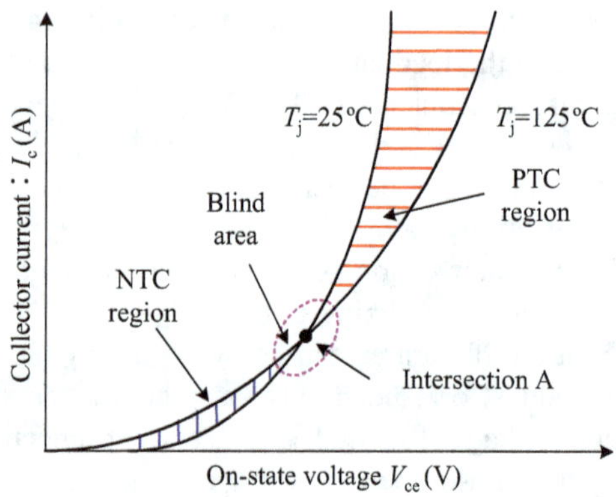

Figure 13.27 Typical IGBT output characteristic curves

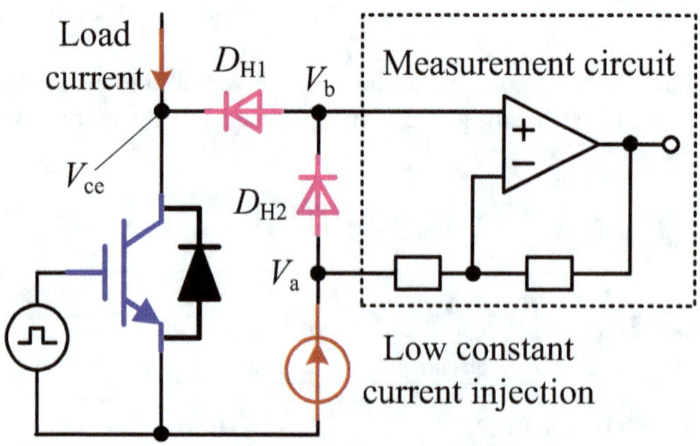

Figure 13.28 On-state V_{ce} measurement method with dual diodes

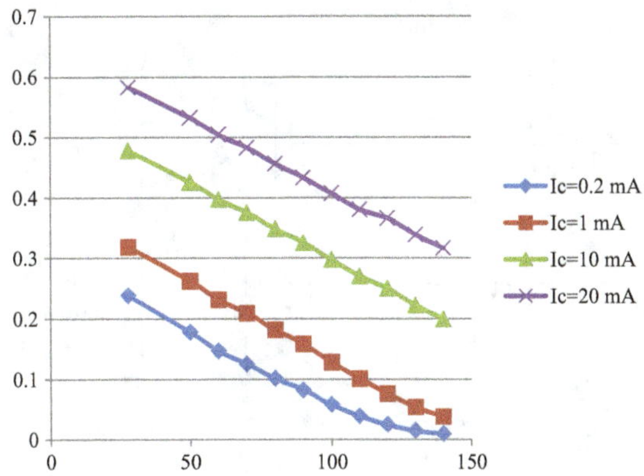

Figure 13.29 Calibration curves under different current injection

V_{sat}–T_j calibration curves at different injection current for IGBT module (IMBI800UG-330) are plotted in Figure 13.29. The measured V_{sat} curves are characterized by excellent linearity, which can simply the calibration procedure and the on-line T_j estimation process.

Referring to the leakage current, which is also proportional to the junction temperature in the case of fixed bus voltage. However, the leak current during the off-state is usually in the range of tens of microamperes, whose value is more vulnerable to the noise interfere. Hence, the leakage current-based TSEP method would not be practical because of low-temperature sensitivity.

13.6.3.2 Dynamic TSEPs

Since the number of d-TSEPs is larger than that of static TSEPs, more and more d-TSEP methods are being developed for T_j estimation recently. Practically, according to the magnitude of electrical parameters, TSEPs can be divided further into gate-related TSEPs and collector-related TSEPs, as depicted in Figure 13.30.

Since the electrical parameters in the gate loop are low-voltage quantities, the gate-related TSEPs can be sampled by low-voltage sampling circuits directly. Some key published gate-related TSEPs are based on peak gate current, integration of gate current, threshold voltage v_{th}, and Miller plateau width. On the other hand, plenty of collector-related TSEPs can be identified, for example, di_c/dt, dv_{ce}/dt, maximum turn-off voltage peak V_{peak}, maximum turn-on current peak I_{peak}, etc.

Compared with the gate-related TSEPs, the collector-related TSEPs usually vary within several hundreds of thousands of voltages or amperes. However, the sampling circuit should be designed to withstand high voltages and/or large currents during converter operations, making the adoption of such a method not convenient or even not feasible at all. Of course, once the collector-related quantities can be measured at a low cost, many new and practical TSEPs candidates can be extracted and developed. Fortunately, a unified extraction method and evaluation system for the collector-related TSEPs has been developed. Benefitting from the

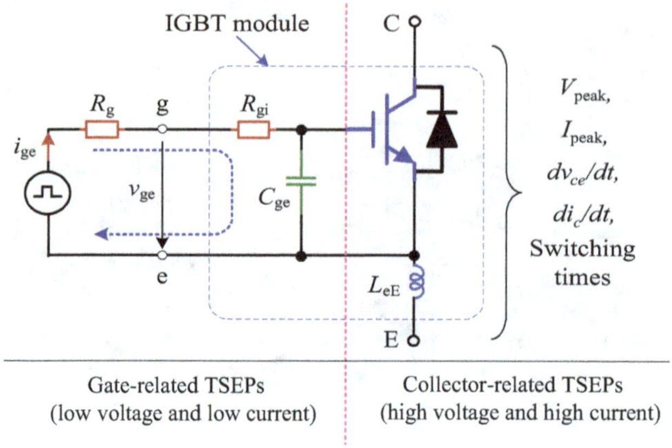

Figure 13.30 Gate-related and collector-related TSEP candidates for IGBT modules

specific package of high-voltage and high-power IGBT modules, most of the collector-related TSEPs can be extracted by using the parasitic inductance between the Kelvin emitter and power emitter terminals. The information of the induced voltage on the internal parasitic inductance provides a cost-effective solution to measure the collector-related TSEPs. The measurable dynamic TSEP candidates are listed in Figure 13.31 [38,39]. According to the different electrical properties, the corresponding measurement broads in terms of are needed to be developed for the on-line T_j estimation.

Generally speaking, practical TSEP-based methods should satisfy the following requirements:

1. Simplified calibration: calibration procedure can be regarded as the prerequisites for TSEP-based methods. For different kinds of TSEPs, the same selected electrical parameters are not only affected by temperature but also affected by the working conditions, in terms of bus voltage and load current. These dependencies of temperature and working conditions should be taken into consideration before the calibration. In principle, the calibration procedure should be as simple as possible to voiding dismounting the converter system.
2. Noninvasive acquisition: the running IGBT is operated in high-frequency switching status according to the control strategies. During the process of normal operation, the protection functions are always in the activation status. From the reliability point of view, the normal operation and protection mechanism should have higher precedence than the TSEP acquisition. Hence, the noninvasive acquisition procedure is necessary for the preferred online TSEP methods.
3. Facilitated integration: as an important interface between IGBT modules and microcontrollers, the gate drivers are an integral part of any power converters. Theoretically, most of the IGBT operation status can be reflected on the gate

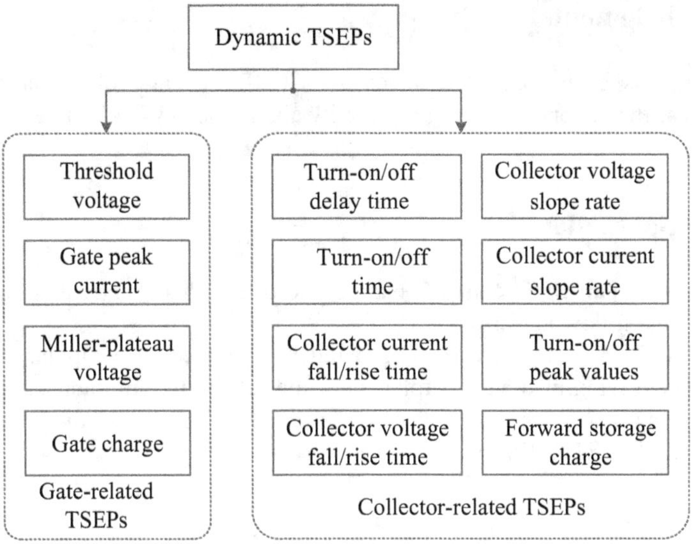

Figure 13.31 Classifications of selected dynamic TSEPs

driver side. For compatibility with the conventional gating and protection functions, the subsequent TSEP acquisition procedure should be implanted and integrated into the gate driver. As a result, both the electrical parameters and T_j knowledge can be extracted and monitored by the intelligent gate driver.

13.7 Summary

This chapter has presented an overview of state-of-the-art advanced gate driver techniques for enhancing the reliability of IGBT modules. Broadly speaking, methods can be classified in detection methods, optimization methods, and protection methods. Additionally, optimization and protection methods can be roughly classified in simple (and cheap) and advanced (even more expensive). Simple methods, like the open-loop and passivity-based ones, perform well in normal applications, but advanced methods, like closed-loop control strategies, even if more expensive, are necessary for special applications, like high-power IGBT modules. In the near future, benefiting from the increase of data processing speed and reducing the cost of digital controllers, the advanced techniques discussed in this chapter could become more and more affordable and popular, even in low-cost applications, for both considerably reducing short-term and long-term reliability issues. In fact, basing on a large number of different failure mechanisms, more and more complex strategies will be needed including most of the mature detection and protection methods toward the so-called reliability-oriented gate driver design.

Acknowledgments

This work has been conducted under the CORPE (Center of Reliable Power Electronics) framework—Aalborg University and funded by The Obel Foundation, Denmark.

Note to the reader

This chapter is part of an edited book. To cite or use this chapter as a reference, please use the following format:

Luo H, Li W, and Iannuzzo F. "Modern insulated gate bipolar transistor (IGBT) gate driving methods for robustness and reliability." In: Iannuzzo F. (ed.), *Modern Power Electronic Devices: Physics, Applications, and Reliability*. Stevenage, UK: IET; 2020. pp. 417–449.

References

[1] Baliga B. J. *Fundamentals of Power Semiconductor Devices*, 2008 edn. New York, NY: Springer, 2008.
[2] Wang H., Liserre M., and Blaabjerg F. "Toward reliable power electronics: Challenges, design tools, and opportunities." *IEEE Ind. Electron. Mag.* 2013, vol. 7, no. 2, pp. 17–26.
[3] Infineon Technologies AG. "Semiconductor & system solutions – Infineon Technologies." [Online]. Available: http://www.infineon.com/cms/en/. [Accessed Jun 20, 2017]
[4] Yang S., Xiang D., Bryant A., Mawby P., Ran L., and Tavner P. "Condition monitoring for device reliability in power electronic converters: A review." *IEEE Trans. Power Electron.* 2010, vol. 25, no. 11, pp. 2734–2752.
[5] Wu R., Reigosa P. D., Iannuzzo F., Smirnova L., Wang H., and Blaabjerg F. "Study on oscillations during short circuit of MW-scale IGBT power modules by means of a 6-kA/1.1-kV nondestructive testing system." *IEEE J. Emerg. Sel. Top. Power Electron.* 2015, vol. 3, no. 3, pp. 756–765.
[6] Smet V., Forest F., Huselstein J. J., *et al.* "Ageing and failure modes of IGBT modules in high-temperature power cycling." *IEEE Trans. Ind. Electron.* 2011, vol. 58, no. 10, pp. 4931–4941.
[7] Durand C., Klingler M., Coutellier D., and Naceur H. "Power cycling reliability of power module: A survey." *IEEE Trans. Device Mater. Reliab.* 2016, vol. 16, no. 1, pp. 80–97.
[8] Holmes D.G. and Lipo T.A., *Pulse Width Modulation for Power Converters: Principles and Practice*. Hoboken, NJ: John Wiley & Sons, 2003.
[9] Eberle W., Zhang Z., Liu Y. F., and Sen P. C. "A current source gate driver achieving switching loss savings and gate energy recovery at 1-MHz." *IEEE Trans. Power Electron.* 2008, vol. 23, no. 2, pp. 678–691.

[10] Huang F. and Flett F. "IGBT fault protection based on di/dt feedback control." 2007 *IEEE Power Electronics Specialists Conference*, Orlando, FL, USA, 17–21 June 2007, pp. 1478–1484.

[11] Musumeci S., Raciti A., Testa A., Galluzzo A., and Melito M. "Switching-behavior improvement of insulated gate-controlled devices." *IEEE Trans. Power Electron.* 1997, vol. 12, no. 4, pp. 645–653.

[12] John V., Suh B.-S., and Lipo T.A. "High-performance active gate drive for high-power IGBT's." *IEEE Trans. Ind. Appl.* 1999, vol. 35, no. 5, pp. 1108–1117.

[13] Idir N., Bausiere R., and Franchaud J. J. "Active gate voltage control of turn-on di/dt and turn-off dv/dt in insulated gate transistors." *IEEE Trans. Power Electron.* 2006, vol. 21, no. 4, pp. 849–855.

[14] Grbovic P. J. "An IGBT gate driver for feed-forward control of turn-on losses and reverse recovery current." *IEEE Trans. Power Electron.* 2008, vol. 23, no. 2, pp. 643–652.

[15] Basler T., Lutz J., Jakob R., and Brückner T. "The influence of asymmetries on the parallel connection of IGBT chips under short-circuit condition." *Proceedings of the 2011 14th European Conference on Power Electronics and Applications*, Birmingham, UK, 30 Aug.–1 Sept. 2011, pp. 1–8.

[16] Kopta A., Rahimo M., Schlapbach U., Kaminski N., and Silber D. "Limitation of the short-circuit ruggedness of high-voltage IGBTs." *2009 21st International Symposium on Power Semiconductor Devices IC's*, Barcelona, Spain, 14–18 June 2009, pp. 33–36.

[17] Ohi T., Iwata A., and Arai K. "Investigation of gate voltage oscillations in an IGBT module under short circuit conditions." 2002 *IEEE 33rd Annual IEEE Power Electronics Specialists Conference. Proceedings* (Cat. No.02CH37289), Cairns, QLD, Australia, 23–27 June 2002, vol. 4, pp. 1758–1763.

[18] Reigosa P. D., Wu R., Iannuzzo F., and Blaabjerg F. "Evidence of gate voltage oscillations during short circuit of commercial 1.7 kV/1 kA IGBT power modules." *Proceedings of PCIM Europe 2015; International Exhibition and Conference for Power Electronics, Intelligent Motion, Renewable Energy and Energy Management*, Nuremberg, Germany, 19–20 May 2015, pp. 1–8.

[19] Wang Z., Shi X., Tolbert L. M., Wang F., and Blalock B. J. "A di/dt feedback-based active gate driver for smart switching and fast overcurrent protection of IGBT modules." *IEEE Trans. Power Electron.* 2014, vol. 29, no. 7, pp. 3720–3732.

[20] Lizama I., Alvarez R., Bernet S., and Wagner M. "A new method for fast short circuit protection of IGBTs." *IECON 2014 – 40th Annual Conference of the IEEE Industrial Electronics Society*, Dallas, TX, USA, 29 Oct.–1 Nov. 2014, pp. 1072–1076.

[21] Horiguchi T., Kinouchi S., Nakayama Y., *et al.* "A high-speed protection circuit for IGBTs subjected to hard-switching faults." *IEEE Trans. Ind. Appl.* 2015, vol. 51, no. 2, pp. 1774–1781.

[22] Rodriguez-Blanco M. A., Claudio-Sánchez A. Theilliol D., *et al.* "A failure-detection strategy for IGBT based on gate-voltage behavior applied to a

motor drive system." *IEEE Trans. Ind. Electron.* 2011, vol. 58, no. 5, pp. 1625–1633.

[23] Park B.G., Lee J.B., and Hyun D.S. "A novel short-circuit detecting scheme using turn-on switching characteristic of IGBT." *2008 IEEE Industry Applications Society Annual Meeting*, Edmonton, AB, Canada, 5–9 Oct. 2008, pp. 1–5.

[24] Lu B. and Sharma S.K. "A literature review of IGBT fault diagnostic and protection methods for power inverters." *IEEE Trans. Ind. Appl.* 2009, vol. 45, no. 5, pp. 1770–1777.

[25] Palmer P. R. and Rajamani H. S. "Active voltage control of IGBTs for high power applications." *IEEE Trans. Power Electron.* 2004, vol. 19, no. 4, pp. 894–901.

[26] Wang Y., Palmer P. R., Bryant A. T., Finney S. J., Abu-Khaizaran M. S., and Li G. "An analysis of high-power IGBT switching under cascade active voltage control." *IEEE Trans. Ind. Appl.* 2009, vol. 45, no. 2, pp. 861–870.

[27] Chen L. and Peng F. Z. "Closed-loop gate drive for high power IGBTs." *2009 Twenty-Fourth Annual IEEE Applied Power Electronics Conference and Exposition*, Washington, DC, USA, 15–19 Feb. 2009, pp. 1331–1337.

[28] Dang L., Kuhn H., and Mertens A. "Digital adaptive driving strategies for high-voltage IGBTs." *IEEE Trans. Ind. Appl.* 2013, vol. 49, no. 4, pp. 1628–1636.

[29] Lobsiger Y. and Kolar J. W. "Closed-loop di/dt and dv/dt IGBT gate driver." *IEEE Trans. Power Electron.* 2015, vol. 30, no. 6, pp. 3402–3417.

[30] Ciappa M. "Selected failure mechanisms of modern power modules." *Microelectron. Reliab.* 2002, vol. 42, no. 4–5, pp. 653–667.

[31] Özkol E., Brem F., and Liu C. "Improving the power cycling performance of IGBT modules by plating the emitter contact." *Microelectron. Reliab.* 2015, vol. 55, no. 3, pp. 552–557.

[32] Held M., Jacob P., Nicoletti G., Scacco P., and Poech M. H. "Fast power cycling test of IGBT modules in traction application." *Proceedings of the Second International Conference on Power Electronics and Drive Systems*, Singapore, 26–29 May 1997, vol. 1, pp. 425–430.

[33] Ma K., Blaabjerg F., and Liserre M. "Thermal analysis of multilevel grid-side converters for 10-MW wind turbines under low-voltage ride through." *IEEE Trans. Ind. Appl.* 2013, vol. 49, no. 2, pp. 909–921.

[34] Luo H., Iannuzzo F., Ma K., Blaabjerg F., Li W., and He X. "Active gate driving method for reliability improvement of IGBTs via junction temperature swing reduction." *2016 IEEE 7th International Symposium on Power Electronics for Distributed Generation Systems (PEDG)*, Vancouver, BC, Canada, 27–30 June 2016, pp. 1–7.

[35] Murdock D. A., Torres J. E. R., Connors J. J., and Lorenz R. D. "Active thermal control of power electronic modules." *IEEE Trans. Ind. Appl.* 2006, vol. 42, no. 2, pp. 552–558.

[36] Andresen M., Buticchi G., and Liserre M. "Study of reliability-efficiency tradeoff of active thermal control for power electronic systems." *Microelectron. Reliab.* 2016, vol. 58, pp. 119–125.

[37] Avenas Y., Dupont L., and Khatir Z. "Temperature measurement of power semiconductor devices by thermo-sensitive electrical parameters: A review." *IEEE Trans. Power Electron.* 2012, vol. 27, no. 6, pp. 3081–3092.

[38] Baker N., Liserre M., Dupont L., and Avenas Y. "Improved reliability of power modules: A review of online junction temperature measurement methods." *IEEE Ind. Electron. Mag.* 2014, vol. 8, no. 3, pp. 17–27.

[39] Niu H. and Lorenz R. D. "Evaluating different implementations of online junction temperature sensing for switching power semiconductors." *IEEE Trans. Ind. Appl.* 2017, vol. 53, no. 1, pp. 391–401.

Chapter 14
Prospects and outlooks in power electronics technology and market

Elena Barbarini[1]

14.1 Global markets figures

When we talk about the power electronics market, we talk about all the solid-state electronics devices and systems which allow us to control and convert electric power. In terms of segments, we can find applications of power devices and systems in automotive, rail and electric vehicle (EV) charging, renewable energy sources such as photovoltaics and wind, UPS (uninterruptible power sources), computing, storage, and motor drivers, consumer electronics, telecommunications, and energy storage systems. Thus in terms of power range, we can go from tens of watts to more than 1 mega-Watt.

The drivers of general electronics are mainly the end-user application. Thus, the parameters to consider for integration in a system are power requirement, performance, reliability and cost. In the power electronics sector, we must also consider the megatrends and government subsidies. Thus, electrical power-conversion optimization is driven by three main factors: electrification of transportation, CO_2 emission reductiongoals, and the development of clean electricity sources.

Transport and energy segments/applications are linked to big projects, depending on global megatrends (CO_2 emission reduction goals, etc.). System-level demand is influenced more by subventions, politics, and environmental decisions than by inverter technology innovations. Linked to the automotive sector, the electric vehicle (EV) charging infrastructure is pushed because of the growing number of EVs on the road. The rail market is boosted by new high-speed lines worldwide, but its dynamics are strongly linked to the global economy and subsidies. Industrial segments are more technology-linked to achieve good efficiencies for power-saving (and thus cost). Here there is high cost-pressure and stiff market competition. Regarding renewables, wind turbine installations will remain rather flat in the coming years, with a demand of 50–60 GW per year. However, the trend is to increase the power of these turbines. On the other hand, photovoltaics is the fastest-growing renewable energy source, with huge untapped potential for

[1]System Plus Consulting/Yole Développement Group, Nantes, France

additional PV installations. Both windmills and photovoltaics are pushing the energy storage market, with many electricity-generation plus storage installations for better integration of generated electricity with the electricity grid. Electricity transmission lines are needed to connect electricity generation sites with the regions that have high energy demand. One example is the high-voltage direct electricity (HVDC) systems needed to transport the electricity generated by large offshore wind installations to inland regions.

The motor drive market is growing thanks to aggressive regulation target impacting DC motor drives. Moreover, industrialization and highly automated manufacturing plants favor the motor drive market with more manufacturing equipment/plants. Increased data storage demand directly affects not just server accounts but also the UPS linked to these systems. Demand for UPS systems was strong during 2018, especially in the United States, Europe, and China. This is due primarily to the boom of data centers, which is UPS's biggest application. However, demand for uninterruptible power source (UPS) in medical applications is also present. The UPS market's main driver is reliability: UPS must be up and running continuously, without the possibility of error, and kick in immediately in case of power failure. Moreover, new levels of energy efficiency standards are needed for power supplies.

In general, technological efforts are made mostly for higher efficiency and higher power-density systems, with minimal thermal losses. This includes new semiconductor dies, packaging materials, and the integration of different components. Thus, all the segments, even with their specificities, can have common synergies and, therefore, share market and technical innovations.

All these factors will be translated into healthy growth of the power electronics industry, which is expected, by Yole Développement, to grow from a \$53.4 billion market for power inverters in 2018 to \$72.6 billion in 2024 and \$17.5 billion for power semiconductor devices in 2018 to \$21.2B billion in 2024. This translates into a continuous increase in insulated-gate bipolar transistors (IGBT) and metal-oxide-semiconductor field-effect transistor (MOSFET) sales, with 2.6% and 3% compound annual growth rate (CAGR) 2018–2024, respectively (Figure 14.1) [1].

If we focus on power semiconductor devices, the biggest market is for discrete (silicon (Si), silicon carbide (SiC), and gallium nitride (GaN) included) with a value of US\$17 billion in 2019 and an expected growth of 3.3% CARG between 2018 and 2014. But, as a key element in power converters and inverters, the growth is expected to be in higher power applications which will require IGBTs or SiC modules to reach the desired efficiency, with an estimated CAGR of 6.6% for the power module market.

At the discrete devices level, the actual biggest market share is for Si MOSFETs, especially at a voltage below 100 V, which has almost 73% of the market share. This is followed by the IGBT market and thyristors. SiC devices account only for 1% in 2018, but the EV segment is expected to push it to 2% in 2024. In the module and intelligent power module (IPM) market, IGBT represents today 88% of the share, but again, thanks to the EV segment is expect that SiC modules will gain 14% of the total module market by 2024. Moreover, the new

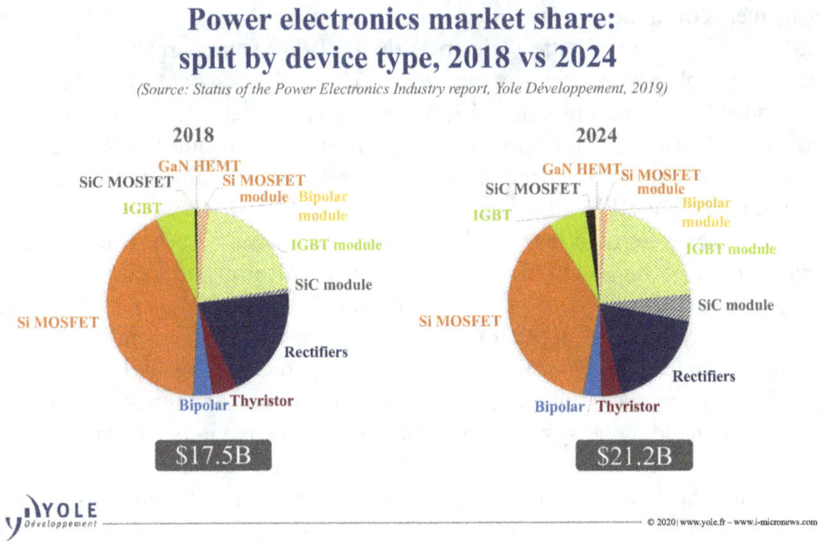

Figure 14.1 Power electronics market trends by device type

applications, such as energy storage, charging Infrastructure and EVs, demand modules with various power levels and reliability requirements, which will lead to a vast choice of power modules in the coming years.

Globally, every power electronics company increased its revenue in 2018 due to the power market's impressive health. 2019 began on a positive note, but demand has dropped because major customers still have device inventory from last year. In the first half of 2019, average selling price (ASP) grew just like in 2018; however, as demand lows and customers begin relying on multiple manufacturers, the ASP felt down in the second half. A market low-down is also expected in 2020, but demand will keep increasing as the EV market continues growing, even some low down has been seen after subvention cuts in China. Thus, we expect a renewed increase of ASP in the coming years due to strong demand for certain applications, that is, EV/hybrid electric vehicle (HEV).

14.2 Impact of EV/HEV sector

Beyond market figures, the most important point is probably the continuous growth showed by the electric and hybrid electric vehicles sector. At the beginning of the electrical revolution the driving application where industrial, then we have seen a major impact of the renewables, and now the main driver of the power market is the automotive sector. EV/HEV market is driven by various incentive mechanisms such as governments' subsidies to push the CO_2 emission reduction targets and engagements of numerous automotive manufacturers' about EV/HEV deployment. In 2018, 1.32 million of battery electric vehicle (BEV) were purchased, along with 0.75 million plug-in hybrid electric vehicle (PHEV)—compared to 0.78 million units and 0.41 million units in 2017, respectively. This equates to year-over-year growth of 68% and 84%, respectively. Moreover, sales of other hybrid cars have also increased. Indeed, over the last years, it has been seen a big push and market increase in the EV/HEV sales, reaching a 29.7% CAGR 2018–2024.

With the electrified car evolution, different trends have been observed, impacting the power electronic technologies and power electronics supply chain. The consumer demand is to have electric cars that have a higher driving range, charges rapidly have an optimum loading capacity and economically affordable. This pushes the car designer to find optimum solutions such as high-power inverters, higher component integration, and sharing of functionalities. To integrate several systems into one system, some solutions can be the e-axle or integration of a DC–DC converter with a battery. Another innovation is the use of more motors per vehicle. The idea is to use one motor to power the front axle and the second one the rear axle. The high-power inverter will use high-end die and packaging to have high performance and reliability, while the low power inverter can still use more conventional packaging and discrete component solutions. Moreover, there are other axes of development, like a 48 V electric system, 800 V batteries, or fuel cell vehicles.

The high integration brings new challenges concerning especially the inverter level and thus the semiconductor power devices. High-power inverters are typically based on high-end semiconductor dice, specific power module designs, and innovative packaging solutions. New design and materials of power module packaging with wide-bandgap (WBG) semiconductors are part of the technology roadmap since the technical developments have shown positive results, including a reduction in size and weight and efficiency improvements.

Today, the semiconductor power devices market for EV/HEV accounts for 23.7% of the total market in 2018 and it is expected to exceed US$3.7 billion in 2024, with 21% CAGR between 2018 and 2024. IGBT modules represent the largest market, with US$1.2 billion, and it is expected to have a 2018–2024 CAGR of 13%. It is followed by the full SiC modules with US$1.2 billion and an expected 2018–2024 CAGR of 48%. It is effective on SiC devices, including diodes, transistors, discrete, and modules, where analysts expect to have the biggest market growth, with a 2018–2024 CAGR of 45%, 48%, and 47.7%, respectively, for diodes, discrete transistors, and modules. The GaN devices market for EV/HEV is expected to have a 2018–2024 CAGR of 75%, from US$3 million to US$76 million, thanks to the possible implementation in the 48V DC/DC of mild hybrid electric vehicles (MHEV) and the onboard chargers (Figure 14.2) [2].

The automotive market is also driving a reshaping of the power electronics industry. Tier 1 companies are particularly exposed to different challenges, such as revenue decrease for internal combustion engine (ICE)-related products. To compensate, Tier 1 companies are increasing their efforts in EV/HEV-related products. However, original equipment manufacturers (OEMs) are becoming more and more

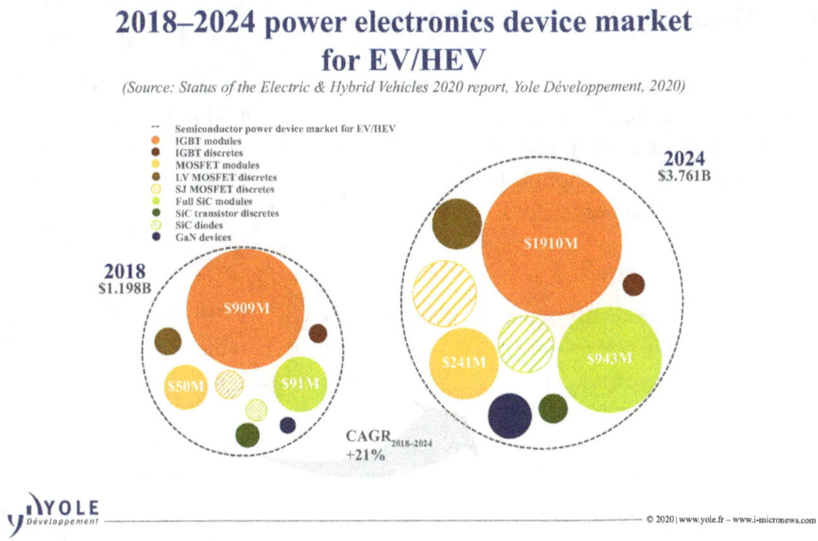

Figure 14.2 Power electronics market trends for EV/HEV by device type

intrusive, particularly in main inverters, to control the key EV/HEV elements. Established semiconductor device suppliers are in a similar situation, on the one hand facing the entrance of some Tiers 1 companies in the device market and addressing challengers from the emergence of WBG devices on the other. At the same time, the increasing demand for semiconductor devices implies the implementation of more 300 mm fabs for silicon devices and 150 mm and later on 200 mm for SiC devices for an increase in production together with cost-saving and technology improvements for better performance and efficiency of the power modules.

The EV/HEV market not only impacts the power ecosystem of the vehicles but also impacts other applications boosted by the need for electrification, such as renewable energy which is pushed by clean driving trends and growing electricity consumption. Charging stations need to be installed for proper geographical coverage, both inside and outside the cities. Considering that big stations on highways will be required to fast charge many cars simultaneously, more grid lines need to be deployed to sustain greater amounts of energy. Energy storage systems also need to be deployed for better distribution of the energy to the grid, and to buffer energy close to charging stations. Other trends such as autonomous driving require more data exchange, more data centers, more radar and LiDAR systems, together with other supporting technology. At the same time, more conservative applications like motor drives will see continued growth thanks to the electrification era.

14.3 Wide-bandgap semiconductors

After many years of research and development, finally, in recent years, we have seen big opportunities for WBG materials to be implemented in power inverters for high-efficiency and high power-density systems. Today, SiC and GaN-on-Si/Sapphire are "mature" enough to penetrate the power market in the automotive, industrial, and consumer sectors.

14.3.1 Silicon carbide

Being able to theoretically reach temperatures above 175–200 °C, SiC is a perfect candidate for high power density systems. Today it is widely accepted that SiC enables performance and cost benefits at the system level. At the device level, SiC is still expensive compared to silicon, but its performance advantages are compelling end-users choose it more often. Until 2018, SiC's main market was centered on diodes used in PFC and PV inverters. After 2018, many applications such as battery charging, as well as onboard DC–DC systems, as well as PV and UPS have proven to benefit widely from the usage of SiC devices, increasing the market share of MOSFET. A continuous increase is expected in SiC for these different markets, with a push toward full SiC modules. Some development is still needed to reach the high voltage and reliability required for applications like rail and wind, but every rail company is studying and evaluating SiC technologies. Besides those large volume applications, numerous smaller segments can be penetrated by SiC MOSFETs, including healthcare applications, welding, high-end air-conditioning, or certain motor drive applications with special requirements on power density and or switching frequency. Concerning the EV/HEV sector, SiC is already used in OBC and but with the implementation of SiC technology for its main inverter application, the market has reached the point of no return.

Ranging from conservative to optimistic, industrial players offer very different forecasts in terms of market value for SiC in the EV/HEV market. These forecasts range from several hundred million dollars to billions of dollars in 2024. Everyone agrees that EV is the market of the highest potential, but perceptions differ regarding how it will grow and how SiC will penetrate the automotive market. These perceptions are based on the data that each player has collected and the arguments that derive from their interpretation of the data. The future adoption of SiC power devices is now undeniable, but today, questionings are more related to how and when they will make it. In 2017, the SiC power semiconductor market's value was about US$300 million, with 39% market share for PFC and 32% for PV applications. By 2024, it is estimated that the market will approach US$2 billion, showing an impressive 29% CAGR2 between 2018 and 2024. The automotive market is undoubtedly the foremost driver with around 50% of total device market share in 2024 and it will be driven mainly by xEV charging infrastructure adoption, with a 2018–2024 CARG of 100% and xEV inverter with a 2018–2024 CARG of 73% (Figure 14.3) [3].

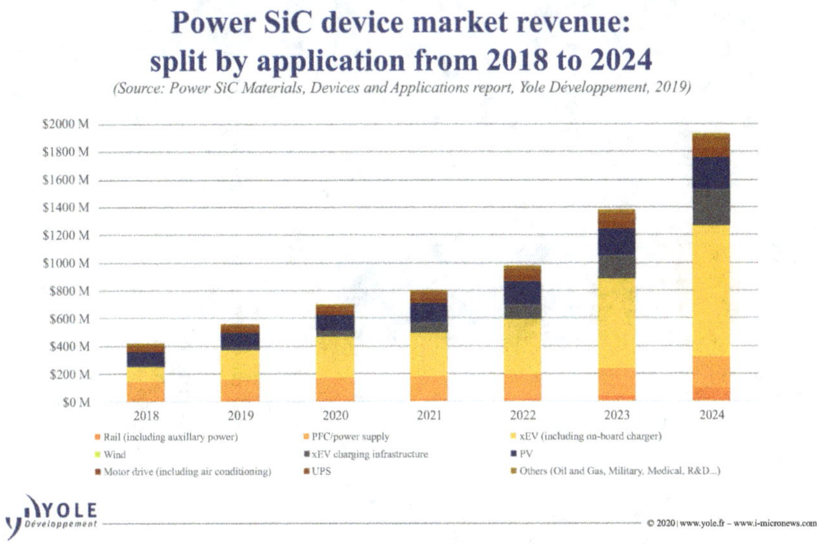

Figure 14.3 Power SiC device market by application

In terms of voltages, for 650 V technologies, diodes will dominate longer, at 1200 V, MOSFETs and diodes co-exist and this segment will be more and more dominated by power module-based solutions; above 1700 V, the value proposition of SiC against silicon-based devices is more pronounced but currently, only a few commercial products are visible. Below 500 V, SiC is facing strong competition from silicon and GaN technology.

Technically speaking, the panorama of the SiC transistor's design is still varying. On the market, we still can find different solutions to die design such as JFET and MOSFET and among the latest, trench and planar structures. The biggest limitation in this phase of increasing demand is the limited supply of 6″ production-grade substrate and SiC-specific equipment like epitaxy reactors. Consequently, only a few players can supply 6-inch wafers and even those players might have some issues if they must provide large volume of reproducible high-quality wafers. Thus substrate prices and equipment costs remain high, with a big impact on final device manufacturing cost (Figure 14.4).

Moreover, behind the adoption of SiC, we need to point out the packaging issue. Today, there is still a limited number of automotive-qualified module suppliers. Since the standard packaging is not yet optimized for SiC higher performances, new designs and materials are continuously introduced. This leads to the entrance of new players, especially outsourced semiconductor assembly and test (OSAT), in the packaging' sector. Another challenge is reliability and ruggedness. There are two areas of concern: the gate oxide and the low defectivity epitaxy processes. Regarding the gate oxide, the major challenge is to avoid drift effect of the threshold voltage under long-term stress. Concerning defectivity, the objective

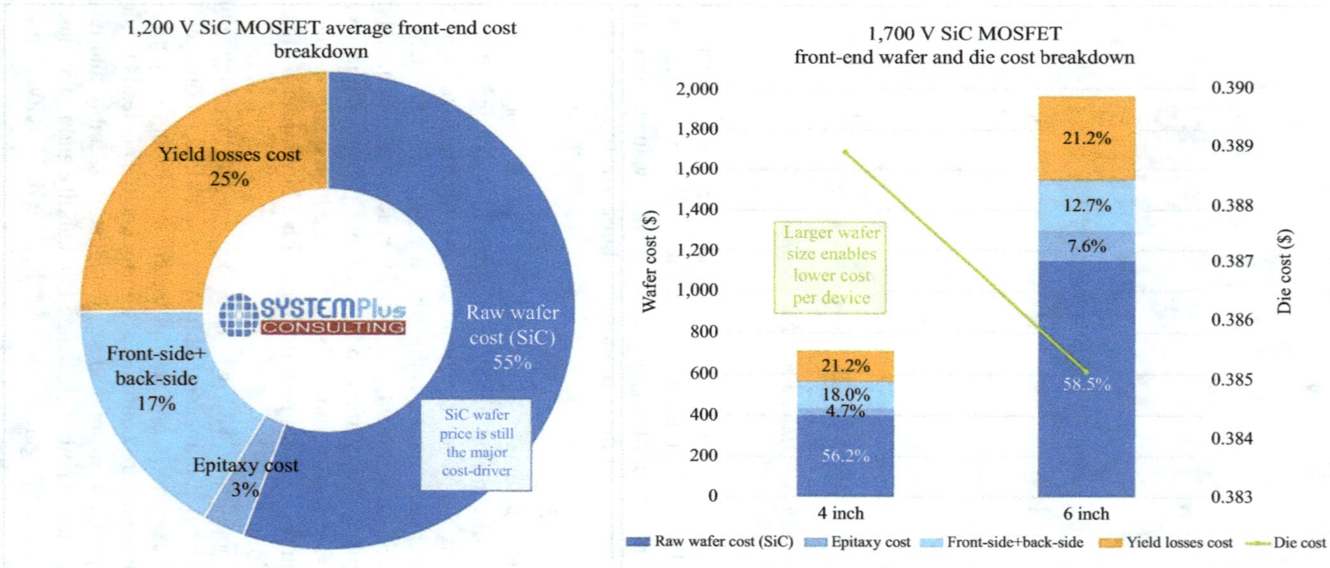

Average 1,200 V MOSFET wafer cost breakdown

Estimation of 1,700 V SiC wafer & die cost breakdown by substrate size

Figure 14.4 Impact of SiC substrate's price on device manufacturing cost is to reduce yield losses, still not comparable with silicon, to allow also larger die sizes at reasonable manufacturing cost. An important challenge is also the thermal and electrical connectivity, between the package and the SiC MOSFET die. SiC MOSFETs were initially released only in three-lead through-hole packages, now several more advanced discrete and module packages are becoming available. In last year we have seen the use of new designs and new materials to reshape the standard modules to improve the thermal dissipation of SiC. The innovations are spread-throughout all manufacturing levels, from the encapsulation to the heatsink. New hard resins, instead of the classic epoxy, have been offered; while in automotive sectors, integrators shifted from case modules to molded single side cooling modules. New substrates have a big impact on thermal dissipation; copper has seen widespread use as material for lead frames while SiN AMB substrates are becoming a standard. Finally, die and substrate attach also plays a role in increasing module reliability especially at higher temperatures; silver sintering is becoming more and more common as an alternative to soldering and can be found in different types of packaging.

In summary, the development of SiC power device technologies will be pushed by the total market growth in specific segments; today and on the mid-term scale, the focus will be on 1200 V components. Even within this voltage class, different applications have different requirements and a one-size-fits-all technology will probably not be enough. The key aspect for chip and package developers will be to provide room for wide enough parametric tradeoffs within a given base technology to realize the application-specific product requirements of a wide range of applications. All these innovations have a cost, thus the improvements by manufacturers must take this factor into account. Innovative technology can have a higher cost by itself, but we always must consider the final impact on the integration and the flexibility of the final system. Initially very expensive, due to the improvements in the base material, die design, and new packaging solutions, we expect that the costs will decline greatly and must continue to do so in the future to access wide market adoption.

14.3.2 Gallium nitride

Over a long period, industrial companies followed up at a distance the development of GaN-based solutions mainly managed by R&D institutes and laboratories. Today the context has changed; a lot of power electronics and compound semiconductor companies including leading players strongly engaged in significant projects of development. Some of them already introduce in their portfolio a GaN product and other finally integrated GaN-based devices in their consumer applications.

From a theoretical point of view, GaN offers fantastic technical advantages over traditional Si MOSFETs; moreover, the lowering of prices can make GaN devices a good competitor of the currently used Si-based power switching transistors. Nevertheless, the improvement of silicon SJ MOSFETs will keep these devices on the market and drive them toward standardization and popularization. Thus, the technical panorama is not clear yet; every manufacturer presents its solution to die design and packaging integration [4].

Regarding technology, we see different trends at different levels. The first technological highlight is at the substrate level. In standard, GaN on Si structure is the high thermal expansion coefficient mismatch between GaN and Si leads to large tensile stress during cooling of epitaxy growing process. Manufacturers propose different solutions such as multiple heterojunction structures or superlattice structures of nanometric layers. Therefore, the innovations in terms of epitaxy deposition and manufacturing process lead to some problems such as very low manufacturing yield and reproducibility issues which have an impact of at least 40% on the final die cost. In 2019, following the product launch and entry of GaN-on-sapphire devices into the consumer fast charger market, GaN-on-sapphire has made plenty of noise. According to industry feedback, the cost of a 4-inch GaN-on-sapphire LED epitaxial (epi) wafer is less than $50, benefiting from a very mature LED industry's capacity and know-how. In our opinion, although it is not the same epi structure, a significantly low epi wafer price is likely in a high-volume market for power applications. Moreover, the GaN-on-sapphire epitaxy represents around 20% of total GaN HEMT wafer cost, with GaN-on-Si epitaxy representing around 40%–50%. GaN-on-sapphire's low cost, easier growth, and lower dislocation density is a crucial advantage. Over the next several years, it can drive many developments for GaN epi houses and LED industry players seeking new market opportunities.

To achieve normally off operation, which is strongly desired for the safe operation, is another technical issue for the GaN power switching transistors. However, it has been very difficult, since the built-in polarization electric field induces extremely high electron density at the heterointerface. Again, manufacturers propose different solutions such as Gate Injection Transistor structure, E-Mode transistor, or cascode co-packaging. Integration is an obvious technological trend in the consumer market. GaN SiP or SoC solutions are mainly used in fast chargers. Some discrete GaN solutions are being proposed in this market, but we expect GaN IC solutions to dominate in the next 5 years, enabling easier solutions for end-users who benefit from the smaller size. On the other hand, GaN discrete solutions are still of interest to higher-power applications like data centers, industrial power supply, and automotive applications. High frequency is one of the main benefits enabled by GaN, but they bring additional challenges for GaN device packaging, such as parasites. To minimize them, companies are following different packaging solutions; from flip-chip packaging to embedded die packages or wafer-level packaging solutions where the connections are made directly on the die through solder balls. All these solutions can impact the choice of the integrator not only for the design of the circuit but also for the cost; in fact, the use of multiples dies to obtain the normally-off operation implies a significant increase of the final cost of the device. This brings to a strong competition which will accelerate technical innovations and lower the prices (Figure 14.5).

Over the last decade, the tiny GaN power market has been driven mostly by high-performance, high-end applications. Moreover, due to its higher cost and end-user reluctance toward GaN, the overall GaN device market has remained very low at $8.8 million and split amongst the few leading players. In 2019, with the integration in several aftermarket chargers and the adoption in inbox fast-chargers, for the first time, GaN power devices entered a high-volume smartphone market. With 5G's arrival in the smartphone market and the significant growth in luxury smartphones, other OEMs who have already announced high power fast charging (>100 W) and look for differentiation with high power chargers, adopting GaN. Nevertheless, >100 W fast charger does not lead to a significant increase in sales and major OEMs remain conservative and first want to gauge the market acceptance and further cost reduction of this new product. However, GaN is not a necessary solution in this context as Si can still do a good job for high power charging (60 W range). In this context, the 2020–2021 period is crucial for GaN-based fast chargers in terms of technical, economic, and sourcing aspects. Bigger market expansion in the consumer applications can be expected, once GaN achieves high maturity and market acceptance as well as cost-competitiveness compared to Si MOSFETs. Today new players are entering the market, but the historical players keep their lead by decreasing production costs as much as possible or by introducing different technologies, thus it is expected that the power GaN device market which will exceed US$350 million by 2024, with 85% CAGR7 between 2018 and 2024 [5].

Concerning the EV/HEV market, GaN is interesting for emerging 48 V DC/DC in MHEV, and onboard chargers in electrified vehicles. Following demonstrations, GaN is expected to outperform Si MOSFET from the light load through to high-current using air-cooling. GaN does not currently have good prospects for immediate large-volume production but on the plus side we expect some volume ramp-up within 4–5 years. At the higher-power level, SiC is currently penetrating while for the lower-power charger, GaN is theoretically well-positioned and could benefit in the long term from the consumer market's ramp-up. The device

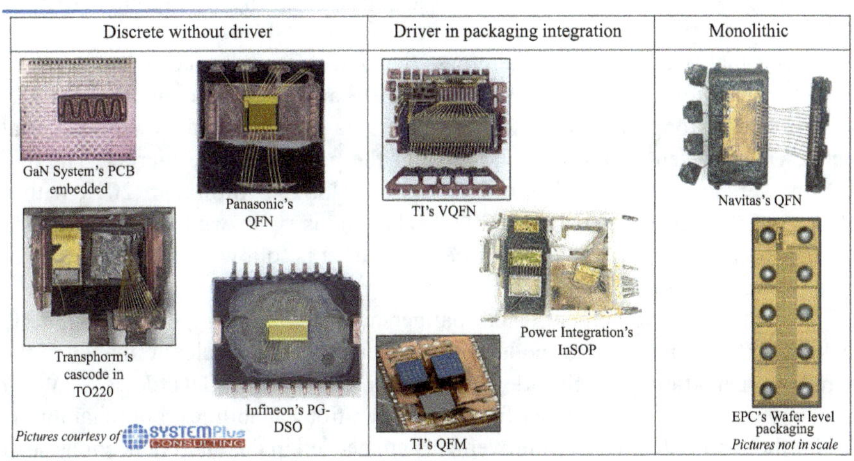

Figure 14.5 Technical solutions for driver integration in GaN devices

manufacturers are working closely with packaging companies to enter the OEM supply chain and enjoy increasing volumes starting in 2023–2024.

GaN is also expected to penetrate industrial and telecom power supply applications including datacom and base stations; with increasing efficiency requirements, data centers will benefit from enhanced GaN device maturity and cost-competitiveness. Other applications, such as LiDAR, envelope tracking, and wireless charging have a significant possibility for GaN adoption, but the market is not yet ready as the end-application is still emerging. As of 2019, analysts expect very little market penetration for industrial applications like UPS and PV, since some technology development is still needed, and this segment competes directly with silicon and SiC.

14.4 Power packaging prospects

Two of the hottest power electronics topics over the last few years are semiconductor performance and thermal management of the packages, to cope with the strong requirement for high efficiency and high-power density with minimal thermal losses. This includes new semiconductor dies, packaging materials, and the integration of different components.

In the past, packaging needs were driven by industrial applications, but today the market has changed, and the future will be different. Electrification of the automotive industry directly impacts the market of power modules and related packaging: now power module packaging is a very dynamic market with constantly reshaping supply chain, continuous innovations, materials' enhancements, and a lot of R&D investments. As a key element in power converters and inverters, according to Yole Développement, the power module market should reach about US$6 billion by 2024 with 6.6% CAGR2 between 2018 and 2024. In parallel, analysts announce a US$2.2 billion power module packaging market in 2024. This market's promising outlook is beneficial for the power module packaging material business. Therefore, the power module packaging material market will achieve a 7.8% CAGR between 2018 and 2024, reaching the US$2.17 billion business opportunity by 2024. Under this dynamic context, the material sector will represent more than one-third of the power module market. The discrete power devices market represents almost US$13.5 billion in 2018 with a 2.9% CAGR between 2018 and 2024. This industry is well established and mature; therefore, the discrete power device market is following the global rise of the power electronic industry.

Technology developments and partnerships are also expected for WBG packaging (both discrete and module). Improvements will accelerate thanks to the increasing penetration of SiC devices and, in the future, GaN devices. WBG semiconductors will allow higher frequency-operation capabilities, but reliability is key for these modules. Thus, improvements are needed in die interconnections such as overlay, silver sintering for die-attach (for higher temperature-junction capability and reliability improvement), and ball bonding (compared to existing wire or ribbon bonding). The embedded die approach enables achieving very low parasitic inductance from interconnections.

14.4.1 Power discrete packaging market

In the discrete power device industry, key characteristics and needs are the low cost per device, the large selection of products and suppliers, and the use of proven, highly standardized products and technologies, including packaging technologies. The discrete power device packaging industry still offers business opportunities, especially for materials suppliers and packaging companies. Packaging technologies for discrete power devices including a lead frame, die-attach, electrical interconnections, and encapsulation should have the aforementioned characteristics; but it is difficult to match the high volume, standardized products and low cost required by device integrators with the acceptable additional costs equated to innovative packaging technologies.

Under this context, the combination of market growth and market size for different packaging solutions gives the complex result with many different variables, including device demand evolution, die size, package type and interconnection method used, device size following downsizing trends, semiconductor content per packaged device, and more. Some of these factors favor a market increase, others a market decrease, participating in rather a flat market evolution. Therefore, the discrete power packaging market's evolution remains flat but is still growing between 2018 and 2024 with a 1.1% CAGR [6].

The adoption of new material on the device level and new markets application pushed the manufacturers to find new packaging technological solutions. Major packaging innovations will be realized on the level of the electrical interconnection, due to the growing adoption of copper clips as a substitute for more conventional wire and ribbon bonding. These innovations came from sectors other than power, such as micro electro-mechanical systems (MEMS) and advanced packaging and this pushed the old supply chain to adapt itself. Discrete device makers can manufacturer power devices internally or by subcontracting the packaging to OSAT companies. Device makers and OSATs are both looking to offer innovative packaging solutions to their customers. Generally, during the first technology adoption phases, innovative products are made internally. Once the demand for units becomes important, the device manufacturers license the technology to other companies or use OSATs, which have huge manufacturing capacities. Die-attach materials, epoxy molding compounds, and interconnection materials are typically supplied by the same materials suppliers, which also provide these solutions to other markets. Lead frames are supplied primarily by numerous Asian materials suppliers since low cost and high volumes are important factors.

With increasing application requirements (i.e., thermal cycling capability in electric vehicles), along with reducing device size and increasing device package-design complexity, players that can provide specific solutions and ensure tight angular and dimensional tolerances in high volume production are increasingly sought by device packaging companies. Advanced packaging companies have significant experience in the packaging of various complex devices for smartphones and microelectronic applications, indeed power electronics, especially at low and mid-power range, represents a real opportunity for them to adapt and transfer their existing advanced packaging solutions to power devices, and thus enlarge their product and customer portfolio. Advanced packaging solutions' highest added value is not in "rather simple" discrete devices, but in devices integrated with a driver, multichip devices, etc.

14.4.2 Power module packaging market

When speaking about power module packaging, the market is becoming extremely competitive. Several new players are arriving from different directions, driven by the EV/HEV market. Moreover, since power module packaging designs and materials are evolving fast, material suppliers will need to adapt to the new industry requirements. In the coming years, power module market leaders will face strong competition from Tier 1 automotive manufacturers and new entrants from China. OSAT companies could also offer services providing advanced packaging technologies to power module manufacturers.

The technology solutions developed for discrete devices are increasingly being applied in power modules and small-size modules. This means that many suppliers of innovative packaging solutions are focusing on other markets which might lead to stronger revenue growth, as compared to a "traditional" discrete power device market. Moreover, with power module packaging designs and materials evolving fast, material suppliers must adapt to new industry requirements.

We see the biggest growth in power semiconductors coming from IGBT modules, which is still motivated by high power efficiency and density requirements from the main power applications. The new applications such as energy storage, charging infrastructure, or EVs demand modules with different power levels and reliability requirements, which will lead to a vast choice of power modules in the coming years. Power modules are being developed with new substrates, die-attach materials, or new semiconductor materials. For all technologies, reducing packaging size is the main trend, with high reliability under harsh environmental conditions.

The need for increased performance as well as the introduction of new materials with different technical properties has pushed manufacturers to find new solutions, not only at the silicon level but also in packaging. In the last year, we have seen the use of new designs and new materials to reshape the standard module, mainly to improve thermal dissipation. The innovations are spread-throughout all manufacturing levels, from the encapsulation to the heatsink. New epoxy resins, instead of the classic silicone gel, have been integrated into automotive and industrial modules and molded double-side cooling modules eliminate the need for a plastic case and enable more compact and highly modular inverters. To reduce inductance caused by connections, there are multiple possible solutions, from the use of copper or aluminum coated copper wires to copper clips or flex printed circuit boards (PCBs). Double side cooling modules and new substrates have had a big impact on thermal dissipation; copper has seen widespread use as material for lead frames while new integrated heatsinks with pin-fin or active metal brazed (AMB) or insulated metal substrate (IMS) substrates are becoming a standard. Finally, solder also plays a role in increasing module reliability especially at higher temperatures; the use of Gold/Tin solders or silver sintering is becoming more and more common and can be found at different levels, either under the die, under the substrate, or the baseplate (Figure 14.6). The traditional power module with a plastic case, all wire bonding, and tin-silver-copper (SAC) solder are still offered by manufacturers, but new solutions allow better integration of the module into the final systems according to the performance requirements [7]. All these innovations have a cost, thus the improvements by manufacturers must take this factor into account. A bigger power module can have a higher cost by itself, but we always must consider the final impact on the integration and the flexibility of assembling. The fast evolution of technology on all design levels and the cost of modules are the two factors that in recent years eliminated the presence of a standard in power module design, and we will expect even more innovation in the future. The request for cost saving during the integration steps pushes design companies to move forward single and more flexible structures, such as one-in-one commutation cells or embedded discrete. As a direct consequence, the power module packaging supply chain is showing strong changes.

Under this dynamic context, the material sector will represent more than one-third of the power module market. In 2018, the largest packaging material segment was baseplates, followed by substrates. The other 32% was represented by die-attach and substrate-attach materials. Thus, major technological choices in these segments can rapidly impact the overall power module packaging market. For example, the market share for silver sintering as a die-attach is increasing, driven

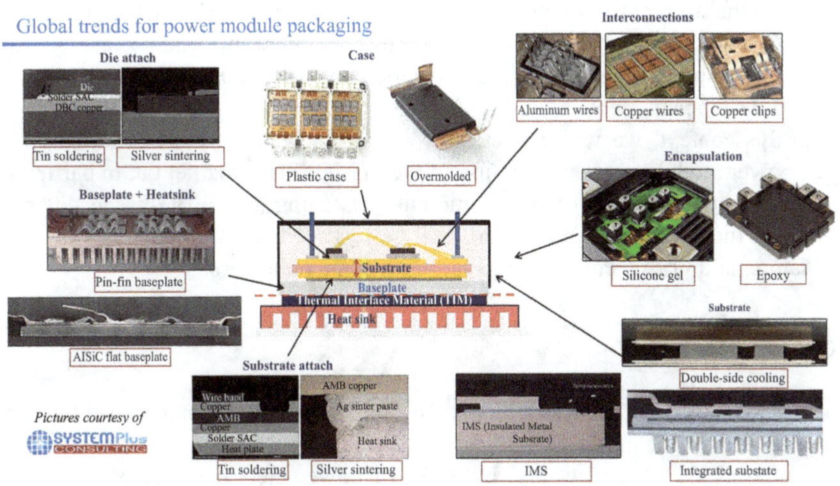

Figure 14.6 Technology trends in power module packaging

especially by EV/HEV. This technology is pricier than more conventional soldering materials, and CAGR for the die-attach market is +10.8% between 2018 and 2024—well higher than for other market segments. The second-highest growth is for interconnections. Yole Développement announces 8.7% CAGR for the 2018–2024 period. This market segment is followed by substrates, with a 2018–2024 CAGR of 8.5% [8].

Besides the materials suppliers, packaging equipment manufacturers (i.e., wire bonders, sintering machines, reflow ovens, and cleaning equipment) will also be positively or negatively impacted by these changes. The power module makers that adopt innovative packaging solutions early on can secure a better market position. Suppliers of packaging solutions for EV/HEV power modules must adapt their strategy, product portfolio, and manufacturing capacities to satisfy strong requirements in terms of costs, manufactured volume, and product reliability.

This is a very challenging task, especially because many players are targeting opportunities offered via rapidly growing EV/HEV demand. To succeed in this competitive environment, new mergers and acquisitions, as well as partnerships, are necessary to help capture new technologies/new customers quickly and increase production capacity. To reduce cost pressure, some companies have already moved or are planning to move at least part of their production capacity to countries with lower production costs.

14.5 Summary

Since its introduction with the first transistors, the power electronics market has been driven by technology innovations, consumer needs, political choices, and environmental issues. After the two first megatrends of industrial and renewable energy sources sectors, the power electronics market of the near future will be driven by the automotive sector. The automotive, and the EV/HEV market, not only will impact the vehicles' ecosystem, but also renewable energy, energy storage, and transportation systems, data centers, radar and LiDAR systems, together with other supporting technologies.

In this context, we will expect, not only an increase of the market share of IGBT power modules and stabilization of the Si MOSFETs market but in particular new design and materials of power module packaging and WBG-based devices. Moreover, the complete supply chain will be reshaped, with new players, system integrators, and OSAT entering the market of semiconductors.

Note to the reader

This chapter is part of an edited book. To cite or use this chapter as a reference, please use the following format:

Barbarini E. "Prospects and outlooks in power electronics technology and market." In: Iannuzzo F (ed.), *Modern Power Electronic Devices: Physics, Applications, and Reliability*. Stevenage, UK: IET; 2020. pp. 451–467.

References

[1] Villamor A. "Status of the power electronics industry 2019." [Online]. Available: https://www.i-micronews.com/products/status-of-the-power-electronics-industry-2019/ [Accessed 1 August 2019].

[2] Rosina M. and Lin H. "Power electronics for electric & hybrid electric vehicles 2020" [Online]. Available: https://www.i-micronews.com/products/power-electronics-for-electric-hybrid-electric-vehicles-2020/ [Accessed 1 January 2020].

[3] Lin H., Dogmus E. and Villamor A. "Power SiC 2019 materials devices and applications" [Online]. Available: https://www.i-micronews.com/products/power-sic-2019-materials-devices-and-applications/ [Accessed 1 July 2019].

[4] Barbarini E. "State of the art of GaN on Si transistors – technology and cost overview." *APEC International Conference Proceedings 2018*, IEEE-Pels, San Antonio, TX, USA, 2018.

[5] Dogmus E., Villamor A. and Lin H. "Power GaN 2019 epitaxy devices applications technology trends" [Online]. Available: https://www.i-micronews.com/products/power-gan-2019-epitaxy-devices-applications-technology-trends/ [Accessed 1 November 2019].

[6] Rosina M. and Agarwal S. "Discrete power device packaging: Materials market and technology trends 2019" [Online]. Available: https://www.i-micronews.com/products/discrete-power-device-packaging-materials-market-and-technology-trends-2019/ [Accessed 1 April 2019].

[7] Barbarini E. and Troadec C. "Power module packaging: Market and technology trends." *15th International Conference on Device Packaging Proceedings, IMAPS 2019*, Phoenix, AZ, USA, 2019.

[8] Agarwal S. and Rosina M. "Status of the power module packaging industry 2019" [Online]. Available: https://www.i-micronews.com/products/status-of-the-power-module-packaging-industry-2019/ [Accessed 1 October 2019].

Index

www.ingramcontent.com/pod-product-compliance
Lightning Source LLC
LaVergne TN
LVHW011551200926
848484LV00001B/304